ADVANCES IN
METAL-ORGANIC CHEMISTRY

Volume 3 • 1994

ADVANCES IN METAL-ORGANIC CHEMISTRY

Editor: LANNY S. LIEBESKIND
Department of Chemistry
Emory University
Atlanta, Georgia

VOLUME 3 • 1994

JAI PRESS INC.

Greenwich, Connecticut *London, England*

55 Old Post Road, No. 2
Greenwich, Connecticut 06836

JAI PRESS LTD.
The Courtyard
28 High Street
Hampton Hill, Middlesex TW12 1PD
England

ISBN: 1-55938-406-9

Transferred to digital printing 2006

CONTENTS

LIST OF CONTRIBUTORS

James W. Herndon — Department of Chemistry and Biochemistry
University of Maryland
College Park, Maryland

Richard C. Larock — Department of Chemistry
Iowa State University
Ames, Iowa

Lyndsay Main — Department of Chemistry
University of Waikato
Hamilton, New Zealand

Yoshiro Masuyama — Department of Chemistry
Sophia University
Tokyo, Japan

Julius J. Matasi — Department of Chemistry and Biochemistry
University of Maryland
College Park, Maryland

Leonard A. McMullen — Department of Chemistry and Biochemistry
University of Maryland
College Park, Maryland

Shun-Ichi Murahashi — Department of Chemistry
Faculty of Engineering Science
Osaka University
Osaka, Japan

Takeshi Naota — Department of Chemistry
Faculty of Engineering Science
Osaka University
Osaka, Japan

Brian K. Nicholson — Department of Chemistry
University of Waikato
Hamilton, New Zealand

Wayne F. K. Schnatter Department of Chemistry and Biochemistry
University of Maryland
College Park, Maryland

Seniz U. Tumer Department of Chemistry and Biochemistry
University of Maryland
College Park, Maryland

INTRODUCTION TO THE SERIES: AN EDITOR'S FOREWORD

The JAI series in chemistry has come of age over the past several years. Each of the volumes already published contains timely chapters by leading exponents in the field who have placed their own contributions in a perspective that provides insight to their long-term research goals. Each contribution focuses on the individual author's own work as well as the studies of others that address related problems. The series is intended to provide the reader with in-depth accounts of important principles as well as insight into the nuances and subtleties of a given area of chemistry. The wide coverage of material should be of interest to graduate students, postdoctoral fellows, industrial chemists and those teaching specialized topics to graduate students. We hope that we will continue to provide you with a sense of stimulation and enjoyment of the various sub-disciplines of chemistry.

Department of Chemistry
Emory University
Atlanta, Georgia

Albert Padwa
Consulting Editor

PREFACE

Volume 3 of *Advances in Metal-Organic Chemistry* brings the series into its fifth year of existence. As with the previous two volumes in the series, an international flavor to the contributions is maintained and both stoichiometric and catalytic metal mediated synthetic organic processes are covered.

Chapter 1, "Orthomanganated Aryl Ketones and Related Compounds in Organic Synthesis" by Lyndsay Main and Brian Nicholson presents the reader with a comprehensive survey of the synthesis and subsequent reactions of compounds possessing an aryl-manganese bond. This chapter serves as a nice introduction to the use of manganese carbonyl derived compounds in organic synthesis; compounds that are easily prepared and handled and undergo a variety of interesting transformations of synthetic value. The incredible facility with which chromium carbene complexes participate in a diversity of synthetic organic transformations is underscored in Chapter 2, "Cyclopropylcarbene-Chromium Complexes: Versatile Reagents for the Synthesis of Five-Membered Rings," a contribution from James ("Pete") Herndon and co-workers. Moving from the stoichiometric to the catalytic regime, Chapter 3 is Richard Larock's treatise on "Palladium-Catalyzed Vinylic Substitution," one of the more important advances in organic synthesis in the last 25 years. This chapter should serve as a resource to those in academia and industry interested in palladium catalyzed transformations of vinyl halides, triflates, and related molecules. The continued development of Japan as a center for important organometallic chemistry and catalysis can be seen in the final two chapters making up Volume 3. In Chapter 4, Shun-Ichi Murahashi and Takeshi Naota have provided a thorough description of "Ruthenium-Catalyzed Oxidative Transformations of Alcohols," one aspect of the synthetically important metal-catalyzed oxidations studied by the Osaka group. Finally, in Chapter 5, Yoshiro Masuyama describes how he and his co-workers have reversed the typical electrophilic reactivity of π-allylpalladium complexes by developing "Palladium-Catalyzed Carbonyl Allylations via π-Allylpalladium Complexes."

Lanny S. Liebeskind
Editor

ORTHOMANGANATED ARYL KETONES AND RELATED COMPOUNDS IN ORGANIC SYNTHESIS

Lyndsay Main and Brian K. Nicholson

Advances in Metal-Organic Chemistry
Volume 3, pages 1–50.

ISBN: 1-55938-406-9

I. INTRODUCTION

Organic synthesis using metal carbonyl complexes of the 3d elements ("the inexpensive metals") is well developed for chromium, iron, cobalt, and nickel in both stoichiometric and catalytic processes.[1] As illustrations we draw attention to the work with (arene)$Cr(CO)_3$ complexes[2] and chromium carbene complexes,[3] the two-volume review on organoiron chemistry,[4] hydroformylation reactions[1] with $Co_2(CO)_8$, alkyne coupling reactions[5] with $CpCo(CO)_2$, and $Ni(CO)_4$-catalyzed carbonylation reactions.[6]

Noticeably absent from this list of elements is manganese. Very few useful syntheses have been reported which make use of manganese carbonyl complexes. For example in Schore's recent review of transition metal mediated cycloaddition reactions of alkynes in organic synthesis,[7] metal carbonyl complexes of the above metals all feature prominently, but manganese is not mentioned at all. While the expense of $Mn_2(CO)_{10}$ (which is the usual starting material) may be a disincentive for work in this area, organomanganese carbonyl compounds have some advantages compared with those of say iron or cobalt; they are usually readily prepared, stored and handled, being only mildly air-sensitive and surviving normal chromatography.

Our attention was focused on manganese when a doctoral student, Lou Gommans, expressed interest in a project using organometallic reagents in organic synthesis. Although some chemistry of (η^5-organo)$Mn(CO)_3$ species was being developed,[8] we noted that σ-organomanganese compounds were attracting only sporadic attention. As a starting point for our own work we were attracted by the reports from Kaesz's group [9–13] in the mid-1970s showing that aromatic ketones such as acetophenone undergo efficient orthometallation reactions with $MeMn(CO)_5$ (Eq. 1).

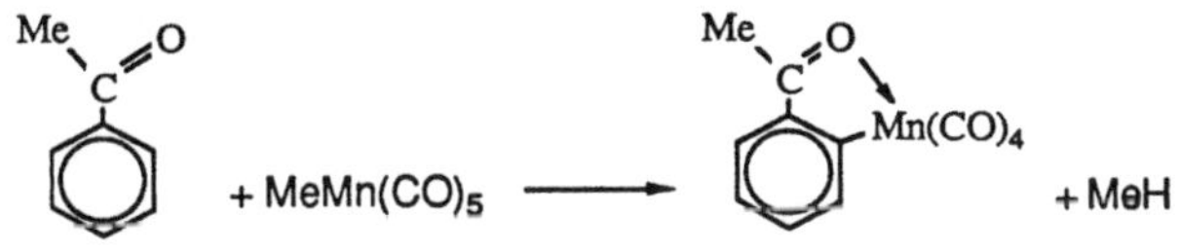

(1)

(1)

While cyclometallation reactions are known for many combinations of metal and organic substrates[14] (and are perhaps best developed for Pd(II) with N-donor substrates[15]), the Kaesz reactions are distinctive: aromatic substrates which lead to an oxygen atom in the metallocyclic ring, as in **1**, are *not* otherwise orthometallated by any of the other metals that react routinely with N, P or S donor substrates.[16,17] Only with rhenium (an expensive manganese mimic) are similar species found.[9] The reasons for the distinctive behavior of manganese towards aromatic ketones is still not clear.

This orthomanganation of aryl ketones has potential in synthesis if the manganese can be subsequently replaced by another metal or nonmetal; coupling to carbon at the *ortho* site is, of course, of prime interest to the organic chemist. Some factors which encouraged us into this area were as follows. First, the carbonyl function is *meta*-directing in electrophilic aromatic substitution, an otherwise useful method for entry into aryl functionalization; electrophiles may react in any case preferentially at the ketone α-carbon. Second, as described above, we could find no other examples of metallation directed *ortho* to ketone functions. Thus the coupling reactions of alkenes at metal-activated aryl carbon, which make aryl complexes of Hg, Pd, Sn, Pb, Tl and other elements so synthetically useful,[18,19] is not applicable to *ortho*-functionalization of aryl ketones. Activation by directed metallation *ortho* to other aryl carbonyl functions is known for some metals, but is inapplicable to ketones. Thus orthothalliation is well known for benzoate esters and benzoic acids, with Pd-promoted coupling of alkenes in the latter case providing a convenient route to isocoumarins; a range of other heterocycles can be similarly obtained.[20,21] Ortholithiation of benzamides is well established[22] and that of benzoic acids is indirectly available through protection of the acid as an oxazoline, the N-atom of which provides the directive effect.[22] Lithiation is inapplicable in the presence of the ketone function. Aryl halogenation also provides activation for Pd(0)-catalyzed coupling with alkenes,[18,23] but the general inaccessibility of *ortho*-halo ketones is a major limitation for *ortho*-functionalization.

We therefore initiated a systematic study of the reactions of orthomanganated ketones with a view to establishing useful synthetic methods which exploited the ready availability of these complexes. During the course of our work other groups[24–26] have independently developed related aspects of this chemistry. While reactivity patterns are by no means fully explored nor synthetic methods optimized, we now review progress with the hope of stimulating greater interest in the versatility of these potentially valuable reagents.

In this survey we cover complexes of the type **2** where X is an oxygen atom. Related examples where X = N, P, As, and S are also well known[17,27–29] and have preparations and properties similar to those of the X = O compounds. However only for the oxygen-donor examples has the reactivity been examined in any detail, so the other types will be discussed only when they illustrate aspects of the chemistry of the oxygen-donor species.

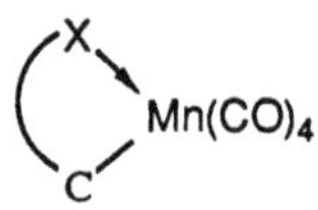

(2)

II. PREPARATION OF CYCLOMANGANATED COMPOUNDS

A. From Aromatic Ketones

Following from the pioneering work from Kaesz's group,[9–13] it is now established that the reaction of Eq. 1 is quite general for aromatic ketones. The only significant modification is the preferred use of $PhCH_2Mn(CO)_5$ as the metallating agent, since it is easier to prepare than $MeMn(CO)_5$ and reacts under less forcing conditions.[28] A typical preparation follows:

Preparation of benzylmanganese pentacarbonyl.[30]

$$Mn_2(CO)_{10} + Na/Hg \rightarrow 2\ Na[Mn(CO)_5]$$

$$Na[Mn(CO)_5] + PhCH_2Br \rightarrow PhCH_2Mn(CO)_5 + NaBr.$$

A 1% sodium amalgam is prepared under nitrogen in a 100 mL Schlenk flask by adding, in small portions, Na metal (0.7 g) to Hg (5 mL, 70 g) (CARE: the dissolution of Na is very exothermic). When the amalgam has cooled, dry deoxygenated tetrahydrofuran (thf) (60 mL) is added by syringe, followed by $Mn_2(CO)_{10}$ (2.0 g, 5.13 mmol). The mixture is stirred vigorously for 2 h. The resulting pale green-grey solution of $Na[Mn(CO)_5]$ is transferred by N_2-flushed syringe, or by decantation under N_2, to a Schlenk flask containing benzyl bromide (1.20 mL, 1.73 g, 10.0 mmol). A white precipitate forms immediately, and after stirring for 10 min the solution is passed through a short column of silica gel (4 × 1 cm, 60–120 mesh), washing with thf (2 × 15 mL). The thf is pumped away under vacuum and the residual yellow oil is sublimed onto a water-cooled cold-finger (1 mm Hg, 50 °C). This yields crystalline $PhCH_2Mn(CO)_5$ (*ca.* 2.2 g, 84%), mp 35–36 °C; ν_{CO} (hexane, cm^{-1}) 2105s, 2010vs, br, 1990 vs; ^{1}H NMR ($CDCl_3$) δ 7.28 (s, 5H, Ph), 2.45 (s, 2H, CH_2). Pure $PhCH_2Mn(CO)_5$ is white, but the pale-yellow product obtained by this route is suitable for further reaction.

Cyclomanganation of p-bromoacetophenone.[31] $PhCH_2Mn(CO)_5$ (0.213 g, 0.75 mmol) and p-bromoacetophenone (0.135 g, 0.68 mmol) are dissolved in redistilled heptane (20 mL). The mixture is refluxed under nitrogen for 2 h, until the reaction appears complete (monitored by t.l.c., or by using solution infrared spectra to follow the relative intensities of the 2105 and the *ca.* 2080 cm^{-1} bands of the starting and product species respectively). The solvent is then removed under vacuum and the residue is chromatographed on preparative-scale silica plates, eluting with diethyl ether/petroleum spirit (1:9). The major yellow band yields yellow crystals of η^2-(2-acetyl-5-bromophenyl)tetracarbonylmanganese (0.239 g, 96%); mp 116–120 °C; ν_{CO} (hexane, cm^{-1}) 2083m, 1996vs, 1949s.

The reaction procedure is identical for a wide range of substrates[10,31] (see Table 1). For larger scale preparations column chromatography using deactivated neutral alumina (activity IV) is more convenient. Elution with petroleum spirit removes traces of unreacted $PhCH_2Mn(CO)_5$, and the polarity of the solvent is gradually increased with CH_2Cl_2 to elute the product as a yellow band.

The direct reaction of Eq. 1 is general for acetophenones and benzophenones, and tolerates a range of substituents on the benzene ring, for example Me, CF_3, OMe, Br, Cl, F, CN, $OSiR_3$. However $-NO_2$ and –OH groups interfere with the metallation reaction[24] unless the hydroxyl hydrogen is intramolecularly H-bonded, as in 9-hydroxyanthraquinone.[10,13] Orthomanganated complexes derived from acetophenones and benzophenones are listed in Table 1, sections (a)–(c). Reaction conditions are similar for all substrates, and yields are generally good to excellent.

A number of unsymmetrically substituted benzophenones have been examined to determine by internal competition which substituents encourage metallation[32,33]. Thus 2,4,5-triethoxybenzophenone gave a mixture of isomers **3a** and **3b** (ratio 2.2:1), showing preferential attack at the unsubstituted aryl ring.

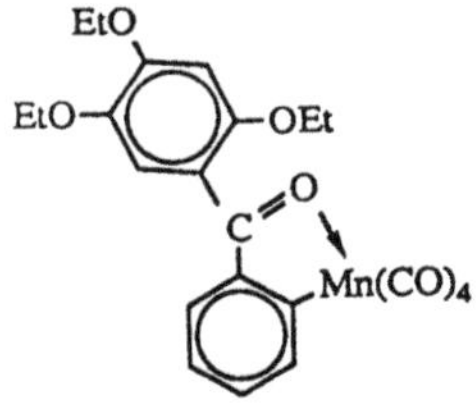

(3a)

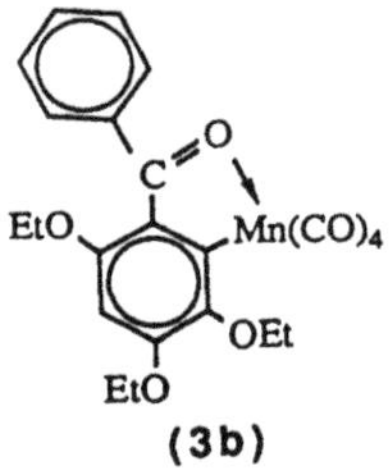

(3b)

With both 3- and 4-CF_3 substituted benzophenones the results were similar; again manganese attack at the unsubstituted ring predominated in an approximately 2:1 ratio. The interpretation of these results is not straightforward, but does indicate that the orthomanganation of benzophenones is less affected by substituents than

Table 1. A Summary of Reported Cyclomanganated Complexes

Compound	Method[a]	Yield(%)	M.Pt(°C)	Ref
(a) Acetophenones				
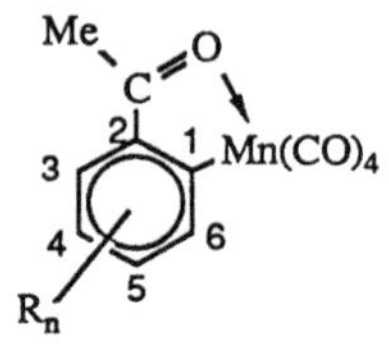				
$R_n = H$	A	68, 85		10,31
	F[c]	84		45
R_n = 4–Me	A	60, 77	122–125	10,24
R_n = 5–Me	A	50–60		10
R_n = 6–Me	A	low		10
R_n = 4–OMe	A	28	110–113	10,24
R_n = 5–OMe	A	50–60		10,31
R_n = 6–OMe	A	63	121–124	10,24
R_n = 5–Cl	A	35–45		10
R_n = 5–Br	A	96	116–120	31
R_n = 4–CF_3	A	89,85	128–131	32,24
R_n = 5–CF_3	A	71	98–100	32
R_n = 4–CN	A	11	109–110	24
R_n = 4–F	A	15	137–139	24
R_n = 6–F	A	70	103–105	24
R_n = 4–Cl	A	42	132–134	24
R_n = 6–Cl	A	44	105–107	24
R_n = 4–Br	A	38	129–130.5	24
R_n = 6–Br	A	27	133–135	24
R_n = 3,5Me_2	A	56	122–124	31
R_n = 3,4,5-$(OMe)_3$	A	80		31
R_n = 4,5,6-$(OMe)_3$	A	86		31
R_n = 4,5-$(OMe)_2$	A	60		31
R_n = 5,6-$(OMe)_2$	A	20		31
R_n = 5,6-(-OCH_2O-)	A	86		35
R_n = 3-$OSiMe_2Bu$	A	86		31
R_n = 3-OCH_2Ph;4,5-$(OMe)_2$	A	90		31
R_n = 3-$OSiMe_2Bu$;4,5-$(OMe)_2$	A	80		31
(b) Other alkyl phenyl ketones				
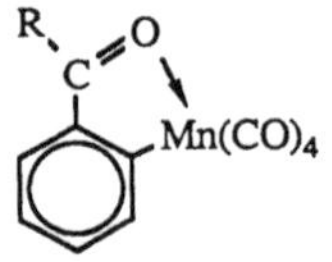				
R = Et	A	76		13
R = Et	F	67		45

(continued)

Table 1. (*continued*)

Compound	*Method*[a]	*Yield(%)*	*M.Pt(°C)*	*Ref*
R = Pr	A	91		13
R = C_6H_{11}	A	54		13
R = cyclopropyl	A	45		13
R = $PhCH_2$	F	66		45
(c) Benzophenones				
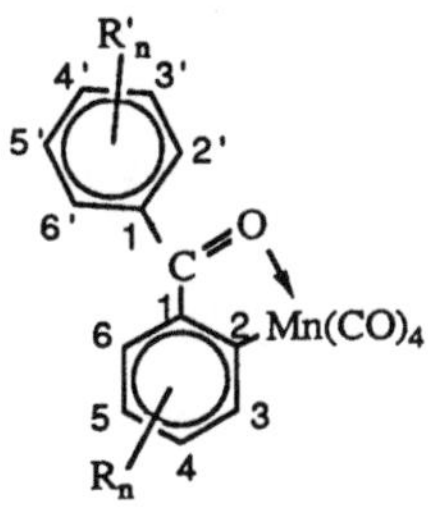				
$R_n = R_n' = H$	A	60		10
	F[d]	80		45
R_n = 4-Me; R_n'=4′-Me	A			10
R_n = 4-OMe; R_n' = 4′-OMe	A			10
R_n = 4-F; R_n' = 4′-F	A			10
R_n = 4-Cl; R_n' = 4′-Cl	A			10
R_n = H; R_n' = 3′-CF_3	A	57		32
R_n = 5-CF_3; R_n'=H	A	19		32
R_n = H; R_n' = 2′,4′,5′-$(OEt)_3$	A	63		35
R_n' = H; R_n = 3,4,6-$(OEt)_3$	A	29		35
(d) Heteroaromatic methyl ketones				
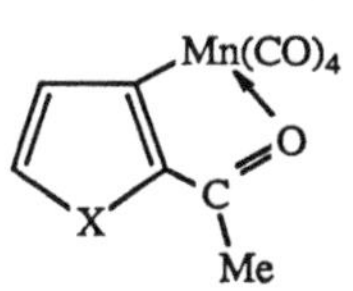				
X = S	A	75		31
X = O	A	18		31
X = NMe	A	28		31
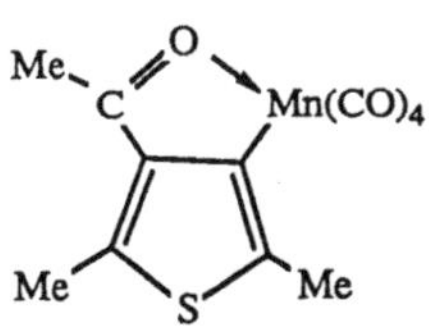				
	A	70		31

(*continued*)

Table 1. *(continued)*

Compound	Method[a]	Yield(%)	M.Pt(°C)	Ref
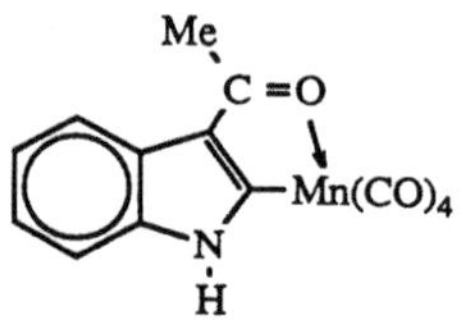	A	90		31
(e) Benzaldehydes				
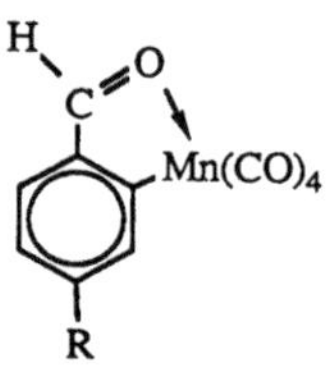				
R = H	A	0		37
R = OMe	A	33		37
R = NMe_2	A	61	124–128	37
(f) Benzoate esters				
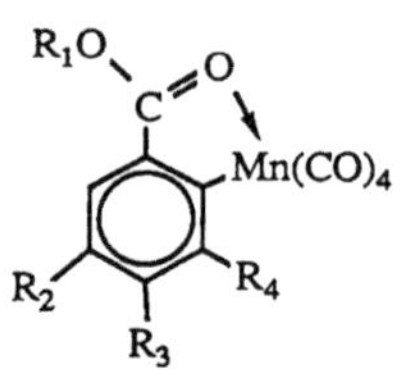				
R_1 = Me; R_2 = R_3 = R_4 = H	A	4		13
R_1 = Me; R_2 = R_3 = R_4 = H	B[e]	57	115–120	39
R_1 = Me; R_2 = R_4 = H; R_3 = OMe	A	74		35
R_1 = Me; R_2 = R_3 = R_4 = OMe	A	73	121–125	35
R_1 = Pr^i; R_2 = R_3 = R_4 = H	A	86		13
(g) Benzamides				
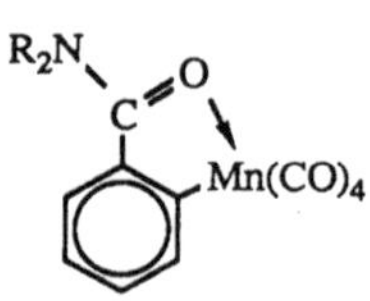				
R_2 = Me_2	A	100	120–125	37
R_2 = Et_2	A	33	96–99	37
R_2 = $(CH_2)_4$	A	77	88–93	37

(continued)

Table 1. (continued)

Compound	Method[a]	Yield(%)	M.Pt(°C)	Ref
(h) N-acyl heteroaromatics				
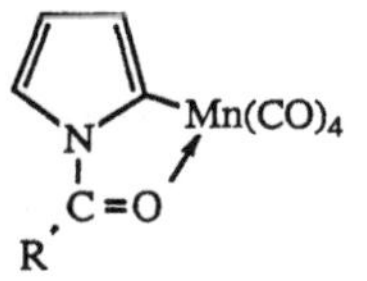				
R = Me	A	36	102	37
R = Ph	A	59		37
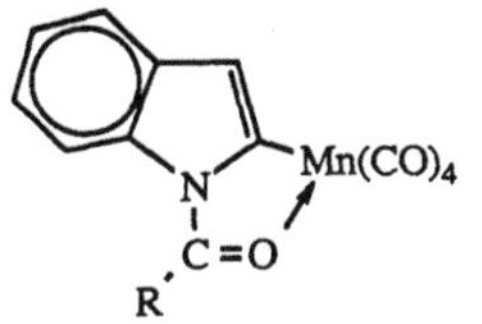				
R = Me	A	68	112	37
R = Ph	A			37
(i) Enones				
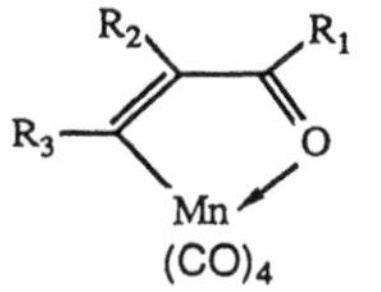				
R_1 = Me; R_2 = CF_3; R_3 = H	C	18	~0	41
R_1 = H; R_2 = CF_3; R_3 = H	C	78		41
R_1 = OMe; R_2 = H; R_3 = COOMe	C[c]	86	47–48	42
R_1 = Me; R_2 = H; R_3 = Ph	C	63(62)[b]	67–68	43
R_1 = Me; R_2 = H; R_3 = Ph	A,C	50, 67	64–65	25,32,44
R_1 = Ph; R_2 = H; R_3 = Ph	A,C	74, 80	65–66	43,44
R_1 = 3,4,5-$(MeO)_3C_6H_2$; R_2 = H; R_3 = Ph	A	90	~110(d)	44
R_1 = Me; R_2 = R_3 = COOMe	C	81(93)[b]	53–54	43
R_1 = Me; R_2 = R_3 = COOH	C	81(97)[b]	~100(d)	43
R_1 = Ph; R_2 = R_3 = COOMe	C	78	89–91	43
R_1 = Me; R_2 = R_3 = COOH	C	73	~150(d)	43
R_1 = Me; R_2 = H; R_3 = COOMe	C	73(77)[b]	55–56	43
R_1 = Me; R_2 = H; R_3 = COOMe	C	93		25
R_1 = Me; R_2 = H; R_3 = COOH	C	70(72)[b]	32–33	43
R_1 = Ph; R_2 = H; R_3 = COOMe	C	91	85–86	43
R_1 = Me; R_2 = H; R_3 = COOH	C	78	117–118	43
R_1 = Me; R_2 = $SiMe_3$; R_3 = Ph	D	72	97–98	25
R_1 = Me; R_2 = Ph; R_3 = COOEt	D	55		25

(continued)

Table 1. *(continued)*

Compound	*Method*[a]	*Yield(%)*	*M.Pt(°C)*	*Ref*
R_1 = Me; R_2 = COOEt; R_3 = Ph	D	32		25
R_1 = CH_2Ph; R_2 = H; R_3 = $CH_2CH_2OCH_2Ph$	D	49		25
R_1 = CH_2Ph; R_2 = H; R_3 = Bu^n	D	63		25
R_1 = CH_2Ph; R_2 = H; R_3 = COOMe	D	94		25
(j) Aliphatic ketones				
R_1 = Me; R_2 = H; R_3 = H; R_4 = COOMe	D	90		25
R_1 = Me; R_2 = Me; R_3 = H; R_4 = COOMe	D	70		25
R_1 = Me; R_2 = H; R_3 = H; R_4 = SO_2Ph	D	90		25
R_1 = Me; R_2 = H; R_3 = Me; R_4 = COOMe	D	25–63		25
R_1 = Me; R_2 = C_6H_{13}; R_3 = H; R_4 = COOMe	D	57		25
R_1 = CH_2Ph; R_2 = H; R_3 = H; R_4 = COOMe	D	88		25
R_1 = CH_2Ph; R_2 = Me; R_3 = H; R_4 = COOMe	D	67		25
R_1 = CH_2Ph; R_2 = H; R_3 = Me; R_4 = COOMe	D	4–55		25
R_1 = CH_2Ph; R_2 = C_6H_{13}; R_4 = COOMe	D	63		25
R_1 = CH = CHMe; R_3 = H; R_4 = COOMe	D	70		25
R_1 = Me; R_2,R_3 = CH_2CH_2O; R_4 = H	E	25	62	48

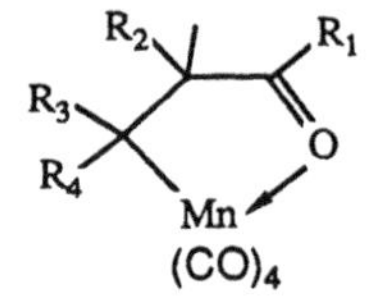

(k) Miscellaneous complexes

The following have all been orthomanganated via Method A: Acetylferrocene, benzoylferrocene[12], 1-tetralone[13,24], benzosuberone[24], flavanone, xanthone, thioxanthen-9-one, thiochroman-4-one, 1,2,3,4-tetrahydro-9-acridanone, 10-methyl-9(10H)-acridone, dibenzosuberone, dibenzosuberenone[35], N-methyl-5,6,7,-trimethoxyisocarbostyril[32], and some carbonyl derivatives of podocarpic acid and dehydroabietic acid.[50]

Notes: [a]*Method A*: Reaction of substrate with $MeMn(CO)_5$ or $PhCH_2Mn(CO)_5$. *Method B*. By salt-elimination between $[Mn(CO)_5]^-$ and ortho-carbomethoxybenzoyl chloride and subsequent bis-decarbonylation. *Method C*: By reaction of an alkyne with $RMn(CO)_5$ (R = H, Me, $PhCH_2$) or with $MeC(O)Mn(CO)_5$. *Method D*: As for C but under 5-10kBar pressure. *Method E*: via a manganese carbene derivative. *Method F*: From reaction of Ph_2Hg with $RMn(CO)_5$.

[b]The second yield in brackets is for $MeC(O)Mn(CO)_5$ as starting complex.

[c]The PPh_3 derivative is also known.

[d]mono-PPh_3, $AsPh_3$, $SbPh_3$ and di-$P(OR)_3$ derivatives are also described.

[e]$P(OCH_2CH_2)_3CEt$ derivative known.

the corresponding reactions with substituted azobenzenes, where much more marked isomer selectivities were found in a similar study.[28,34]

For acetophenone substituted at the 3′-position two isomers are possible depending on whether metallation occurs at the 2′- or the 6′-position (Eq. 2).

For X = OMe the more sterically crowded 2′-isomer **4a** (X = OMe) is the dominant one formed,[10,31] indicating an electronic directing effect from the OMe

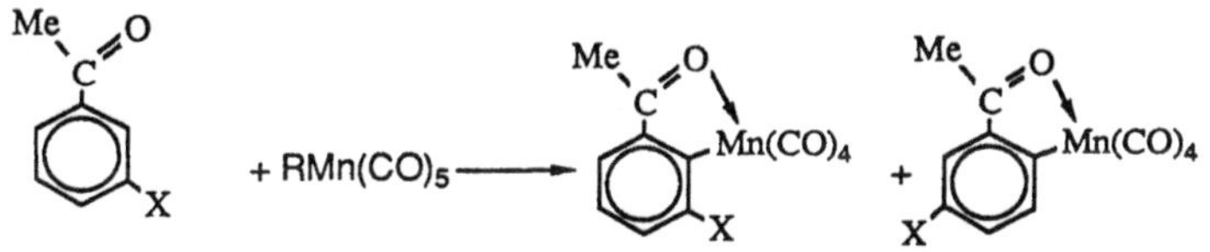

(2)

(4a) (4b)

substituent, possibly *via* lone-pair coordination to the entering metal. However, for 3′,4′-dimethoxyacetophenone metallation preferentially takes place (in a ratio of 3:1) at the less-hindered 6′-site, probably because the 4′-OMe group now sterically prevents the 3′-OMe rotating to direct a lone pair towards the 2′-position.[31] This steric interpretation is supported by results with the 3′,4′-(methylenedioxy)acetophenone, where attack reverts predominantly to the 2′ site, giving isomer **5** with only traces of the other possible isomer.[35]

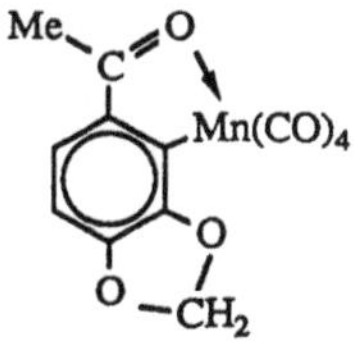

(5)

With 3′-haloacetophenones the same behavior is observed, in that there is an electronic directing effect favoring metallation at the 2′-site which diminishes with increasing size. For the reaction of Eq. 2 the ratio of isomers **4a:4b** is 4.5:1 (X = F), 1.05:1 (X = Cl) and 0.69:1 (X = Br). For X = Me or CF_3 only isomer **4b** is found.[24]

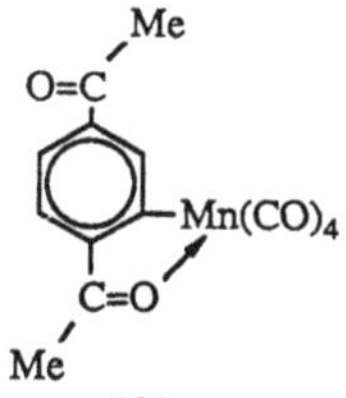

(6)

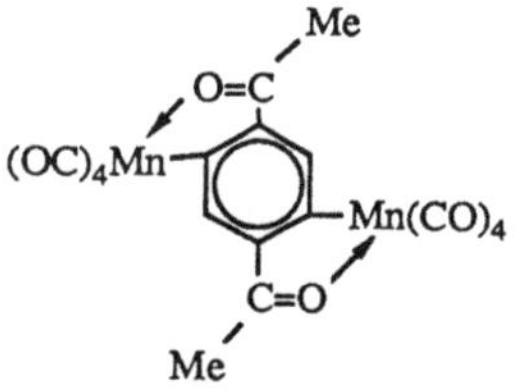

(7)

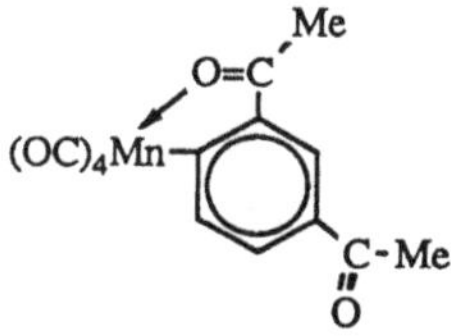

(8)

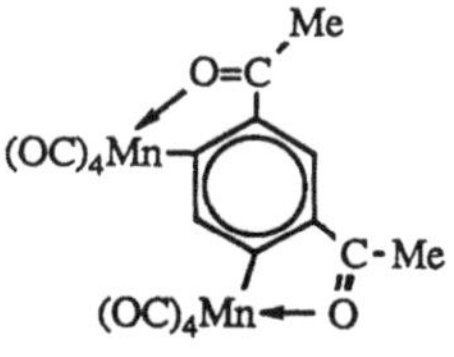

(9)

For 1,3-diacetylbenzene or 1,4-diacetylbenzene, mono- and di-manganated derivatives can be obtained depending on the stoichiometry of the reaction mixture (e.g. **6–9**),[32] although even with a ratio of $PhCH_2Mn(CO)_5$:diacetylbenzene of 1:1 some disubstituted product forms. In each case only the sterically least hindered isomer is found.

There are very few other examples in the literature of polymetallated benzene rings, so these species should prove to be especially useful in synthesis.

With 1,3,5-triacetylbenzene no trimanganated species was observed, which is not surprising given the sterically protected nature of potential *ortho* sites for attack by manganese.[32]

When the C=O group is incorporated into a fused ring, orthometallation still proceeds normally; for example, with chromone,[33] compound **10** is formed.

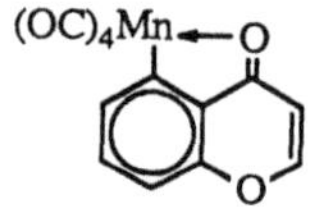

(10)

Other substrates that have given good yields of orthomanganated products include 1-tetralone,[13,24] benzosuberone,[24] flavanone, xanthone, thioxanthen-9-one, thiochroman-4-one, 1,2,3,4-tetrahydro-9-acridanone, 10-methyl-9(10H)-acridone, dibenzosuberone, and dibenzosuberenone.[33,35]

B. From Quinones

With quinones such as anthraquinone or duroquinone, reaction with $MeMn(CO)_5$ gave highly colored and unstable products which could not be characterized.[10] It is possible[36] these are radical adducts related to those formed between $Mn_2(CO)_{10}$ and quinones. However $MeRe(CO)_5$ and anthraquinone did give[10] orthometallated complexes such as (**11a**) and (**11b**), and $PhCH_2Mn(CO)_5$ gave the manganese analogues, albeit in low yields.[13]

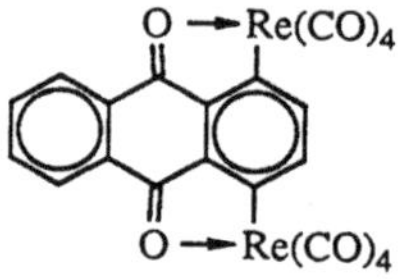

(11a)

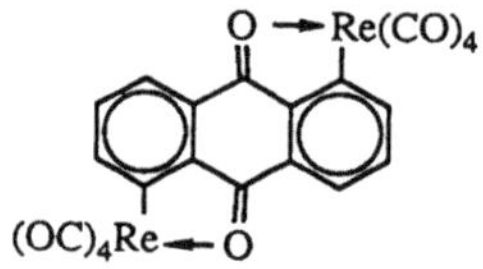

(11b)

C. From Heteroaromatic Ketones

Various acetyl-thiophenes, -furans and -*N*-methylpyrroles can be orthomanganated [see Table 1, section (d)], although only with the thiophenes are the yields up to the usual satisfactory levels, possibly because of the lower aromaticity of the other heterocyclic rings.[37] 2-Acetylpyrrole reacts with $PhCH_2Mn(CO)_5$, possibly by cyclometallation at the N–H bond, but no products have been properly characterized. With 2-acetyl-*N*-methylpyrrole normal manganation does occur in the 3-position, and 3-acetylindole gives **12** in good yield, probably because there is now no possibility of forming a chelate ring involving the C=O group and the N atom.[37.]

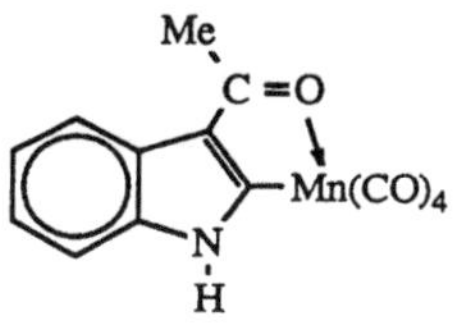

(12)

D. From Aromatic Aldehydes

Attempts to orthomanganate benzaldehyde have met with no success.[13,37] However the presence of OMe or NMe_2 substituents in the *para* position allows for direct metallation in good yields.[37] The reasons for this are not clear but the electron-releasing substituents enhance the donor strength of the aldehyde O atom [as is evident from the lowering of the C=O stretching frequencies of the substituted benzaldehydes to 1685 and 1660 cm^{-1} respectively from that (1700 cm^{-1}) of the parent molecule]. This would enhance complexing of the donor O atom to $PhCH_2Mn(CO)_5$, and the metallation step would occur from a complex in which there is increased electron density on the manganese, which, at least for an oxidative addition mechanism,[38] may provide activation. Mechanisms are further considered in Section J.

Both 2-formyl- and 3-formyl-thiophene can be cyclometallated with $PhCH_2Mn(CO)_5$ under normal conditions[32] but yields are only 10–20%.

E. From Aromatic Esters

Methyl benzoate gives only trace amounts of the orthomanganated derivative **13** under the usual conditions.[13,33] This is not due to any instability of the product since Casey and Bunnell[39] have prepared **13** by an indirect route, outlined in Eq. 3, and once formed **13** has similar properties to other cyclomanganated species.

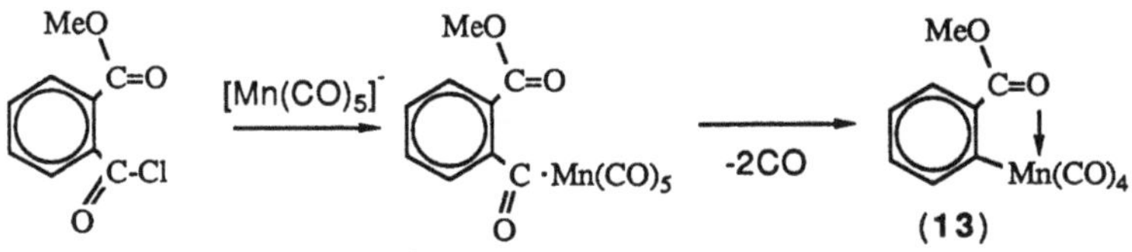

(13) (3)

As with benzaldehyde, a 4-OMe substituent on the aromatic ring of benzoate esters does allow metallation to proceed in good yield; both methyl 4-methoxybenzoate and methyl 3,4,5-trimethoxybenzoate give the expected derivatives on treatment with $PhCH_2Mn(CO)_5$.[33,40] However, there is still unpredictability in this area since methyl 3,5-dimethoxybenzoate is also effectively orthometallated in spite of having no 4-OMe group. Furthermore, isopropyl benzoate *does* give[13] good yields of the orthomanganated complex with $PhCH_2Mn(CO)_5I$. There is no obvious explanation for the difference in behavior between the methyl or isopropyl esters.

F. From Benzamides

Benzamide does not undergo orthomanganation. Alternative reactions involving the NH_2 group may be responsible since with *N,N*-dialkylamides normal reaction takes place to give species such as **14**, R = Me, Et, but not for R = Pr^i, Ph, possibly for steric reasons.[37] There is no sign in these reactions of the *N*-donor isomer **15**, analogous to those formed by orthomanganating the amine $PhCH_2NMe_2$ or the imine PhCH═NMe.[17]

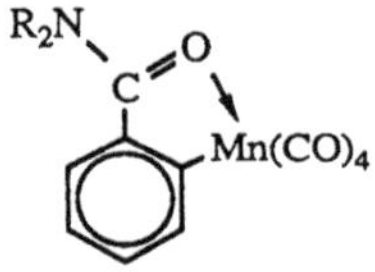

(14)

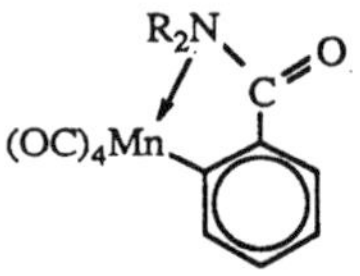

(15)

G. From *N*-Acyl Heteroaromatics

N-acetyl-pyrrole and -indole give **16** and **17** in good yield.[37] These differ significantly from the complexes formed from the other substrates discussed above in that a nitrogen atom is now incorporated into the five-membered metallocyclic ring, and this has a distinctive effect on the spectroscopic properties of the complexes and leads to different types of products in their reactions (see below). For *N*-benzoylindole three possible isomers can be envisaged since metallation could occur at the five-membered ring at C(2) to give the analogue of **17**, at the fused benzenoid ring at C(7), or at the phenyl ring of the benzoyl group (cf. benzamides). However only the first alternative is realized, despite this leading to strain in the fusion of two five-membered rings.[37]

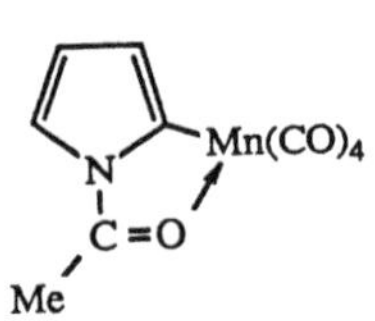

(16)

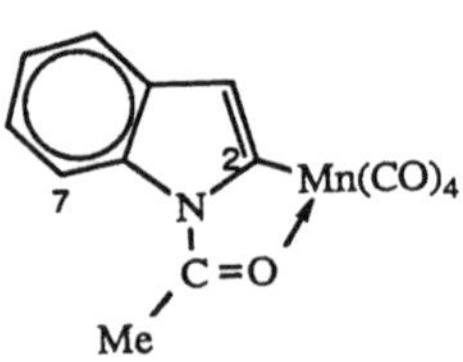

(17)

H. From Non-Aromatic Manganocycles

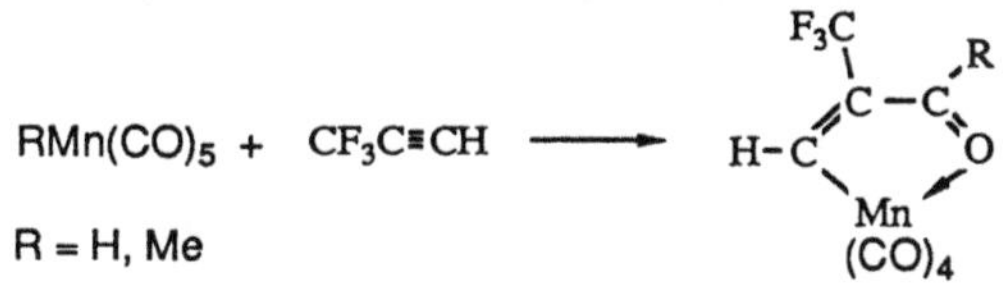

(4)

In 1968 Harbourne and Stone[41] reported that $HMn(CO)_5$ or $MeMn(CO)_5$ reacted with the alkyne F_3CCCH according to Eq. 4 to generate species with five-membered chelate rings directly analogous to those discussed above. This was then extended in a detailed study by Booth and Hargreaves,[42,43] who showed the reaction of $HMn(CO)_5$ or $RMn(CO)_5$ (R = Me, Ph) with alkynes was quite general and a range of cyclometallated α,β-unsaturated ketones **18** could be thus obtained. Acyl manganese pentacarbonyl complexes such as $MeC(O)Mn(CO)_5$ could also be used and sometimes gave improved yields.

Booth and Hargreaves proposed the reaction sequence given in Eq. 5 to explain the products obtained.

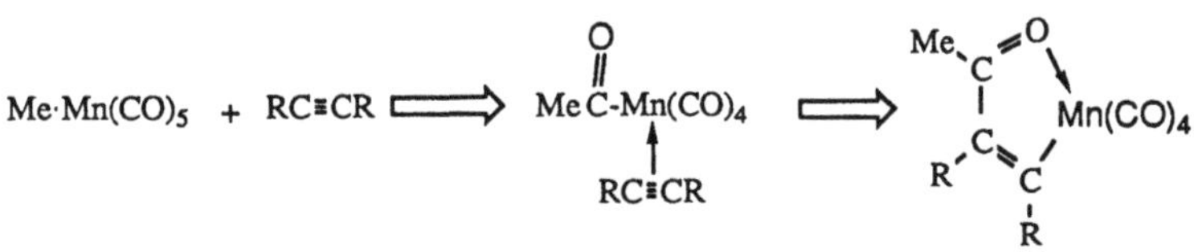

(5)

(18)

A by-product of these reactions was assigned structure **19** on spectroscopic evidence,[43] and this has been recently confirmed by the X-ray structural characterization of **19** (R = Ph).[44] These η^5 species presumably arise through double alkyne insertion and cyclization.

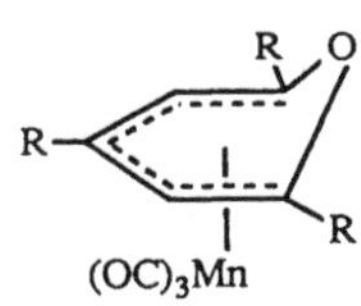

(19)

More recently this method of synthesis of cyclomanganated complexes of type **18** has been extended in a major study by DeShong and co-workers[25] by including $PhCH_2Mn(CO)_5$ as the starting complex and using a wider range of alkynes. For the less reactive manganese complexes pressures in the range 2–10 kbar were found to be an advantage.

It is noteworthy that for unsymmetrical alkynes the insertion usually occurs regiospecifically to give only one isomer of the manganocycles, although the reaction of $MeMn(CO)_5$ with PhCCCOOEt gave both isomers **20** and **21** with the former slightly more favored.[25]

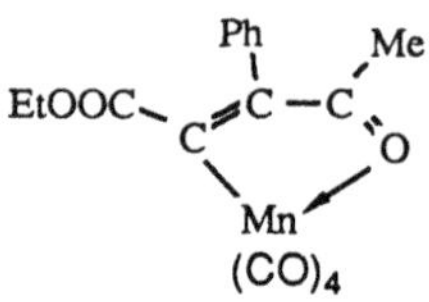

(20)

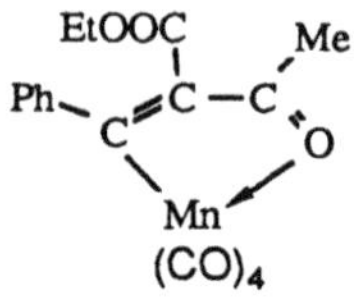

(21)

There are some examples of the preparation of these non-aromatic complexes by the direct metallation route, such as Eqs. 6 and 7.[13,32,44]

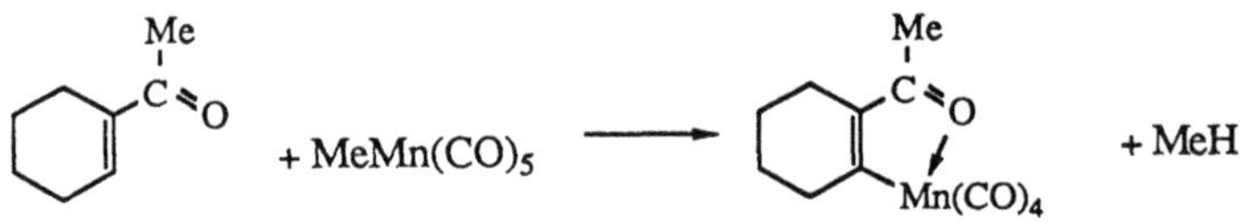

(6)

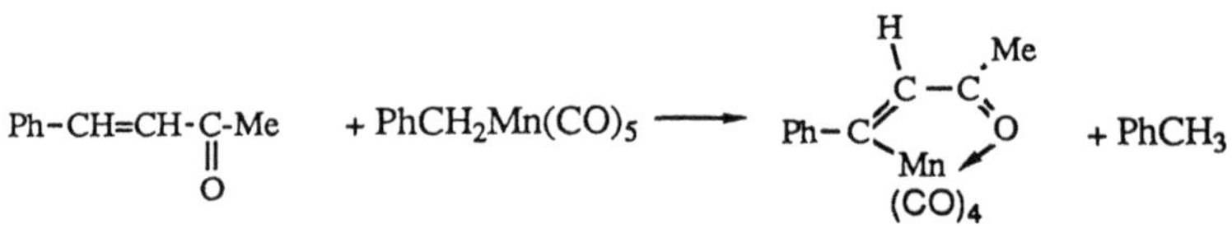

(7)

The direct method has some advantages in that α,β-unsaturated ketones are inexpensive reagents compared to alkynes. Further work is needed in light of the potential uses for complexes such as **18** as reagents in organic synthesis. The entries in Table 1, section (i) indicate the wide range of compounds of this type that is accessible.

DeShong's group has also investigated the corresponding reactions of $RMn(CO)_5$ (R = Me, $PhCH_2$) with *alkenes*.[25,26] Here, high pressures were essential for reaction but conditions were established which lead generally to complexes of type **22** which contain manganacyclic rings, but differ from those above in that there is no C=C bond incorporated, that is they are saturated analogues of the examples discussed above. There are also other examples of this type prepared by less direct routes [see Table 1, section (j)].

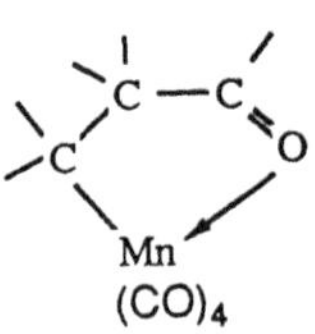

(22)

I. From Diphenylmercury

Haupt and co-workers[45] have recently described the preparation of a range of orthomanganated aryl ketones from the reaction of diphenylmercury with alkylmanganesepentacarbonyl compounds, according to Eq. 8.

$$RMn(CO)_5 + (C_6H_5)_2Hg \rightarrow RC(O)C_6H_4Mn(CO)_4 + Hg + C_6H_6 \qquad (8)$$

($R = Me$, Et, Pr, $PhCH_2$; refluxing toluene, 1 h; yields 65–80%)

Related complexes with EPh_3 (E = P, As, Sb) or $P(OR)_3$ replacing CO ligands were also prepared by the same method, starting with the appropriately substituted manganese complex. This reaction is an interesting alternative to the more common direct metallation route.

J. Mechanistic Considerations

Little can be deduced about mechanisms of orthomanganation from the paucity of data available. Aryl substituent effects are variable. For ketones, yields are not strongly dependent on electron-donating or -withdrawing properties of substituents or on their ring positions, while for aldehydes, as discussed in Section D, the picture is different. Selectivity between rings in benzophenones is not strongly dependent on substituents.

In his 1977 review, Bruce[14] discussed ring substituent effects on the regioselectivity of orthomanganation of azobenzenes, which indicated nucleophilic character for the manganese. Recently, Ryabov[38] reviewed mechanisms of intramolecular metallation of C–H bonds in transition metal complexes and although he did not discuss orthomanganation reactions, he related enhanced electron density on the attacking metal center to facilitation of an oxidative addition reaction. He did point out, however, the ambiguity of interpretation arising when a substituent may influence electron density at both the metal and the C–H bond, as applies for ring substituents in orthomanganation of acetophenones for instance.

Some results[46] which avoid this ambiguity were provided by comparison of reactions of $PhCH_2Mn(CO)_5$ and $PhCH_2Mn(CO)_4PPh_3$. To obtain what amounts to kinetic data free of contribution from steps preceding the actual metallation, we measured the product ratio of *ortho*-D and *ortho*-H manganated ketone (Eq. 9) for the reaction of mono-*ortho*-deuteriated acetophenone **25** with $PhCH_2Mn(CO)_5$.

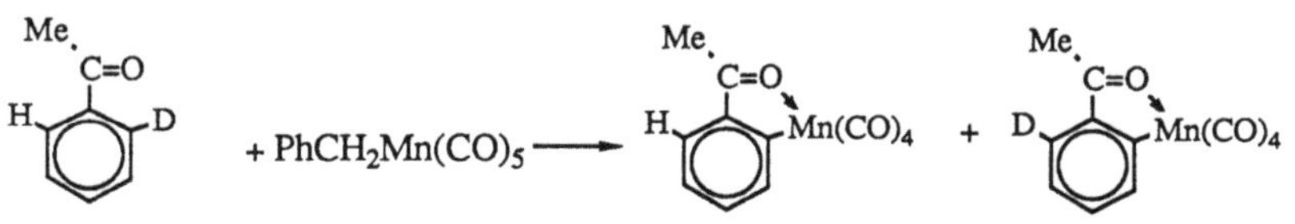

(9)

(25) 24% 76%

As the ratio shows, there is a threefold rate preference for substitution of H over D *in the C–H bond-breaking step* of the reaction, that is, there is a kinetic isotope effect (k_H/k_D) of about 3 when Mn replaces H at the aryl carbon. Such selectivity disappears for $PhCH_2Mn(CO)_4PPh_3$ however: $k_H/k_D \sim 1.1$. If C–Mn bond-formation is synchronous with C–H bond-breaking, the coordinated phosphine must therefore increase the reactivity of the metal center so much that it no longer discriminates between C–H and C–D. Such activation by electron-donation from the phosphine ligand is consistent with the demands of an oxidative addition mechanism.[38] However more data across a range of reactions is required for reliable interpretation and the ease of preparation of mono-*ortho*-deuteriated aryl carbonyl compounds from orthomanganated precursors (see Section IV–H) opens up this area for study.

III. PROPERTIES OF CYCLOMANGANATED COMPLEXES

A. General

All of the cyclomanganated complexes discussed in this review are easily handled. They are mostly yellow-to-orange crystalline solids with sharp melting points, although some can only be isolated as oils. They are not particularly susceptible to oxidation and can be safely handled in air, although they are best refrigerated for long-term storage, and reactions with them are generally carried out under an inert atmosphere. The complexes tend to dissolve readily in all organic solvents which means there are no constraints for spectroscopy and reactivity studies, while chromatography on silica or alumina requires no special precautions. Compared with many organometallic reagents, orthomanganated compounds are a dream to work with!

B. Infrared Spectra

The cyclometallated complexes are readily identified by the characteristic pattern of the CO ligands of the $Mn(CO)_4$ group.[10,25,31,3] For a *cis*-$L_2Mn(CO)_4$ molecule four bands are predicted in the ν_{CO} region of the spectrum. There is usually a medium intensity band at around 2080 cm^{-1}, two intense bands near 2000 cm^{-1} (which are often accidentally degenerate and not resolved) and a strong band around 1945 cm^{-1}. There is surprisingly little variation in band positions with different organic fragments, although the *N*-acetyl complexes such as **16** or **17** have bands shifted by approx. 15 cm^{-1} to higher frequencies, presumably because of the electronegative nitrogen atom in the metallocyclic ring.[37] Similarly the saturated species **22**, with no C$=$C in the ring, have bands that are shifted to lower frequencies,[25] presumably because there is less electron density removed from the manganese by delocalization in the metallocyclic ring (cf. discussion of X-ray crystal structures below).

The aromatic ketone complexes show two medium intensity bands in the 1500–1600 cm^{-1} region, which are associated with vibrations involving the C=O group and the aromatic rings, with the lower energy one of the pair being assigned mainly to the carbonyl stretch. Coordination lowers the frequency by about 150 cm^{-1} compared to the value for the free ligand.[10,31] For the enone complexes of type **18** a band in the region 1500–1600 cm^{-1} was tentatively assigned to $\nu_{C=O}$,[43] but the later study reports values that suggest the peaks in the 1400–1500 cm^{-1} region may be more probable for this stretch.[25] For all of these species, coupling with modes associated with the other multiple bonds in the molecules will preclude any absolute assignment. For the saturated complexes **22** the bands observed in the 1620–1650 cm^{-1} region can perhaps be more definitely associated with the coordinated ketone group.[25]

C. NMR Spectra

^{1}H spectra are routinely available and are useful for purity checks and for determining regiospecificity. ^{13}C data have been reported for a limited range of complexes involving aromatic ketones,[31,37] and some generalizations can be made. There are four signals around δ 200 ppm: three broad ones of intensity 1:1:2 which can be assigned to the terminal CO groups, and the fourth, a sharp signal to the carbon of the ketone group, having been shifted by 12–21 ppm to lower field on coordination. The aryl carbon atom bonded to manganese is normally found in the range δ 190–210 ppm, although it can be as low as 166 ppm; when compared with the free ketones it appears coordination shifts the signal of the metallated carbon by 60–75 ppm to lower field, those on either side by 9–14 ppm, with only small shifts for the remaining aryl carbon signals.[31]

More recently, ^{13}C data have been reported for the non-aromatic complexes of type **18** and **22**. The same pattern is generally observed although with a better spectrometer and lower symmetry molecules the lowfield signals from the CO ligands were fully resolved. Detailed assignments were not made.[25]

De Shong's group[47] has also reported some ^{55}Mn NMR data for cyclometallated nonaromatic ketones of type **18** and **22**. Chemical shifts of about –680 ppm found for "saturated" species **22** and about –810 ppm for "unsaturated" species **18**, were very different from those of $RMn(CO)_5$ species.

D. X-ray Crystal Structures

A range of cyclomanganated complexes has been structurally characterized, and the structural data for some representative examples are given in Table 2. It is immediately apparent that the cyclometallated ring has a remarkably constant geometry in a wide range of derivatives. The metallocyclic ring is always essentially

Table 2. Selected Bond Lengths (Å) for Some Cyclomanganated Compounds[a]

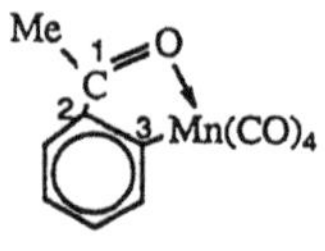
(T1)

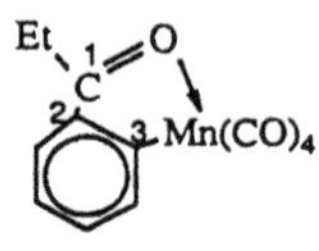
(T2)

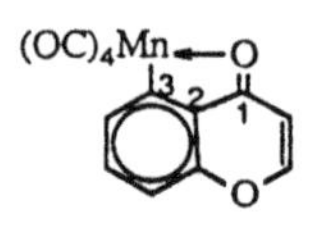
(T3)

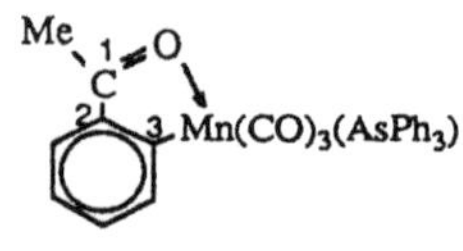
(T4)

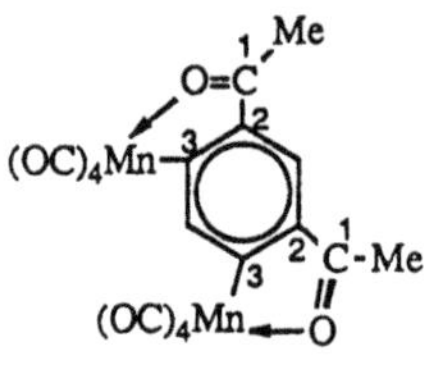
(T5)

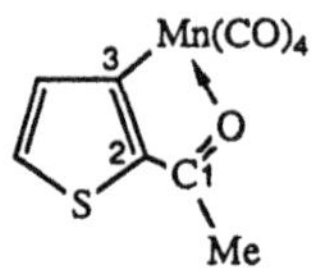
(T6)

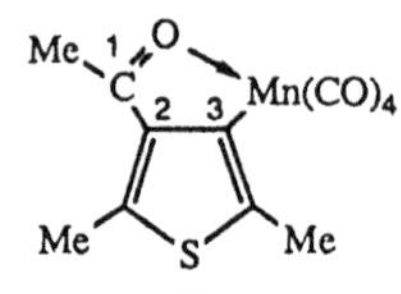
(T7)

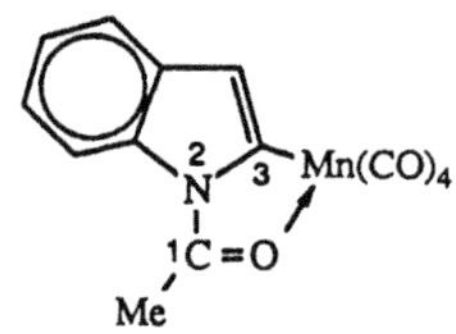
(T8)

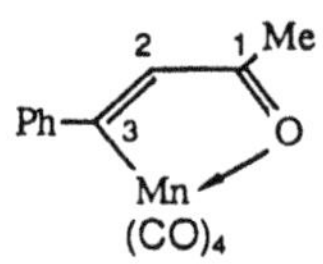
(T9)

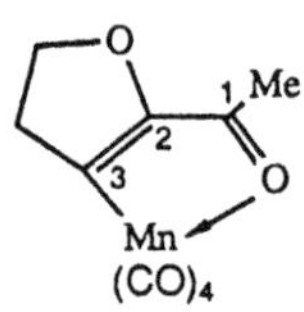
(T10)

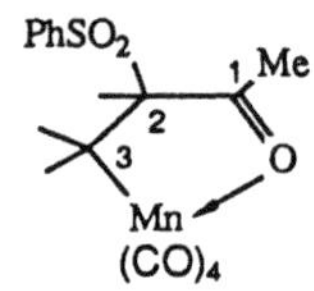
(T11)

Compound	*Mn-O*	$C_1{=}O$	C_1-C_2	C_2-C_3	*Mn*-C_3	*Ref.*
T1	2.055(2)	1.244(3)	1.455(3)	1.406(3)	2.042(2)	10
T2	2.064(3)	1.248(4)	1.444(4)		2.074(2)	45
T3	2.080(2)	1.258(3)	1.413(4)	1.426(3)	2.039(3)	35
T4	2.05(2)	1.20(3)	1.45(3)		2.07(2)	45
T5	2.055(3)	1.252(4)	1.455(5)	1.420(5)	2.036(3)	32
T6	2.069(3)	1.263(4)	1.418(5)	1.392(5)	2.030(4)	31
T7	2.055(2)	1.251(3)		1.454(3)	2.039(2)	31
T8	2.054(2)	1.249(3)	1.353(3)[b]	1.443(3)[b]	2.012(2)	37
T9	2.053	1.251	1.410	1.360	2.040	25,47
T10	2.06	1.29			1.99	48
T11	2.055	1.242	1.486	1.517	2.117	25,47

Notes: [a]These are all the relevant examples to date with "simple" ligands. Others incorporating cyclic phosphine ligands and ferrocenyl groups have not been included.
[b]C-N bonds.

planar with Mn–O and Mn–C(sp^2) distances in the ranges 2.05–2.08Å and 2.00–2.07Å respectively. Both of these are shorter than the sum of the covalent radii of the atoms involved, suggesting some multiple bonding. The C═O bonds range over 1.24–1.26Å (excluding the less accurate data), slightly longer than expected for a true double bond, while the C_1-C_2 distances are again short. These features are especially marked for compound **T8** (Table 2), which also incorporates a nitrogen atom into the metallocyclic ring. All of this data is consistent with delocalized π-bonding over the whole ring, and this has led to the suggestion that the ring is best viewed as a manganafuran species.[25] The only exception to this generalization is compound **T11**, where the Mn–C bond is longer because C_3 is sp^3 hybridized and delocalized bonding is no longer viable. These interpretations are all consistent with the results of Fenske-Hall MO calculations.[47]

Other general features that have been noted include the classification of the CO groups, according to the Mn-C distances. The longest Mn–CO bonds are those to the carbonyl ligands *trans* to each other, the intermediate one to the carbonyl group *trans* to the C(aryl), and the shortest to the carbonyl *trans* to the O donor atom. This is readily understood in terms of the relative π-bonding properties of the atoms around the manganese atom[10,31,37,48,49] and is well-correlated with the ^{13}C NMR shifts of the CO ligands.[31]

IV. CYCLOMANGANATED COMPOUNDS IN ORGANIC SYNTHESIS

A. Introduction

Although our work has now extended to other cyclomanganated compounds, it was apparent initially that the ease of preparation and the stability of orthomanganated ketones might be usefully exploited in their *ortho*-functionalization, particularly if they could be made to couple with alkenes in the presence of Pd(II), as in Heck reactions of aryl-mercury[18,51] and -halo[18,23] compounds. There was also some potential within an intermediate of the type proposed for such reactions[18,51a] to undergo cyclization by the palladiated carbon in the special case where an *ortho*-ketone function is adjacent (Eq. 10).

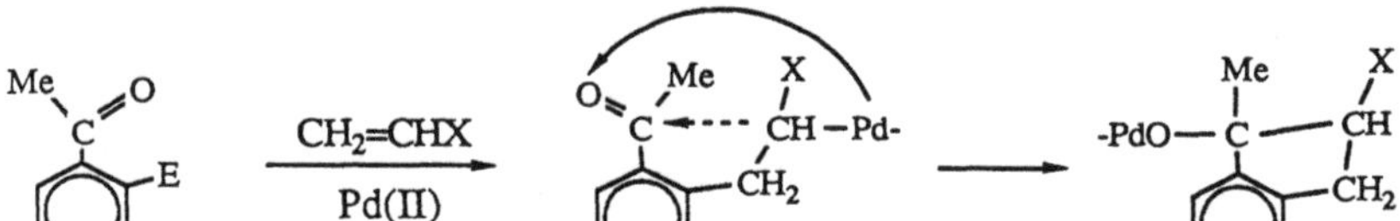

(10)

A literature search revealed only one reaction with the potential to give such cyclization: the Pd(0)-promoted coupling of 2-bromoacetophenone with methyl acrylate.[52] However, this reaction gave a high yield of the normal Heck product, the arylated alkene, and there was no cyclization. Our early results[53] revealed, however, that with orthomanganated precursors, cyclized products can be formed in large excess over arylalkenes under suitable conditions.

B. Reactions of Orthomanganated Aryl Carbonyl Compounds with Alkenes in the Presence of Pd(II)

From the restricted range of reactions so far studied, there is sufficient diversity in reactivity to suggest considerable potential for these reactions in organic synthesis. The coupling reactions normally generate in excellent yield as Eq.11 illustrates,[35] three main coupling products, arylalkene (**25**), arylalkane (**26**) and indene (**27**), whose ratio varies with reaction conditions and with substrate. Often a fourth minor product is the parent ketone formed by demetallation at the aryl carbon.

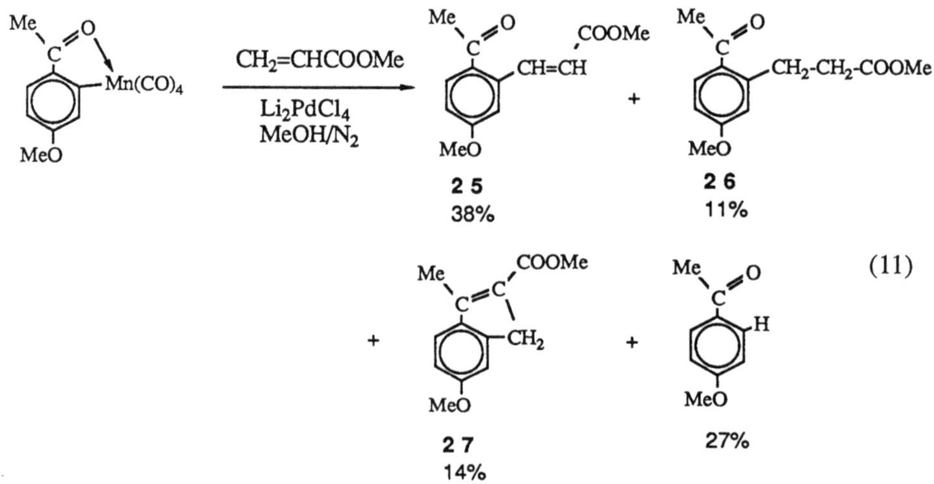

(11)

For reactions promoted by $PdCl_2$, study has so far been centered on methyl acrylate (methyl propenoate) and methyl vinyl ketone (3-propen-2-one) (Section 1 below), but a survey has been made of other alkenes so as to provide a contrast with the known types of products obtained in coupling reactions with the major Heck reaction precursors, aryl-mercury[18,51] and -halo compounds [18,23](Section 2). Coupling reactions catalytic in Pd(II) or with activation by trimethylamine oxide are covered in Sections 5 and 6.

1. Methyl Acrylate and Methyl Vinyl Ketone: General Procedures for Reactions of Alkenes with Orthomanganated Ketones in the Presence of Li_2PdCl_4.

Reactions have been routinely carried out[53] either in dry methanol under N_2 at room temperature or in dry acetonitrile under reflux. The latter conditions favor cyclization to indene products.[53] Palladium chloride is solubilized as Li_2PdCl_4 by the addition of LiCl and other reagents are then added. Precipitation of palladium usually soon starts, but because of apparent variations in reactivity, the reactions are normally continued with stirring for a few hours or overnight. Standard methods are illustrated by reference to the reaction[35] in Eq. 11.

Reaction of η^2-(2-acetyl-5-methoxyphenyl)tetracarbonylmanganese with methyl acrylate and Li_2PdCl_4 in MeOH. $PdCl_2$ (112 mg, 0.63 mmol) and LiCl (107 mg, 2.5 mmol) were added to dry methanol (10 mL) and stirred under N_2 for 3 h. The orthomanganated ketone (200 mg, 0.63 mmol) and methyl acrylate (270 mg, 3.1 mmol) were added. Precipitation of Pd was evident after 30 min, but stirring was continued at room temperature overnight. The mixture was then filtered to remove metal, the filtrate run through a short silica column, and the solvent evaporated. The residue was chromatographed on a silica layer (400 × 200 × 2 mm) using a mixture (8:1 v/v) of petroleum spirit (b.p. 60–80 °C) and ethyl acetate as eluent. Four successive developments were required for complete separation. The bands in order of decreasing R_f gave the following compounds after extraction and solvent removal: unreacted orthomanganated ketone (9%); the cyclic product, methyl 3-methyl-6-methoxyindene-2-carboxylate (**27**), white plates, mp 91–2 °C (14%); the parent ketone, 4′-methoxyacetophenone (27%); the arylalkane, methyl 3-(2-acetyl-5-methoxyphenyl)-propanoate (**26**), an oil (11%); and the arylalkene, methyl (*E*)-3-(2-acetyl-5-methoxyphenyl)-propenoate (**25**), an oil (38%).

The reaction in dry MeCN as solvent[53] was carried out identically except that the reaction mixture was refluxed without exclusion of air for 3 h after addition of the organic reagents. The only products separated in significant yield were methyl 3-methyl-6-methoxyindene-2-carboxylate (**27**) (76%) and methyl (*E*)-3-(2-acetyl-5-methoxyphenyl)-propenoate (**25**) (8%).

In Table 3 are listed yields for the above and related reactions[33,35,40,53] of other orthomanganated acetophenones and of 2-acetylthiophene. The results for orthomanganated 3-acetyl-2,5-dimethylthiophene[35,40] are excluded here because they appear in Table 5. Major products, particularly the indenes, could probably be isolated by crystallization rather than chromatography in many cases.

Some significant points about product yields are:

1. Indene formation is generally better achieved in MeCN under reflux than in MeOH. That the 2-acetylthiophene reactions produce no indene-type products at all may reflect the strain of fusing two 5-membered rings, although this is not prohibitive in the 3-acetyl-2,5-dimethylthiophene case (see Table 5);
2. Arylalkenes of *E* rather than *Z* configuration are formed as is consistent with similar couplings with arylmercury compounds.[18,51] An exception noted later (Scheme 1) is the result for acrylonitrile which gives[35] both isomers;

Table 3. *Yields (%)[a]* of Products From Li_2PdCl_4-Promoted Coupling of Orthomanganated Ketones with Methyl Acrylate and Methyl Vinyl Ketone in MeOH and MeCN

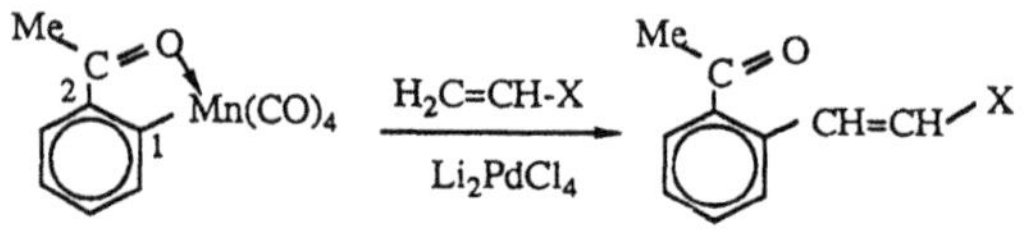

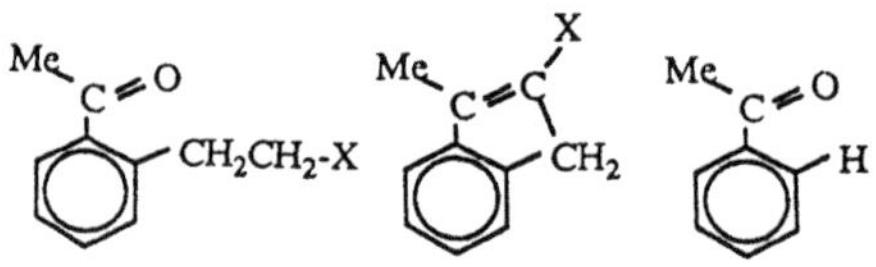

Orthomanganated Ketone Ring substituents [Parent ketone]	*X*	*Solvent[b]*	*E-Arylalkene*	*Arylalkane*	*Indene*	*Parent Ketone*
None	COOMe	MeOH	75			
[Acetophenone]	COOMe	MeCN	34	10	8	
	COMe	MeCN	14		61	
5-OMe	COOMe	MeOH	38	11	14	27
[4′-Methoxyaceto-	COOMe	MeCN	8		76	
phenone]	COMe	MeCN	10		46	21
3,4,5-(OMe)3	COOMe	MeOH	36		22	23
[2′,3′,4′-	COOMe	MeCN	12		66	
Trimethoxyaceto-phe-none]	COMe	MeCN	12	3	80	
4,5,6-(OMe)3	COOMe	MeOH				90
[3′,4′,5′-Trimethoxyaceto-phe-none]	COOMe	MeCN			89 (5)[c]	
[2-Acetylthiophene][d]	COOMe	MeOH	63e			
	COOMe	MeCN	60[e]	16[e]		3[e]
	COMe	MeCN[f]	40[e]	56[e]		
	COMe	MeCN[g]	20[e]	60e		

Notes: [a]There are gaps in the Table where yields are low and undetermined or where products have not been detected at all.

[b]Reactions in MeOH at room temperature; in MeCN under reflux (see text).

[c]Without $PdCl_2$ and LiCl; other unstable product material decomposed on silica in attempted chromatographic purification.

[d]No cyclization product has been found for this orthomanganated ketone; see Scheme 1 for other alkenes in MeOH.

[e]Corresponding structures to those of the acetophenone derivatives at the head of the Table.

[f]Stirred at room temperature for 90 min.

[g]Refluxed for 5 h.

3. For the arylalkane products there is as yet no general correlation of yields with structure or reaction conditions, although yields appear to be higher in the 2-acetylthiophene case when indene yields are low (perhaps because of difficulty of cyclization);
4. Demetallation to form the parent ketone occurs to a variable degree (final column of Table 3), but is a major reaction only for one attempted coupling, that of methyl acrylate at the sterically crowded center of orthomanganated 3′,4′,5′-trimethoxyacetophenone and then only in MeOH, not MeCN under reflux; and
5. Structures of other unknown but minor products have not generally been pursued. The possibility of traces of products derived from carbonyl insertion into an aryl-metal bond is suggested, however, by the observation (g.c.m.s.) of a common minor (<3%) product, assigned the vinyl phthalide structure **28**, formed during the coupling reaction of both methyl acrylate and methyl vinyl ketone with orthomanganated 2′,3′,4′-trimethoxyacetophenone.[35,40] Carbonyl insertion is more predominant in the reaction of orthomanganated acetophenone with excess Br_2 (Section E-2) which gives a high yield of the dibromide of the corresponding vinyl phthalide.

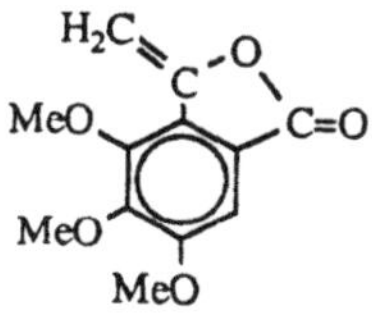

28

2. Coupling Reactions of an Orthomanganated Ketone with Other Alkenes

η^2-(2-Acetylthienyl)tetracarbonylmanganese was chosen to determine the types of aryl functionalization obtainable from aryl-manganese precursors with a range of alkenes for comparison with those previously reported for Heck reactions.[18,23,51] Any other orthomanganated ketone would have made comparison difficult because of formation of cyclic products which are not possible for simple aryl-mercury or -halo compounds (but see subsequent work described in Section 3 below).

The reactions with alkenes were carried out in methanol using the standard conditions described above. The results are depicted in Scheme 1. The results for those alkenes which introduce functions differing from those in arylmercury coupling reactions[18,51] are tabulated for comparison in Table 4; aryl-halo compounds are not included because as far as we are aware they always give[23] the same product types as the mercury compounds. The comparison is not an exact one because of differences in reaction conditions, but Table 4 reveals some synthetically exploitable variation in the nature of the coupling products in the Mn as opposed

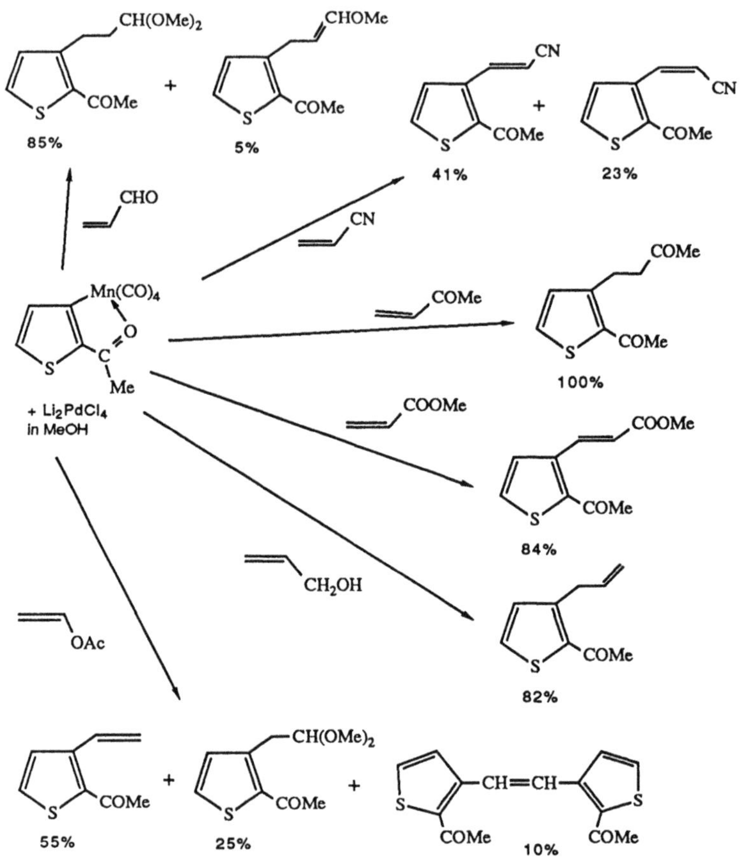

Scheme 1.

Table 4. Major Functional Groups Introduced by Pd(II)-Promoted Coupling of Various Alkenes at Metallated Aryl Carbon: Differences Between Manganese[a] and Mercury[b]

Alkene	$H_2C = CHCOMe$	$H_2C = CHCHO$	$H_2C = CHCH_2OH$
Ar-Mn[a]	$-CH_2CH_2COMe$	$-CH_2CH_2CH(OMe)_2$	$-CH_2CH = CH_2$
Ar-Hg[b]	$-CH = CHCOMe$	$-CH = CHCHO$[c]	$-CH_2CH_2CHO$[e]
	$[-CH_2CH_2COMe]$[d]		

Notes: [a]Based on the results in Scheme 1 for reaction in MeOH. Other alkenes in Scheme 1 give the same coupling product types as arylmercury compounds so are excluded from this table.

[b]For arylmercuric chloride and diarylmercury compounds, commonly in MeCN and MeOH, in some cases with $CuCl_2$.

[c]Reference 51a and reference 18, p. 268.

[d]Reference 54 and reference 18, p. 272; a two-phase CH_2Cl_2/HCl(aq.) system with $(n\text{-}Bu)_4NCl$ is used [see text: Section 3(iii)].

[e]Reference 51b and reference 18, p. 276.

to the Hg (or halogen) cases. Possible explanations for these reactivity differences are considered in Section 3.

3. A Possible Role for Displaced Manganese in Cyclization and Other Coupling Reactions with Alkenes

In this section, general mechanistic aspects are considered. These provide the rationale for undertaking the experiments which showed a marked difference in the extent of indene formation from an orthomanganated thienyl ketone as compared with its orthomercurated analogue. The results of these experiments are rationalized, along with others from Section 2 which also indicate a significant difference in alkene-coupling reactivity between Mn and Hg precursors.

General Mechanistic Considerations. The formation of arylalkene, arylalkane and indene products can be envisaged for any aryl-element precursor according to the following generalized equations [E = $Mn(CO)_4$, HgCl or other element functional group; X = COOMe for methyl acrylate; $PdCl_2$ and -PdCl are used for simplicity, but Pd will be further complexed by Cl^- or another donor under our reaction conditions]. Equations 12–14 are as proposed by Heck [23,51] for arylation of alkenes. The elimination of metal hydride (Eq. 13) gives the (*E*)-arylalkene. The mechanistic and stereochemical aspect of these reactions (*syn*-addition and *syn*-elimination) have been discussed.[23]

$$\mathrm{ArE} \xrightarrow[\mathrm{PdCl_2}]{\mathrm{CH_2{=}CHX}} \mathrm{ArCH_2\text{-}\underset{\displaystyle PdCl}{\underset{|}{C}H}\text{-}X} \qquad (12)$$

$$\mathrm{Ar\text{-}\underset{\displaystyle H}{\underset{|}{C}H}\text{-}\underset{\displaystyle PdCl}{\underset{|}{C}H}\text{-}X} \longrightarrow \mathrm{ArCH{=}CHX + HPdCl} \qquad (13)$$

$$\mathrm{HPdCl} \longrightarrow \mathrm{Pd + HCl} \qquad (14)$$

Formation of arylalkane products is likely to occur through protonolysis (Eq. 15), possibly mediated when X is a carbonyl function by initial protonation at carbonyl oxygen, with subsequent ketonization of enolic product.

$$\mathrm{ArCH_2\text{-}\underset{\displaystyle PdCl}{\underset{|}{C}H}\text{-}X + HCl} \longrightarrow \mathrm{ArCH_2\text{-}CH_2X + PdCl_2} \qquad (15)$$

This reaction releases Pd in an unchanged oxidation state, but it nevertheless depends on the prior formation of HCl (Eqs. 13 and 14). It is clear that the amount of arylalkane cannot exceed the amount of arylalkene unless protic solvents (e.g.

MeOH) or contaminants (e.g H_2O, possibly not until workup) take the place of HCl in Eq. 15. The source of the additional protons when the yield of arylalkane exceeds that of arylalkene deserves study and is further discussed below.

For indene formation, envisaged as in Eq. 16 (X = H, PdCl or $Mn(CO)_4L$; L = Cl^- or solvent), a proton is released (as HOX) in the elimination step, so the amount of indene is not limited by the extent of formation of HCl with arylalkene (Eqs. 13 and 14) under otherwise aprotic conditions. The absence of indene formation in the Pd(0)-promoted coupling of *ortho*-bromoacetophenone and methyl acrylate[52] referred to above suggested that the cyclized intermediate in Eq. 16 with X = Pd(II) either is not formed or is formed reversibly and does not lead to indene product, at least under the conditions concerned. We were therefore encouraged to check quantitatively the extent to which such cyclization and consequent indene formation are specific to manganese.

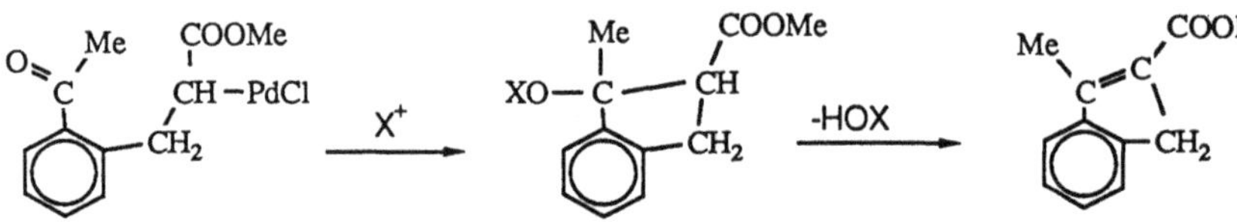

(16)

Experimental Study of the Specificity of Manganese in Promoting the Indene-Forming Cyclization Reaction of Orthomanganated Ketones. If Pd-promoted coupling reactions of alkenes with aryl-metal precursors involve the formation of the alkene-inserted palladiated intermediate (Eq. 12), it would be expected in the case of orthometallated ketones that any potential for the cyclization of the palladiated carbon to the ketone carbonyl (Eq. 16) would be realized irrespective of the metal in the precursor, that is, unless the displaced metal group or a product derived from it could itself promote the cyclization (e.g. as X^+ in Eq. 16).

The availability of orthomercurated aryl ketones *via* transmetallation of orthomanganated ketones with mercuric chloride (Section E-1) allowed us to determine the product distribution for methyl acrylate coupling with an orthomercurated precursor and compare it with that for the corresponding orthomanganated precursor under identical conditions. Orthomercurated and orthomanganated 3-acetyl-2,5-dimethylthiophene in acetonitrile as solvent were used for this study because we knew there was a good balance between the cyclized indene-type product and the arylalkene in the manganese case, so that any product ratio difference for the Hg case could be readily detected.

Conditions employed are in a footnote to Table 5, which shows cyclization to be much more strongly favored in the Mn than the Hg case. This suggests that the displaced manganese group must be effective at promoting cyclization in a way that the displaced mercury group is not. Some possible interpretations are considered in the next section.

Table 5. Product Ratio[a] for the Li_2PdCl_4-Promoted Reaction[b] in MeCN of Methyl Acrylate with Orthomanganated and Orthomercurated 3-Acetyl-2,5-dimethylthiophene

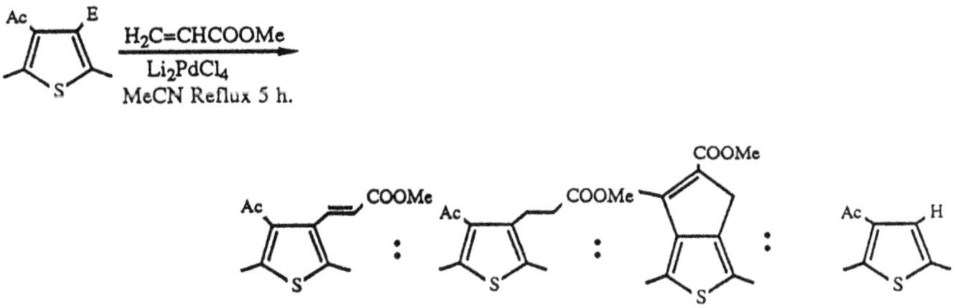

E = $Mn(CO)_4$	35[c]	5	58	2
E = HgCl	85[c]	3	3	9

Notes: [a]Product ratio, determined by g.c. separation and ion-trap detection without response factor calibration, is semiquantitative only. In the Hg, but not the Mn case, there were a number of other very minor (<2%) unidentified products.

[b]Reaction conditions: Orthometallated ketone (0.77 mmol), $PdCl_2$ (0.77 mmol), LiCl (3.1 mmol), H_2C = CHCOOMe (3.8 mmol) in dry MeCN (10 mL) refluxed for 5 h. (drying tube to exclude atmospheric moisture).

[c]A very small amount (<0.5%) of a product with the same mass and fragmentation factor but shorter retention time is assumed to be the *Z*-isomer.

A Specific Role for Displaced Manganese in Coupling Reactions Leading to Indenes and Other Products Alternative to Those from Arylmercury Compounds. Although the state of Mn when it is displaced by the electrophile $PdCl_2$ is not known with certainty, by analogy with electrophilic bromodemetallation using Br_2 in CCl_4 (Section E-2) for which $[BrMn(CO)_4]_2$ is a major product, displacement as $ClMn(CO)_4$ (or a ligand exchange product of this species) seems likely. It may remain coordinated or become recoordinated to the donor ketone carbonyl oxygen and by electron-withdrawal increase the susceptibility of the ketone carbonyl to addition by the palladiated carbon (Eq. 16; X = $Mn(CO)_4Cl$ or $Mn(CO)_4L^+$; L = solvent; the Mn function is abbreviated in the following discussion to [Mn]). If [Mn] is more effective at coordinating to oxygen and polarizing the carbonyl bond than displaced $HgCl_2$ from an orthomercurated precursor, the relatively high yield of indene in the manganese case is rationalized.

Similar polarization by [Mn] may explain why, for an orthomanganated precursor, allyl alcohol couples to introduce the allyl group (Scheme 1). Elimination of ClPdOH in the palladiated intermediate **29** may be preferred over elimination of ClPdH to form the enol of the saturated aldehyde which (Table 4) is formed[51b] from arylmercury precursors.

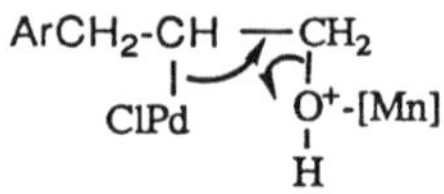

29

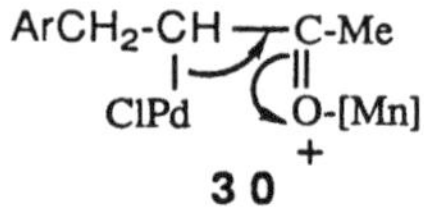

30

A similar activated elimination (**30**), with subsequent solvolysis of the product *O*-manganated enol, may account for the quantitative yield of arylalkane (saturated ketone) formed from methyl vinyl ketone in MeOH (Scheme 1). It is not easy, however, to reconcile this with the absence (Scheme 1) of arylalkane formation from the ester, methyl acrylate, for which activated elimination equivalent to that in **30** must be available. We have as yet made no comparative study with arylmercury compounds in MeOH to see whether they react with enones to give arylalkane products, though Heck reported[51a] one reaction—that of acrylaldehyde in MeOH with di-(2-naphthyl)-mercury. It gave arylalkene but the yield was only 8% and it is not clear whether arylalkane would have been detected if formed. There is one report[54] of arylmercury (and aryltin) compounds giving arylalkanes with enones instead of the normal arylalkenes. This involves Pd(II)-promoted coupling in a two-phase CH_2Cl_2/HCl(aq) system incorporating Bu_4NCl. Here the added acid present, rather than the displaced element, may be involved in directing the coupling to the arylalkane product.

With orthomanganated precursors, any traces of adventitious acid present, or produced through unknown side-reactions, could activate elimination (**29** and **30** with H in place of [Mn]). In the aprotic solvent MeCN, however, acid would have to be present at a stoichiometric level matching the amount of arylalkane formed, unless protonation does not occur until workup. The high yield of arylalkane formed in the reaction of orthomanganated 2-acetylthiophene with methyl vinyl ketone in acetonitrile (Table 3; last two lines) is a case in question. Clearly, further study is required to establish the proton source. It is interesting that methanol or small amounts of water are stated[23] not to interfere in the Pd(0)-catalyzed coupling reactions of aryl halides with alkenes to form arylalkenes, there being no reports of arylalkane by-products.

4. *Coupling Reactions of Orthomanganated Esters*

If cyclizations analogous to those with orthomanganated ketones could be made to occur orthomanganated esters would give indanones by ester substitution [cf. the facile indenone formation from orthomanganated esters and amides with alkynes (Section IV-C)]. However, only a few studies of Pd(II)-promoted reactions of orthomanganated esters with alkenes have so far been made and cyclization is not yet known.

The reaction of the methyl 3,4,5-trimethoxybenzoate derivative with methyl vinyl ketone in MeCN gave[35] neither indanone nor arylalkene (Eq. 17).

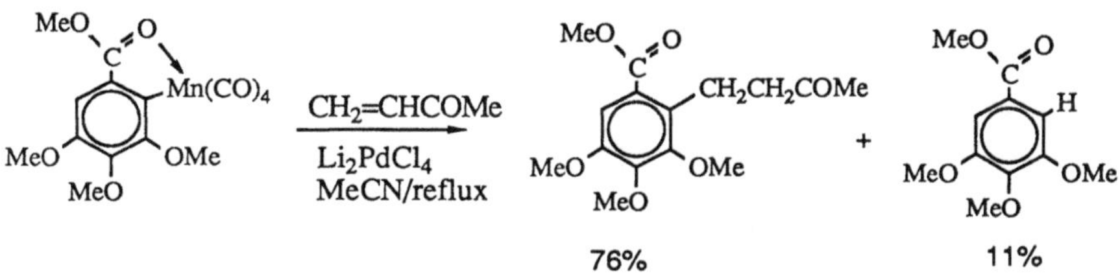

(17)

Palladium acetate has been uscd[35] in stoichiometric amount to promote the reaction of orthomanganated methyl 4-methoxybenzoate with methyl acrylate in acetonitrile. The reaction was done in the dark. Protiodemetallation was the main reaction (56%), with low yields of arylalkane (22%) and arylalkene (15%), suggesting that palladium acetate is not as useful as palladium chloride in promoting coupling. The same reaction, but using palladium chloride in catalytic amount, is described in the next section.

Two orthomanganated esters from the podocarpic acid field have also been studied[50] with methyl acrylate in MeCN, but with activation by Me_3NO (cf. Section 6 below). Arylalkane products are formed in good yield, with arylalkene as a minor byproduct in one case but not another.

5. Coupling Reactions with Catalytic Quantities of Pd(II)

Little study has so far been made in this area.

Addition of cupric reagents to reoxidize Pd(0) has been employed by Heck[51a] to make coupling reactions of arylmercury compounds catalytic in Pd. We have attempted one similar reaction, that of orthomanganated methyl 4-methoxybenzoate and methyl acrylate in MeCN, using only 1 mol% Li_2PdCl_4 with $CuBr_2$ as the oxidant.The result (Eq. 18) was arylalkane formation in low yield. The extensive demetallation, which is not the result of photolysis as the reaction was carried out in the dark, is a limitation that must be overcome if a catalytic method is to be developed.

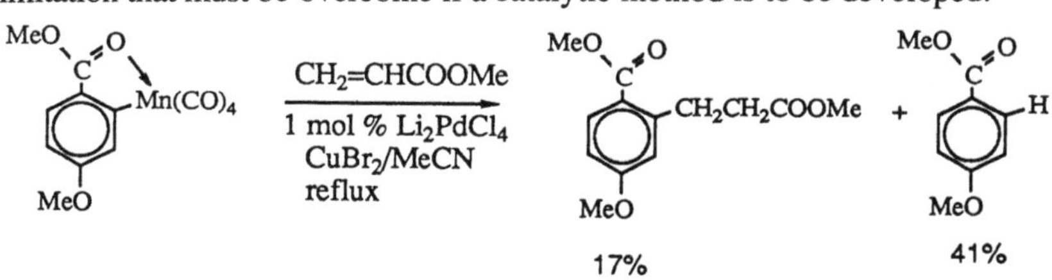

(18)

Although it formed part of a study aimed at achieving indene-type cyclizations,[50] the reaction in Eq.19 does give a high yield of arylalkane coupling product. It involves 10 mol % $Pd(OAc)_2$ under conditions commonly used for Pd(0)-catalyzed coupling of aryl halides,[23] so the role of the various reagents is not apparent.

The difficulty of indene formation from tetralone type precursors of the type in Eq. 19 may result from the carbonyl group not being free to rotate to expose its antibonding π-molecular orbital to the neighboring reactant metallated center (cf.

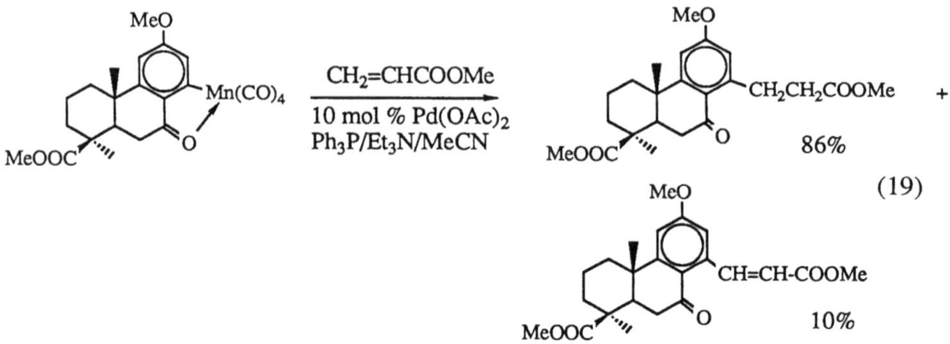

(19)

Eq. 10). It is interesting to note in this context that Liebeskind found no cyclization product from orthomanganated tetralone with alkynes, whereas with increased flexibility for twisting the carbonyl group in the 7-membered ring of benzosuberone, cyclization to form indenol did occur[24] (Section C-1).

6. *Coupling Reactions in the Absence of Pd(II)*

It was noted earlier (Table 3; footnote[c]) that coupling to form an indene product is possible[40] in the absence of Pd(II). The yield, however, was only low and the attempted separation of other unstable products was not achieved.

Trimethylamine oxide has been successfully used by Liebeskind[24] to promote coupling of alkynes with orthomanganated ketones to form cyclic indenol products (Section C) and this method has been adopted[50] for alkene coupling reactions in an attempt to synthesize indene analogues from podocarpic and dehydroabietic acid precursors. Again, however, arylalkane products predominate with tetralone-like precursors, the arylalkane and arylalkene products given in Eq. 19 being formed in yields of 59% and 29% respectively when Me_3NO (1.5 mol equiv.) is used in place of $Pd(OAc)_2$ with methyl acrylate in MeCN. There are similar results[50] for other precursors, with an indene derivative as a by-product for only one orthomanganated tetralone-type precursor in which a bulky isopropyl group adjacent to the C–Mn bond would promote cyclization. A better yield of indene derivative was obtained, however, with an exocyclic ketone (Eq. 20) suggesting this reaction may be more generally useful when there are no unfavorable orientation restrictions on the ketone carbonyl.

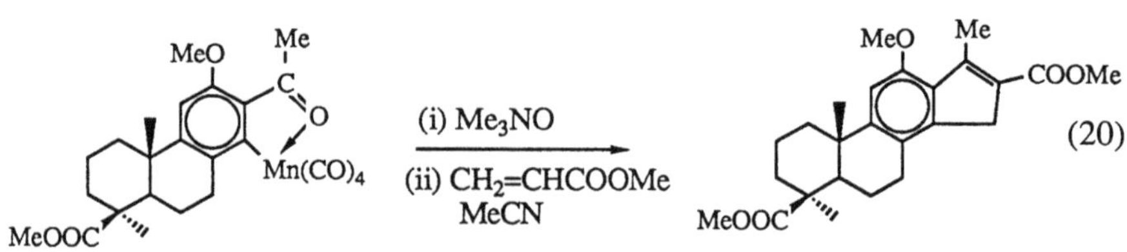

(20)

C. Reactions of Cyclomanganated Compounds with Alkynes

The reactions of organometallic complexes with alkynes have been well-studied, and a recent review summarizes the use of these reactions in cycloaddition processes.[7] However the use of manganese carbonyl complexes in this area has again been neglected until very recently, when two groups independently reported that high yields of cyclopenta-annulated products could be produced from reactions of alkynes with orthomanganated substrates. Liebeskind and co-workers[24] made use of Me_3NO to activate the manganese complex while we[55] used a thermally-induced reaction.

1. Synthesis of Inden-1-ols

Thermal Method. If a mixture of **1** and diphenyl acetylene is refluxed together in benzene for 8 hours, a quantitative yield of the inden-1-ol **31** can be obtained as in Eq. 21.[32,55]

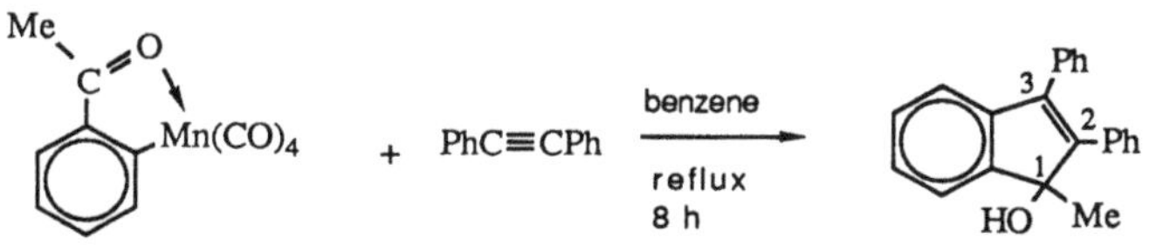

1 31 (21)

Preparation of 1-methyl-2,3-diphenylinden-1-ol (**31**). Nitrogen is bubbled through benzene (15 mL) in a Schlenk flask. Diphenylacetylene (0.067g, 0.376 mmol) and η^2-(2-acetylphenyl)tetracarbonylmanganese (**1**, 0.104g, 0.364 mmol) are added, and the mixture is refluxed under nitrogen for 8 h. At this stage t.l.c. shows that all starting manganese complex has been consumed. Solvent is removed under vacuum and the residue is chromatographed on silica gel plates, eluting with CH_2Cl_2/petroleum spirit (1:10). The upper, yellow band is $Mn_2(CO)_{10}$ (0.026g, 46%), while the extract from the second yellow band is recrystallized from petroleum spirits to give **31** (0.107g, 98%), m.p. 145–146 °C.

The solvent choice for this reaction is critical. With refluxing methanol the reaction is slower and the yield of **31** is diminished to about 50%, with appreciable quantities of acetophenone being produced in a competing demetallation process. In refluxing heptane no inden-1-ol product is observed at all after 4 h, although all starting material is consumed.

The corresponding inden-1-ol products are also formed in variable yield from **1** under similar conditions with other symmetrical alkynes; namely, C_2H_2 (64%), Pr^nCCPr^n (51%), and $Me_3SiCCSiMe_3$ (9%), (**32a–c**) but not with dimethyl acetylenedicarboxylate, MeO_2CCCCO_2Me.[32,55]

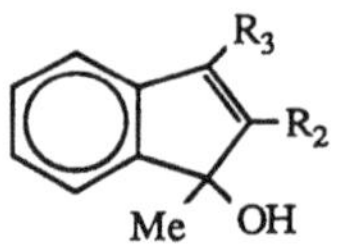

32

a. $R_2=R_3=H$ d. $R_2=Ph$, $R_3=H$
b. $R_2=R_3={}^nPr$ e. $R_2=SiMe_3$, $R_3=H$
c. $R_2=R_3=SiMe_3$ f. $R_2={}^nPr$, $R_3=Me$

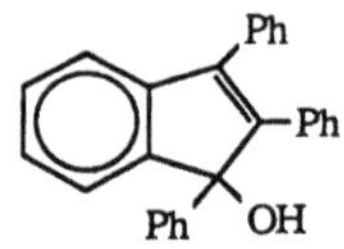

33

Good yields were obtained also with the unsymmetrical alkynes PhCCH (84%), Me_3SiCCH (82%) and Pr^nCCMe (32%), the preferred regioisomers having the bulky group at the 2-position of the inden-1-ol (**32d–f**). Note that **32e** is a vinyl silane, a class of compound which is readily elaborated by a number of well established reactions.[56]

Orthomanganated benzophenone also undergoes reaction with PhCCPh under the standard conditions to give inden-1-ol **33** in 85% yield, so that a variety of indene products with different substituents at the 1, 2, or 3 positions will be generated with the appropriate choice of starting ketone and alkyne.[32]

Me_3NO Method. In these reactions the orthometallated complex is first briefly treated in acetonitrile solution with Me_3NO to give a red-orange solution, which is then treated with the alkyne over a further 8–16 h (Eq. 22).[24]

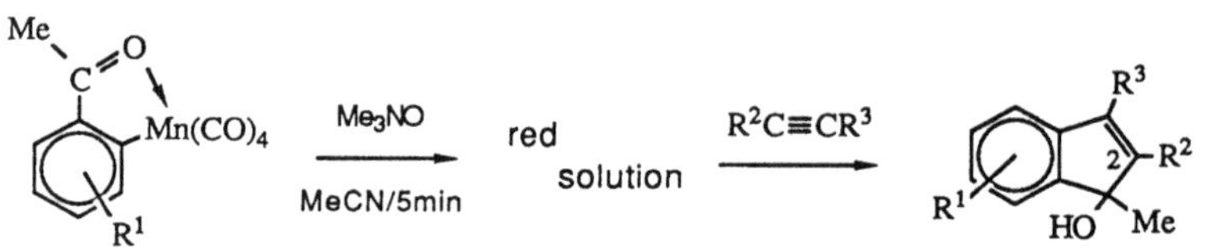

(22)

R^1=H, F, Cl, CF_3
R^2=H,Et,Bu,Ph,C_6H_{11},$SiMe_3$,OEt,CO_2Et
CH_2SiMe_3, $CH_2CH{=}CH_2$
R^3=H,Me,Et,Bu,Ph,C_6H_{11},OEt,$CH_2CH{=}CH_2$
(Not all combinations)

The extensive series of symmetrical and unsymmetrical alkynes used by Liebeskind's group[24] emphasize the generality of the reaction. Terminal, internal, electron-rich and electron-poor alkynes all participate and yields are good to excellent. The regiospecificity of the reaction is the same as in the thermal route, favoring the isomer in which the bulkiest group of the alkyne ends up at the 2-position of the indenol. Only with the alkynes $RCCCO_2Et$ was there any indication of a possible electronic directing effect in that the ester group was always incorporated at the 2-position, independent of the steric bulk of the R group.

A number of acetophenones substituted with F, Cl or CF_3 groups were also orthomanganated and then treated with Me_3NO/alkyne reagents under the same conditions to provide the corresponding inden-1-ols (Eq. 22). The compound **34** (n = 2) was prepared similarly, starting from benzosuberone and hex-3-yne, but no **34** (n = 1) could be found when orthomanganated α-tetralone was used.[24]

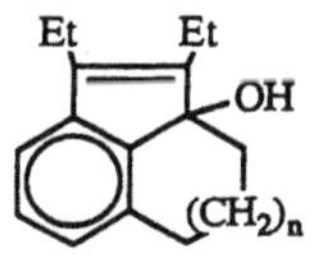

34

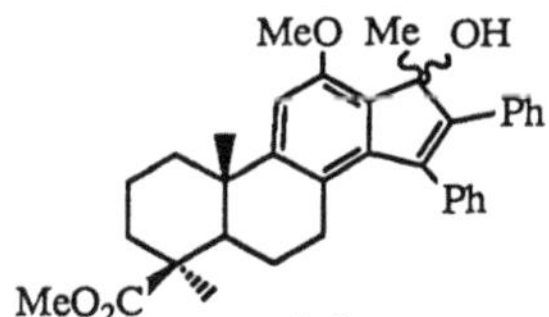

35

Another application of this reaction is the formation of **35** from the appropriate manganated precursor.[50]

Reaction of Manganated N-acetylindole with DMAD. In an extension of the thermally-induced method, **36** was reacted with dimethyl acetylenedicarboxylate.[32] The tricyclic carbinolamine **37**, characterized by an X-ray crystal structure analysis, was formed in 17% yield.

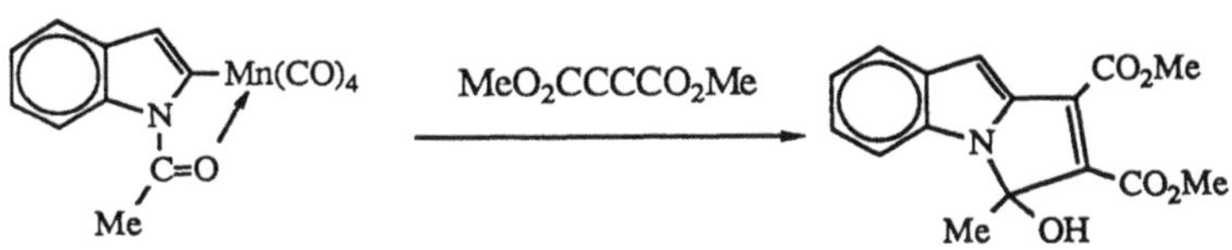

36 **37** (23)

The product **37** has the core of the mitomycin class of antibiotics,[57] so this system may merit a more detailed investigation.

Mechanism of Annulation Reactions. A reasonable sequence can be proposed based on related reactions of alkynes with organometallic substrates (Scheme 2).

The first step is loss of CO to provide a vacant coordination site, brought about either by thermally-induced dissociation[32,55] or by Me_3NO conversion[24] of a CO ligand to CO_2. The alkyne then becomes η^2-coordinated, being orientated such that the bulkiest substituent is directed away from the orthometallated arene. A regiospecific migration of the aryl group to the nearest alkyne carbon generates the insertion product, which may exist as a seven-membered metallocyclic ring. Addition of C–Mn bond across C=O would give a manganese alkoxide species which would generate the inden-1-ol product on displacement of the manganese group by H^+, either from adventitious water or on workup. Support for this sequence comes from evidence in other reactions that the insertion of unsymmetrical alkynes into M–C

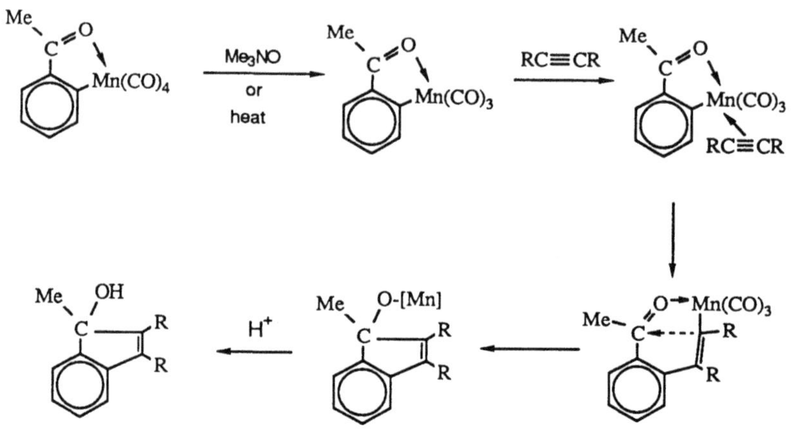

Scheme 2.

bonds usually gives the isomer with the bulkiest group adjacent to the metal,[58] and from the isolation of a seven-membered ring species **38** from the insertion of Ph_2C_2 into the Mn–C bond of orthomanganated triphenylphosphite.[59]

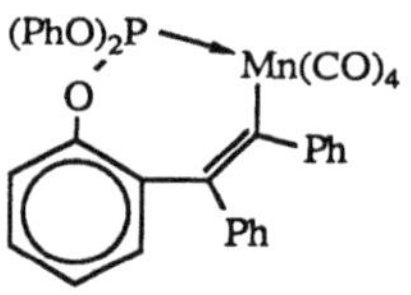

38

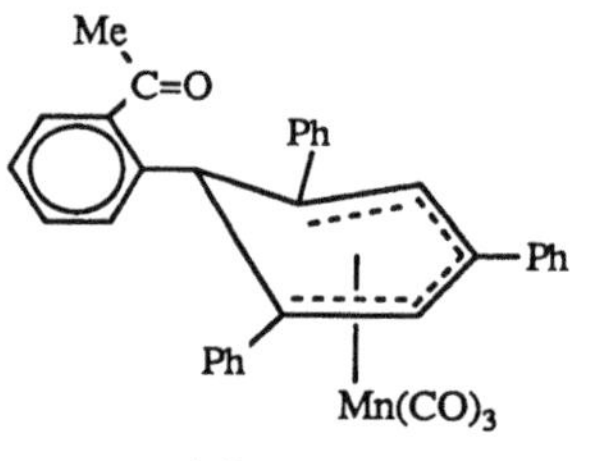

39

Competing Alkyne Cyclotrimerizations. Most of the annulation reactions are specific, with few by-products. However from the thermal reaction of **1** with PhCCH in benzene a second product was isolated in 3% yield, and identified as the η^5-arene manganese tricarbonyl complex **39**. This has apparently arisen from multiple insertion of three acetylene fragments into the Mn–C bond and subsequent cyclization. Consistent with this, the yield of **39** is increased to 14% if excess PhCCH is used in the reaction.

2. *Synthesis of Indenones*

The thermally-induced reaction of Ph_2C_2 with orthomanganated *N,N*-dimethylbenzamide gives 2,3-diphenylindenone (56% yield), as in Eq. 24.[32,55] The appropriately substituted indenones are also formed from reaction of Ph_2C_2 with the orthomanganated derivative of methyl 3,4,5-trimethoxybenzoate in quantitative yield, and from reaction with the *p*-dimethylaminobenzaldehyde manganese complex in 49% yield. Similarly the manganated derivative of *p*-dimethylaminobenzaldehyde reacts with C_2H_2 to form the appropriate indenone, although yields were only 20%.

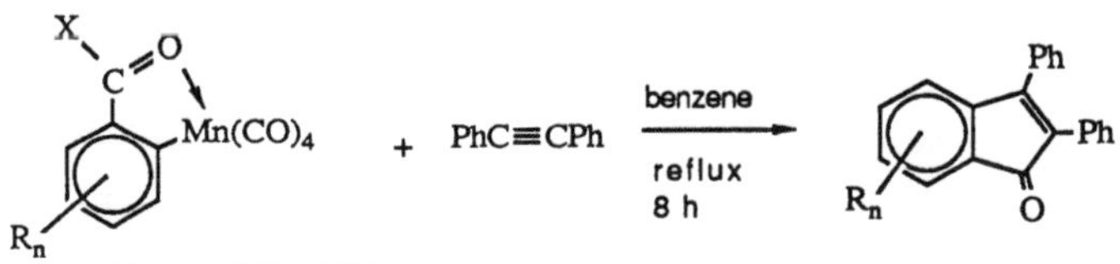

(24)

X = H, OMe, NMe_2

The formation of the indenone products can be rationalized by the sequence given for the inden-1-ol reactions (Scheme 2), but with the final protonolysis replaced by elimination of X[Mn] Eq. 25.

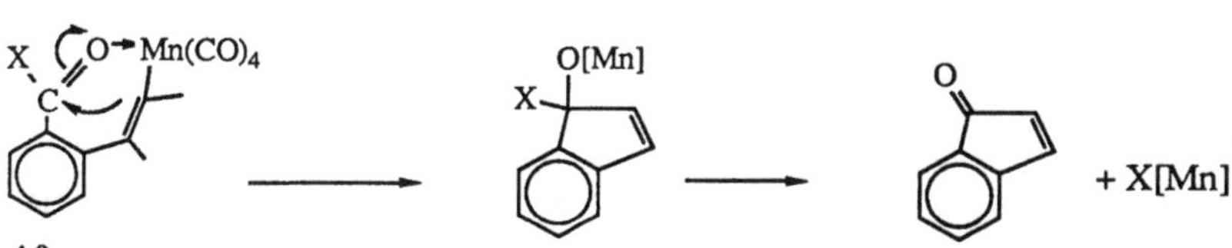

(25)

40

With C_2H_2 (in excess) the cyclometallated *N,N*-dimethylbenzamide gave an η^5-hexadienyl complex **41** in 55% yield, rather than an indenone, and the analogous complex was formed from orthomanganated methyl 3,4,5-trimethoxybenzoate (36%) and C_2H_2.[32]

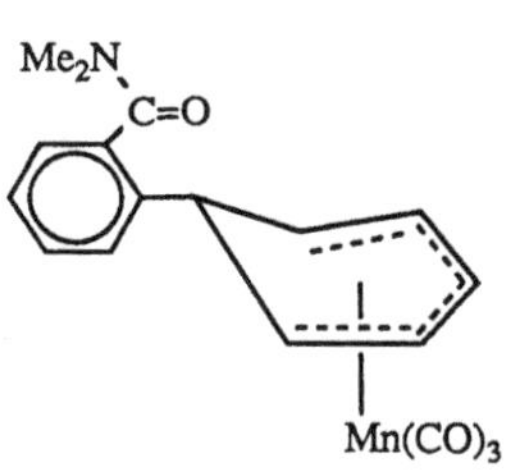

41

Presumably with C_2H_2 there is reduced steric hindrance at the metal center of the mono-insertion product (cf. **40**) to π-complexing of a second alkyne, and subsequent insertion can compete with cyclization. Ring substituents also promote intramolecular cyclization reactions, and those of substituted alkynes (e.g., the phenyl groups of Ph_2C_2) are likely to promote indenone formation here at the expense of further alkyne insertion.

3. Reactions of Cyclomanganated Enones

The reactions of cyclomanganated enones such as **42** with alkynes follow a different pathway from those of manganated aryl ketones. Thus a mixture of cyclomanganated 3′,4′,5′-trimethoxychalcone **42** and Me_3SiCCH refluxed in benzene gives as the main product the complex **43a**, with an η^5-pyranyl ligand as in Eq. 26[44].

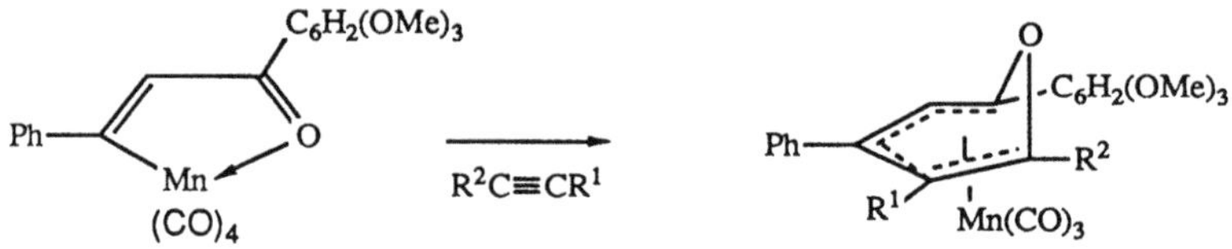

(26)

42

43a. R^1 = H; R^2 = $SiMe_3$
43b. R^1 = Ph; R^2 = Ph

Preparation of η^5*-pyranyltricarbonylmanganese*[44]. A mixture of cyclomanganated 3′,4′,5′-trimethoxychalcone (**42**) (0.30g, 0.64 mmol) and trimethylsilylacetylene (0.08g, 0.82 mmol) in benzene (25 mL) is refluxed for 2 h, by which time t.l.c. shows all starting manganese compound has been used. The solvent is removed in vacuum and the residue chromatographed on an alumina column (activity IV, 15 × 2 cm). Elution with petroleum spirits provides, as the first band, the pyranyl complex **43a** (0.14g, 41%), ν_{CO} 2010m, 1948 vs, 1929m.

Similar complexes are formed with Ph_2C_2, and with other cyclometallated enones.[44] These have presumably formed by initial insertion of the alkyne into the Mn–C bond of **42**, to give an intermediate of the type **44**.

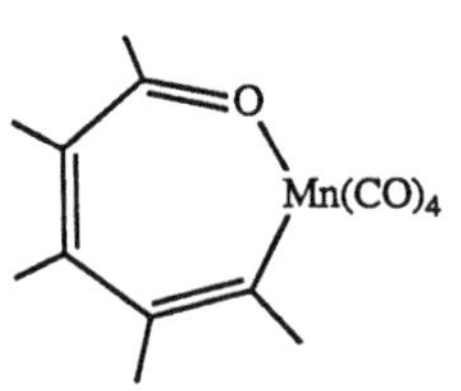

44

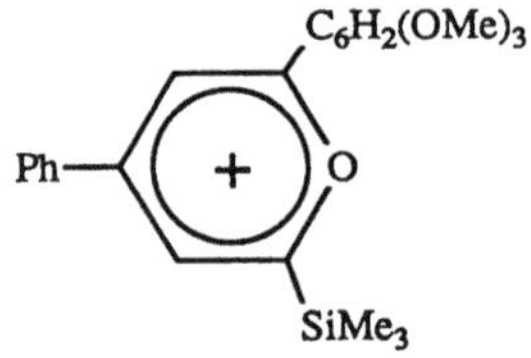

45

Closure of the two ends of the six-membered chain, loss of CO from Mn and a change from η^2 to η^5 coordination would give the product **43**. Complexes of the type **43** were originally found as by-products in the preparation of cyclomanganated compounds from $RMn(CO)_5$ and alkynes,[42–44] but the synthesis by the route described here gives better yields. Compounds of the type **43** may have application in organic synthesis in that the pyranyl ligand can be displaced under oxidative conditions to give pyrylium salts. Thus treatment of **43a** with $AgPF_6$ (1:1) in CH_3CN gives **45** as the PF_6^- salt.[44]

D. Reactions of Orthomanganated Ketones with Isocyanates

The reactions of orthomanganated aryl ketones with isocyanates has recently been reported from Liebeskind's group.[60] The reactions follow Eq. 27, giving as the initial products 3-methylidene phthalimidines **46**, which could be isolated when R = *p*-tolyl.

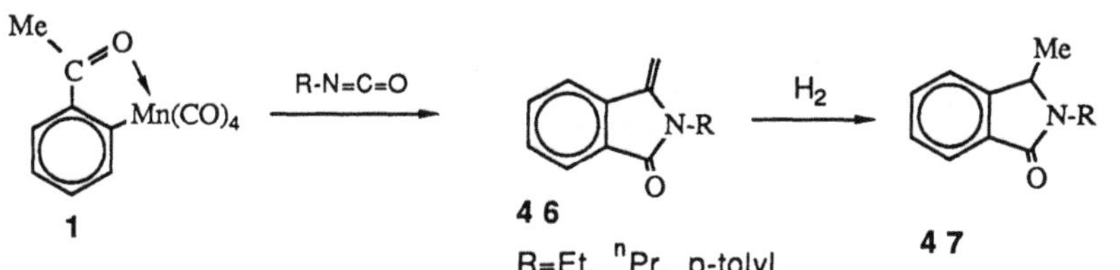

1 46 R=Et, ⁿPr, p-tolyl 47 R=Et, ⁿPr (27)

When R = Et or Pr^n, the phthalimidenes **46** are unstable so were rapidly hydrogenated to give the related stable phthalimidines **47**. Again various substituents (F, Cl, CF_3, MeO) on the original acetophenone did not interfere with the reaction. Yields were generally moderate to good so these reactions should be generally valuable in the synthesis of 3-alkylidene phthalimidines, which incorporate the core unit of a number of isoindole-derived alkaloids.

E. Reactions of Orthomanganated Ketones with Electrophiles

1. *Reactions with* $HgCl_2$

When orthomanganated acetophenone (**1**) and an equimolar quantity of $HgCl_2$ are refluxed in dry methanol, transmetallation occurs to give the corresponding orthomercurated product **48** in good yield, as follows:[61]

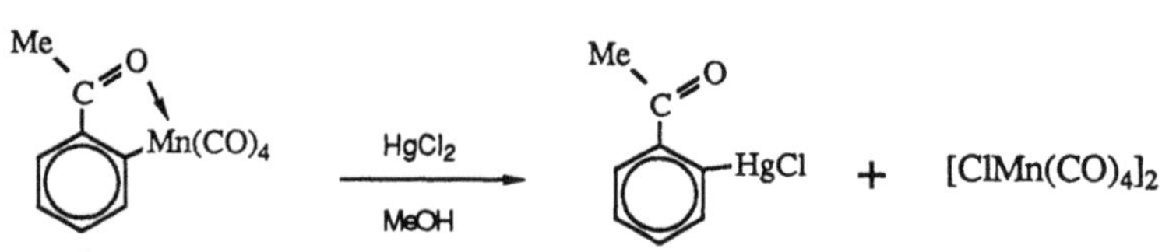

1 48 (28)

Preparation of 2-acetylphenylmercury(II) chloride:[61] A solution of **1** (0.586g, 2.04 mmol) and $HgCl_2$ (0.554g, 2.04 mmol) in dry methanol (15 mL) was refluxed for 3 h. The mixture was allowed to cool and the precipitate was collected by filtration, washed with cold methanol and air-dried to give 2-MeC(O)C_6H_4HgCl, **48**, (0.347g, 48%), mp 214.5–216 °C.

The mercury complex **48** is *not* a cyclometallated compound, since a crystal structure analysis shows negligible interaction between the oxygen atom and the mercury atom.[61] The mercuration reaction should prove to be a valuable addition to the organometallic chemists' armory, since mercurated acetophenones are not otherwise available; direct mercuration gives attack at the methyl group, and preparations via aryl lithium or Grignard reagents are precluded by the ketone group.[61] The transmetallation reaction is quite general for orthomanganated complexes, giving mercurated acetophenones, benzophenones, methyl benzoates, and acetylthiophenes by use of the appropriate orthomanganated species.[61] Non-aromatic species are also suitable substrates, since the cyclometallated enone **49** gives 70% yields of **50** on reaction with $HgCl_2$[44] under the same conditions.

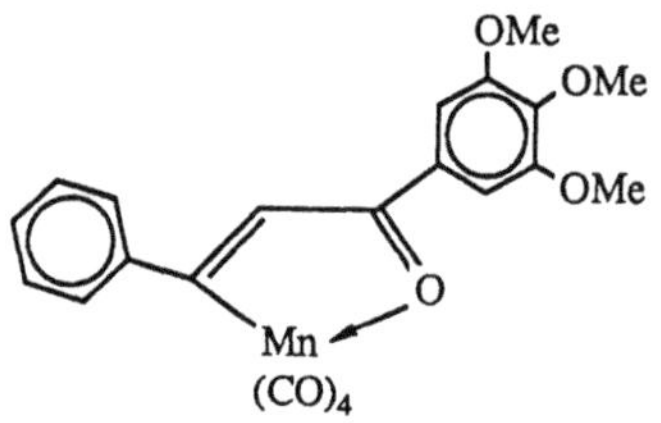

49

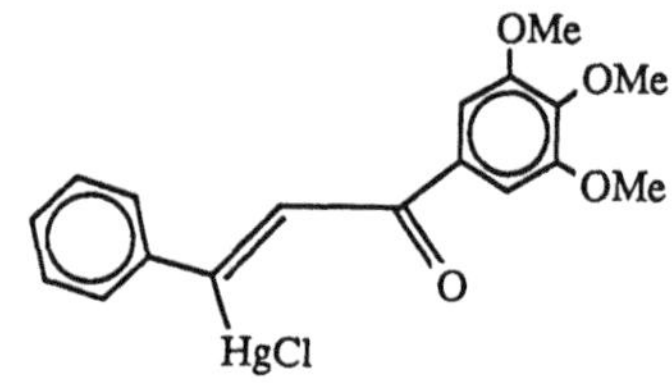

50

2. Reactions with Br_2

Crawford first reported[62] the use of bromine in carbon tetrachloride for bromodemanganation. Orthomanganated benzophenone gave, with a 3 molar excess of Br_2 and after workup involving aqueous alkaline extraction, 2-bromobenzophenone and 2-benzoylbenzoic acid. The latter was envisaged (Scheme 3) as derived from the acyl bromide formed by bromodemetallation at acyl carbon after carbonyl insertion into the C–Mn bond. Also detected were $BrMn(CO)_5$, $[BrMn(CO)_4]_2$, and hydrated $MnBr_2$.

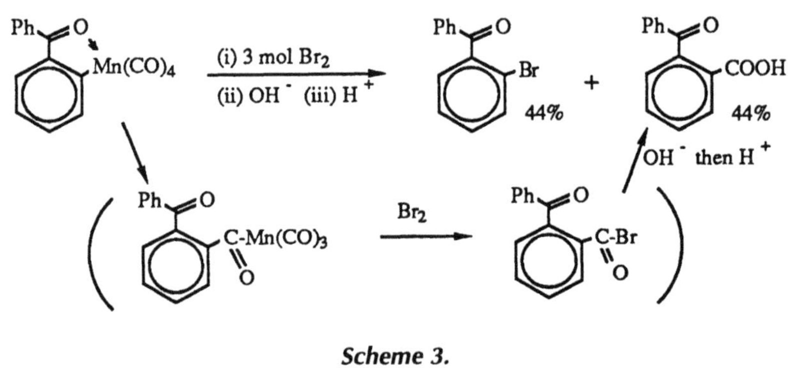

Scheme 3.

Repetition[62] of the reaction with orthomanganated acetophenone and again a 3 molar excess of Br_2, gave no evidence for *o*-bromoacetophenone. The major product (56%) was the dibromide **52** of the vinyl phthalide **51**, the latter probably (Eq. 29) being formed from the acyl bromide. When we recently[40] repeated this reaction using one mole of Br_2, at least five products were obtained, among which the vinyl phthalide **51** and *o*-bromoacetophenone were the only ones identifiable, but the yields were poor. For acetophenone, therefore, the synthetic value of Br_2 for bromodemanganation is apparently limited by diversion through carbonyl insertion to unwanted products.

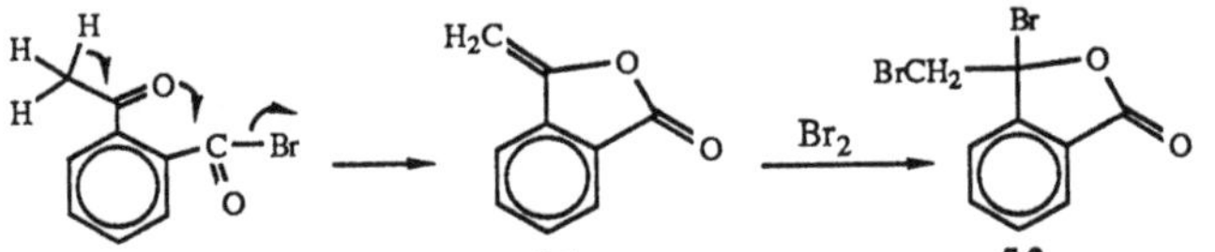

51 52 (29)

When, however, electrophilic attack at the manganated aryl carbon is activated by electron-donating methoxy groups, good yields of orthobromo ketones are obtained, as the following preparation shows.

Preparation of 2′-bromo-3′,4′,5′-trimethoxyacetophenone.[46] A solution of Br_2 in CCl_4 (0.34 mol L^{-1}, 0.78 mL, 0.27 mmol) is added dropwise under nitrogen to a solution of η^2-(2-acetyl-4,5,6-trimethoxyphenyl)tetracarbonylmanganese (100 mg, 0.27 mmol) in CCl_4. A thick orange precipitate forms at once and the reaction is complete in 5 min. The orange solid, identified as $[BrMn(CO)_4]_2$ by infrared spectroscopy, is removed by filtration and the solution is evaporated to give an oil which is chromatographed on a silica layer using 3:7 ether/petroleum spirit as eluent. The first yellow band is a small amount of $BrMn(CO)_5$, while the other colorless band (R_f 0.4) yields a nearly quantitative amount of 2′-bromo product containing a trace of impurity (probably the 2,2′-dibromo analogue by ^{1}H-NMR). Recrystallization from 5:1 hexane/ether gives white crystals of pure 2′-bromo-3′,4′,5′-trimethoxyacetophenone, mp 44–46 °C.

Methyl 2-bromo-3,4,5-trimethoxybenzoate can be prepared in 84% yield from the orthomanganated ester by the same method.[35] In spite of the absence of activating groups, 3-bromo-2-acetylthiophene is similarly prepared[35] in 44% yield along with other unidentified products, a somewhat more successful synthesis than that for 2′-bromoacetophenone described above.

For orthomanganated 2′-*t*-butyldimethylsiloxy-3′,4′-dimethoxyacetophenone, the *o,p*-directing oxy groups would activate the unsubstituted ring position to electrophilic attack moreso than the manganated ring position. Bromodeprotonation to form **53** (Scheme 4) is apparently much faster than bromodemanganation.[40] The HBr formed is probably responsible for the protiodemanganation products **54** and **55** from the manganated 5-bromo and parent acetophenones respectively.

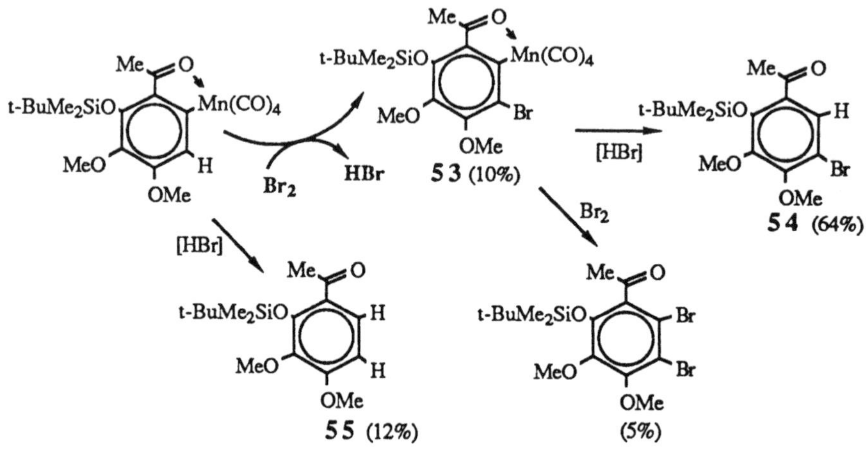

Scheme 4.

Orthomanganated 2′,3′,4′-trimethoxyacetophenone reacts similarly,[40] giving 5′-bromo-2′,3′,4′-trimethoxyacetophenone (68%) and the parent ketone (17%), although the intermediate (cf. **53**) was not isolated in this case. The resistance of the C–Mn group to substitution in these latter cases is a synthetic limitation in that the Br group is not introduced *ortho* to ketone function, the whole point of the initial metallation. Direct bromination of the parent ketones without manganation would probably give the same products in better yield.

A similar resistance to substitution of manganese is observed with nonaromatic cyclometallated species. With manganated benzylidene acetone (Eq. 30), bromination at the C–H bond of the unsaturated cyclometallated ring took place,[25] but no cleavage of the C–Mn bond was found even with excess *N*-bromosuccinimide. Bromination occurs[47] similarly with Br_2 in CH_2Cl_2, but Br_2 gives no reaction with the corresponding saturated manganocycles such as **22**.

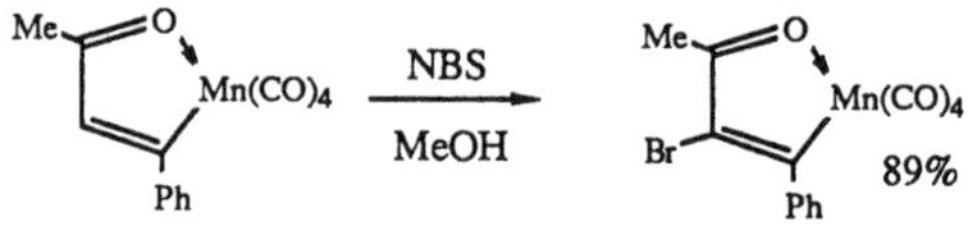

(30)

3. *Reactions with ICl*

Iodine appears to give no reaction with orthomanganated ketones at all, but iodine chloride in carbon tetrachloride reacts cleanly even though it is much less reactive than bromine. It appears to have the advantage over bromine that it reacts exclusively to substitute the metal rather than the proton in the orthomanganated ketone in Scheme 4, as the following preparation indicates.

Preparation[40] *of 2′-t-butyldimethylsiloxy-3′,4′-dimethoxy-6′-iodoacetophenone (**56**).* η²-(2-Acetyl-3-*t*-butyldimethylsiloxy-4,5-dimethoxyphenyl) tetracarbonylmanganese (112 mg, 0.235 mmol) was dissolved in 2 mL CCl_4 and the solution saturated with nitrogen. ICl (35 mg, 0.237 mmol) dissolved in 2 mL nitrogen-saturated CCl_4 was added. The flask was stoppered and left at room temperature. After 4 days, the color of the ICl had disappeared. The bright orange precipitate which had formed was filtered. In CCl_4 it had ν_{CO} 2099, 2046, 2009 and 1972 cm^{-1} matching closely frequencies reported for $[ClMn(CO)_4]_2$. The filtrate solution was evaporated and the solid separated on a silica layer using 3:7 ether/petroleum spirit as eluent. This gave slightly impure **56** (100 mg, 100%), which was purified by crystallization from hexane as white crystals, mp 115–117 °C.

The other following *ortho*-iodinated compounds were prepared likewise[33,40] but in variable yield, in some cases as a result of the reaction being incomplete (unreacted ICl) when the reaction was worked up after the period indicated.

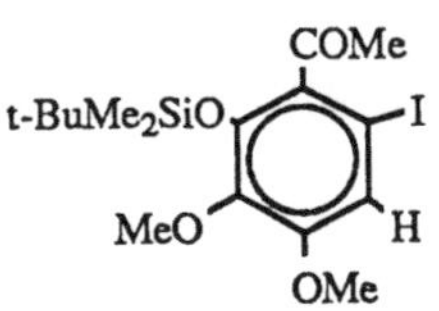

56
100% (4 days)

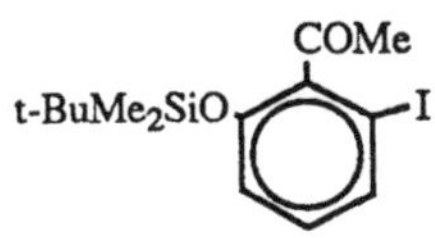

57
5% (1 day)

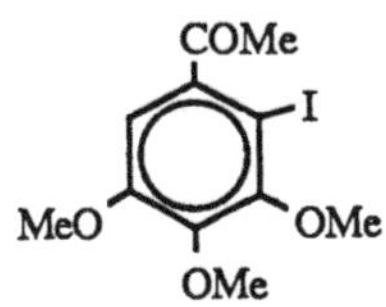

58
93% (4 days)

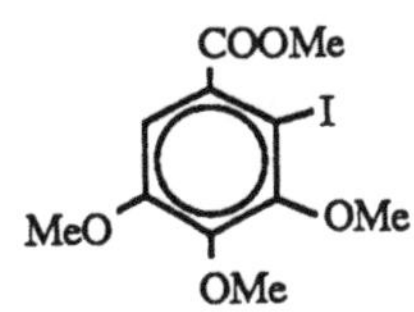

37% (5 days)

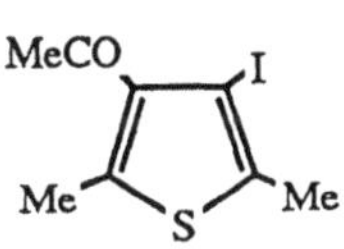

100% (1 day)

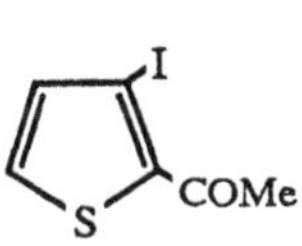

75% (1 day)

In the case of 2-(*t*-butyldimethylsiloxy)-6-iodoacetophenone (**57**) the yield (5% after 1 day) was improved to about 40% after 6 h if the reaction solution was exposed to bright sunlight.[40] Iodination was also observed during an attempted[40] *alkylation* of orthomanganated 3′,4′,5′-trimethoxyacetophenone with iodomethane in CCl_4. After 3 weeks at room temperature, 2′-iodo-3′,4′,5′-trimethoxyacetophenone (**58**) was formed in 55% yield along with 33% demanganated ketone, methyl iodide succeeding as an iodination reagent where I_2 fails. Another attempted[40] methylation using CH_3I and $Pd(OAc)_2$ in MeCN under reflux for 2 h gave neither alkylation nor iodination of orthomanganated 4′-methoxyacetophenone, only demetallation (72%).

4. *Reaction with Ph_2PCl*

Orthomanganated acetophenone was refluxed with Ph_2PCl in petroleum spirit (bp 40–60 °C) to see if transfer of the organic group to phosphorus could be achieved.[63] Initially a phosphine-substituted compound **59** was formed [as shown by the characteristic infrared spectrum: ν_{CO} (cm^{-1}) 2016s, 1949s, 1910s], but prolonged reflux gave a mixture from which the dimanganese complex **60** was isolated in 17% yield. It was characterized by X-ray crystallography.[64]

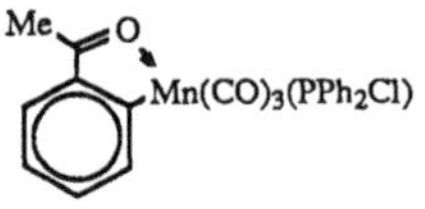

59

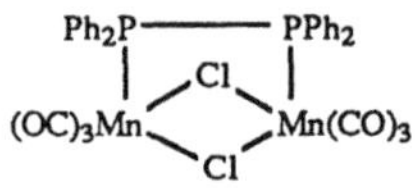

60

The reaction is obviously complicated, and the co-products have not been identified. However, the unusual *in situ* formation of a P–P bond indicates that cyclomanganated compounds have unexplored potential in the preparation of organometallic as well as organic species.

F. Reactions with HCl

Demetallation of manganated complexes of enones with HCl in MeCN has been reported[25] from DeShong's group (Scheme 5).

The competing pathways were explained by initial protonation of the ligand to give the cationic complex **61**, which can undergo either hydrogen migration and loss of the Mn moiety to give enone, or CO insertion and cyclization leading to a butenolide product. Double bond migration under the acid conditions gives the isomeric butenolide which, however, may be obtained in better yield by reduction using DIBAL-H/n-BuLi (Section G).

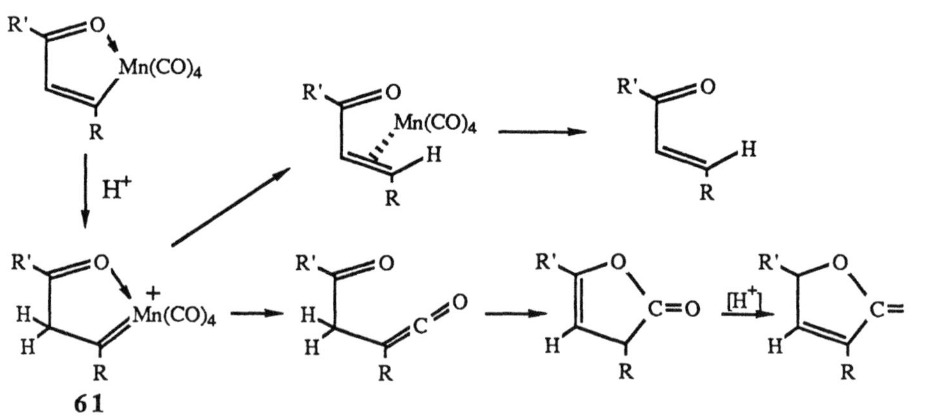

Scheme 5.

Demetallation of orthomanganated aryl carbonyl compounds has not been systematically studied though it is recognized as the source of parent ketone byproduct in a number of the syntheses reported above.

G. Reducing Agents

Very little study has been carried out in this area.

Hydrogen. Orthomanganated acetophenone gives[62] no reaction in heptane at room temperature with H_2 at 12 atm. pressure. However, with D_2 (7 atm at room temperature, then heated to 90 °C for 19 h) a mixture of d_1- and d_2-acetophenone and d_2- and d_3-1-phenylethanol was obtained, indicative of *ortho*-deuteriation at one and both *ortho* positions of acetophenone, likewise in the alcohol but with additional deuteriation at C1.

Li[BEt$_3$H] and DIBAL-H/n-BuLi. DeShong has reported[26a] the following reduction of a cyclometallated saturated ketone (Eq. 31). The CO-insertion-derived product is akin to the dibromide of the vinyl phthalide (**52**; Section IV-E-2) and the analogous product from the Ce(IV) reaction (Eq. 34; Section IV-H). Presumably the alkoxide formed from hydride addition to the carbonyl group cyclizes to displace [Mn] from a carbonyl insertion moiety.

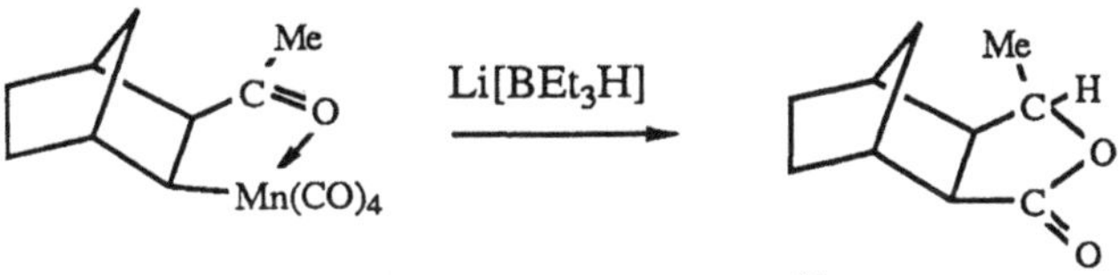

(31)

Cyclometalled unsaturated ketones likewise give[25] the corresponding unsaturated lactones (furanones) by reaction with $LiBEt_3H$ and with DIBAL-H, but a significant amount of lactol is formed as well. A better reagent is DIBAL-H/n-BuLi complex. Reactions such as that in Eq. 32 proceed in 50–60% yield. The nonconjugated isomer of the product in Eq. 32 is formed as the major product in the reaction with HCl (cf. Scheme 5).

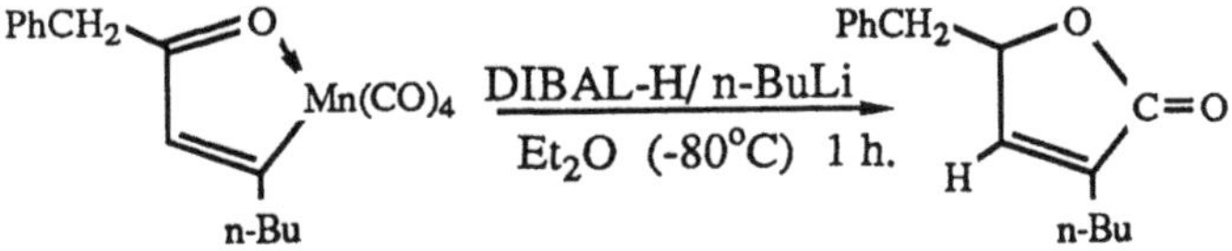

(32)

H. Reactions with the Oxidizing Agents Cerium (IV) Ammonium Nitrate (CAN) and Manganese (III) Acetate

Oxidation with these reagents results generally in demetallation. Orthomanganated acetophenone gives about 50% acetophenone on reaction with 0.5 mol

CAN in acetic acid, along with several other unknown minor products. Orthomanganated 3′,4′,5′-trimethoxyacetophenone gives 3′,4′,5′-trimethoxyacetophenone (83%) by reaction with 1 mol CAN in acetone, and the same product (93%) by reaction with 1.05 mol of $Mn(OAc)_3.2H_2O$ in MeOH. Such reactions can be applied to the synthesis of specifically mono-*ortho*-deuteriated acetophenones by carrying out the reaction in a deuterio solvent as the following example[40] (Eq. 33) in CD_3COOD shows.

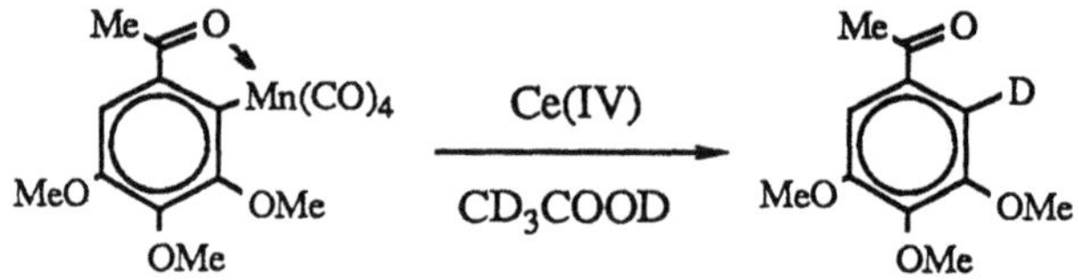

(33)

Preparation of 2′-deuterio-3′,4′,5′-trimethoxyacetophenone. A solution of η^2-(2-acetyl-3,4,5-trimethoxyphenyl)tetracarbonylmanganese (0.239 g, 0.63 mmol) in CD_3COOD (>99.5% D) was treated with oven-dried cerium(IV) ammonium nitrate (0.348 g, 0.65 mmol), added in small portions over 30 min. The solution was poured into a large excess of water and extracted with ether (3 × 20 mL). The combined extract was thoroughly washed with sodium bicarbonate solution and dried ($MgSO_4$). After filtration, the solvent was removed and the residual oil chromatographed on a silica layer, eluting with 1:4 ether/petroleum spirit. The major colorless band (see below for other products) was identified as 2′-deuterio-3′,4′,5′-trimethoxyacetophenone (50% yield), 99% isotopically pure by ^{1}H-, ^{2}H- and ^{13}C-NMR.

Equally specific deuteriation of acetophenone and of methyl 3,4,5-trimethoxybenzoate were similarly achieved.[40,46] The advantage of the method is that it involves neither strongly acidic or basic conditions, which for other deuteriation methods can be responsible for random exchange at alternative sites. The monodeuteriated compounds have value as orthomanganation substrates in the determination of hydrogen isotope effects[46] (Section II-J).

In the case of orthomanganated 3′,4′,5′-trimethoxyacetophenone, two other significant products are identified in the reaction with CAN in acetic acid as Eq. 34 shows.

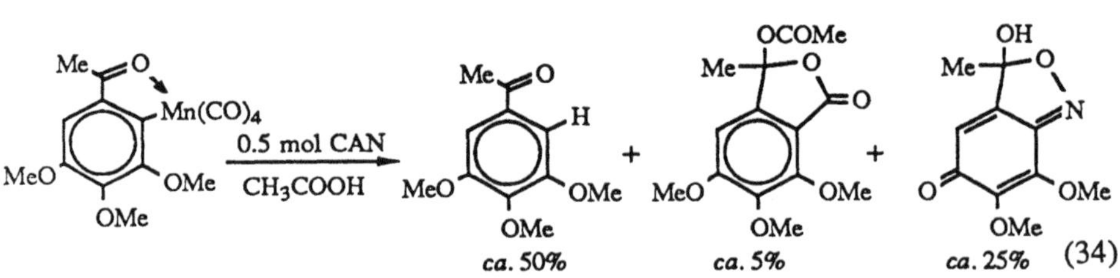

(34)

The minor product, whose structure was assigned[40] by NMR and HRMS, is a carbonyl insertion product reminiscent of the dibromide of the vinyl phthalide formed with Br_2 (Scheme 4; Section IV-E-2). In the present case, acetic acid rather than Br_2 has added across the double bond. The other quinoid compound, formed in 25% yield and identified[40] by X-ray crystallography, has resulted from demethylation, possibly from a nitroso precursor (**62**).

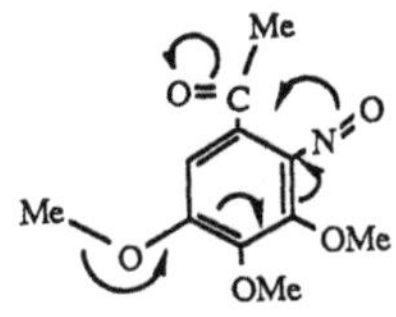

62

We are unaware of CAN leading to nitrosation although it is known to nitrate activated aromatic rings. Current study[65] shows that $NOBF_4$ as a nitrosation reagent in CH_2Cl_2 reacts with orthomanganated 3′,4′,5′-trimethoxyacetophenone to give about 20% of the quinoid compound. A much better yield (ca. 90%) is obtained with the orthochloromercurated precursor and $NOBF_4$.

I. Photolysis

With a 350-nm light source, cyclometallated saturated ketones undergo photolytic demetallation in degassed acetonitrile.[25] The proton replacing the metal must be derived from water during workup: there is no deuterium incorporation from CD_3CN, but there is from D_2O if it is added to quench the reaction. The reaction proceeds readily at low pressures, slowly at atmospheric pressure (Ar atmosphere), and is stopped completely under an atmosphere of CO, thus indicating that the first step in the reaction is loss of CO. Photolytic demetallation of the corresponding unsaturated cyclomanganated ketones was not reported,[25] but was achieved with HCl or with reducing reagents as discussed in the appropriate sections.

NOTE ADDED IN PROOF

Since this review was written, coupling reactions of some orthomanganated diterpenoid compounds with alkenes, alkynes, and electrophiles have been reported in detail by Woodgate's group at Auchland.[66]

ACKNOWLEDGMENTS

It is with great pleasure that we acknowledge the enthusiasm and laboratory skills of the research students who have contributed to the development of organo-manganese carbonyl chemistry at Waikato—Lou Gommans, Nick Robinson, Janine Cooney, Meto Leach, Warren Tully, Les Arnold, Alastair Boyd, Warren Grigsby, and Rene De Wit. Financial support from the New Zealand Universities Grants Committee is gratefully acknowledged.

REFERENCES AND NOTES

1. Wender, I.; Pino, P. *Organic Syntheses via Metal Carbonyls*. J. Wiley and Sons, New York, 1968, Vol. 1, 1977, Vol 2; Collman, J. P.; Hegedus, L. S.; Norton, J. R.; Finke, R. G. *Principles and Applications of Organotransition Metal Chemistry*, University Science Books: Mill Valley, CA, 1987; Wilkinson, G; Stone, F. G. A.; Abel, E. W., Eds. *Comprehensive Organometallic Chemistry*. Pergamon Press: Oxford, 1982; Vol. 8; Davies, S. G. *Organotransition Metal Chemistry: Applications to Organic Synthesis*. Pergamon Press: Oxford, 1982.
2. Muetterties, E. L.; Bleeke, J. R.; Wucherer, E. J.; Albright, T. A. *Chem. Rev.* **1982**, *82*, 499; Solladie-Cavallo, A. In *Advances in Metal-Organic Chemistry*; Liebeskind, L. S., Ed.; JAI Press: Greenwich, 1988; Vol. 1, p 99.
3. Wulff, W. D. In *Advances in Metal-Organic Chemistry*; Liebeskind, L. S., Ed.; JAI Press: Greenwich, 1988; Vol. 1, p. 209.
4. Von Gustorf, E. A. K.; Grevels, F. W.; Fischler, I. *The Organic Chemistry of Iron*. Academic Press, 1978; Vol. 1, 1981; Vol. 2.
5. Vollhardt, K. P. C. *Angew. Chem., Int. Ed. Engl.*, **1984**, *23*, 539.
6. Jolly, P. W. *Comprehensive Organometallic Chemistry* Wilkinson, G.; Stone, F. G. A.; Abel, E. W., Eds.; Pergamon Press, 1982, Chapter 56.
7. Schore, N. E. *Chem. Rev.* **1988**, *88*, 1081.
8. Treichel, P. M. *Comprehensive Organometallic Chemistry*; Wilkinson, G.; Stone, F. G. A.; Abel, E. W. Eds.; Pergamon Press, 1982.
9. McKinney, R. J.; Hoxmeier, R.; Kaesz, H. D. *J. Amer. Chem. Soc.*, **1975,** *97*, 3059; McKinney, R. J.; Kaesz, H. D. *J. Amer. Chem. Soc.*, **1975,** *97*, 3066.
10. McKinney, R. J.; Firestein, G.; Kaesz, H. D. *Inorg. Chem.* **1975**, *14*, 2057.
11. Knobler, C. B.; Crawford, S. S.; Kaesz, H. D. *Inorg. Chem.* **1975**, *14*, 2062. See also McKinney, R. J.; Crawford, S. S. *Inorg. Syntheses* **1989**, *26*, 155.
12. Crawford, S. S.; Kaesz, H. D. *Inorg. Chem.* **1977**, *16*, 3193.
13. A. W. Cabral, Ph.D. Thesis, University of California, Los Angeles, 1981.
14. Bruce, M. I. *Angew. Chem. Int. Ed. Engl.* **1977**, *16*, 73; Omae, I. *Intramolecular-Coordination Compounds*; Elsevier: Amsterdam, 1986 (*J. Organomet. Chem. Library, 18*).
15. Constable, E. C. *Polyhedron* **1984**, *3*, 1037.
16. Trofimenko, S. *Inorg. Chem.* **1973**, *12*, 1215.
17. Bruce, M. I.; Liddell, M. J.; Snow, M. R.; Tiekink, E. R. T. *Aust. J. Chem.* **1988**, *41*, 1407; Bennett, R. L.; Bruce, M. I.; Goodall, B. L.; Stone, F. G. A. *Aust. J. Chem.* **1974**, *27*, 2131; Bennett, R. L.; Bruce, M. I.; Matsuda, I. *Aust. J. Chem.* **1975**, *28*, 1265 and refs therein; Little, R. G.; Doedens, R. J. *Inorg. Chem.* **1973**, *12*, 840; 844.
18. Heck, R. F. *Palladium Reagents in Organic Synthesis;* Academic Press: London, 1985, Ch. 6.
19. Trost, B. M.; Verhoeven, T. R. In *Comprehensive Organometallic Chemistry*; Wilkinson, G.; Stone, F. G. A.; Abel, E., Eds.; Pergamon Press: Oxford, 1982; Vol. 8, p. 854.
20. Larock, R. C.; Varaprath, S.; Law, H. H.; Fellows, C. A. *J. Amer. Chem. Soc.* **1984**, *100*, 5274.
21. Larock, R. C.; Liu, C. L.; Law, H. H.; Varaprath, S. *Tetrahedron Lett.* **1984**, *25*, 4459.
22. Snieckus, V. *Bull. Soc. Chim. France* **1988**, *67*.
23. Heck, R. F. *Org. Reactions* **1982,** *27*, 345.
24. Liebeskind, L. S.; Gasdaska, J. R.; McCallum, J. S.; Tremont, S. J. *J. Org. Chem.* **1989,** *54*, 669.
25. DeShong, P.; Sidler, D. R.; Rybczynski, P. J.; Slough, G. A.; Rheingold, A. L. *J. Amer. Chem. Soc.* **1988,** *110*, 2575.

26. (a) DeShong, P.; Slough, G. A. *Organometallics* **1984**, *4*, 636; (b) Deshong, P.; Slough, G. A.; Elango, V. *J. Amer. Chem. Soc.* **1985**, *107*, 7788; (c) Deshong, P.; Slough, G. A.; Rheingold, A. L. *Tetrahedron Lett.* **1987** *28*, 2229; (d) Deshong, P.; Sidler, D. R.; Slough, G. A. *Tetrahedron Lett.* **1987** *28*, 2232.
27. Darensbourg, M. Y.; Darensbourg, D. J.; Drew, D. *J. Organometal. Chem.* **1974**, *73*, C25.
28. Bennett, R. L.; Bruce, M. I.; Stone, F. G. A. *J. Organometal. Chem.* **1975**, *94*, 65; Bennett, R. L.; Bruce, M. I.; Matsuda, I.; Doedens, R. J.; Little, R. G.; Veal, J. T. *J. Organometal. Chem.* **1974**, *67*, C72; Doedens, R. J.; Little, R. G.; Veal, J. T. *Inorg. Chem.* **1975**, *14*, 1138.
29. Suarez, A.; Vila, J. M.; Pereira, M. T.; Gayoso, E.; Gayoso, M. *J. Organometal. Chem.* **1987**, *335*, 359; Ceder, R. M.; Sales, J.; Solans, X.; Font-Altaba, M. *J. Chem. Soc., Dalton Trans.* **1986**, 1351.
30. cf Closson, R. D.; Kozikowski, J.; Coffield, T. H. *J. Org. Chem.* **1957**, *22*, 598; see also Bruce, M. I.; Liddell, M. J.; Pain, G. N. *Inorg. Syntheses* **1989**, *26*, 172.
31. Cooney, J. M.; Gommans, L. H. P.; Main, L.; Nicholson, B. K. *J. Organometal. Chem.* **1988**, *349*, 197.
32. Robinson, N. P. D. Phil. Thesis, University of Waikato, 1989.
33. Cooney, J. M. M. Sc. Thesis, University of Waikato, 1988.
34. Bruce, M. I.; Goodall, B. L.; Stone, F. G. A. *J. Chem. Soc., Dalton Trans.* **1978**, 687.
35. Cooney, J. M. unpublished data, University of Waikato, 1989.
36. Bowmaker, G. A.; Campbell, G. K. *Aust. J. Chem.* **1979**, *32*, 1897.
37. Robinson, N. P.; Main, L.; Nicholson, B. K. *J. Organometal. Chem.* **1988**, *349*, 209.
38. Ryabov, A. D. *Chem. Rev.* **1990**, *90*, 403.
39. Casey, C. P.; Bunnell, C. A. *Inorg. Chem.* **1975**, *14*, 796.
40. Gommans, L. H. P. D. Phil. Thesis, University of Waikato, 1986.
41. Harbourne, D. A.; Stone, F. G. A. *J. Chem. Soc. (A)* **1968**, 1765.
42. Booth, B. L.; Hargreaves, R. G. *J. Chem. Soc. (A)* **1969**, 2766.
43. Booth, B. L.; Hargreaves, R. G. *J. Chem. Soc. (A)* **1970**, 308.
44. Tully, W. M. Sc. Thesis, University of Waikato, 1990.
45. Haupt, H. J.; Lohmann, G.; Florke, U. *Z. Anorg. Allg. Chemie* **1985**, *526*, 103.
46. Gommans, L. H. P.; Main, L.; Nicholson, B. K. *J. Chem. Soc., Chem. Commun.* **1986**, 12.
47. DeShong, P.; Slough, G. A.; Sidler, D. R.; Rybczynski, P. J.; von Philipsborn, W.; Kunz, R. W.; Bursten, B. E.; Clayton, T. W. *Organometallics* **1989**, *8*, 1381.
48. Casey, C. P.; Boggs, R. A.; Marten, D. F.; Calabrese, J. C. *J. Chem. Soc., Chem. Commun.* **1973**, 243.
49. McKinney, R. J.; Huie, B. T.; Knobler, C. B.; Kaesz, H. D. *J. Amer. Chem. Soc.* **1973**, *95*, 633.
50. Cambie, R. C.; Metzler, M. R.; Rutledge, P. S.; Woodgate, P. D. *J. Organometal. Chem.* **1990**, *381*, C26.
51. Heck, R. F. *J. Amer. Chem. Soc.* **1968**, *90*, (a) 5518; (b) 5526; (c) 5531; (d) 5535; (e) 5538.
52. Terpko, M. O.; Heck, R. F. unpublished results quoted in reference 23.
53. Gommans, L. H. P.; Main, L.; Nicholson, B. K. *J. Chem. Soc., Chem. Commun.* **1987**, 761.
54. Cacchi, S.; Misiti, D.; Palmieri, G. *Tetrahedron* **1981**, *37*, 2941.
55. Robinson, N. P.; Main, L.; Nicholson, B. K. *J. Organomet. Chem.* **1989**, *364,* C37.
56. Colvin, E. W. *Silicon Reagents in Organic Synthesis*; Academic Press: London, 1988.
57. Verboom, W.; Reinhoudt, D. N. *Rev. Trav. Chim. Pays-Bas* **1986**, *105*, 199.
58. e.g. Maassarani, F.; Pfeffer, M.; Le Borgne, G. *Organometallics* **1987**, *6*, 2029.
59. Grigsby, W. J.; Main, L.; Nicholson, B. K. *Bull. Chem. Soc. Japan* **1990**, *63*, 649.
60. Liebeskind, L. S.; Johnson, S. A.; McCallum, J. S. *Tetrahearon Letters* **1990**, *31*, 4397.
61. Cooney, J. M.; Gommans, L. H. P.; Main, L.; Nicholson, B. K. *J. Organomet. Chem.* **1987**, *336*, 293.
62. Crawford, S. A. Ph.D. Thesis, University of California, Los Angeles, 1975.
63. Arnold, L. D. Phil. Thesis, University of Waikato, 1992.

64. Arnold, L.; Main, L.; Nicholson, B. K. *Applied Organometal. Chem.* **1990**, *4*, 503.
65. Leach, M. M. Sc. Thesis, University of Waikato, 1991.
66. Cambie, R. C.; Metzler, M. R.; Rutledge, P. S.; Woodgate, P. D. *J. Organometal. Chem.* **1992**, *429*, 41, 59; and references therein.

CYCLOPROPYLCARBENE-CHROMIUM COMPLEXES:

VERSATILE REAGENTS FOR THE SYNTHESIS OF FIVE-MEMBERED RINGS

James W. Herndon, Seniz U. Tumer, Leonard A. McMullen, Julius J. Matasi, and Wayne F.K. Schnatter

Advances in Metal-Organic Chemistry
Volume 3, pages 51–95.

ISBN: 1-55938-406-9

I. INTRODUCTION

Reactions involving alkynes and Fischer carbene-chromium complexes have emerged as a powerful tool for synthetic organic chemistry, allowing for the direct synthesis of a variety of structural types including cyclobutenones, furans, cyclopentenones, aromatic rings, and various bicyclic structures.[1,2] The most well-developed of the many versions of this reaction is the coupling of alkynes with α,β-unsaturated carbene-chromium complexes to produce highly substituted aromatic rings (Scheme 1). A diverse array of natural products have been synthesized using this aromatic ring-forming reaction, including anthraquinone anticancer agents,[3] naphthoquinone antibiotics,[4] vitamin E,[5] Khellin,[6] and psoralen derivatives.[7] In this reaction, a net [3+2+1] cycloaddition reaction occurs, whereby carbons a–c of α,β-unsaturated carbene-chromium complex **1**, the two alkyne carbons, and a carbon monoxide ligand combine to form the new ring. The newly formed ring is initially a cyclohexadienone derivative **2**, which undergoes enolization to form the corresponding phenol **3**.

In the organic literature, there is a notable lack of cycloaddition reactions available for the synthesis of carbocyclic seven-membered rings.[8] This is a highly important ring system in a number of medicinally-important compounds, most notably the colchicine derivatives,[9] guaianolides and pseudoguaianolides,[10] and phorbol esters.[11] As part of a program to develop a general cycloaddition reaction for the synthesis of carbocyclic seven-membered rings, we have examined the reaction of cyclopropylcarbene-chromium complexes with alkynes. In many organic reactions, the ability of a cyclopropane ring to mimic the chemistry of an alkene is well-documented, and thus if complexes **1** and **4** react similarly with alkynes (Scheme 2), then cycloheptadienone derivatives such as compound **3** would be produced. Limited success has been reported by others in the use of cyclopropane analogs to divert classical six-membered ring forming reactions into seven-membered ring-forming reactions.[12]

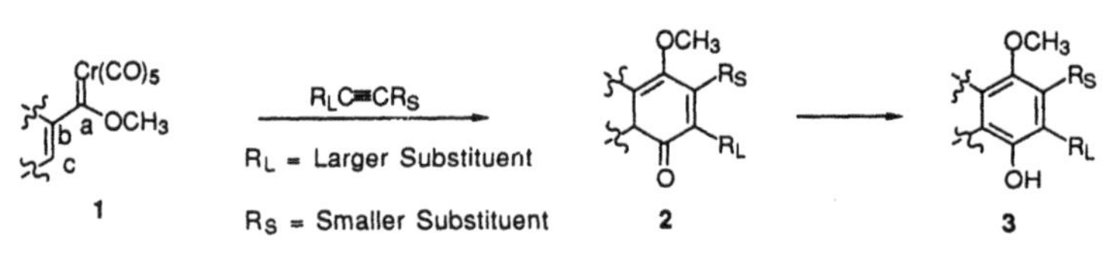

Scheme 1.

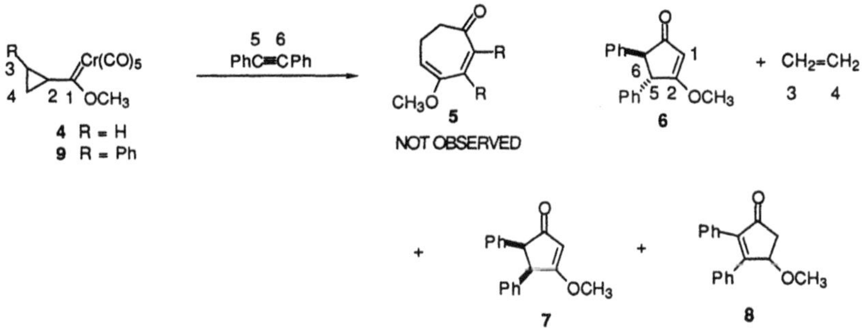

Scheme 2.

As we recently reported,[13] the reaction of cyclopropylcarbene complex **4** and diphenylacetylene did not produce cycloheptadienone **5**, but gave exclusively the cyclopentenone derivatives **6–8**, plus ethylene. This reaction was found to be general for a variety of alkynes. The conversion of complex **4** and alkynes to cyclopentenone derivatives allows for the construction of highly functionalized five-membered rings from readily available components. The development of cycloaddition approaches for the construction of carbocyclic five-membered rings has become a very active area of research in recent years,[14] with only a few one-pot reactions actually being developed. As such, the overall conversion represented by this reaction is a potentially powerful tool for the synthetic organic chemist. In this paper, we report on the scope, limitation, and mechanism of this highly useful new five-membered ring forming reaction.

II. REACTION OF ALKYNES WITH CYCLOPROPYLCARBENE-CHROMIUM COMPLEXES

A. Scope and Limitations

When complex **4** was allowed to react with diphenylacetylene (0.1 M solution of both components in THF) at 65 °C under nitrogen, no identifiable products were obtained, with the exception of insoluble aromatic tars, tentatively identified as diphenylactetylene polymers.[15] Polymerization of the alkyne can probably be suppressed if the concentration of the alkyne can be kept to a minimum throughout the reaction.[16] When diphenylacetylene was slowly added to a solution of complex **4** in THF at 65 °C, compound **6** was obtained in 34% yield. The minor products **7** (6%) and **8** (2%) were also obtained. Compound **6** was formed exclusively (41%) when the reaction was performed in refluxing dioxane (100 °C). The yield was not improved significantly by the addition of excess alkyne.

Scheme 3.

The most reasonable correlation between compound **6** and the reactants is outlined in Scheme 2. The carbonyl carbon arises from a CO-ligand of the carbene complex, and carbons 1 and 2 of cyclopentenone **6** originate from carbons 1 and 2 of the carbene complex. Carbons 5 and 6 correspond to the alkyne carbons of diphenylacetylene. This implies that carbons 3 and 4 of the carbene complex are lost as a two-carbon fragment. When complex **9** was reacted with diphenylacetylene under identical conditions, compound **6** (44%) was obtained along with styrene (41%). To balance the equation for formation of **6** from **4** or **9**, an external source of hydrogen must be involved. When the reaction was performed in the presence of external hydrogen sources such as water, methanol, triethylsilane,[17] 1,3-cyclohexadiene, or hydrogen, the yield of formation of **6** increased. The best yields of **6** (79%) were obtained using 1% aqueous dioxane as the solvent, and with the carbene complex and the alkyne added to a solution of refluxing aqueous dioxane over a period of 2–6 h. Apparently, water and low-oxidation state chromium interact to provide the source of hydrogen. When deuterium oxide was substituted for water in the reaction, compound **6** was obtained with deuterium atoms at carbons 1 (20%), 5 (90%), and 6 (90%). During the course of the reaction, a dark green color appeared. This color is suggestive of chromium(III) species.

A list of various alkynes reacting with complex **4** is presented below in Table 1. As can be seen from this table, the cycloaddition reaction is general for a wide variety of alkynes. With monosubstituted alkynes (entries B, H–N), the only regioisomer obtained was that in which the substituent on the alkyne was alpha to the carbonyl group in the product. When using monosubstituted alkynes in the reaction, it is critical that no more than 1.3 equivalents of alkyne per carbene complex is used. When large excesses of monosubstituted alkyne were used, cyclopentenones such as **9** (Scheme 3) were obtained, which result from a Reppe-type coupling of the alkynes with CO.[18] With 1-phenylpropyne (entry C) and 2-methyl-2-pentyne (Entry F), both unsymmetrically-disubstituted alkynes, the major products obtained from reaction with complex **4** were those where the larger substituent was alpha to the carbonyl group. The reaction appears to proceed in a way such that the larger substituent will end up alpha to the carbonyl group in the final product. Further support that the reaction is controlled by sterics and not electronics can be seen in entry D, where an alkyne having similar steric properties at both carbons, but different electronic properties gave little selectivity, providing a 57:43 mixture of the two regioisomers.

Table 1. Reaction of Complex **4** with Alkynes in Refluxing Aqueous Dioxane[a,b]

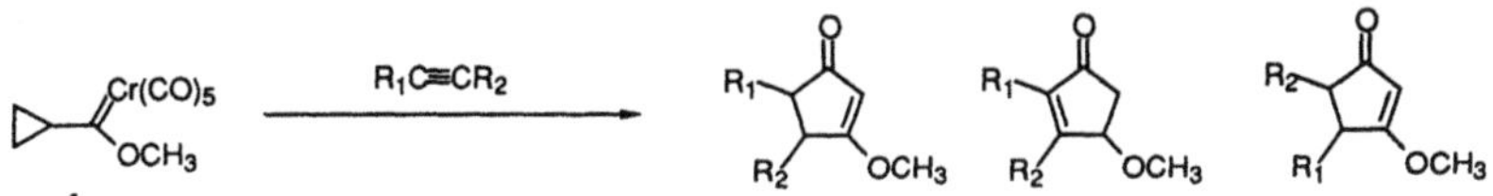

4 **10** **11** **12**

Entry[b]	*R_1*	*R_2*	*Yield 10(%)*	*trans:cis 10*	*Yield 11 (%)*	*Yield12 (%)*
A	Ph	Ph	79	24:1	4	
B	Ph	H	62		0	0
C	Ph	CH_3	73	9:1		12[c]
D	Ph	p-$PhOCH_3$	39	1.1 :1[d]	0	39[d]
E	n-C_3H_7	n-C_3H_7	60	1 :1[d]	0	
F	i-C_3H_7	CH_3	48	1.1 :1[d]		14[e]
G	n-C_3H_7	H	68		0	0
H	c-C_3H_5	H	53		0	0
I	-$(CH_2)_4OH$	H	68		0	0
J	- $(CH_2)_4$OTBDMS[f]	H	58		0	0
K	-$(CH_2)_4OSO_2CH_3$	H	38[g]		0	0
L	-CH_2OCH_3	H	0[h]		0	0
M	*cis*-CH=$CHOCH_3$	H	42[i]		0	0
N	-COOEt	H	0		0	0
O	-$Si(CH_3)_3$	-$Si(CH_3)_3$	54[j]		0	0

Notes: [a] In all cases, the alkyne and complex **4** (0.1 M in dioxane) were added to a refluxing solution of aqueous dioxane via syringe pump over a period of 2 h.

[b] Entry letters define R_1 and R_2 for compound **10**.

[c] Only the compound having the double bond between the phenyl and methyl substituents was obtained.

[d] The *trans:cis* ratio was determined by examination of the crude ^{1}H NMR spectrum; these compounds were isomerized to the pure *trans* isomer by treatment with sodium methoxide and methanol prior to characterization.

[e] The *trans:cis* ratio was 9.1:1.

[f] TBDMS = *t*-butyldimethylsilyl.

[g] A 16% yield of compound **10I** was also obtained from the reaction.

[h] Only the methoxy elimination product 6-methyl-3-methoxy-2-cyclopentenone was obtained in 16% yield.

[i] The product was obtained as a 3:2 *trans:cis* mixture about the enol ether double bond.

j Cyclopentadienone **15** was obtained in 14% yield, accompanied by 3-methoxy-2-cyclopentenone in 54% yield, presumably due to protiodesilylation of **10O**.

The scope and limitations of the reaction has been explored using a variety of functionalized alkynes. The reaction is tolerant of alcohol functionality (entry I), and similar yields were obtained if the alcohol functionality was protected (entry J). Even mesylate functionality could be tolerated in the reaction; hex-5-yn-1-yl mesylate coupled with carbene complex **4** to give a mixture of the expected mesylate (**10K**, R = –$(CH_2)_4OSO_2CH_3$) and the corresponding alcohol (**10I**, R = –$(CH_2)_4OH$), which presumably results from a solvolysis of the mesylate under the

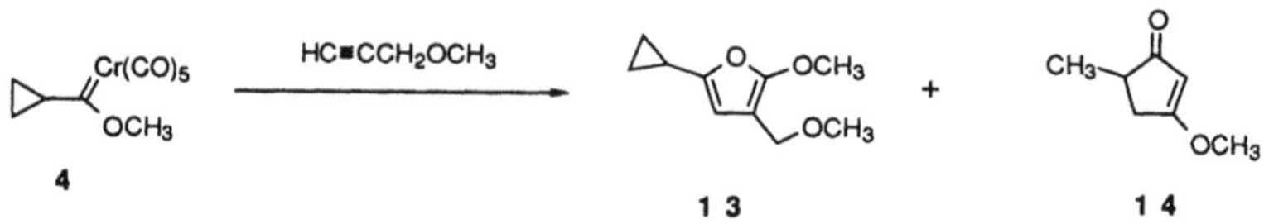

Scheme 4.

conditions of the reaction.[19] In three cases, the reaction of carbene complex **4** with alkynes failed to produce any of cyclopentenones **10–12**. The reaction of complex **4** with methyl propargyl ether resulted in the formation of the furan **13** and the cyclopentenone **14** (Scheme 4). The five-membered ring-forming reaction did not proceed using the electron-deficient alkyne ethyl propiolate.[20] The reaction of bis-trimethylsilylacetylene with complex **4** afforded cyclopentadienone derivative **15** in 14% yield (Scheme 5), accompanied by 3-methoxy-2-cyclopentenone, presumably the result of protiodesilylation of cyclopentenone **10O**.

In most cases, the major alkene isomer obtained from the reaction mixture was the thermodynamically more stable vinylogous esters **10** or **12**[21] having the original alkyne substituents with the *trans* relative stereochemistry. At shorter reaction times and lower reaction temperatures, other alkene isomers such as **11**, as well as significant amounts of the *cis* isomer, were obtained from the reaction. Both base and chromium hexacarbonyl are effective alkene isomerization catalysts for conversion of **7** and **8** to **6**. Most of the products in the Table could be rearranged to essentially pure *trans* isomers by the action of sodium methoxide and methanol.

We have briefly examined the reaction of other cyclopropylcarbene-chromium complexes with diphenylacetylene. As can be seen in Table 2, the simple cyclopropylcarbene complex **4** was the optimal carbene complex for use in the cyclization reaction. Lower overall yields of the cyclopentenone **6** were obtained with the phenyl-substituted carbene complexes **9**, and the sterically-encumbered tetramethyl-substituted carbene complex **17** was not reactive with diphenylacetylene at 100 °C. The vinyl-substituted carbene complex **16** reacted with diphenylacetylene to afford a mixture of cyclopentenones **6** and **20** (Scheme 6). Cyclopentenone **20** results from a rearrangement-carbonylation reaction of starting carbene complex **16**. Compound **20** was also obtained from the thermolysis of complex **16** in the absence of an alkyne, and this reaction was found to be general for a wide variety

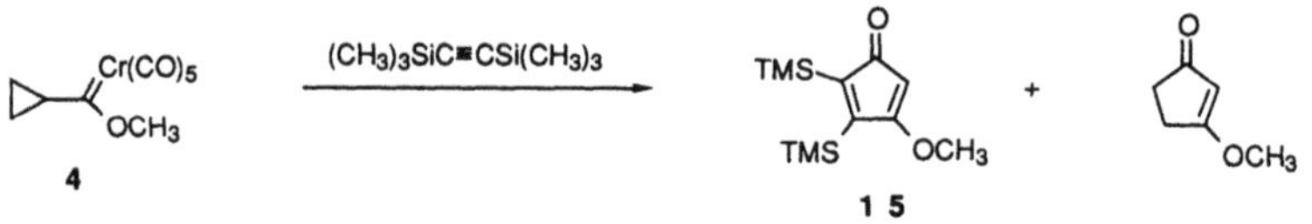

Scheme 5.

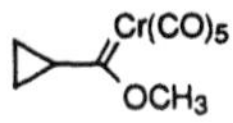

4

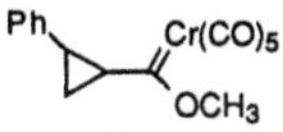

9

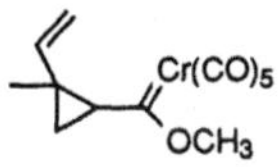

16

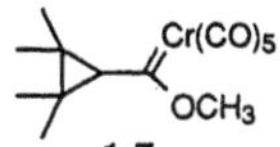

17

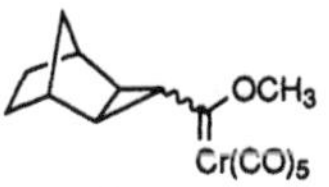

18

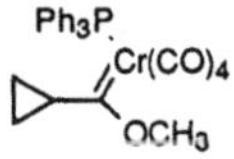

19

Table 2. Yield of Cyclopentenones **6–8** from Reaction of Diphenylacetylene with Various Cyclopropylcarbene-Chromium Complexes in Refluxing Aqueous Dioxane

Complex #	*Yield Cyclopentenones* **6–8**
4	83
9	53
16[a]	14
17	0
18	42[b]
19	60

Notes: [a] A 44% yield of cyclopentenone **20** was also obtained.
[b] No water was used in this reaction; a 20% yield of cyclopentadienone **35** was also obtained.

of 2-vinylcyclopropylcarbene-chromium complexes (see ahead). The triphenylphosphine-substituted carbene complex **19** was also reactive with diphenylacetylene, giving cyclopentenone **6** as the major product of the reaction. Although reactions from this species seemed to be much more facile,[22] the overall yield from this complex was disappointing. The norbornyl-fused cyclopropylcarbene complex **18** also provided cyclopentenone **6** as the major product of the reaction, in spite of the extra energy associated with formation of the highly strained alkene norbornene.[23]

Preparation of carbene complex **4**. To a solution of cyclopropyl bromide (4.80 g, 39.7 mmol) under nitrogen at –78 °C in diethyl ether (90 mL) was

16 6 20

Scheme 6.

added by syringe *t*-butyllithium (36.4 mL of a 2.2 M hexane solution 80.1 mmol) over a period of 5–15 min. This solution was stirred at –78 °C for a period of 30 min., and then transferred by cannula to a suspension of chromium hexacarbonyl (8.80 g, 40.0 mmol) in diethyl ether (100 mL) at 0 °C. This mixture was allowed to stir 1 h at 0 °C, after which time methyl triflate (25.2 g, 153 mmol) was added. The reaction mixture was warmed to 25 °C, and then stirred 3 h at 25 °C. The mixture was placed in a separatory funnel and washed with saturated aqueous sodium bicarbonate and saturated aqueous sodium chloride, and then dried over magnesium sulfate. After the solvent was removed on a rotary evaporator, the residue was subjected to flash chromatography on silica gel using hexane as the eluent. After solvent removal, a yellow solid (6.52 g, 59%) was obtained. ^{1}H NMR (C_6D_6): δ 3.78 (s,3H), 3.01 (tt, 1H, J = 7.6, 3.2 Hz), 0.85 (m, 2H), 0.48 (m, 2H); (in $CDCl_3$): δ 4.65 (s, 3H), 3.46 (m, 1H), 1.36 (m, 2H), 1.18 (m, 2H). The spectral data were in agreement with those reported previously for this compound.[28]

Reaction of carbene complex **4** *with diphenylacetylene.* To a three-neck round-bottom flask equipped with reflux condenser, stopper, and septum, under nitrogen, was added 20 mL of 1% aqueous dioxane solution. The dioxane solution was heated to reflux. To this refluxing solution, a solution of carbene complex **4** (0.083 g, 0.30 mmol) and diphenylacetylene (0.107 g, 0.6 mmol) in dioxane (25 mL) was added by syringe pump over a period of 3 h. After the addition was complete, the mixture was allowed to reflux for a period of 6 h. The mixture was then allowed to cool to 25 °C, and the solvent was removed on a rotary evaporator. Ethyl acetate (50 mL) was added and the solution was filtered through silica gel. The solvent was removed on a rotary evaporator, and final purification was achieved through flash chromatography on silica gel, using 3:2 hexane:ethyl acetate as the eluent. The product in the first fraction (0.003 g, 4%) was identified as compound **8**. ^{1}H NMR ($CDCl_3$): δ 7.15–7.36 (m, 10 H), 5.04 (dd, 1 H, J = 5.9, 2.0 Hz), 3.39 (s, 3H), 2.96 (dd, 1 H, J = 18.3, 5.9 Hz), 2.66 (dd, 1 H, J = 18.3, 2.0 Hz); IR ($CDCl_3$): 3060 (m), 2930 (m), 1705 (s), 1600 (m), 1440 (m), 1350 (m), 1260 (m), 1210 (m), 1180 (s) cm^{-1}. The product in the second fraction (0.061 g, 76%) was identified as compound **6**. ^{1}H NMR ($CDCl_3$): δ 7.2 (m, 6 H), 7.0 (m, 4 H), 5.58 (d, 1 H, J = 1.1 Hz), 3.98 (dd, 1 H, J = 3.3, 1.1 Hz), 3.83 (s, 1 H), 3.59 (d, 1 H, J = 3.3 Hz); ^{13}C NMR ($CDCl_3$): δ 203.7 (s), 190.0 (s), 139.3 (s), 138.8 (s), 128.9, 128.7, 127.6, 127.4, 127.2, 127.0 (overlapping in SFORD spectrum), 104.5 (d), 62.4 (d), 59.0 (q), 56.4 (d); IR ($CDCl_3$): 3040 (m), 2945 (m), 1694 (s), 1598 (s), 1500 (m), 1457 (m), 1443 (m), 1358 (s), 1347 (s), 1170 (s) cm^{-1}; MS (EI): 264 (parent), 233, 205, 187, 159, 128, 115, 102, 91, 69, HRMS; calcd for $C_{18}H_{16}O_2$ 264.1150, observed 264.1150. The product in the third fraction (0.002 g, 3%) was identified as compound **7**. ^{1}H NMR ($CDCl_3$): δ 6.75–7.03 (m, 5H), 5.75 (d, 1 H, J = 0.7 Hz), 4.48 (d, 1H,

J = 7.8 Hz), 4.25 (d, 1H, J = 7.8 Hz), 3.89 (s, 3H); IR ($CDCl_3$): 3068 (w), 2942 (m), 1691 (s), 1599 (m), 1455 (m), 1354 (m), 1335 (m), 1232 (m), 1133 (s), 1101 (s), 996 (m) cm^{-1}.

B. Intramolecular Reactions

The cycloaddition reaction between alkynes and cyclopropylcarbene-chromium complexes is regioselective only with monosubstituted alkynes. The overall synthetic utility of the cycloaddition reaction would be enhanced substantially if a regioselective version of the above reaction could be developed. Among the more important five-membered ring-containing synthetic targets are the prostaglandins (e.g., PGA_2, **21**) and prostacyclins (e.g., PGI_2, **22**), which are potentially important cardiovascular drugs.[24] Both of these compounds have as their central nucleus unsymmetrically-substituted cyclopentane rings, and thus regioselectivity problems would be encountered during attempts to construct these ring systems using the cyclopropylcarbene-chromium complex-alkyne cycloaddition reaction.

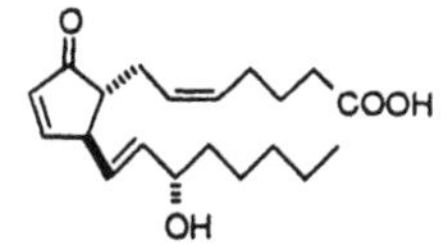

PGA_2 **21**

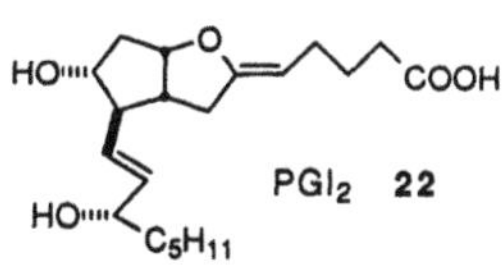

PGI_2 **22**

One possible method to control the regioselectivity would be through an intramolecular version of the cycloaddition reaction (Scheme 7).[25] This approach has been used in the reaction of alkynes with arylcarbene-chromium complexes.[7,26] Thermolysis of an alkynyl-containing carbene-chromium complex such as **23** in the presence of water should lead primarily to fused ring compound **24**. Bridged-bicyclic compound **25** is an anti-Bredt alkene and should not form provided that the length of tether is reasonably small. Compound **24** is a vinylogous ester, and in theory is susceptible to a wide variety of synthetic manipulations.[27] Thus subsequent ring opening reactions could then provide cyclopentenone derivatives of well-defined regiochemistry.

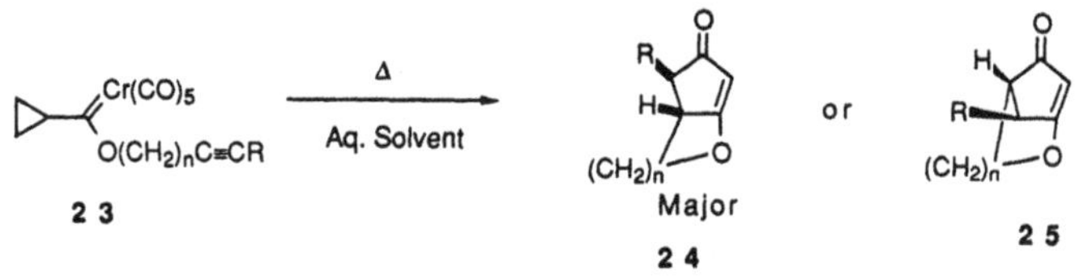

Scheme 7.

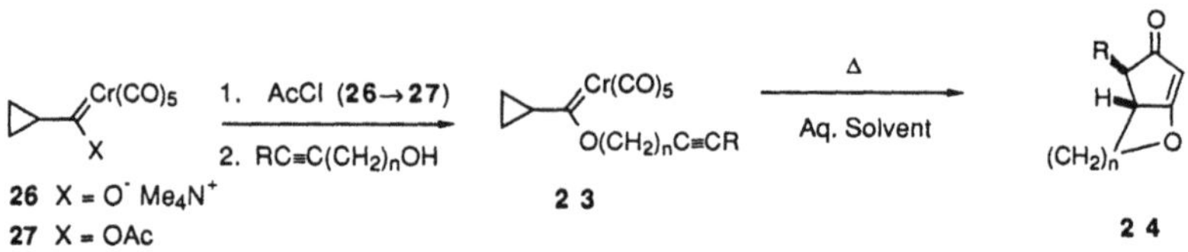

Scheme 8.

In the first phase of these studies, methods for the synthesis of alkyne-containing carbene complexes were explored. The reaction of acetylenic alcohols with the anhydride-like complex **27**[26] (generated *in situ* from complex **26** and acetyl chloride) would appear to be the most obvious synthetic route to these complexes (Scheme 8). Acetoxycarbene complex **27** has been reported.[28] Although this complex decomposes below 25 °C, it could be trapped with azide ion to produce the corresponding isonitrile complex **28** in what is formally a metallo-Schmidt reaction (Scheme 9).

The method in Scheme 8 does in fact provide the desired carbene complexes, but the reaction was very sensitive to temperature due to competing decomposition of **27**.[28] Primary alcohols worked well in this reaction, providing the desired carbene complexes **23** in high yield. Secondary alcohols afforded much lower yields of the desired complexes, but the problem could be partially overcome by use of excesses of the acetoxycarbene complex. The yields for formation of a variety of these carbene complexes using the acetoxycarbene route is presented below in Table 3. The other obvious route to these complexes, reaction of acylate salt **26** with alkyne-containing triflates provided only low yields of the desired carbene complexes.

It was initially felt that high dilution conditions would not be required for these intramolecular cycloaddition reactions since they are now intramolecular. When a 0.1 M solution of carbene complex **23G** in 1% aqueous dioxane was heated at 80 °C for a period of 2 h, the desired compound **24G** (*cis* and *trans*) was obtained only in trace amounts along with the alcohol **30** and possible dimeric structure **31** (Scheme 10). Compound **30** is the result of a hydrolysis reaction of the enol ether functionality in cyclopentadienone intermediate **29**. Compound **31** can result from dimerization of the same cyclopentadienone.[29] If the reaction was performed under high dilution conditions, then compound **24G** (*cis* and *trans*) was obtained in higher

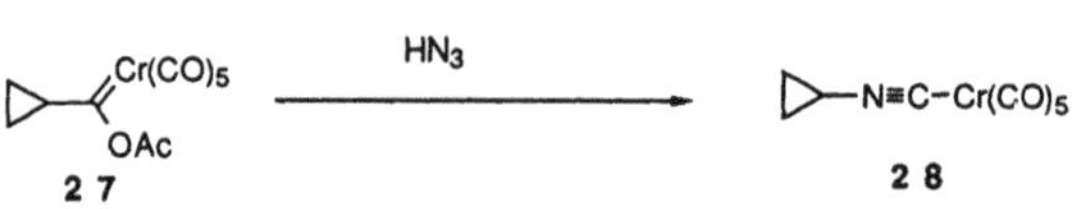

Scheme 9.

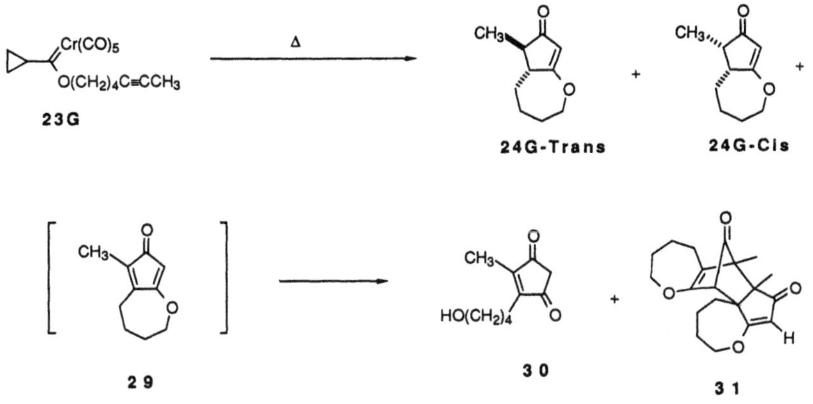

Scheme 10.

yield (26%), along with the alcohol **30** (9%) and the cyclopentadienone dimer **31** (20%). In order to assist with the reduction of cyclopentadienone **9**, chromium hexacarbonyl was added to the reaction, giving the expected products **24G** in higher yield (38%), but the dimerization and hydrolysis products were still formed in similar yields. The yields of compound **24G** were considerably higher in toluene (70%), and none of dimer **31** was formed using this solvent. The reason for this might be the poorer ligation ability of toluene relative to dioxane, and thus the cyclopentadienones were more likely to exist as their chromium complexes.

The results from a systematic study of the intramolecular cyclization reaction are summarized in Table 3. As expected, the reaction was highly regioselective, giving only fused-ring isomers and no bridged-ring isomers. With internal alkynes, the reaction proceeds efficiently when forming heterocyclic rings of sizes 5–7. If both of the substituents are alkyl, then there was a problem with dimerization and hydrolysis when the reaction was conducted in dioxane, however this problem was largely be overcome if the reaction was performed in toluene. If one of the substituents is aryl then no such problems exist, regardless of the solvent used. The reaction was general for internal alkynes, and reasonable yields of the expected products **24** were formed in all cases. Only in the reactions in Entries D and G were significant amounts of dimerization and cyclopentenedione products obtained. The formation of the cyclopenta[*a*]furan ring system (entry A), which is present in prostacyclins and in many prostaglandin synthesis intermediates, proceeded in 51% yield. Addition of chromium hexacarbonyl to the reaction in Entry C led to an increase in the *trans*:*cis* ratio, presumably due to a thermodynamic equilibration by the chromium species. The *trans-cis* selectivity appears to be greater in dioxane than in toluene although the yields are considerable better in toluene.

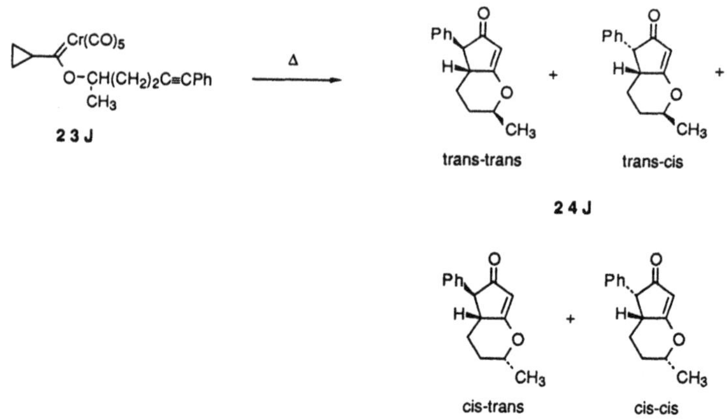

Scheme 11.

Thermolysis of complex **23J** in aqueous dioxane leads to a 40:35:14:11 mixture of four stereoisomers in 61% yield (Scheme 11). The major isomer (*trans-trans*) differs from the second-most abundant one (*trans-cis*) only in the configuration of the substituents on the five-membered ring. The methyl group apparently controls the overall stereochemistry of the reaction since the six-membered ring substituents

Table 3. Preparation and Thermolysis of Alkyne-Containing Cyclopropylcarbene Complexes

Entry[a]	*n*	*R*	*Yield* **23**	*Yield* **24**	*trans:cis* **24**
A	2	CH_3	66	51	65:35
B	2	H	75	Trace	
C	3	C_6H_5	73	74	93 :7[b]
D	3	n-C_3H_7	82	31	61:49[c]
E	3	2-Thienyl	68	51	80:20[b,d]
F	3	H	70	26	
G	4	CH_3	63	38	66:34[b,c]
H	4	H	69	60	
I	5	H	73	30	
J	Ph-C≡C$(CH_2)_2$CHOHCH_3		56[e]	61	f
K	Compound **23D** in 1% aq. toluene			60	40:60
L	Compound **23G** in 1% aq. toluene			70	77:23

Notes: [a] Table entry letters define R and n for compounds **23** and **24**, all reactions were performed in 1% aqueous dioxane unless otherwise noted.
[b] Chromium hexacarbonyl (1 eq.) was present during the reaction.
[c] Cyclopentadienone dimers and cyclopentenediones were also obtained.
[d] A 3% yield of the compound having the double bond at the carbon containing R was also obtained. This compound is the natural plant growth regulator chrycorin.[30]
[e] An excess of **27** (2 eq.) was used.
[f] See text.

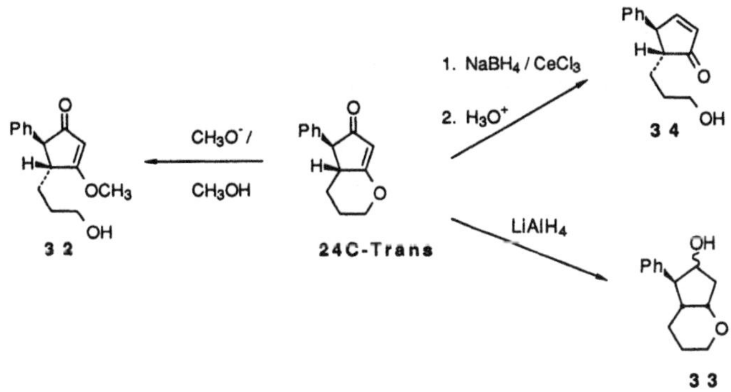

Scheme 12.

are diequatorial in both of the two major isomers. Groups larger than methyl might lead to even greater stereoselectivity in the reaction.

Reactions involving terminal alkynes (Entries B,E,H,I) were less efficient, and highly sensitive to ring size, giving reasonable yields of cycloadducts only in Entry H. A similar dependence with regard to length of tether has been observed in the intramolecular reaction of terminal alkynes with arylcarbene-chromium complexes,[7,26] and has been attributed to electronic effects in the complexation of the alkyne with the chromium atom.

Although the stereoselectivity of the reaction was somewhat disappointing, the possibility exists for epimerization, since all carbons of the newly-formed five-membered ring are epimerizable. Treatment with sodium methoxide-methanol did not accomplish the *cis-trans* isomerization, but rather led to ring opening of the vinylogous lactone **24C** to the vinylogous ester **32** (Scheme 12). Triethylamine was an effective reagent for epimerization of compound **24C** but failed for the less acidic compounds **24D** and **24G**. Potassium *t*-butoxide effected the desired epimerization reaction in the less acidic compounds.

The vinylogous ester functionality was susceptible to further synthetic manipulation, which will now allow for the conversion to monocyclic compounds. Surprisingly, lithium aluminum hydride reduction of the enone **24C** provided only alcohol **33** in 66% yield. The desired reduction of the ketone could be effected by reaction of compound **24C** (*trans*) with sodium borohydride/cerium trichloride,[31] followed by aqueous acid, leading to the enone alcohol **34** (*trans*) in 68% yield (93% based on recovered starting material). The corresponding *cis* isomer underwent the same conversion to give compound **34** as a 5:1 *cis*:*trans* mixture. Treatment of compound **24C** (*trans*) with sodium methoxide/methanol led to the vinylogous ester alcohol **32**; both the *cis* and the *trans* isomers of **24C** led to the *trans* compound

32. Compound **32** is the product that would have been obtained from a regioselective reaction between carbene complex **4** and 5-phenyl-4-pentynol.

Preparation of carbene complex **23C.** To a solution of acylate salt **26**[28] (0.405 g, 1.2 mmol) in dichloromethane (20 mL) at 0 °C under nitrogen was added via syringe acetyl chloride (0.080 mL, 1.2 mmol), followed by immediate addition of 5-phenyl-4-pentyn-1-ol[85] (0.193 g, 1.2 mmol). The reaction mixture was stirred at this temperature for about 1 h and then warmed to 25 °C and stirred for 20 min. The solvent was removed on a rotary evaporator. Flash chromatography of the residue after evaporation using hexane as the eluent gave pure carbene complex **23C** (0.354 g, 73%) as a yellow oil after solvent removal. ^{1}H NMR ($CDCl_3$): δ 7.20 (m, 5H), 5.05 (t, 2H, J = 6.1), 3.50 (m, 1H), 2.60 (t, 2H, J = 6.9), 2.15 (tt, 2H, J = 6.9, 6.1),1.40 (m, 2H), 1.20 (m, 2H). IR (CH_2Cl_2): 3051, 2970, 2061, 1979, 1938, 1365, 1040, 960 cm^{-1}. ^{13}C NMR($CDCl_3$): δ 352.0, 224.0, 216.7, 131.6, 128.2, 127.9, 123.3, 87.4, 82.0, 78.9, 41.5, 29.7, 28.4, 17.8, 16.3. MS (CI): m/e 404 (M, 0.80), 337 (0.30), 297 (0.80), 264 (0.70), 229 (18.3),171 (2.8),135 (6.2), 91 (15.5), 69 (100),67 (55.2). High res. MS: Calcd. for $C_{20}H_{16}CrO_6$: 404.0352. Found: 404.0333.

Thermolysis of carbene complex **23C**; *Preparation of compound* **24C**. To a three-neck round-bottom flask equipped with a reflux condenser and rubber septum, under nitrogen was added 1,4-dioxane (100 mL) and water (1 mL), and the solution was heated to reflux. To this refluxing solution was added a solution of carbene complex **23C** (0.354 g, 0.88 mmol) and chromium hexacarbonyl (0.190 g, 0.9 mmol) in 1,4-dioxane (30 mL) via syringe pump over a period of 4 h. After the addition was complete, the mixture was heated at reflux an additional 20 h and then cooled to 25 °C. Next, the green mixture was filtered through Celite and the solvent was removed on a rotary evaporator. The final purification was achieved by flash chromatography on silica gel, using hexane/ethyl acetate (4/1) as the eluent. Two fractions were obtained. The compound in the first fraction (0.130 g, 69%) was identified as the *trans* isomer of compound **24C**. ^{1}H NMR ($CDCl_3$): δ 7.07–7.32 (m, 5H), 5.37 (d, 1H, J = 1.0), 4.1–4.37 (m, 2H), 3.24 (d, 1H, J = 3.8), 2.80–2.93 (m, 1H), 2.15–2.31 (m, 1H), 1.78–2.00 (m, 2H), 1.45–1.65 (m, 2H). ^{13}C NMR ($CDCl_3$): δ 203.5, 189.9, 138.5, 128.7, 128.0, 127.0, 106.1, 68.6, 59.4, 44.8, 24.3, 22.5. IR (CH_2Cl_2): 3061(w), 2970(m), 1685(vs), 1603(vs), 1495(w), 1460(w), 1395(m), 1335(m), 1300(m), 1168(s), 930(m) cm^{-1}. MS (EI): m/e 214 (M, 100),197 (3.4),186 (19.1), 158 (10.3), 144 (9.3), 131 (20.2), 115 (19.8), 103 (12.2), 91 (11.1), 84 (12.8), 77 (8.9), 69 (15.1). High res. MS: Calcd. for $C_{14}H_{15}O_2$: 215.1072. Found: 215.1070. This compound was assigned as the *trans* isomer based on the smaller coupling (3.8 Hz) between the benzylic hydrogen and the allylic hydrogen. The product in the second fraction (0.010 g, 5%) was assigned as the *cis* isomer of compound **24C**. ^{1}H

NMR ($CDCl_3$): δ 7.04–7.32 (m, 5H), 5.56 (d, 1H, J = 1.6), 4.32–4.43 (m, 1H), 4.07–4.19 (m, 1H), 3.90 (d, 1H, J = 7.3), 3.17–3.31 (m, 1H), 1.80–1.91 (m, 2H), 1.41–1.58 (m, 1H), 0.97–1.11 (m, 1H). ^{13}C NMR ($CDCl_3$): δ 205.0, 191.5, 137.4, 129.2, 128.4, 127.0, 108.0, 68.9, 55.4, 41.0, 40.1, 22.7, 21.0. IR (CH_2Cl_2): 3060(w), 2930(m), 1689(s), 1607(vs), 1504(w), 1462(w), 1410(m), 1304(m), 1170(m), 940(m) cm^{-1}. MS (EI): m/e 214 (M,30.2), 197 (1.5), 186 (8.0), 158 (5.5), 131 (13.1), 115 (17.0), 91 (10.5), 86 (65.9), 84 (100), 77 (9.7), 69 (23.3), 51 (48.0). The compound was assigned as the *cis* isomer based on the larger coupling (7.3 Hz) between the benzylic hydrogen and the allylic hydrogen.

C. Mechanism of the Cyclopentenone-Forming Reaction

1. Reduction of Cyclopentadienones to Cyclopentenones

Before an actual mechanistic scheme can be proposed, the role of the added proton source must be determined. Approximately two protons are provided by the proton source as determined from the deuterium labelling study. Since chromium is oxidized during the five-membered ring forming reaction, a plausible origin of the redox process is the net reduction of two protons to H_2 concomitant with the oxidation of chromium. The most obvious precursor to the cyclopentenone products would be the corresponding cyclopentadienone derivative, which was actually obtained from the reaction between bis-trimethylsilylacetylene and complex **4**. The ability of bulky trimethylsilyl groups to protect highly reactive cyclopentadienones from reactions such as dimerization is well-documented.[32] By eliminating water from the reaction between complex **4** and diphenylacetylene in THF at reflux, and keeping the reaction times to a minimum, cyclopentadienone **35** could be obtained from the reaction in 24% yield (Scheme 13). The combination of chromium(0) (from chromium hexacarbonyl) and water is an effective reagent for conversion of compound **35** to cyclopentenone **6**. In fact, chromium hexacarbony/water is specific for the reduction of cyclopentadienones. Cyclopentadienones **35** (65%), **15** (21%), and tetraphenylcyclopentadienone (91%)[33] were reduced to the corresponding cyclopentenone derivatives with no overreduction products obtained. Neither

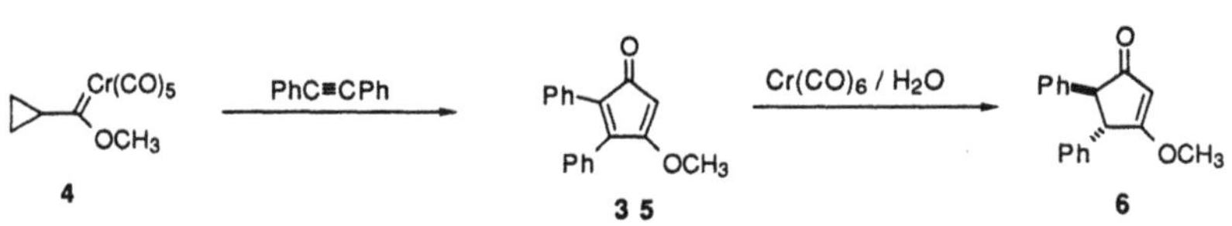

Scheme 13.

cyclohexenone nor 4,4-dimethyl-2,5-cyclohexadienone were reduced under these conditions.

We have briefly probed the mechanism of the cyclopentadienone reduction. If the reaction employing diphenylacetylene in the presence of D_2O was performed while hydrogen gas was bubbled through the system, then only deuterium incorporation was observed. A minimal amount of CO_2 was formed during the reduction of tetraphenylcyclopentadienone, confirming that chromium is the reducing agent and not CO. A possible mechanism involves electron transfer, since the one-electron reduction potential for cyclopentadienone derivatives is quite low.[34] This mechanism would also explain the reductive loss of methoxy in the reaction between complex **4** and methyl propargyl ether, which could occur via intermediate **41** in Scheme 14. This step is similar to the step where benzylic ethers are cleaved during the Birch reduction.[35] The proposed mechanism implies that chromium (II) and not chromium (III) would form in the reaction. Chromium (II) chloride was less effective for the reduction of tetraphenylcyclopentadienone (31%), although it did proceed to some extent. In the absence of an added proton source, the starting carbene complex could serve as the proton donor due to its high acidity.[36]

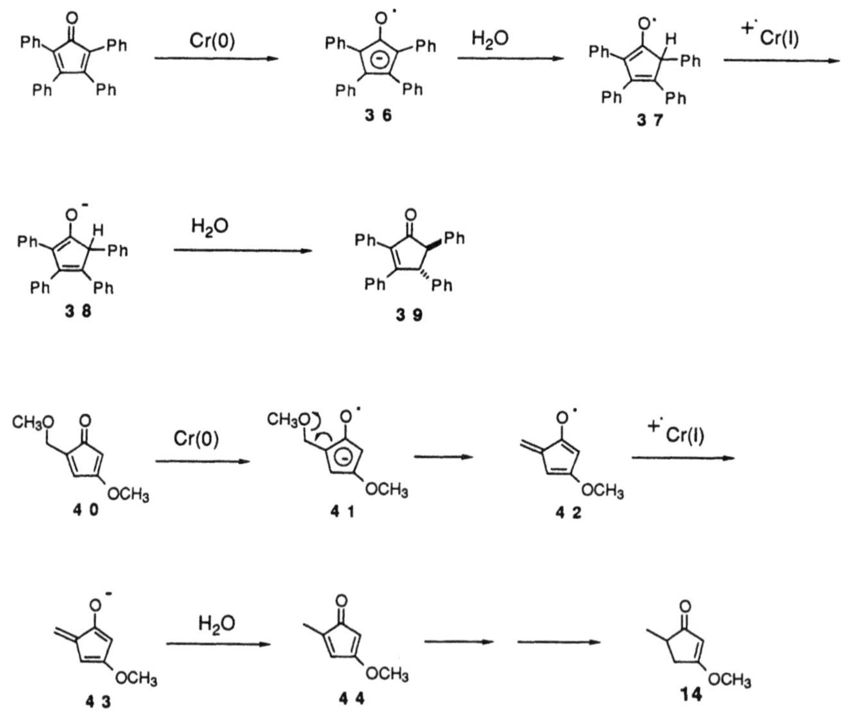

Scheme 14.

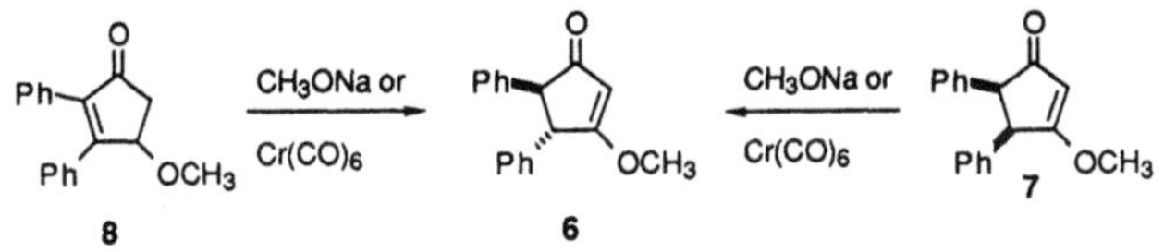

Scheme 15.

A control experiment was performed to establish that deuterium incorporation occurs during the reduction and not as a result of epimerization. To a refluxing solution of compound **6** (deuterium labelled) in 1% aqueous dioxane was added a solution of complex **4** and 4-octyne in dioxane. After final purification, compound **6** was recovered from the reaction with 93% of the original deuterium intact. Compound **10E** was obtained from the reaction with no deuterium incorporation, thus confirming that H-D exchange does not account for the deuterium incorporation when the cycloaddition reaction was performed in the presence of D_2O.

What is the kinetic site of reduction of the 3-methoxycyclopentadienone intermediates? The *trans* vinylogous esters appear to be the result of thermodynamic isomerizations and not kinetic preferences. Although no conditions could be found where compounds **7** or **8** were major products of the reaction between complex **4** and diphenylacetylene, these compounds were more abundant when the reaction was performed at lower temperatures and at shorter reflux times. Chromium hexacarbonyl was an effective reagent for the isomerization of *cis*-cyclopentenone **7** to the corresponding vinylogous ester **6** (Scheme 15). During the isomerization, arenechromium tricarbonyl complexes were formed which provided **6** after treatment with iodine. Since the isomerization reaction proceeds without H-D exchange, chromium (0) appears to be responsible for the isomerization reaction, and not simple enolization-reprotonation.

2. *Mechanism of the Cycloaddition Reaction*

Having established that cyclopentadienones are the penultimate intermediate of the reaction, plausible classes of mechanisms can now be proposed for the cycloaddition reaction (Schemes 16–18). In Schemes 16, 17 the alkyne first complexes, then inserts into carbene complex **4** to form the carbene complex **46** according to the mechanism of Doetz.[37] Next, the cyclopropane ring of **46** opens to give the metallacycloheptadiene complex **47**; this conversion is formally a 1,5-alkyl shift. CO-insertion then occurs to give the metallacyclooctadienone complex **48**, which upon simple reductive elimination would give the originally-expected cycloheptadienone **5**. Instead, we propose that intramolecular alkene insertion in **48** occurs to give the metallocyclopentane **49**, which can fragment with loss of ethylene[38] to give the cyclopentadienone **50**. A model suggests that the geometry required for

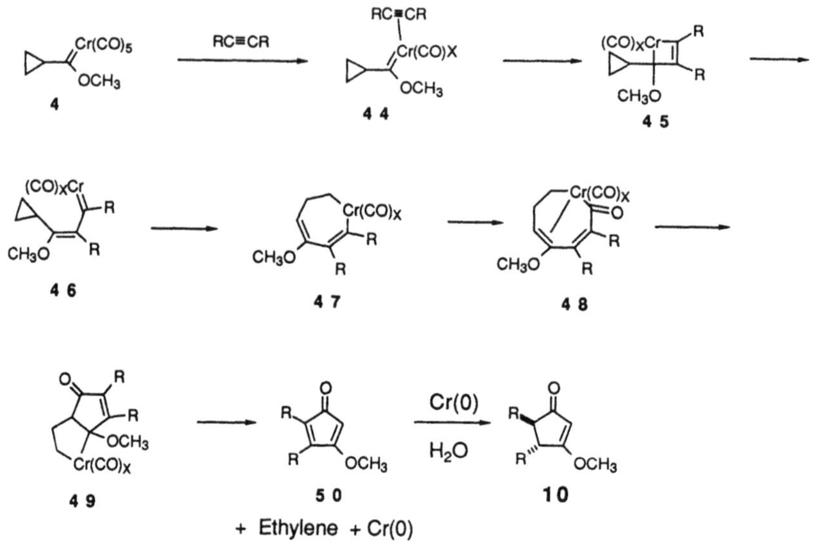

Scheme 16.

conversion of **48** to **49** is difficult to attain. The alternate insertion regiochemistry would provide metallacyclobutane intermediate **51** (Scheme 17), which can fragment with loss of ethylene to produce carbene complex intermediate **52**.[39] Unstabilized carbene complex **52** should be very electrophilic, and thus susceptible to a ring-expansion reaction which provides **53**.[40] Loss of chromium would then provide cyclopentadienone **50**. Reduction of the cyclopentadienone with water and Cr(0) leads to the observed cyclopentenone derivatives **10–12**.

Alternatively, cyclopentadienone **50** could be formed by the mechanism outlined in Scheme 18. Ring enlargement of the starting carbene complex **4** to metallacyclopentene complex **54**, followed by fragmentation[38] leads to the mixed alkyne-alkene complex **55**. Ligand exchange with diphenylacetylene concomitant with loss of ethylene leads to the bis alkyne complex **56**. Coupling of the alkyne ligands[41] would lead to the chromacyclopentadiene **57**, which would provide cyclopentadienone **40** after CO-insertion and reductive elimination. Alkyne coupling of this type

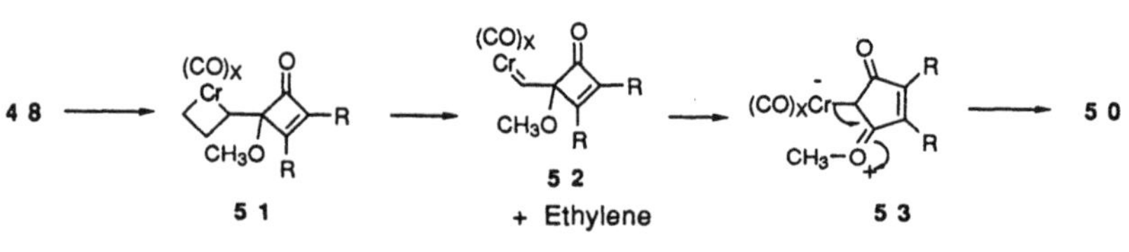

Scheme 17.

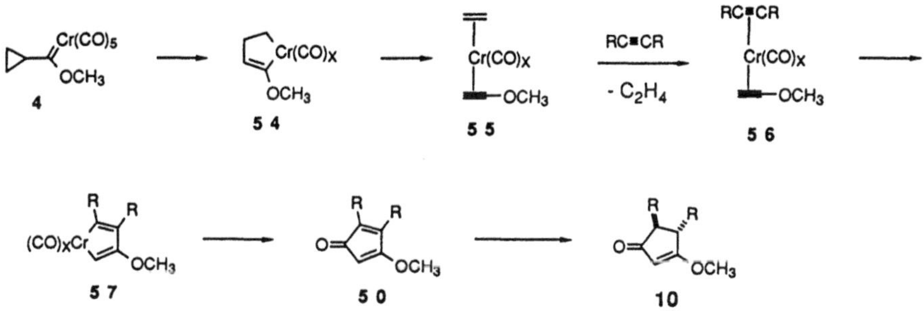

Scheme 18.

leading to the formation of chromacyclopentadienes is precedented for organochromium species.[42]

Evidence to date supports the mechanisms of Schemes 16 or 17, but not the mechanism in Scheme 18. The steps involved in the conversion of **4** to **46** have been proposed for virtually every reaction involving a chromium-carbene complex and an alkyne.[43] The key difference in the two mechanistic schemes (16, 17 vs 18) is the timing of the cyclopropane ring opening step. This occurs in the very first step of Scheme 18, and after the alkyne insertion in Schemes 16,17. Thermolysis of carbene complex **4** at 100 °C produces no new products, suggesting that conversion of **4** to **54** is not the first step of the reaction unless this step is heavily influenced by the presence of the alkyne. Furan **13** was formed as a byproduct in the reaction of carbene complex **4** and methyl propargyl ether.[44] The mechanisms proposed for furan formation also involve similar steps to those suggested for the conversion of **4** to **46**. According to Schemes 16,17, the cyclopropane ring opening step is a simple 1,5-alkyl shift. Thus if the formation of intermediate **46** is prevented, then the cyclopropane ring opening reaction will not place. This is exactly the situation created when an alkyne is replaced by an alkene in the reaction. Treatment of complex **4** with methyl acrylate leads to the cyclopropylcyclopropane derivative **59** (46:54 *trans*:*cis*)(for a more thorough discussion of this reaction see ahead).[45] This reaction presumably proceeds via the intermediacy of metallacyclobutanes (e.g., **58**, Scheme 19).[46] This result strongly suggests that the [2+2] cycloaddition

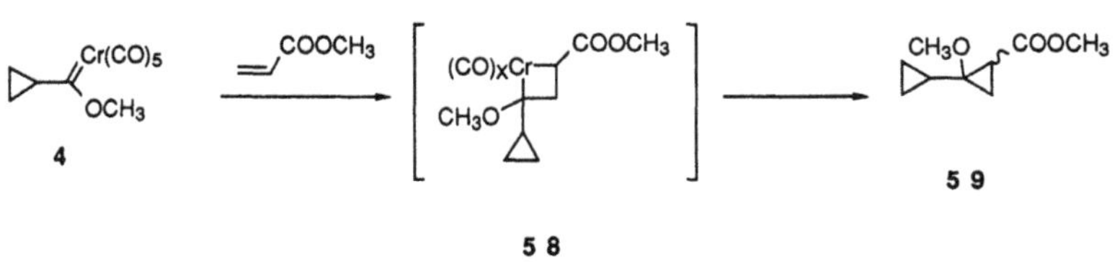

Scheme 19.

between the metal-carbene and a C–C π-bond can occur before ring opening of the cyclopropane ring, which is required for the mechanisms of Scheme 16,17.

3. Regiochemistry

The overall regiochemistry of the reaction is set in the metallacyclobutene ring forming step (**44** → **45**) of the reaction. Investigations by others in this area has shown that the larger group ends up alpha to the chromium in the metallacyclobutene.[47] Following these substituents through the mechanism in Schemes 16,17, it is predicted that the larger substituent will be alpha to the carbonyl group in the final product. This is exactly the case when monosubstituted alkynes and unsymmetrically-disubstituted alkynes are used in the reaction. That these processes are controlled by steric effects can be seen by viewing the examples in Table 1 (entries B, C, F–M), where the larger group does end up alpha to the carbonyl group in the product. Also, little regioselectivity is observed in the reaction of complex **4** with phenyl(*p*-methoxyphenyl)acetylene, indicating that there is only a minor[48] electronic effect in the reaction. The reaction is completely regioselective with monosubstituted alkynes, but mixtures are usually obtained when unsymmetrically-disubstituted alkynes were used in the reaction. A comparison of regioselectivities observed in this reaction with those observed between the reaction of arylcarbene-chromium complexes and alkynes reveals that the regioselectivity in the cyclopentenone-forming reaction is slightly lower.

III. REACTION OF CYCLOPROPYLCARBENE-CHROMIUM COMPLEXES WITH ALKENES

A. Synthesis of Cyclopropylcyclopropanes

From the above studies, it is apparent that the cyclopropane ring in complex **4** is reluctant to undergo ring-opening processes. In some cases, such as when the ring is activated by the presence of alkenyl substituents (compound **16**), a facile ring opening reaction occurred and cyclopentenone **20** was formed (Scheme 6). Since the cyclopropane rings in simple complexes such as **4** are somewhat robust, these complexes are potentially useful for cyclopropylcarbene-transfer processes.[49] Free cyclopropylcarbenes are often unstable and undergo ring-opening and/or ring-expansion processes,[50] and typically do not undergo cyclopropanation reactions with alkenes. Cationic cyclopropylcarbene-iron complexes undergo cyclopropanation reactions with electron-rich alkenes with no complications from opening of the cyclopropane ring.[51] As a possible complement to this method, we have investigated the reaction of cyclopropylcarbene-chromium complexes with electron-deficient alkenes.[45] If carbene complex **4** functions as a cyclopropanating reagent, donor-acceptor substituted cyclopropanes (e.g., **59**) would be provided

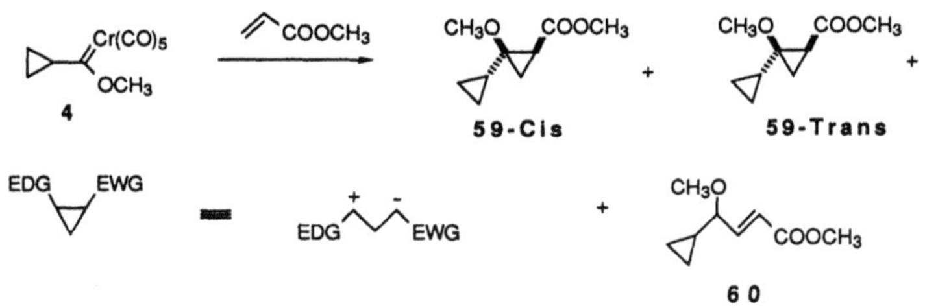

Scheme 20.

(Scheme 20). The reaction of pentacarbonyl[phenyl(methoxy)methylene]chromium with electron-deficient alkenes leads to donor-acceptor substituted cyclopropanes in good yields.[52] Donor-acceptor substituted cyclopropanes are synthetically useful compounds[53] and can serve as a synthon for an all-carbon 1,3-dipole. Compound **59** contains two carbocation-stabilizing substituents at the donor carbon, a methoxy and a cyclopropyl group. The presence of an additional donor substituent at this carbon might enhance the donor aspect of these compounds, and the possibility for opening of the unactivated cyclopropane ring might lead to new types of products from these reactions.

In initial studies, the reaction between cyclopropylcarbene complex **4** and methyl acrylate was investigated. Under all conditions examined (Table 4), a mixture of the expected cyclopropylcyclopropanes **59**-*trans* and **6**-*cis*, and the C–H insertion product **60** was obtained. The reaction of vinylcyclopropylcarbene complex **16** with methyl acrylate afforded only cyclopentenone **20** (see Scheme 6) in 48% yield; apparently this ring expansion process occurs considerably faster than the cyclopropanation reaction. In general, the yield of **59** was slightly higher if the reaction was conducted at 65 °C instead of 101 °C. The yield was not significantly affected by concentration, but was slightly higher if the reaction was conducted under high dilution conditions, achieved by syringe pump addition of a solution of the alkene and carbene complex over a period of 5 h to refluxing THF. The *trans*:*cis*[54] ratio of the products was virtually 1:1 except when the reaction was conducted at 65 °C in

Table 4. Reaction of Complex **4** with Methyl Acrylate

Entry	*Solvent*	*Temp.*	*Additive*	*Yield* **59**	*trans:cis* **59**	*Yield* **60**
1[a]	THF	65 °C	None	64%	46:54	5
2	THF	65 °C	None	63%	48:52	9
3	THF	65 °C	PPh_3	39%	26:74	7
4	Dioxane	101 °C	None	52%	48:52	7
5	Dioxane	101 °C	PPh_3	58%	47:53	12

Notes: [a] Methyl acrylate was added to a refluxing solution of carbene complex **4** in THF.

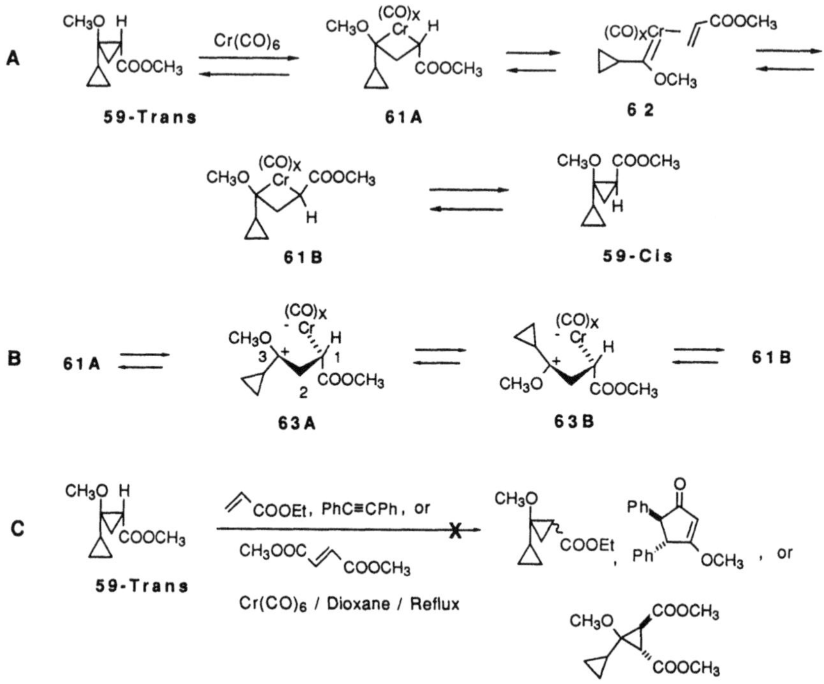

Scheme 21.

the presence of triphenylphosphine. Treatment of cyclopropylcyclopropane **59-*trans*** with chromium hexacarbonyl (a byproduct of the cyclopropanation reaction) in refluxing dioxane led to *trans-cis* isomerization, and compound **60** was detected in trace amounts. There was no isomerization in the absence of chromium hexacarbonyl. The stereochemical preferences in the cycloaddition reaction itself cannot be assessed, since product formation is most likely accompanied by isomerization. In one case (Table 4, entry 3) the reaction leads to primarily the *cis* isomer. In this case, the triphenylphosphine could tie up a coordination site in the chromium (0) by-product, preventing the *trans-cis* isomerization.

Plausible mechanisms for the *trans-cis* isomerization by chromium are outlined above in Scheme 21. The donor-acceptor activated C–C bond of the cyclopropane is weak and might be susceptible to oxidative addition reactions,[55] leading to metallacyclobutane **61**. In the mechanism in Scheme 21A, retrometallacyclobutane formation gives intermediate carbene complex **62** which, after rotation of the alkene ligand, can then reform the metallacyclobutane with the opposite stereochemistry. In the mechanism in Scheme 21B, metallacyclobutane **61A** converts to the zwitterionic species **63A**, which after rotation about the C_2–C_3 bond (giving **63B**) can provide metallacyclobutane **61B**, which is the precursor to **59-*trans***. The mecha-

Table 5. Reaction of Complex **4** with Alkenes in Refluxing THF

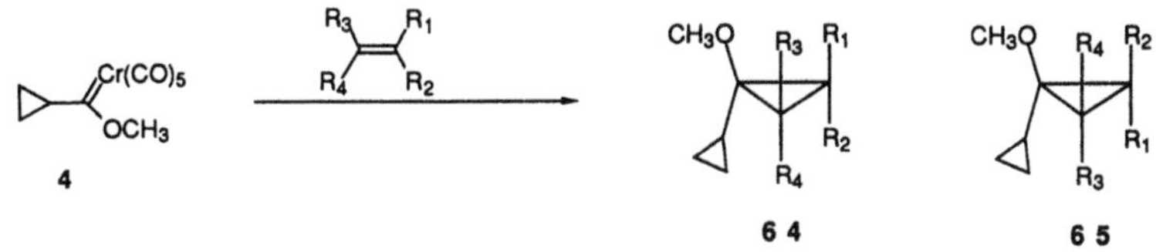

Entry[a]	*R1*	*R2*	*R3*	*R4*	*Yield* *(64+65)*[b]	**65:64**
A	$COOCH_3$	H	H	H	63[c]	48:52
B	$COOCH_3$	H	H	$COOCH_3$	62[d]	
C	$COOCH_3$	H	$COOCH_3$	H	37[e]	
D	$COOCH_3$	CH_3	H	H	69	22:78
E	$COOCH_3$	H	H	CH_3	66	>95% **64**[f]
F	$CON(CH_3)_2$	H	H	H	81	21:79
G	1-cyclopenten-1-yl	H	H	H	66	40:60[g]
H	$COOCH_3$	H	H	Ph	5	0:100
I	$COOCH_3$	H	CH_3	CH3	0[h]	
J	$COOCH_3$	CH_3	H	CH_3	50	14:86

Notes: [a] Table entry letters define substituents for compounds **64** and **65**.
[b] The yield refers to compounds which are pure by combustion analysis and/or spectroscopic data.
[c] A 9% yield of compound **60** was also obtained.
[d] A 18% yield of compound **67** was also obtained.
[e] A 12% yield of compound **67** was also obtained; the total yield refers to an 89:11 mixture of **64C**: **64B**.
[f] The relative stereochemistry of the methyl and carbomethoxy groups was *trans*.
[g] The stereochemistry could not be assigned reliably; the major product has been suggested as *cis* by analogy with the other systems.
[h] A 7% yield of C-H insertion product **66** was obtained.

nism in Scheme 21B implies that cyclopropanation by Fischer-carbene complexes is reversible. If a carbene complex such as **62** were present in the reaction, then it might be possible to exchange the alkene ligand with external ligands. Reaction of **59**-*trans* with large excesses of ethyl acrylate, diphenylacetylene, or dimethyl fumarate failed to provide any of the expected exchange products (Scheme 21C). This result suggests that the mechanism in Scheme 21A does not account for the *cis-trans* isomerization, however the possibility that **62** converts to **61** faster than any exchange with external ligands cannot be ruled out.

The reactivity of complex **4** with a variety of alkenes was examined. As can be seen in Table 5 the reaction appears to be quite general for ester-substituted alkenes. The reaction appears to be stereoselective;[49b] crotonate, cinnamate, tiglate, and fumarate esters result in products that retain the stereochemistry of the original alkene substituents. In one reaction, dimethyl maleate with complex **4** a mixture of stereoisomers was obtained: both cyclopropane **64B** and the expected cyclopropane **64C**, in which the ester groups are in the *cis* relationship were obtained. Maleate

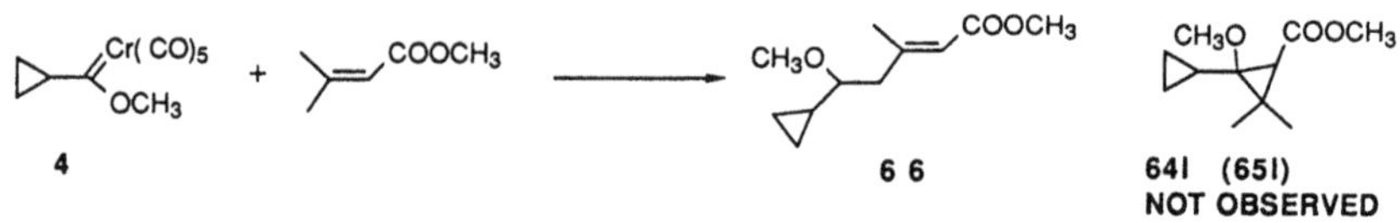

Scheme 22.

to fumarate isomerization occurs under the conditions of the reaction. ^{1}H NMR analysis of the mixture before completion (entry C) showed that both dimethyl maleate and dimethyl fumarate were present. Presumably this accounts for the lack of stereoselectivity in the reaction of **4** with dimethyl maleate. In addition, treatment of cyclopropylcyclopropane **64C** with chromium hexacarbonyl leads to isomerization to cyclopropylcyclopropane **64B**. The reaction appears to be quite sensitive to steric effects, since the product yields are reduced as the number of substituents on the double bond increases. Also, substituents on the β-carbon of the alkene are more detrimental to the yield than are substituents on the α-carbon. For example, the product yields are greater for the reaction of **4** with methyl tiglate (entry J) than for the corresponding reaction with methyl 3,3-dimethylacrylate (entry 1). The reaction of carbene complex **4** with methyl 3,3-dimethylacrylate afforded only the allylic C–H insertion product 66 in 7% yield (Scheme 22). The reaction of methyl cinnamate with carbene complex **4** led to a low yield of the expected cyclopropylcyclopropane (**64H**) accompanied by η^6-(methyl cinnamate)tricarbonylchromium (0) in 5% yield.

In reactions employing maleate and fumarate esters, the reduction products dimethyl succinate and compound **67**, in which both of the cyclopropane rings have been opened, were also observed (Scheme 23). The extra hydrogens in the open-

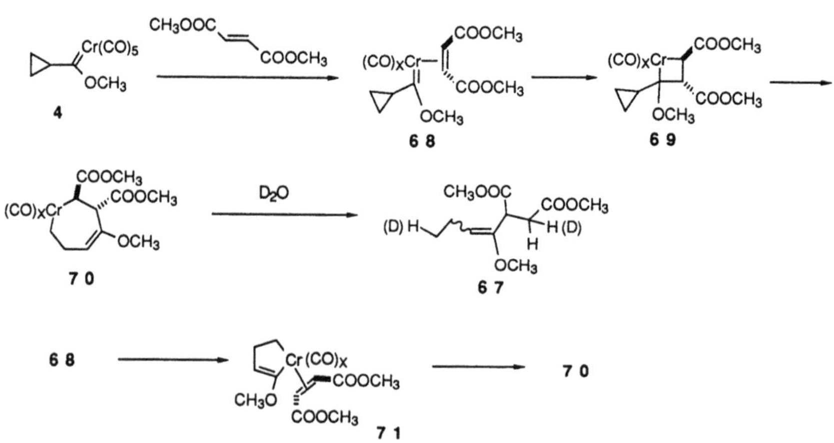

Scheme 23.

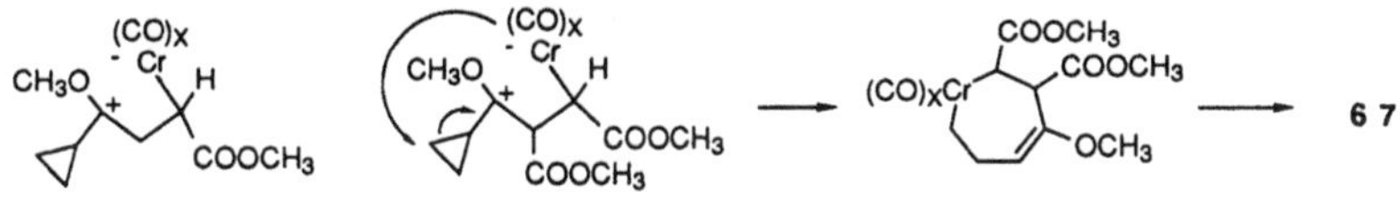

6 3   7 2   7 0

Scheme 24.

chain product **67** both come from proton sources. When the reaction between complex **4** and dimethyl fumarate was conducted in THF/$D_2$0 99:1, compound **67** was obtained with deuteriums at the 2- and 7-positions. No deuterium incorporation was detected in compound **64B**, confirming that epimerization is not due to deprotonation. In the absence of the added proton source, the cyclopropylcarbene complex itself might function as the proton source since the alpha hydrogen would be quite acidic.[36] The proposed mechanism for formation of **67** is outlined in Scheme 23. Protonolysis of the chromium-carbon bonds in intermediate **70** provides compound **67**.

Why is compound **67** only formed in reactions involving 1,2-diesters? Perhaps this is due to the greater backbonding ability of these alkenes relative to simple α,β-unsaturated esters. In complex **68**, the fumarate would be more strongly bound to the chromium than in the analogous complex containing an acrylate ester. This slows the step involving metallacyclobutane formation and perhaps allows other processes such as cyclopropane ring opening (e.g., **68** → **71**, Scheme 23) to compete. Alternatively, the formation of **67** might be a reflection of the relative stabilities of carbocation intermediates. A comparison of zwitterions **63** and **72** (Scheme 24) suggests that the carbocation portion of **72** is inductively destabilized by the presence of the carbomethoxy group.[56] Thus the carbocation in **72** will require more electron donation by the methoxy and cyclopropyl groups, thus activating the cyclopropane with respect to ring-opening reactions, via the cyclopropylcarbinyl cation.[57] A final point of divergence involves the regiochemistry of metallacyclobutane ring formation. Recently, it has been shown that when alkynes having strongly electron-withdrawing groups react whh carbene complexes, there is enhanced control of the regiochemistry by electronic effects.[48] Thus for alkenes with a single electron-withdrawing group, the predicted regiochemistry should be as shown in Scheme 25, and metallacyclobutane **73** should be the electronically-favored intermediate in the reaction. Metallacyclobutane **74** differs from **73** in that an electron-withdrawing group is present at the chromium-containing carbon. Thus, the carbene-chromium bond is stabilized by this interaction and should be less likely to undergo reductive elimination reactions. This might then allow ring opening processes to compete with reductive elimination to form cyclopropanes. Compound **60** provides a possible clue as to the regiochemistry of metallacyclobutane formation; C–H insertion product **60** can arise via β-hydride elimination and reductive elimination from a metallacyclobutane such as **74**, but not from **73**. Compounds such as **75** have never been reported from the reaction of chromium(0) carbene

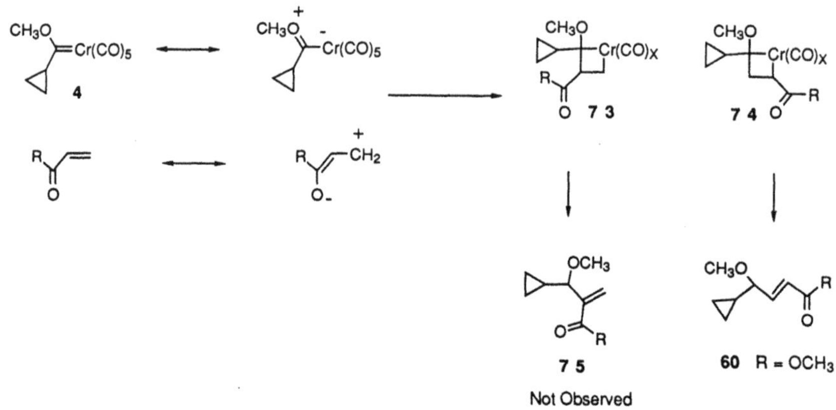

Scheme 25.

complexes and electron-deficient alkenes. We thus suggest that the differences lie in the different ligating abilities of fumarates (maleates) vs. acryaltes or are due to inductive destabilization of carbocation intermediates.

Cyclopropylcyclopropanes were also obtained from the reaction of complex **4** with α,β-unsaturated amides. Complex reaction mixtures were obtained from the reaction of **4** with α,β-unsaturated ketones. Although none of the desired cyclopropanation products could be isolated from the reaction of complex **4** and methyl vinyl ketone, we were able to isolate 2,7-octanedione from the reaction in low yield. The cyclopropanation reaction appears to be restricted to activated alkenes since norbornene, a highly strained alkene which complexes readily to metals,[58] and styrene both fail to react with complex **4**. Interestingly, 1-vinylcyclopentene couples readily with carbene complex **4**, giving the bicyclopropane **64(65)G** in 66% yield. Previously only electron-deficient and highly electron-rich dienes were reported to couple with chromium-carbene complexes.[46,49] We are currently examining the generality of the reaction of chromium carbene complexes with simple 1,3-dienes.

Stereochemical assignments were made on the basis of nOe studies or through the use of chemical shift additivity predictions established by Reissig[59] (Table 6). In compound **64C** there is an 11% enhancement of the signals at δ 0.14 (H_B) when the signal at δ 1.83 (H_A) was irradiated (Scheme 26). A comparison of compound

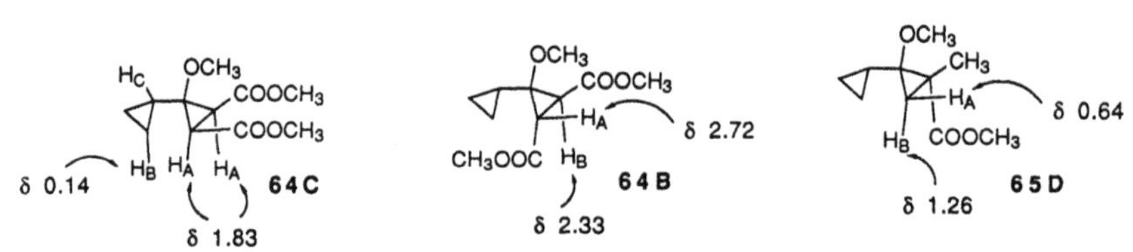

Scheme 26.

Table 6. Chemical Shifts for Protons on the Donor-Acceptor Substituted Cyclopropane Rings

Compound #	R_1	$R_2\delta$	$R_3\delta$	$R_4\delta$
59-*cis* (**64A**)	$COOCH_3$	H (1.46)	H (1.63)	H (0.81)
59-*trans* (**65A**)	$COOCH_3$	H (1.97)	H (1.15)	H (1.15)
64B	$COOCH_3$	H (2.33)	H (2.72)	$COOCH_3$
64C	$COOCH_3$	H (1.83)	$COOCH_3$	H (1.83)
64D[a]	$COOCH_3$	CH_3	H (1.63)	H (0.45)
65D	$COOCH_3$	CH_3	H (1.26)	H (0.64)
64E[b]	$COOCH_3$	H (1.16)	H (1.89)	CH_3
64F	$COON(CH_3)_2$	H (1.43)	H (1.66)	H (0.49)
65F	$COON(CH_3)_2$	H (1.98)	H (1.07)	H (1.07)
64H[c]	$COOCH_3$	H (2.06)	H (3.25)	C_6H_5
64J	$COOCH_3$	CH_3	H (2.12)	CH_3
65J	$COOCH_3$	CH_3	H (1.51)	CH_3

Notes: [a] An exact assignment between R_3 and R_4 can not be made. The chemical shift difference between these protons should be greater in **64D** (1.18 ppm) than in **65D** (0.62 ppm).

[b] Only this isomer was obtained; the unusually low chemical shift for R_2 is consistent only with the *cis* isomer (**64E**) (compare with the same proton in **59**-*cis* and **64F**).

[c] Only this isomer was obtained; the stereochemistry has been suggested as *cis* by analogy to the reaction in Table 5, Entry E.

64C and **64B** reveals an application of these rules, where alkoxy groups and carbonyl groups induce a strong downfield shift to a proton *cis* on the cyclopropane ring, while alkyl groups induce an upfield shift for a *cis* proton. In **64C**, the chemical shift of H_A is δ 1.83; this proton is not *cis* to the methoxy or the ester group. In compound **64B**, the analogous protons, H_A and H_B, exhibit higher chemical shifts since each is *cis* either to a carbomethoxy group and/or a methoxy group. Interestingly, there is a strong nOe interaction between protons H_A and H_B in **64C**, but no detectable nOe between H_A and H_C. This suggests that compound **64C** exists predominately in the conformation depicted in Scheme 26. Similarly, the NOESY spectrum for compound **65D** shows nOe interactions between H_A and methoxy, and between H_B and carbomethoxy, which is consistent with the stereochemical assignment determined from chemical shift effects. A greater chemical shift difference is predicted between H_A and H_B in **64D** ($\Delta\delta = 1.18$) than in **65D** ($\Delta\delta = 0.62$). A similar application of these rules has been used for other structures in this paper. As can be seen in Table 6, in all cases where R_2 and $R_4 = H$, the chemical shift for these protons is higher in compound **65** than in compound **64** (R_1 is the carbonyl group). In compound **65** these protons are always *cis* to methoxy, whereas they are *cis* neither to methoxy nor carbomethoxy in **64**. In cases where $R_3 = H$, the chemical shift for this proton is higher in compound **64** than in compound **65**. In compound **64**, this proton is *cis* to methoxy and carbomethoxy, while it is *cis* only to carbomethoxy in compound **65**.

*Reaction of complex **4** with methyl methacrylate (Table 5, entry D).* Complex **4** (0.276 g, 1.00 mmol) and methyl methacrylate (0.200 g, 2.00 mmol) were dissolved in THF (20 mL). This solution was added dropwise via syringe pump over a period of 5 h to refluxing THF (100 mL) under nitrogen, was refluxed for an additional 12 h after the addition was complete, and the solvent was removed on a rotary evaporator. The residue after evaporation was then dissolved in ethyl acetate (10 mL) and the resulting suspension was filtered through Celite. The solvent was removed on a rotary evaporator, and the residue after evaporation was distilled (Kugelrohr, 25–45 °C, 0.025 mmHg). The distillate was purified by flash chromatography (silica gel, 9:1 hexane:ethyl acetate). The product in the first fraction (R_f 0.59 in 4:1 hexane:ethyl acetate) was identified as methyl 1α-methoxy-2α-methyl-[1,1′-bicyclopropyl]-2β-carboxylate **65D** (0.028 g, 15%). ^1H NMR ($CDCl_3$): δ 3.63 (s, 3 H), 3.35 (s, 3 H), 1.38 (s, 3 H), 1.26 (d, 1 H, J = 5.9 Hz), 1.09 (m, 1 H), 0.64 (d, 1 H, J = 5.9 Hz), 0.27–0.59 (m, 4 H); ^{13}C NMR ($CDCl_3$): δ 173.3, 70.2, 54.6, 51.5, 32.8, 22.4,15.4,10.7, 5.6, 3.2; IR ($CDCl_3$) 3090 (w), 3008 (m), 2953 (m), 2833 (w), 1720 (vs), 1461 (m), 1436 (s), 1320 (s), 1303 (s), 1210 (s), 1197 (s), 1154 (s), 1135 (vs), 1086 (m), 1040 (m) cm^{-1}; MS(EI) m/z (relative intensity) 184 (M^+), 137, 129, 111, 91, 87, 83, 74 (100), 69, 59; HRMS (EI) calcd for $C_{10}H_{16}O_3$ 184.1099, found 184.1095. The product in the second fraction (R_f 0.51 in 4:1 hexane:ethyl acetate) was identified as methyl 1α-methoxy-2β-methyl-[1,1′-bicyclopropyl]-2α-carboxylate **64D** (0.099 g, 54%) ^1H NMR ($CDCl_3$): δ 3.67 (s, 3 H), 3.25 (s, 3 H), 1.63 (d, 1 H, J = 6.0 Hz), 1.34 (s, 3 H), 1.22 (m, 1 H), 0.47–0.75 (m, 3 H), 0.45 (d, 1 H, J = 6.0 Hz), 0.18 (m, 1 H); ^{13}C NMR ($CDCl_3$): δ 172.2, 70.7, 54.4, 51.5, 32.0, 21.3,16.3,10.6, 4.7, 3.1; IR ($CDCl_3$): 3088 (w), 3011 (m), 2953 (m), 2832 (w), 1721 (vs), 1462 (m), 1437 (s), 1336 (m), 1316 (m), 1296 (s), 1232 (m), 1197 (s), 1184 (m), 1146 (vs), 1072 (m), 1041 (m) cm^{-1}; MS (Cl) m/z (relative intensity) 185 (M+1, 100), 169,153,141,137, 125,109, 97, 93, 69. Anal. Calcd. for $C_{10}H_{16}O_3$: C, 65.19; H, 8.75. Found: C, 65.07; H, 8.60.

B. Reactions of Cyclopropylcyclopropanes

The reactivity of donor-acceptor substituted cyclopropanes has recently been reviewed.[53] As predicted,[60] the cyclopropanes **64** and **65** are very susceptible to acid-catalyzed hydrolysis reactions, which give the compounds (**76**) shown in

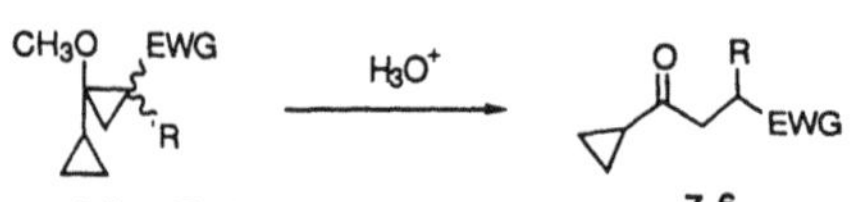

64 (65)

76

A. EWG = $CON(CH_3)_2$, R = H: 100%

B. EWG = $COOCH_3$, R = H: 93%

C. EWG = $COOCH_3$, R = CH_3: 88%

Scheme 27.

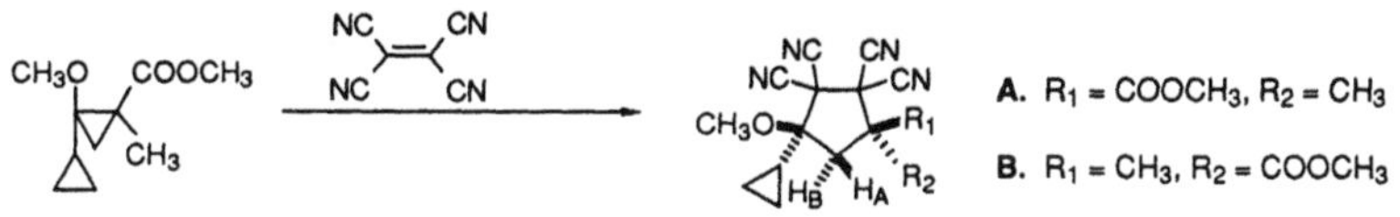

64B (65B)

77

Scheme 28.

Scheme 27. In this reaction, only the donor-acceptor activated cyclopropane opens, while the unactivated cyclopropane ring is untouched. The product of the reaction is a 1,4-dicarbonyl compound. Thus carbene complex **4** can be viewed as a synthon for the cyclopropylcarbonyl anion. The cyclopropylketone functionality is further susceptible to a variety of synthetically useful transformations.[61]

Donor-acceptor substituted cyclopropanes have been reported to undergo [3+2] cycloaddition reactions with some electron-deficient alkenes.[62,53] Compound **64D** reacts with the electron-deficient alkene tetracyanoethylene in refluxing acetonitrile to give the five-membered ring containing compounds **77A** and **B** as a 72:28 mixture (Scheme 28). This is presumably a stepwise reaction, and both of the cyclopropane stereoisomers give identical ratios of the five-membered ring adducts. The minor stereoisomer obtained was the one depicted by structure **77A**. In this compound the methyl and cyclopropyl substituents are on the same side. Since these substituents produce a strong upfield shift for *cis* hydrogens vicinal on the cyclopentane ring,[63] a greater difference in the chemical shifts of H_A and H_B would be predicted. In the major isomer these shifts are δ 2.46 and 1.86 ($\Delta\delta = 0.60$), and in the minor isomer these are δ 3.14 and δ 1.43 ($\Delta\delta = 1.71$). In this conversion as well, there is no complication from a ring opening reaction of the unactivated cyclopropane ring. Less electrophilic alkenes such as dimethyl fumarate did not undergo this conversion.

IV. THERMAL REACTIVITY OF CYCLOPROPYLCARBENE-CHROMIUM COMPLEXES

A. Thermolysis of 2-Vinyl-2-methylcyclopropylcarbene Complex 16

As reported previously, simple cyclopropylcarbene-chromium complex **4** is not thermally reactive. The proposed rearrangement of chromium carbene complex **4** to metallacyclopentene **54** (Scheme 29) can be viewed as an analogue of the vinyl cyclopropane rearrangement,[64] where the chromium-carbon double bond has replaced a carbon–carbon double bond. The classical vinylcyclopropane rearrangement is typically very sluggish and often requires high temperatures for initiation. If the conversion of **4** to **54** is related to the vinylcyclopropane rearrangement, the sluggishness of this conversion should come as no surprise. A related electrocyclic

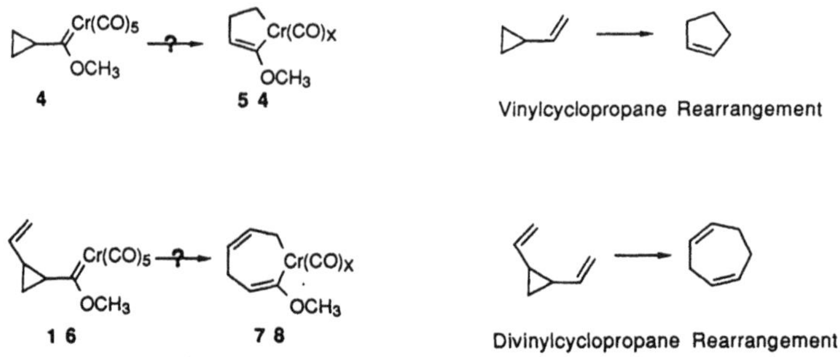

Scheme 29.

cyclopropane ring-opening reaction, the divinylcyclopropane rearrangement, is quite facile and often occurs at a much lower temperatures.[65] We therefore initiated a study of the thermal reactivity of 2-alkenylcyclopropylcarbene-chromium complexes (e.g., **16**), which are organometallic analogs of divinylcyclopropane. It had previously been suggested that electrocyclic-like processes might occur for the carbon–chromium double bond.[66] Thus the 2-alkenylcyclopropylcarbene-chromium complexes might undergo facile cyclopropane ring-opening reactions (e.g., **16** → **78**).[67]

The vinylcyclopropylcarbene complex **16** was synthesized from isoprene[68] as an inseparable 58:42 mixture of *cis*:*trans* stereoisomers. This carbene complex is quite thermally reactive. The thermolysis of complex **16** in refluxing THF for a period of 3 h led to the formation of the 5-vinyl-2-cyclopentenone derivative **20** in 30% yield (Scheme 30). The thermolysis of carbene complex **16** in dioxane at 100 °C in a pressure bottle for 3 h led to the formation of the expected cyclopentenone derivative **20** in a 48% yield. Also obtained from the reaction in refluxing THF was unreacted starting carbene complex **16** in 15% yield. Analysis of recovered **16** revealed enrichment in one of the two possible stereoisomers, suggesting that one of the stereoisomers is more reactive.

The stereochemistry of the two stereoisomers of **16** was assigned by nuclear Overhauser enhancement (nOe) studies (Scheme 31). Irradiation of a 60:40 *cis*:*trans* mixture of carbene complex **16** at δ 1.50 produced an 18% enhancement of the resonance at δ 3.65; however irradiation of the protons at δ 1.15 produced

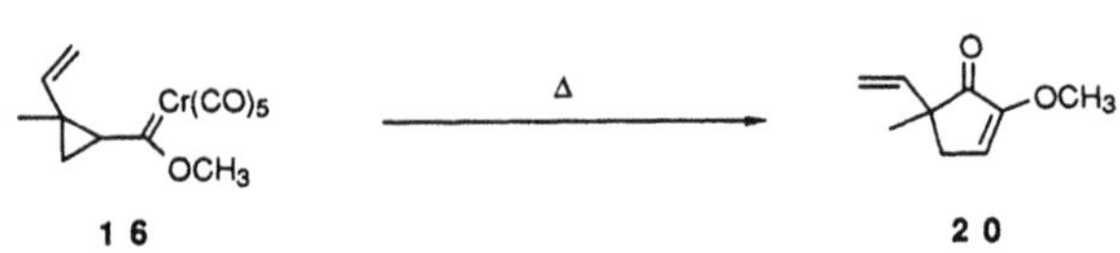

Scheme 30.

16-Cis **16-Trans**

Scheme 31.

no observable enhancement in the resonance at δ 3.75. From this study it was determined that the major isomer was that one having the chromium-carbene and alkene substituents *cis*. Throughout the remainder of this manuscript, the terms *cis* and *trans* will refer to the relative stereochemistry of the alkene and carbene substituents. The recovered carbene complex was enriched in the *trans* isomer, consequently the *cis* isomer of carbene complex **16** was converted to product faster than the *trans* isomer. At 52 °C, the first order rate constant for the disappearance of the *cis* isomer was $(3.0 + 0.1) \times 10^{-5}$ sec^{-1}, while that for the *trans* was $(0.83 \pm 0.06) \times 10^{-5}$ sec^{-1}.

The same order of reactivity for *cis*/*trans* stereoisomers has been observed in the divinylcyclopropane rearrangement, although to a greater extent.[65] Schneider and Rau have shown that *cis*-divinylcyclopropane rearranges at 30 °C to 1,4-cycloheptadiene, and the *trans* isomer requires temperatures in excess of 160 °C to undergo the same rearrangement. It was suggested that rearrangement of the *trans* isomer to cycloheptadiene occurs only after isomerization to the *cis* isomer.

Based upon the above observations several mechanisms can be proposed for the transformation of chromium carbene complex **16** to the 5-vinylcyclopentenone derivative **20**. The mechanisms proposed in Scheme 32 are analogous to the divinylcyclopropane rearrangement. In path a, chromium carbene complex **16** undergoes a divinylcyclopropane rearrangement to give the metallacycloheptadiene **80**, which undergoes σ→π-allyl interconversion[69] to yield the π-allyl-metal complex **81**. Carbonyl insertion affords the corresponding π-allyl(acyl)metal complex **82**, which upon reductive elimination gives the observed 5-vinylcyclopentenone derivative **20**. Reductive elimination could conceivably provide compound cycloheptadienone **83**,[70] however this process was not observed.

Alternatively, carbene complex **16** can undergo CO-dissociation followed by alkene complexation, resulting in complex **84** (Scheme 32, path b). Divinylcyclopropane-like rearrangement of complex **84** would lead directly to the π-allyl-metal complex complex **81**, which is then converted to cyclopentenone **20** as in path a.

In addition to mechanisms based on the divinylcyclopropane rearrangement, other possibilities exist. The vinyl substituent might be activating the cyclopropane system by weakening the carbon–carbon sigma bond of the cyclopropane ring, since sigma bonds to allylic or benzylic carbons are weaker than their saturated

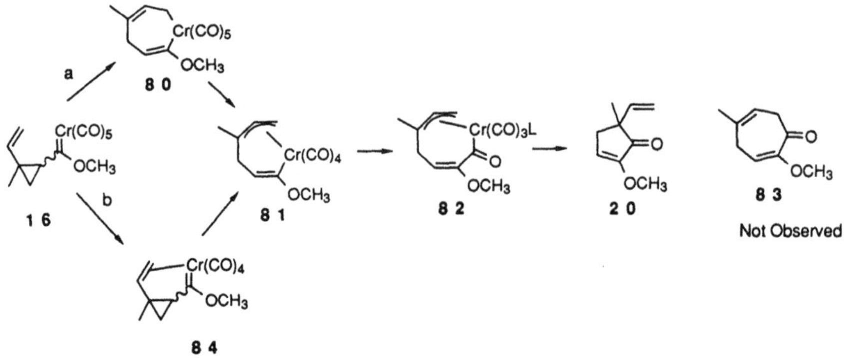

Scheme 32.

counterparts.[71] For example the cleavage of a carbon–carbon bond of propane requires 88 kcal mol, whereas the cleavage of carbons C-3 and C-4 in 1-butene requires only 72 kcal/ mol. A similar effect is seen in ethyl benzene where the cleavage of the carbon–carbon sigma bond between the two alkyl carbons also requires 72 kcal/mol. Thus ring expansion of the carbene complex **16** to metallacyclopentene **85** (Scheme 33a) should be considerably more energetically favorable than conversion of **4** to **54** which leads to the formation of metallacyclopentene **75**. Carbonyl insertion and subsequent reductive elimination would provide complex **20**. Alternatively this reaction might proceed through a discreet dipolar intermediate such as zwitterion **86** (Scheme 33b). The formation of zwitterion **86** results in a stable tertiary allyl cation and a carbanion which is directly conjugated to the electron deficient carbene-chromium π bond.[36] Thus carbene complex **16** is formally a donor-acceptor-substituted cyclopropane.[53] The zwitterion can then collapse to form metallacyclopentene **85**, which proceeds to the observed product **20**.

The mechanism proposed in Scheme 32b involves the dissociation of a carbonyl ligand as the initiating step of the reaction. Evidence to substantiate this presumption is obtained from the thermolysis of carbene complex **16** under carbon monoxide pressure (3 atm.). These conditions result in greatly increased reaction times (18 h) and reduced yields. Consequently, the presence of excess carbon monoxide inhibits the reaction, suggesting that the dissociation of a carbonyl ligand is an important step in the overall pathway to the observed product.

An aromatic ring should be similar to a vinyl substituent in its ability to activate the cyclopropane ring since its affect on bond strengths[71] and carbocation stability[72] is similar. However, it is known that aromatic rings are much more reluctant to participate in Cope type rearrangements. It has been shown that the rearrangement of the *cis*-1-aryl-2-vinylcyclopropane requires temperatures in excess of 200 °C.[73] If the mechanisms of Scheme 32 are operative, then carbene complexes analogous to **16**, where vinyl has been replaced by phenyl (e.g., **87,88**, Scheme 34), should

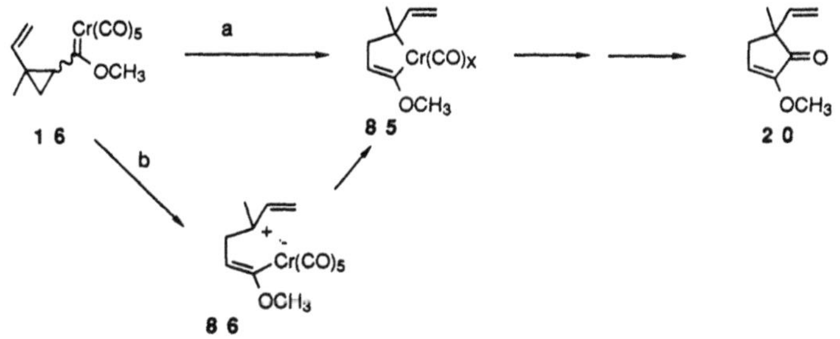

Scheme 33.

undergo conversion to cyclpentenones **89,90** at a much slower rate than conversion of **16** to **20**. If the mechanisms of Scheme 33 are operative, there should be very little difference between complexes **87,88** and **16**. When the 2-phenyl-2-methylcyclopropylcarbene chromium complex **88** was heated at 95 °C for a period of 6 h there was no indication of conversion to **90**. However, if this complex was heated in refluxing dioxane for an extended period of time (36 h) it slowly converted to 5-arylcyclopentenone derivative **90** in 27% yield. TLC analysis revealed that a significant amount of carbene complex **88** was still present. The 2-phenylcyclopropylcarbene chromium complex **88** demonstrated similar reactivity, and provided a 22% yield of the corresponding cyclopentenone derivative **89** after 36 h at 25 °C.

The above results suggest that the mechanisms proposed in Scheme 33 are unlikely pathways. The aryl substituted carbene complexes **87,88** exhibited sluggish reactivity and resulted in reduced yields of cyclopentenone products relative to the vinyl substituted carbene complex **16**. This suggests a more intimate role for the alkene, as is the case with the mechanisms in Scheme 32, where the function of the alkene is as a ligand or to provide a π-component for the Cope rearrangement. Additionally, it is not clear how the relative stereochemistry of the vinyl or aryl substituents relative to the carbene complex would affect the reaction if it proceeded by the mechanism proposed in Schemes 33. A preference for the *cis* isomer has been observed in the rearrangement of **16** as discussed previously. In the mechanisms of Scheme 33, there is no obvious reason for the enhanced reactivity of one stereoisomer of complex **16**.

Further evidence for the mechanisms proposed in Scheme 32 was obtained from the thermolysis of 2-vinylcyclopropylcarbene chromium complex **16** for 1 h in DMF at 100 °C . The expected cyclopentenone derivative **16** was obtained, however the yield had increased from 43% in dioxane to 68%, while the overall reaction time had decreased. It was apparent that the DMF solvent was somehow facilitating the reaction. At first glance, this affect could be attributed to the more polar solvent facilitating the zwitterionic mechanism (Scheme 33b) pathway. Also, DMF is known to be a weakly basic solvent, while chromium carbene complexes, in

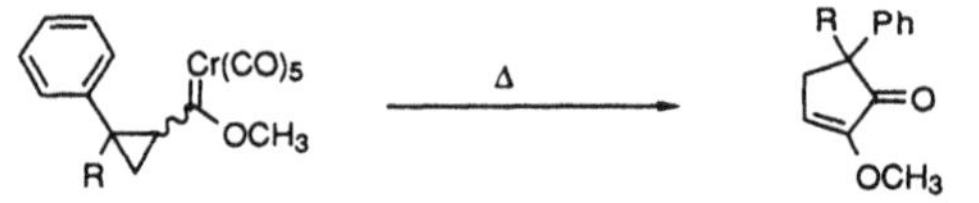

87 R = H
88 R = CH_3

89 R = H
90 R = CH_3

Scheme 34.

general, are known to have fairly acidic alpha hydrogens.[36] Therefore, DMF might also be inducing an equilibration between the *cis* and *trans* isomers of carbene complex **16**. The rate enhancement could thus be due to isomerization of the less reactive *trans* isomer to the more reactive *cis* isomer. Deuterium labeling experiments were conducted to test for the possibility of this equilibration. Carbene complex **16** underwent thermolysis at 100 °C in 99:1 DMF/D_2O affording the expected 5-vinylcyclopentenone derivative **20** completely deuterated at C-3 (Scheme 35). A second experiment was performed with carbene complex **16** in 99:1 THF/D_2O. Thermolysis of the reaction mixture was carried out in a pressure bottle at 100 °C and yielded vinylcyclopentenone derivative **16** with no deuterium incorporation. Consequently, it appears that a deprotonation-protonation process is able to occur in DMF, but not in THF. This deprotonation process apparently facilitates the reaction by allowing the less reactive *trans* isomer of carbene complex **16** isomerize to the more reactive *cis* isomer, which ultimately leads to higher yields and reduced reaction times.

In view of all the experimental evidence provided in preceding results. It seems that the thermal rearrangement of 2-alkenylcyclopropylcarbene chromium complexes is proceeding by a pericyclic-like process. This process appears to be similar to a divinylcyclopropane rearrangement.

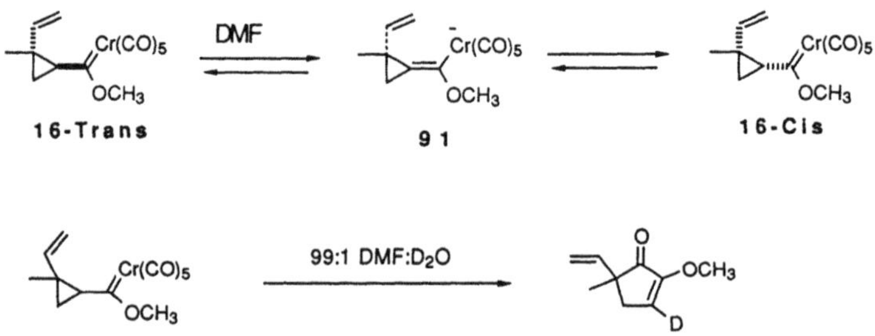

16-Trans **91** **16-Cis**

16 **20** (3-D_1)

Scheme 35.

Thermolysis of carbene complex **16.** A solution of the carbene complex **16** (0.42 g, 1.3 mmol) in DMF (20 mL) was heated to 80 °C for 4 h. The solution was filtered through celite, and poured into a mixture of water and pentane in a separatory funnel. The pentane layer was washed two times with water and once with saturated aqueous sodium chloride solution. The pentane layer was dried over magnesium sulfate and the solvent was removed on a rotary evaporator. Final purification was achieved by flash chromatography on silica gel, using 4:1 hexane:ethyl acetate as the eluent. A colorless oil identified as compound **20** (0.134 g, 68% yield) was obtained. ^{1}H NMR($CDCl_3$): δ 6.33(t, 1H, J = 3.4); 5.78(dd, 1H, J = 17.6,10.9); 5.08 (d, 1H, J = 17.6); 5.05 (d, 1H, J = 10.9); 3.70 (s, 3H); 2.60 (dd, 1H, J = 18.9, 3.4); 2.39 (dd, 1H, J = 18.9, 3.4); 1.21 (s, 3H). ^{13}C NMR ($CDCl_3$): δ 203.5; 155.5;140.5,123.8;113.8; 56.8; 48.8; 37.5; 23.4. IR($CDCl_3$): 3100 (w); 2960 (m); 2910 (s); 2820 (w); 1720 (s); 1645 (s); 1470 (m); 1380 (m); 1360 (m); 1305 (m); 1275 (s); 1140 (m); 1060 (s); 1020 (m) cm^{-1}. Mass Spec(EI): 152 (M, 46);137 (19); 124 (36); 123 (42); 110 (52); 109 (90); 91 (38); 84 (57); 81 (100)77 (33). High Res. Mass Spec: calcd. for $C_9H_{12}O_2$ - 152.0837; found - 152.0837. Analysis calcd. for $C_8H_{10}O_2$: C - 71.05 %; H - 7.89 %; found: C - 70.95 %; H - 7.87%.

B. Scope and Limitations of the Thermal Rearrangement of 2-Alkenylcyclopropylcarbene Complexes

The scope and limitations of the thermal rearrangement of a variety of 2-alkenylcyclopropylcarbene complexes has been explored as shown in Table 7. The reaction appears to be general for a wide variety of these complexes. In cases where R_2 is H, the alkene isomerization product **99** was also obtained.

Previous investigators have examined other γ, δ-unsaturated carbene complexes, however these complexes do not tend to undergo Cope rearrangements. For example, when the γ,δ-unsaturated amino-tungsten carbene complex **92** underwent thermolysis at 80 °C the only product formed was the metal carbene-olefin complex **93**,[74] and there was no evidence of any type of product resulting from a Cope rearrangement. One example of a reaction reminiscent of the Claisen rearrangement has been observed. The allyloxycarbene-tungsten complex **94** produced a complex mixture of products upon thermolysis at 50 °C, of which the major product was the aromatic ketone **96**.[75] One of the proposed mechanisms for the formation of **96**

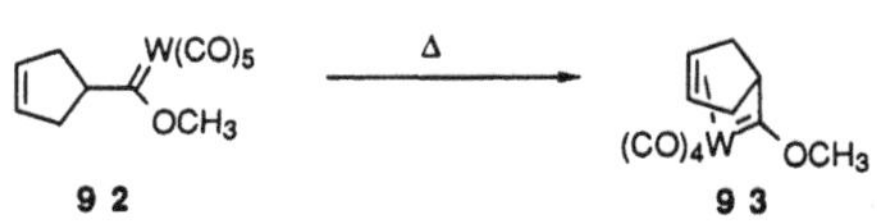

Scheme 36.

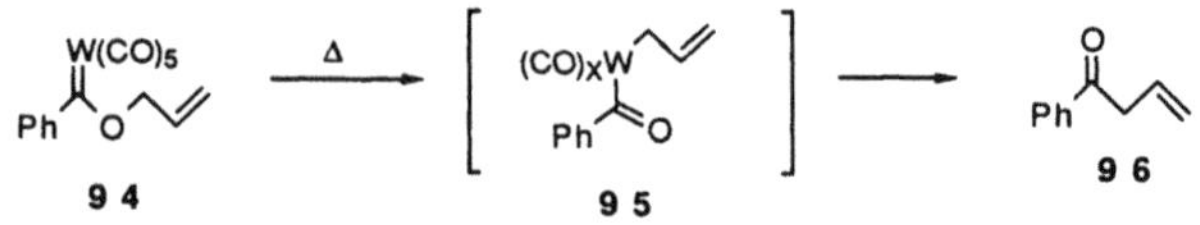

Scheme 37.

involves a migration of the allyl group to the tungsten by a mechanism similar to the Claisen rearrangement, giving intermediate **95**, which upon reductive elimination yields **96**. No additional evidence was provided to support this mechanism.

Presumably, ring opening of the cyclopropane provides the necessary driving force to induce the Cope-like process in the thermal rearrangements of 2-vinylcyclopropylcarbene chromium complexes. Only in one case does the thermal rearrangement of 2-alkenylcyclopropylcarbene complexes fail (Table 7, entry E). The

Table 7. Thermal Reactions of Cyclopropylcarbene-Chromium Complexes

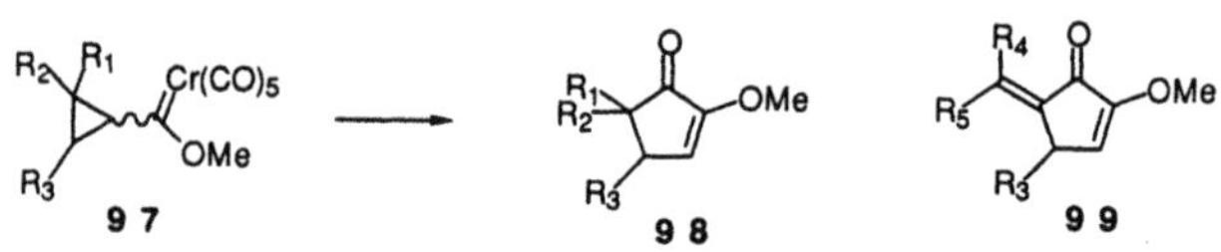

Entry[a]	R_1	R_2	R_3	R_4	*trans:cis Complex*[b]	*Reaction time/temp.*[c,d]	*Yield* **98**[e]
A	vinyl	CH_3	H		42:58	3h/101 °C	43
B	Ph	H	H		46:54	14h/ 101°C[f]	22
C	Ph	CH_3	H		68:32	14h/101 °C[f]	27
D[k]	1-cyclopentenyl	H	H	$-(CH_2)_4-$	—	1 h/101 °C	69
E	H	R_2,R_3=-CH=CH$(CH_2)_2$-			100:0	4.5h/101 °C[f]	0
F	1-propenyl	H	CH_3	CH_2CH_3	53:479	1 h/ 100 °C	53[g,h,i]
G	vinyl	H	H	CH_3	49:51	2.0 h /100 °C	49[i,j]
H	vinyl	$R_2,R_3=-(CH_2)_4-$			41:59	1.5h/ 101 °C	67
I	vinyl	$R_2,R_3=-(CH_2)_5-$			78:22	1.5h/ 100 °C	57[h]

Notes: [a] Table entry letters define R_1, R_2, R_3 for compounds **97–99**.

[b] The *trans:cis* designation refers to the relative stereochemistry of the chromium carbene and the alkenyl (aryl) substituents in the starting carbene complex.

[c] The reaction was performed in anhydrous p-dioxane unless otherwise noted.

[d] All reactions were allowed to proceed until the carbene complex had completely disappeared unless otherwise noted.

[e] The yields refer to products obtained after chromatographic purification.

[f] The starting material had not disappeared when the reaction was stopped.

[g] The relative stereochemistry of the methyl and propenyl groups is *trans*.

[h] The reaction was conducted in DMF.

[i] The yield refers to the product isolated after isomerization of the double bond (compound 99).

[j] The reaction was conducted in THF in a pressure bottle.

[k] The starting carbene complex was stereoisomerically pure, a *cis* or *trans* assignment was not possible.

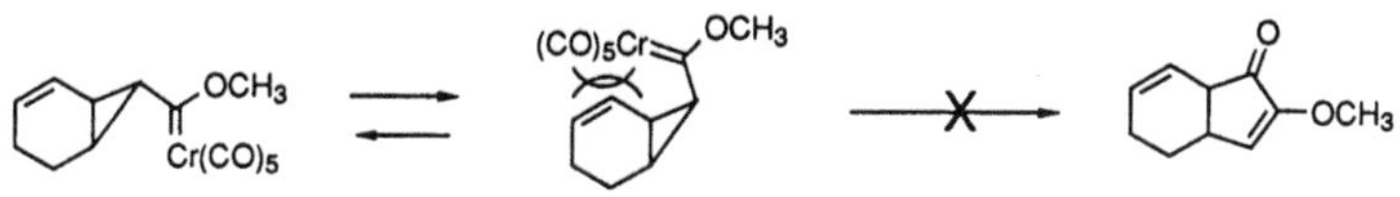

97E

98E

Scheme 38.

bicyclic carbene complex **97E** was synthesized from 1,3-cyclohexadiene exclusively as the *exo* isomer. Upon thermolysis of carbene complex **97E** in refluxing dioxane for 3 h, there was no indication of any product formation and complex **97E** was still present. If the *exo* (*trans*) isomer of carbene complex **97E** was not able to equilibrate to the more reactive *endo* (*cis*) isomer, then a base-induced isomerization might facilitate the reaction. Consequently, the thermolysis reaction was continued in the presence of triethylamine, which resulted only in the decomposition of carbene complex **97E** without any evidence of product formation. Attempts to perform the reaction in DMF also resulted in decomposition of the carbene complex **97E**. The lack of reactivity of bicyclic carbene complex **97E** was attributed to the inability of the chromium carbene double bond to adopt the *cis* conformation due to the strong steric interaction between the bulky chromium carbonyl group and the six membered ring containing the alkene (Scheme 38). From these results it seems that the ability of a carbene complex to attain the *cis* relative orientation of alkene and carbene complex substituents determines the ability of that carbene to convert to the corresponding cyclopentenone product.

In the limited number of examples studied, the reaction proceeds with retention of configuration (Table 7, entries F, H, I). In entries F and H, the product of the reaction is the thermodynamically most stable isomer,[76] while the product of entry 1 (**98I**) is the thermodynamically less stable isomer.[77] Note that the mechanism of Scheme 32a predicts a loss of stereochemistry, while that of 32b predicts retention of stereochemistry. The results of entries H and I do not necessarily imply that the reaction proceeds with retention of stereochemistry through the mechanism in Scheme 32b. The mechanism of Scheme 32a would lead to intermediate **100**, which upon $\sigma \rightarrow \pi$-allyl interconversion could lead to either intermediate **101** or **102** (Scheme 39). Intermediate **101** should be kinetically preferred, which upon CO-insertion and reductive elimination would provide compound **98H,I**. The observed retention of stereochemistry might therefore be the result of a kinetic preference and cannot be considered as conclusive evidence for the mechanism in Scheme 32b over Scheme 32a. The determination of the stereochemical fate of the thermolysis reaction must await more sophisticated studies.

C. Synthesis of Nine-Membered Rings

Methods for the construction of nine-membered rings are somewhat difficult due to the strain and conformational flexibility present in this ring system.[78] Unfortu-

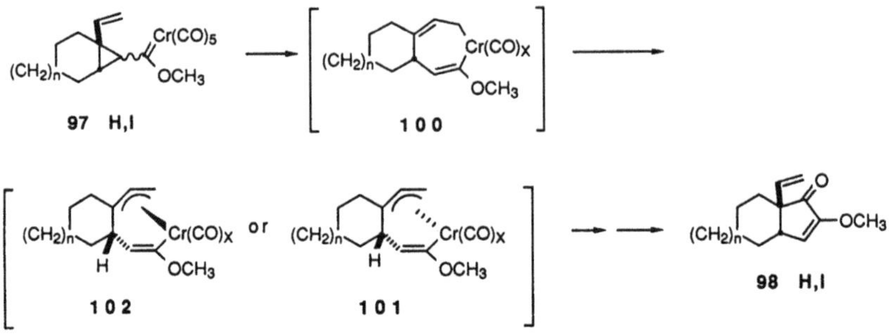

Scheme 39.

nately, simple Cope rearrangements of 1,2-divinylcyclopentane derivatives typically do not afford the cyclononadiene derivatives upon thermolysis.[79] In fact 1,5-cyclononadiene derivatives typically revert by way of a Cope rearrangement to 1,2-divinylcyclopentane derivatives. For example, cyclononadiene **105** undergoes a Cope rearrangement to yield the corresponding divinylcyclopentane **103** (Scheme 40).[80] Oxy-Cope rearrangements[81] and anionic oxy-Cope rearrangements[82] of 1,2-divinylcyclopentane derivatives (e.g., **104**) are effective methods for the construction of these ring systems. The anionic oxy-Cope rearrangement provides mild conditions for the synthesis of nine-membered rings, however the synthesis of the 1,2-divinylcyclopentanols themselves is often a tedious process.

Thermal rearrangement of 2-alkenylcyclopropylcarbene chromium complexes such as **16(97)** leads to the formation of 5-vinyl-2-cyclopentenone derivatives **20(98)** (Table 7). This transformation provides a facile route to highly functionalized cyclopentenones, which can be converted to the corresponding 1,2-divinylcyclopentenol upon treatment with a vinyl-metal reagent (Scheme 41). Therefore, the cyclopentenone derivatives **20(98)** are potential precursors to nine-membered ring containing compounds. The flexibility associated with the cyclopropylcarbene thermolysis reaction allows for the synthesis of a variety of substitution patterns on a nine-membered ring.

The treatment of cyclopentenone derivative **20** with vinylmagnesium bromide resulted in the formation of the divinylcyclopentenol **107A**, accompanied by the

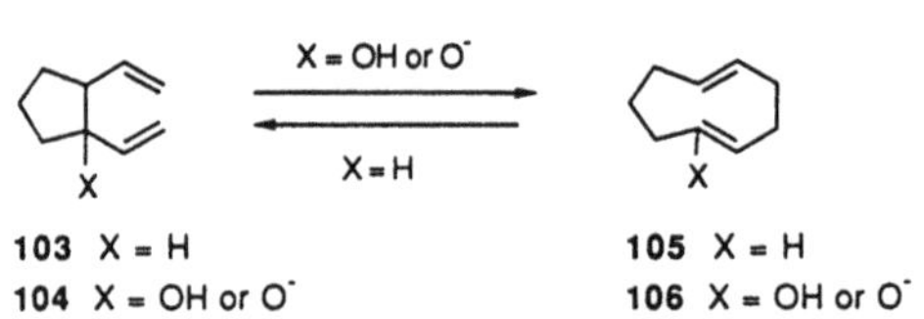

Scheme 40.

presumed conjugate addition product **108A** as diagnosed by infrared spectroscopy (a carbonyl stretch was present at 1737 cm^{-1}, suggestive of a cyclopentanone). Compound **107A** was not stable to chromatographic purification, and was used in the next step without purification. When the mixture of **107A** and **108A** was treated with potassium hydride[83] and 18-crown-6 the corresponding nine-membered ring **109A** was formed in a 11% yield. Presumably, the low yield is due to the inefficiency in the vinyl anion addition step. The conjugate addition side reaction can be suppressed through the use of organocerium reagents, formed by the addition of vinyl Grignard or lithium reagents to an anhydrous solution of cerium trichloride.[84] When cyclopentenone derivative **16** was treated with the vinyl cerium reagent derived from vinylmagnesium bromide, none of the conjugate addition product is observed. Additionally, when the anionic oxy-Cope rearrangement was performed with this crude alcohol **107A**, a 45% yield of the nine-membered ring derivative **109A** was obtained. The generality of this four-carbon ring expansion has been investigated with a variety of variously substituted vinyl cerium reagents and 5-vinylcyclopentenone derivatives. The results are shown in Table 8. The vinyl

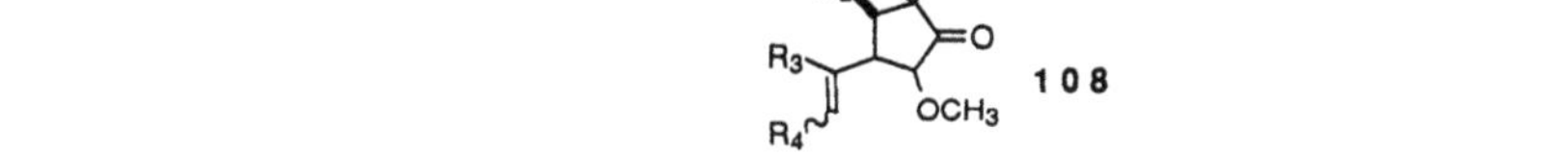

Scheme 41.

Table 8. Reaction of 2-Alkenylcyclopentanones with Alkenylcerium Reagents Followed by Anionic Oxy-Cope Rearrangements

Entry[a]	*R1*	*R2*	*R3*	*R4*	*Yield* **107**	*Yield* **109**	*E/Z* **109**
A	CH_3	H	H	H	85%	50%	100:0
B	CH_3	H	CH_3	H	95%	53%	100:0
C	CH_3	H	H	CH_3	93%	57%	100:0
D	CH_3	H	H	Ph	91%	51%	100:0
E	R_1, R_2=-$(CH_2)_4$-		CH_3	H	89%	47%	100:0
F	R_1, R_2=-$(CH_2)_4$-		H	H	87%	67%	100:0
G	R_1, R_2=-$(CH_2)_5$-		H	H	89%	47%	100:0

Notes: [a] Table entry letters define substituents for **107–109**.

anion addition reactions appear to be insensitive to steric effects, since good yields of adducts were obtained in all cases. The stereoselectivity of the vinyl anion addition reaction could not be determined, since the alcohols were unstable to chromatographic purification. In all cases the Cope rearrangement reaction proceeded in 40–60% yield, regardless of the substitution pattern on the vinyl group.

Upon examination of the 1,2-divinylcyclopentenol intermediate **107A** it is apparent that two possible Cope rearrangements can occur (Scheme 42). The alcohol can rearrange to either the observed nine-membered ring **109A** or to the cyclopentene derivative **110**. In no case were products analogous to compound **110** obtained, presumably because the norbornene-like geometry required for this process is difficult to attain.

In all cases a single alkene stereoisomer was formed. The assignment of the stereochemistry about the C-5,C-6 alkene was accomplished through nOe studies. The proton spectrum of compound **109B** contains a triplet and doublet of doublets in the vinylic region at δ 5.15 and δ 4.95 respectively. The resonance at δ 5.15 was significantly broadened, presumably due to long range coupling interactions. Additionally, decoupling of the doubly allylic proton at δ 3.05 causes the vinylic proton resonance at δ 4.95 to collapse. Consequently, the proton resonance at δ 5.15 was assigned as the vinylic proton (H_A) at the C-6 position (Scheme 43). Irradiation of the doubly allylic proton resonance at δ 3.05 (H_B) produces an 18% enhancement in the vinylic resonance at δ 5.15. Also irradiation of the other doubly allylic proton at δ 2.45 (H_C) produces an 8% enhancement of the same vinylic proton resonance. This is very suggestive of the (*E*)-configuration about the C-5, C-6 double bond. The formation of the (*E*)-configuration about the newly formed double bond suggests that the reaction is proceeding through a chair-like transition state (e.g., **111**) or a boat-like transition state where the vinyl groups are in a position exo to the ring (e.g., **112**).

Reaction of cyclopentenone **20** *with 2-propen-2-yl magnesium bromide/cerium trichloride.* A 25 mL round bottom flask fitted with a rubber septum, containing cerium trichloride heptahydrate (0.611 g,1.60 mmol) was heated to 150 °C under vacuum (0.005 mmHg) for a period of 24 h. Heating for shorter periods of time led to irreproducible yield of **107B**. The flask was then cooled to 0 °C and THF (6 mL) was added. The solution was allowed to stir at 25 °C for 2 h, the solution was then cooled to –78°C, and a solution

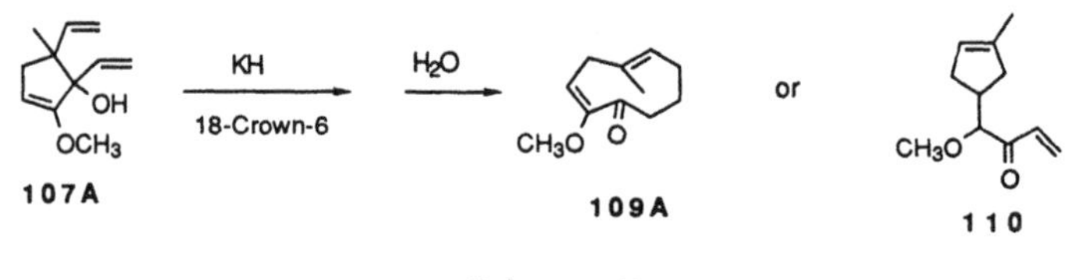

Scheme 42.

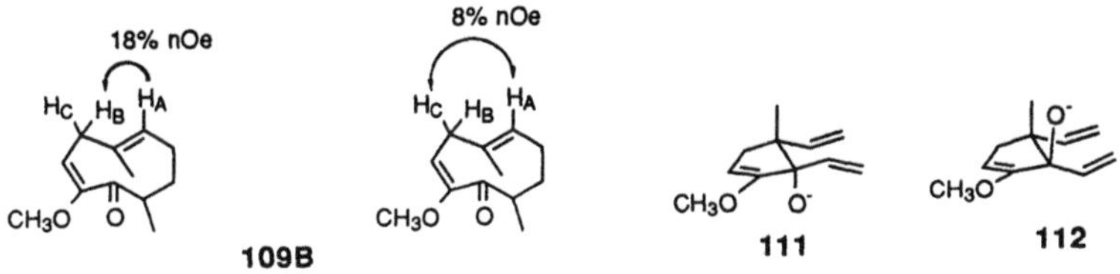

109B 111 112

Scheme 43.

of 2-propenylmagnesium bromide (2.1 mL of an 0.7M THF solution, 1.5 mmol) was added via syringe. The mixture was stirred for 1.5 h at –78 °C and a solution of compound **20** (0.050 g, 0.33 mmol) in THF (8 mL) was added dropwise via syringe over a 15 min period. The solution was stirred at –78 °C for 1 h and then poured into a 4% solution of acetic acid in water in a separatory funnel. The mixture was extracted with ether and the combined ether layers were washed with saturated aqueous sodium bicarbonate and saturated aqueous sodium chloride solutions. The ether layer was dried over magnesium sulfate and the solvent was removed on a rotary evaporator. The residue after evaporation (0.061 g, 95%) was used immediately in the Cope rearrangement reaction without further purification.

Preparation of cyclononadienone derivative **109B**. Pentane (10 mL) was added to a flask fitted with a septum under nitrogen containing potassium hydride (0.180 g of a 35% mineral oil dispersion, 1.60 mmol). The mixture was stirred vigorously for 5 min and then the solid was allowed to settle. The pentane was decanted, and the above process was repeated two times. THF (10 mL) was added, and to the resulting suspension was added via syringe a solution of compound **107B** (0.061 g, 0.31 mmol) and 18-crown-6 (0.42 g, 1.6 mmol) in THF (5 mL). The solution was stirred for 30 min at 25 °C and then cooled to –78 °C and ethanol (1.0 mL) was added very slowly. The mixture was poured into a mixture of saturated ammonium chloride and petroleum ether in a separatory funnel. The organic layer was washed with water and saturated aqueous sodium chloride, dried over magnesium sulfate, and the solvent was removed on a rotary evaporator. The residue after evaporation was purified by flash chromatography on silica gel using 9:1 hexane:ethyl acetate as the eluent. After removal of the solvent, a colorless oil (0.033 g, 53%) identified as compound **109B** was obtained. ^{1}H NMR ($CDCl_3$): δ 5.15 (t, 1H, J = 8.3 Hz); 4.95 (dd, 1 H, J = 9.3, 5.1 Hz); 3.55 (s, 3H); 3.05 (dd, 1 H, J = 15.0, 5.1 Hz); 2.85 (quintet of doublets, 1H, J = 7.0, 3.2 Hz); 2.45 (dd, 1H, J = 15.0, 9.3 Hz); 2.20 –1.50 (m, 4H); 1.60 (br s, 3H); 1.00 (d, 3H, J = 7.0 Hz). ^{13}C NMR ($CDCl_3$): δ 208.7, 157.5, 138.2, 124.2, 97.1, 55.2, 46.1, 33.8, 26.8, 26.4, 22.9,18.7. IR ($CDCl_3$): 3038 (w), 2985 (s), 2945 (vs), 2895 (m), 2880 (m), 1710 (vs), 1655 (m), 1470 (s), 1390 (m), 1245 (m), 1210 (s), 1170 (w), 1128 (m), 1105 (m), 1090 (m), 1075 (m), 1045 (m) cm^{-1}. MS (EI): 194 (M, 20),179 (6),166 (7),151(45), 137 (31),124 (100),109 (99). High res MS: calcd. for $C_{12}H_{18}O_2$: 194.1307. Found: 194.1312.

V. CONCLUSIONS

We have studied the scope, limitation, and mechanism of the reaction between alkynes and cyclopropylcarbene-chromium complexes. The overall conversion represents a versatile one-step method for the synthesis of highly-substituted five-membered rings. The simplicity and availability of the reacting components, alkynes and cyclopropylcarbene-chromium complexes, coupled with the high regioselectivity and functional group tolerability makes this reaction a powerful method for carbocyclic five-membered ring construction. We have also examined thermolysis reactions of cyclopropylcarbene-chromium complexes. When substituted by an alkenyl substituent at the 2-position, these complexes provide 5-vinyl-2-cyclopentenone derivatives in good yield via a mechanism reminiscent of the divinylcyclopropane rearrangement. These complexes can be converted to nine-membered ring-containing compounds by treatment with vinylcerium reagents followed by anionic oxy-Cope rearrangement of the resulting trienols. Further exploration into the scope, limitations, mechanism, and synthetic applications is underway in our laboratories.

ACKNOWLEDGMENTS

We thank the National Institutes of Health (GM-40777), the Petroleum Research Fund, administered by the American Chemical Society, and the University of Maryland for financial support of this research. We are grateful to the following undergraduate researchers for experimental assistance: Jonathan Hahm, Carina Tan, Steven Strong, Charles Daitch, and Ming Chen.

REFERENCES AND NOTES

1. For a review of metal-carbene complexes, see: (a) Dötz, K. H. *Angew. Chem., Int. Ed. Engl.* **1984**, *23*, 587–608. (b) Wulff, W.D. In *Advances in Metal-Organic Chemistry*; Liebeskind, L. S., Ed; JAI Press: Greenwich, CT, 1989; Vol. 1.
2. For a recent advance in the synthesis of bicyclo[x.1.0] ring systems and furanones using chromium-carbene complexes and enynes, see: Hoye, T. R.; Rehberg, G. M. *J. Am. Chem. Soc.* **1990**, *112*, 2841–2842.
3. Wulff, W. D.; Xu, Y.-C. *J. Am. Chem. Soc.* **1988**, *110*, 2312–2314.
4. Semmelhack, M. F.; Zask, A. *J. Am. Chem. Soc.* **1983**, *105*, 2034–2043.
5. Dötz, K. H.; Kuhn, W. *Angew. Chem., Int. Ed. Engl.* **1983**, *22*, 732.
6. Yamashita, A.; Toy, A.; Scahill, T. A. *J. Org. Chem.* **1989**, *54*, 3625-3634.
7. Wulff, W. D.; McCallum, J. S.; Kunng, F.-A. *J. Am. Chem. Soc.* **1988**, *110*, 7419–7434.

8. (a) Davies, H. M. L.; McAfee, M. J.; Oldenburg, C. E. M. *J. Org. Chem.* **1989**, *54*, 930–936. (b) Noyori, R. *Acc. Chem. Res.* **1979**, *12*, 61–66. (c) Hoffmann, H. M. R. *Angew. Chem., Int. Ed. Engl.* **1984**, *23*, 1–19. (d) Wulff, W. D.; Yang, D. C.; Murray, C. K. *J. Am. Chem. Soc.* **1988**, *110*, 2653–2655. (e) Trost, B. M.; MacPherson, D. T. *J. Am. Chem. Soc.* **1987**, *109*, 3483–3484. (f) Sammes, P. G. *Gazz. Chim. Ital.* **1986**, *119*, 109–114. (g) Boger, D. L.; Brotherton, C. E. *J. Org. Chem.* **1985**, 3425–3427.
9. Suffness, M.; Cordell, G. A. In *The Alkaloids*; Brossi, A., Ed.; Academic Press: San Diego, 1985; Vol. 25, pp 69–73.
10. Lee, K. H.; Mar, E. C.; Okamoto, M.; Hall, I. H. *J. Med. Chem.* **1978**, *21*, 819–822.
11. See Wender, P. A.; McDonald, F. E. *J. Am. Chem. Soc.* **1990**, *112*, 4956–4958, and references therein.
12. Tsuji, T.; Nishida, S. *Acc. Chem. Res.* **1984**, *17*, 56–61. For an example where this approach was successful, see reference 8g.
13. (a) Herndon, J. W.; Tumer, S. U.; Schnatter, W. F. K. *J. Am. Chem. Soc.* **1988**, *110*, 3334–3335. (b) Herndon, J. W.; Tumer, S. U.; McMullen, L. A.; Matasi, J. J.; Schnatter, W. F. K.; Daitch, C. E. *Comm. Inorg. Chem.* **1990**, *10*, 1–24.
14. For a recent review, see: Paquette, L. A.; Doherty, A. M. *Polyquinane Chemistry*; Springer: Berlin, 1987.
15. The ability of alkylcarbene-chromium complexes to polymerize alkynes is well-documented. Katz, T. J. Lee, S. J. *J. Am. Chem. Soc.* **1980**, *102*, 422–424.
16. Wulff, W. D.; Kaesler, R. W.; Peterson, G. A.; Tang, P.-C. *J. Am. Chem. Soc.* **1985**, *107*, 1060–1062.
17. Triethylsilane / $Cr(CO)_6$ is an effective reagent for the reduction of ketones and enones. Keinan, E. H.; Perez, D. *J. Org. Chem.* **1987**, *52*, 2576–2580.
18. A similar type of side reaction has been previously suggested. Yamashita, A.; Timko, J.M.; Watt, W. *Tetrahedron Lett.* **1988**, *29*, 2513–2516. For a similar reaction in cobalt chemistry, see: Doyama, K.; Fujiwara, K.; Joh, T.; Maeshima, K. S. *Chem Lett.* **1988**, 901–904.
19. Schadt, F. L.; Bentley, T. W.; Schleyer, P. vR. *J. Am. Chem. Soc.* **1976**, *98*, 7667–7679.
20. The formation of abnormal products from the reaction of arylcarbene-chromium complexes and propiolate esters is precedented. Yamashita, A.; Scahill, T. A. *Tetrahedron Lett.* **1982**, *23*, 3765–3768.
21. Taskinen, E.; Mukkala, V. M. *Tetrahedron* **1982**, *38*, 613–616.
22. For a similar phosphine-substitution effect, see Xu, Y.-C.; Wulff, W. D. *J. Am. Chem. Soc.* **1987**, *52*, 3263–3275.
23. Liebmann, J. F.; Greenburg, A. *Chem. Rev.* **1976**, *76*, 311–365.
24. (a) For prostaglandins, see Johnson, R. A. In *Advances in Prostaglandin, Thromboxane, and Leukotriene Research*; Pike, J.E.; Morton, D. R., Ed; Raven Press: New York, 1985; Vol. 14, pp 131–154. (b) For prostacyclins, see Caton, M. P. L.; Hart, T. W. In *Advances in Prostaglandin, Thromboxane, and Leukotriene Research*; Pike, J. E.; Morton, D. R., Ed; Raven Press: New York, 1985; Vol. 14, pp 73–129.
25. Herndon, J. W.; Matasi, J. J. *J. Org. Chem.* **1990**, *55*, 786–788.
26. Semmelhack, M. F.; Bozell, J. J. *Tetrahedron Lett.* **1982**, *23*, 2931–2934.
27. Stork, G.; Danheiser, R. L. *J. Org. Chem.* **1973**, *38*, 1775–1776.
28. Connor, J. A.; Jones, E. M. *J. Chem. Soc., Dalton Trans.* **1973**, 2119–2124.
29. Ogliarusso, M. A.; Romanelli, M. G.; Becker, E. I. *Chem. Rev.* **1965**, *65*, 261–367.
30. Tad, M.; Chiba, K. *Agric. Biol. Chem.* **1984**, *48*, 1367–1369.
31. Danishefsky, S. J.; Maring, C. J. *J. Am. Chem. Soc.* **1985**, *107*, 1269–1274.
32. Volhardt, K. P. C. *Angew. Chem., Int. Ed. Engl.* **1984**, *23*, 539–556.
33. This reduction reaction is precedented: Brown, D. A.; Hargaden, J. P.; McMullin, C. M.; Gogan, N.; Sloan, H. *J. Chem. Soc.* **1963**, 4914–4918.
34. Oestrich, S.; Broser, W.; Kurreck, H. *Z. Naturforsch., B.* **1977**, *32B*, 686–692.

35. Birch, A. J.; Subba Rao, G. *Adv. Org. Chem.* **1972**, *94*, 8947–8948.
36. Casey, C. P.; Boggs, R. A.; Anderson, R. L. *J. Am. Chem. Soc.* **1972**, *94*, 8947–8948.
37. Fischer, H. Munlemeier, J.; Markl, R. Dötz, K. H. *Chem. Ber.* **1982**, *115*, 1355–1362.
38. Chappell, S. D.; Cole-Hamilton, D. J. *Polyhedron*, **1982**, *1*, 739–777.
39. Casey, C. P.; Burkhardt, T. J. *J. Am. Chem. Soc.* **1974**, *96*, 7808–7809.
40. For a ring expansion of this type, see Liebeskind, L. S.; Mitchell, D.; Foster, B. S. *J. Am. Chem. Soc.* **1987**, *109*, 7908–7910.
41. Wakatsuki, Y.; Nomura, O.; Kitaura, K.; Morokuma, H.; Yamazaki, H. *J. Am. Chem. Soc.* **1983**, *105*, 1907–1912.
42. (a) Knox, S. A. R.; Stansfield, R. F. D.; Stone, F. G. A.; Winter, M. J.; Woodard, P. *J. Chem. Soc., Dalton Trans.* **1982**, 173–185. (b) Michman, M.; Zeiss, H. H. *J. Organomet. Chem.* **1970**, *25*, 161–166.
43. It has recently been suggested that chelated vinylcarbene-metal complexes are more reasonable structures than metallacyclobutenes in these reactions. Hofmann, P.; Hämmerle, M. *Angew. Chem., Int. Ed. Engl.* **1989,** *28*, 908–909.
44. For another recent case where methyl propargyl ether behaves abnormally in the reactions with chromium-carbene complexes, see Semmelhack, M. F.; Jeong, N.; Lee, G. R. *Tetrahedron Lett.* **1990**, *31*, 609–610.
45. (a) Herndon, J. W.; Tumer, S. U. *Tetrahedron Lett.* **1989**, *30*, 4771–4774. (b) Full paper: Herndon, J. W.; Tumer, S. U. *J. Org. Chem.* **1991**, *56*, 286–294.
46. Reactions between carbene complexes and electron-deficient alkenes have been reported to be stereospecific and proceed through metallacyclobutane intermediates. See references 49 a,b,e. Reactions involving electron-rich alkenes are not always stereospecific, and proceed via zwitterionic intermediates. a. Casey, C. P.; Hornung, N. L.; Kosar, W. P. *J. Am. Chem. Soc.* **1987**, *109*, 4908–4916. b. Wulff, W. D.; Yang, D. C.; Murray, C. K. *J. Am. Chem. Soc.* **1988**, *110*, 2653–2655.
47. Wulff, W. D.; Tang, P.-C.; McCallum, J. S. *J. Am. Chem. Soc.* **1981**, *103*, 7677–7678.
48. In some cases, there is an electronic effect on the regioselectivity of the reaction between alkynes and cyclopropylcarbene-chromium complexes. Brandvold, T. A.; Wulff, W. D.; Rheingold, A.L. *J. Am. Chem. Soc.* **1990**, *112*, 1645–1647.
49. (a) Dötz, K. H.; Fischer, E. O. *Chem. Ber.* **1970,** *103*, 1273–1278. (b) Dötz, K. H.; Fischer, E. O. *Chem. Ber.* **1972**, *105*, 1356–1367. (c) Dötz, K. H.; Fischer, E. O. *Chem. Ber.* **1972**, *105*, 3966–3973. (d) Dorrer, B.; Fischer, E. O.; Kalbfus, W. *J. Organomet. Chem.* **1974**, *81*, C20–C22. (e) Cooke, M. D.; Fischer, E. O. *J. Organomet. Chem.* **1973**, *56*, 279–284. (f) For a review, see Brookhart, M.; Studabaker, W. B. *Chem. Rev.* **1987**, *87*, 411–432. (g) For a comparison of group 6 metal-carbenes in the cyclopropanation reaction, see Harvey, D. F.; Brown, M. F. *Tetrahedron Lett.* **1990**, *31*, 2529–2532.
50. Ho, G.-J.; Krogh-Jesperson, K.; Moss, R. A.; Shen, S.; Sheridan, R. S.; Subramanian, R. *J. Am. Chem. Soc.* **1989**, *111*, 6875–6877.
51. Brookhart, M.; Studabaker, W. B.; Husk, R. G. *Organometallics* **1987**, *6*, 1141–1145.
52. (a) Wienand, A.; Reissig, H. U. *Tetrahedron Lett.* **1988**, *29*, 2315–2318. (b) Buchert, M.; Reissig, H. U. *Tetrahedron Lett.* **1988**, *29*, 2319–2320.
53. Reissig, H.-U. *Top. Curr. Chem.* **1988**, *144*, 73–136.
54. These compounds have been designated as *cis* or *trans* to conform with IUPAC nomenclature. Thus these compounds are derivatives of bicyclopropyl, and the *cis-trans* designation refers to the relative stereochemistry between the ester and methoxy substituents.
55. Doyle, M. P.; van Leusen, D. *J. Org. Chem.* **1982**, *47*, 5326–5339.
56. Grob, C. A. *Angew. Chem., Int. Ed. Engl.* **1982**, *21*, 87–154.
57. Wiberg, K. B.; Hess, B. A.; Ashe, A. J. III In *Carbonium Ions*; Olah, G. A., Schleyer, P. vR., Eds; John Wiley and Sons: New York, 1982; Vol. 2, pp 1295–1346.

58. For example, norbornene is one of only a few alkenes which participate efficiently in the intermolecular Pauson-Khand cyclization. Khand, I. U.; Knox, G. R.; Pauson, P. L.; Watts, W. E.; Foreman, M. E. *J. Chem. Soc., Perkin Trans.* **1973**, 977–981.
59. (a) Reichelt, I.; Reissig, H. U. *Chem. Ber.* **1983**, *116*, 3895–3914. b. Kunkel, E.; Reichelt, I.; Reissig, H. U. *Liebigs Ann. Chem.* **1984**, 512–530.
60. Wenkert, E.; Buckwalter, B. E.; Craveiro, A. A.; Sanchez, E. L.; Sathe, S. S. *J. Am. Chem. Soc.* **1978**, *100*, 1267–1273.
61. For a review of the chemistry of cyclopropanes, see Wong, H. N. C.; Hon, M.-Y.; Tse, C.-W.; Yip, Y.-C.; Tanko, J.-M.; Hudlicky, T. *Chem. Rev.* **1989**, *89*, 165–198.
62. Marino, J. P.; Laborde, E. *J. Org. Chem.* **1987**, *52*, 1–10.
63. (a) Herndon, J. W.; Wu, C. *Tetrahedron Lett.* **1989**, *30*, 5745–5748. (b) Melchiorre, C.; Gianella, M.; Giardina, D.; Gaultieri, F. *Synth. Commun.* **1975**, *5*, 95–100.
64. Hudlicky, T.; Kutchan, T. M.; Naqvi, S.M. *Org. Reactions* **1985**, *33*, 247–335.
65. Schneider, M.P.; Rau, A. *J. Am. Chem. Soc.* **1979**, *101*, 4424–4427.
66. (a) Casey, C.P.; Shusterman, A.J.; Vollendorf, N.W.; Haller, K.J. *J. Am. Chem. Soc.* **1982**, 104, 2417–2423. (b) Casey, C. P.; Vollendorf, N. W.; Haller, K. J. *J. Am. Chem. Soc.* **1984**, *106*, 3754–3764. (c) Casey, C.P.; Shusterman, A. J. *Organometallics* **1985**, ***4***, 736–744.
67. Herndon, J. W.; McMullen, L. A. *J. Am. Chem. Soc.* **1989**, *111*, 6854–6856.
68. Compound **16** was synthesized according to the following sequence: isoprene + dibromocarbene → 2-methyl-2-vinyl-1,1-dibromocyclopropane (A); A + Bu_3SnH → 2-methyl-2-vinyl-1-bromocyclopropane (B); B + *t*-BuLi, then $Cr(CO)_6$, then $CH_3OSO_2CF_3$ → **16**.
69. For a review of the chemistry of allylmetal complexes, see Consiglio, G.; Waymouth, R. M. *Chem. Rev.* **1989**, *89*, 257–276.
70. Reductive elimination from these types of complexes typically gives five- rather than seven-membered rings. For an example, see Trost, B. M.; Walchli, R. A. *J. Am. Chem. Soc.* **1987**, *109*, 3487–3489.
71. Generally, sigma bonds to allylic or benzylic carbons are weaker than their non-benzylic counterparts. For example, see: Gordon, A. J.; Ford, R. A. *The Chemist's Companion*; John Wiley and Sons: New York, 1972; p 113.
72. For a table of heterolytic bond dissociation energies, see Morrison, R.T.; Boyd, R.A, *Organic Chemistry,* 5th ed., Allyl and Bacon: New York, 1987, p 22.
73. Marrel, E. C.; Lin, C. *J. Am. Chem. Soc.* **1978**, *100*, 877–883.
74. Parlier, A.; Rudler, H.; Daran, J. C.; Alvarez, C.; Reyes, F. D. *J. Organomet. Chem* **1987**, *327*, 339–356.
75. (a) Casey, C. P.; Shusterman, A. J.; Vollendorf, N. W.; Haller, K. J. *J. Am. Chem. Soc.* **1982**, *104*, 2417–2423. (b) Casey, C. P.; Vollendorf, N. W.; Haller, K. J. *J. Am. Chem. Soc.* **1984**, *106*, 3754–3764. (c) Casey, C. P.; Shusterman, A .J. *Organometallics* **1985**, *4*, 736–744.
76. Jones, T. K.; Denmark, S. E. *Helv. Chim. Acta* **1983**, *66*, 2377–2396.
77. Heathcock; C. H.; Graham, S. L.; Pirrung, M. C.; Plavac, F.; White, C. T. in *The Total Synthesis of Natural Products*; ApSimon; J., Ed.; Wiley: New York; 1983; Vol.5, pp 333–377. See also reference 76.
78. Engler, E. M.; Androse, J. D.; Schleyer, P. V. R. *J. Am. Chem. Soc.* **1973**, *95*, 8005–8025.
79. Barton, D. H. R.; Jackman, L. M.; Hahn, L. R.; Sutherland, J. K. *J. Chem. Soc.* **1965**,1772–1778.
80. Barton, D. H. R.; Sutherland, J. K. *J. Chem. Soc.* **1965**,1769–1771.
81. Kato, T.; Kondo, H.; Nishino, M.; Tanaka, M.; Harta, G.; Miyake, A. *Bull. Chem. Soc. Jpn.* **1980**, *53*, 2958–2961.
82. Paquette, L. A.; DeRussy, D. T.; Pegg, N. A.; Taylor, R. T.; Zydowsky, T. M. *J. Org. Chem.* **1989**, *54*, 4576–4581.
83. MacDonald, T. L.; Natalie, K. J. Jr.; Prasad, G.; Sawyer, J. S. *J. Org. Chem.* **1986**, *51*,1124–1126.
84. Imamoto, T.; Takiyama, N.; Nakamura, K. *Tetrahedron Lett.* **1985**, *26,* 4763–4766.
85. Yamaguchi, M.; Nobayashi, Y.; Hirao, I. *Tetrahedron* **1984,** *40*, 4261–4266.

PALLADIUM-CATALYZED VINYLIC SUBSTITUTION

Richard C. Larock

Advances in Metal-Organic Chemistry
Volume 3, pages 97–224.

ISBN: 1-55938-406-9

I. INTRODUCTION

One of the more important advances in organic synthesis in the last 25 years has been the development of palladium-catalyzed vinylic substitution reactions (Eq. 1). These reactions which generally cannot be affected via simple nucleophilic dis-

$$\text{RCH=CHX} \xrightarrow{\text{catalytic Pd}} \text{RCH=CHY} \qquad (1)$$

X = halogen, sulfonate, sulfonyl, phosphate, metal (Li, Mg, Zr, Cu, Zn, Hg, B, Al, Si, Sn, Te)

Y = H, R, $CH_2CH{=}CH_2$, CH_2Ar, CH=CHR, CH=C=CHR, C≡CR, $CH_2C{\equiv}CR$, Ar, CHO, COR, CO_2R, $CONR_2$, CN, halogen, OR, SR, SiR_3, SnR_3, etc.

placement processes often proceed readily in the presence of only catalytic amounts of palladium. These reactions will be reviewed according to the group X displaced and the nature of the new group Y being introduced, starting with vinylic halides. The chemistry of vinylic palladium compounds generated by additions to alkynes, although a very important area of organopalladium chemistry, will not be discussed here.

II. VINYLIC HALIDES

Oxidative Addition

Vinylic halides readily undergo oxidative addition to palladium(O) reagents (Eq. 2). Phosphine complexes (L = PPh_3, Ph_2PMe, $PhPMe_2$, PEt_3) containing vinylic groups bearing one or more fluorine,[1,2] chlorine,[3,4] aryl,[5] ketone,[6] acid,[7] ester,[7,8] amide[7] and nitrile[7] groups have been isolated and well characterized. Analogous perfluorovinyl complexes bearing arsine ($PhAsMe_2$)[9] and isocyanide (*t*-BuNC, 2,6-$Me_2C_6H_3NC$)[9,10] ligands have also been prepared. Oxidative addition of vinyl chloride to $Pd(CO)(PPh_3)_3$ results in carbonyl insertion to form the corresponding acyl complex (Eq. 3).[11,12] These reactions are the key first step in a vast majority of the palladium-catalyzed substitution processes covered in this review.

$$\underset{\text{X = Cl, Br, I}}{\text{RCH=CHX}} + \text{PdL}_4 \longrightarrow \text{(RCH=CH)(L)Pd(L)(X)} \qquad (2)$$

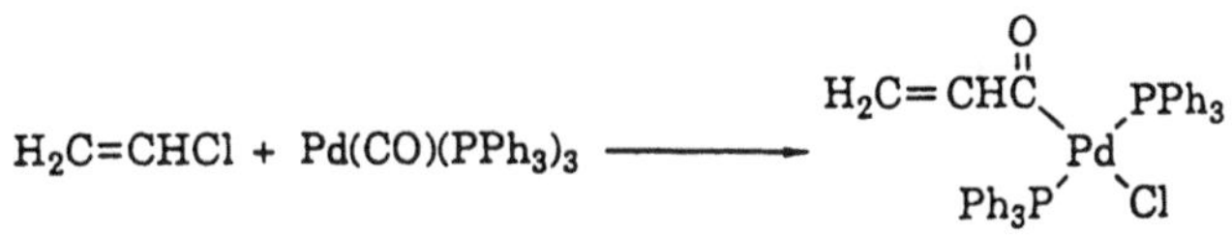

(3)

The mechanism of oxidative addition has only been explored using (*E*)- and (*Z*)-1,2-dichloroethylene, where the reaction is rather complex, involving reversible steps in which both retention and inversion of stereochemistry are observed.[13]

Thermolysis

The majority of vinylic palladium complexes which have been isolated appear to be quite thermally stable. While thermolysis of $(Ph_3P)_2PdX(CH{=}CHCN)$ gave none of the dimer,[14] the corresponding ester has been reported to both rearrange[15] and dimerize[16] (Eq. 4).

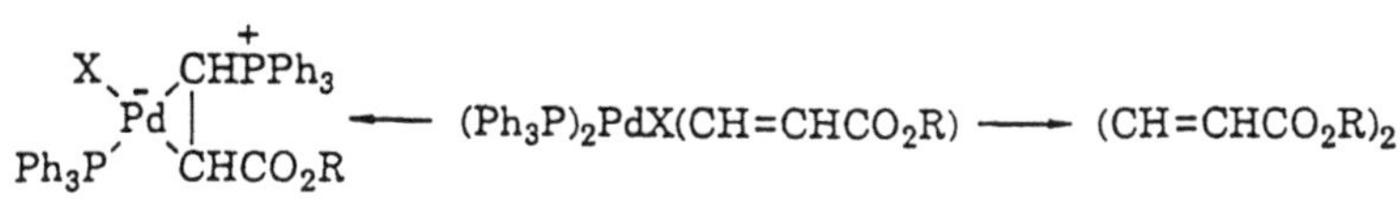

(4)

Hydrogen Substitution

While simple chlorovinyl[4] and acrylonitrile[7] complexes of palladium undergo straightforward protonolysis or deuterolysis to the corresponding alkene by HCl or DCl (Eq. 5), the analogous acrylate esters react with HCl or HBF_4 to afford dimers (Eq. 6).[16,17] On the other hand, the related complex **1** reacts stereoselectively with deuterotrifluoroacetic acid to give the corresponding deuterated ester in high yield (Eq. 7).[8]

$$L_2PdX(CH{=}CHY) \xrightarrow{HCl} H_2C{=}CHY \qquad Y = Cl,\ CN \tag{5}$$

$$L_2PdX(CH{=}CHCO_2R) \xrightarrow{HCl\ or\ HBF_4} RO_2CCH_2CH{=}CHCH_2CO_2R \tag{6}$$

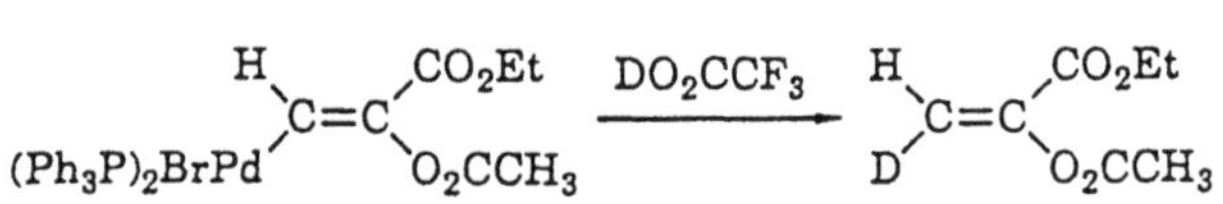

1

(7)

Several palladium-catalyzed processes are available for the reduction of vinylic halides to alkenes. The use of $Pd(PPh_3)_4$ and MgH_2 stereospecifically reduces (*E*)- and (*Z*)-styryl bromide in good yields (Eq. 8).[18] Only catalytic amounts of

$$\mathrm{PhCH{=}CHBr} \xrightarrow[\mathrm{Pd(PPh_3)_4}]{\mathrm{MgH_2}} \underset{65\%}{\mathrm{PhCH{=}CH_2}} \qquad (8)$$

$Pd(PPh_3)_4$ are needed for the reduction of styryl bromide, when sodium formate or polymethylhydrosiloxane are employed.[19] However, the yields are only 35% and 37% respectively.

The hydrogenation of 3-halo-2-alkenals in the presence of catalytic amounts of palladium on charcoal provides excellent yields of the corresponding unsaturated aldehydes (Eq. 9).[20]

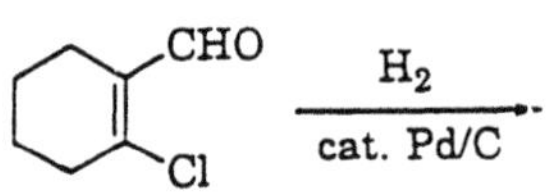

(9)

Tetra-*n*-butylstannane in the presence of catalytic amounts of tetravinylstannane and $PdCl_2(CH_3CN)_2$ has been used to reduce a vinylic halide, but this may not be a general process.[21]

All of these reduction processes probably involve initial oxidative addition to form a vinylic palladium species which is subsequently reduced. These same reagents also reduce a variety of other organic halides.

Nitrile Formation

The real power of the palladium-catalyzed vinylic substitution reactions lies in the ease with which a wide variety of new carbon-carbon bonds can be formed using this methodology. For example, catalytic amounts of $Pd(PPh_3)_4$ in the presence of 18-crown-6 allow the facile synthesis of α,β-unsaturated nitriles from vinylic chlorides and bromides (Eq. 10).[22]

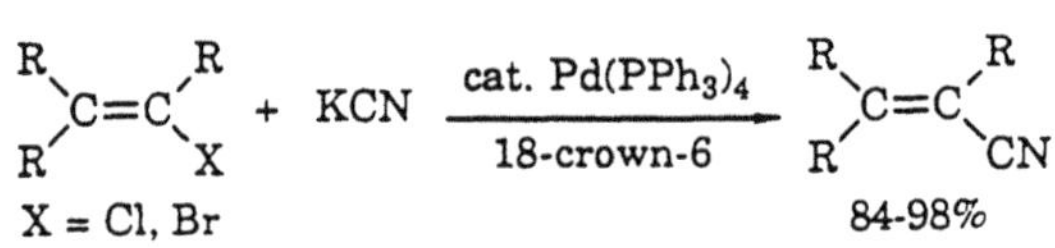

(10)

Carbonylation

The palladium-catalyzed carbonylation of vinylic halides affords a convenient route to a variety of α,β-unsaturated carbonyl compounds. Thus, carbonylation in the presence of H_2[23] or *n*-Bu_3SnH[24–26] affords aldehydes (Eq. 11). Like many of the transformations we will cover, the latter process accommodates additional functionality (Eq. 12), but certain alkenals isomerized during work-up, and reduc-

$$\underset{X = Br, I}{RCH{=}CHX} \xrightarrow[\text{cat. } (Ph_3P)_2PdI_2 \text{ or } Pd(PPh_3)_4]{CO,\ H_2 \text{ or } n\text{-}Bu_3SnH} RCH{=}CH\overset{O}{\overset{\|}{C}}H \qquad (11)$$

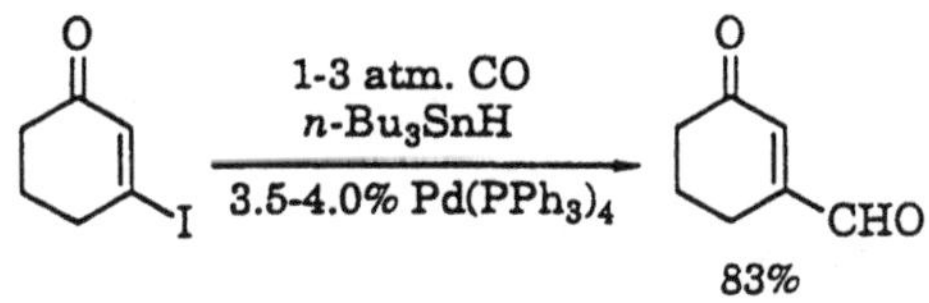

(12)

tion of the vinylic halide to the corresponding alkene is an occasional problem. These reactions undoubtedly proceed via CO insertion into the initial vinylic palladium complex, followed by reduction of the resulting acyl complex.

Carbonylation and subsequent cyclization onto stabilized carbanions has also been reported (Eq. 13).[27]

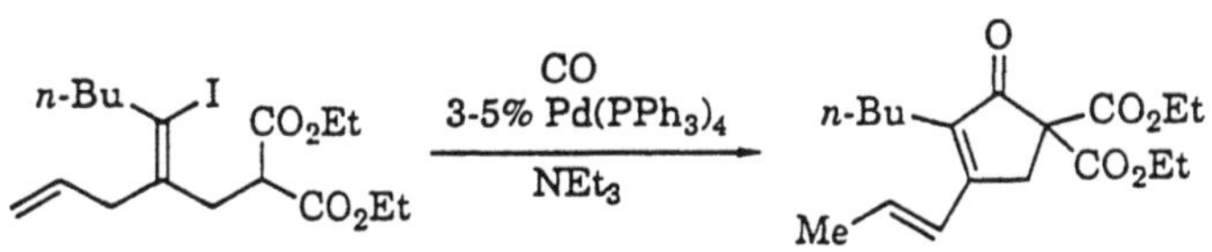

(13)

While carbonylation in the presence of an organoaluminum compound has been reported to produce a modest yield of allylic alcohol (Eq. 14),[28] organotin compounds often afford excellent yields of ketones (Eq. 15).[26,29,30] Alkyl,[31] vinylic,[32] alkynyl,[32] aryl[32] and heterocyclic[33] tin compounds have been successfully employed in this reaction. Mechanistically, the reaction likely proceeds via alkylation of an acylpalladium intermediate, followed by reductive elimination (Eq. 16).

$$E\text{-}PhCH{=}CHBr \xrightarrow[i\text{-}Bu_3Al]{CO} \underset{50\%}{E\text{-}PhCH{=}CH\overset{OH}{\overset{|}{C}}HCH_2CH(CH_3)_2} \qquad (14)$$

$$RCH{=}CHX + CO + SnR_4' \text{ or } n\text{-}Bu_3SnR' \xrightarrow{\text{cat. Pd}} RCH{=}CH\overset{O}{\overset{\|}{C}}R' \qquad (15)$$

$$RCH{=}CH\overset{O}{\overset{\|}{C}}PdX \xrightarrow{SnR_4'} RCH{=}CH\overset{O}{\overset{\|}{C}}PdR' \longrightarrow RCH{=}CH\overset{O}{\overset{\|}{C}}R' \qquad (16)$$

Acylpalladium species generated by carbonylation can also undergo intramolecular alkene insertion, followed by either palladium β-hydride elimination

(Eq. 17) or a second carbonylation (Eqs. 18 and 19). While the former process (Eq. 17) seems to require a stoichiometric amount of palladium, the latter two are catalytic in palladium.

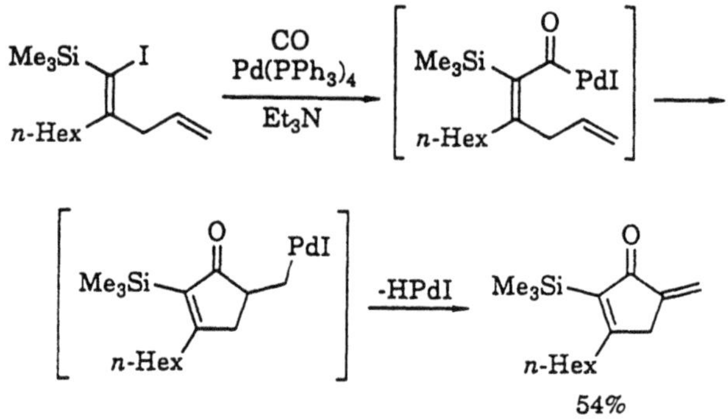

(17)[34]

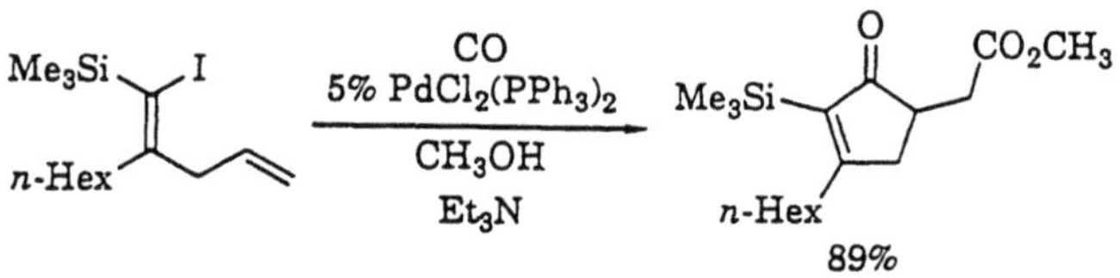

(18)[35]

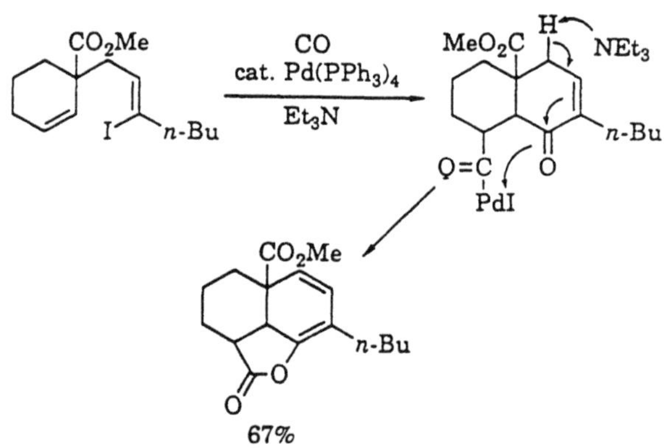

(19)[36]

Carbonylation in the presence of terminal alkynes affords another route to enynones (Eq. 20).[37]

$$\mathrm{PhCH{=}CHBr + CO + HC{\equiv}CPh \xrightarrow{\text{cat. } PdCl_2(dppf)} PhCH{=}CH\overset{O}{\overset{\|}{C}}C{\equiv}CPh} \quad (20)$$

81%

dppf = 1,1'-bis(diphenylphosphino)ferrocene

The carbonylation of vinylic halides in the presence of water, alcohols and amines provides a useful route to the corresponding α,β-unsaturated carboxylic acids, esters and amides respectively. The acids seem to be best prepared using phase transfer conditions, either: (1) 5 atm. CO, 30% aq. NaOH, cat. $PdCl_2(PPh_3)_2$, cat. *n*-Bu_4NI in xylene at 95 °C;[38] or (2) 1 atm. CO, 5N NaOH, cat. $Pd(diphos)_2$ or $Pd(PPh_3)_4$, cat. $(PhCH_2NEt_3)Cl$, in *t*-AmOH-CH_2Cl_2 at room temperature[39] (Eq. 21).

$$\mathrm{PhCH{=}CHBr} \xrightarrow[\text{cat. Pd}]{\text{CO, NaOH}} \mathrm{PhCH{=}CHC({=}O)ONa} \qquad (21)$$

The carbonylation of vinylic dibromides affords different products depending on the reaction conditions.[40] Using *tert*-amyl alcohol as the solvent, one obtains malonic acids (Eq. 22). In benzene, however, either α,β-unsaturated acids or diynes are formed depending on the nature of the substituent on the carbon-carbon double bond (Eq. 23).

$$\underset{\text{R = alkyl, aryl}}{\mathrm{RCH{=}CBr_2}} \xrightarrow[\text{cat. Pd(diphos)}_2\text{, (PhCH}_2\text{NEt}_3\text{)Cl, }t\text{-AmOH}]{\text{CO, 5N NaOH}} \underset{28\text{-}93\%}{\mathrm{RCH{=}C(CO_2H)_2}} \qquad (22)$$

$$\mathrm{RCH{=}CHCO_2H} \xleftarrow{\text{R = alkyl}} \mathrm{RCH{=}CBr_2} \xrightarrow{\text{R = aryl}} \mathrm{RC{\equiv}CC{\equiv}CR} \qquad (23)$$

While the double carbonylation (150 atm. CO, cat. $PdCl_2(PPh_3)_2$, H_2O, Et_3N) of aryl halides proceeds in modest yields, vinylic halides only afford the single insertion product (Eq. 24).[41]

$$\mathrm{RCO_2H} \xleftarrow{\text{R = vinylic}} \mathrm{RI} \xrightarrow{\text{R = aryl}} \mathrm{RC({=}O)CO_2H} \qquad (24)$$

While several procedures have been reported for the conversion of vinylic halides to α,β-unsaturated esters,[42–47] the best involve 1 atmosphere of carbon monoxide in the presence of an alcohol and tri-*n*-butylamine,[42] or aluminum,[46] boron,[47] titanium[44] or zirconium[44] alkoxides (Eq. 25). With the aluminum, boron and zirconium alkoxides, catalytic amounts of (1,5-hexadiene)rhodium(I) chloride

$$\mathrm{RCH{=}CHCO_2R'} \xleftarrow[\text{R'OH, }n\text{-Bu}_3\text{N; X = Br, I}]{\text{CO, cat. PdX}_2\text{(PPh}_3\text{)}_2} \mathrm{RCH{=}CHX} \xrightarrow[\text{metal alkoxide; X = Br}]{\text{CO, cat. Pd(PPh}_3\text{)}_4} \mathrm{RCH{=}CHCO_2R'} \qquad (25)$$

dimer are also required. The alcohol procedure exhibits some loss of carbon-carbon double bond stereochemistry, while the metal alkoxide approaches have not been examined in that regard.

The alcohol approach has been applied to a number of simple vinylic halides,[42] as well as the following more highly functionalized systems (Eqs. 26–28).

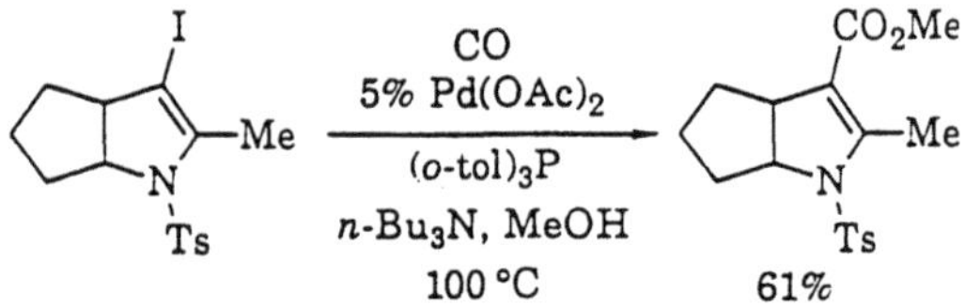

(26)[48]

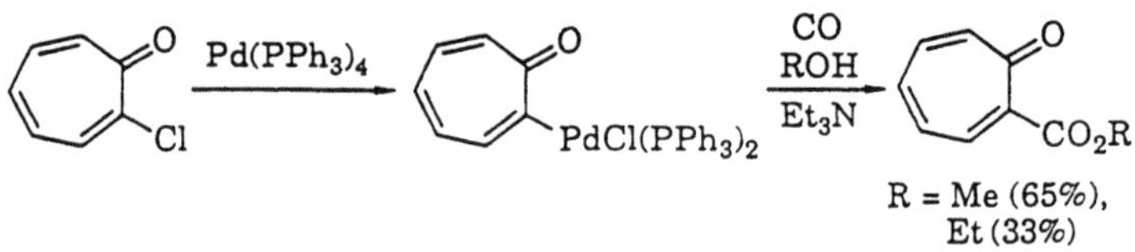

(27)[49]

$$n\text{-}C_8F_{17}CH{=}C(I)SiMe_3 \xrightarrow[Et_3N]{CO,\ \text{cat. } PdCl_2(PPh_3)_2} n\text{-}C_8F_{17}CH{=}CHCO_2Et \quad 61\%$$

(28)[50]

The carbonylation of vinylic halides bearing alcohol functionality provides a convenient route to unsaturated lactones (Eqs. 29–33). A variation of this last reaction has recently been used in the synthesis of germacranolide[55,56] and cembranoid[57] lactones (Eqs. 34 and 35).

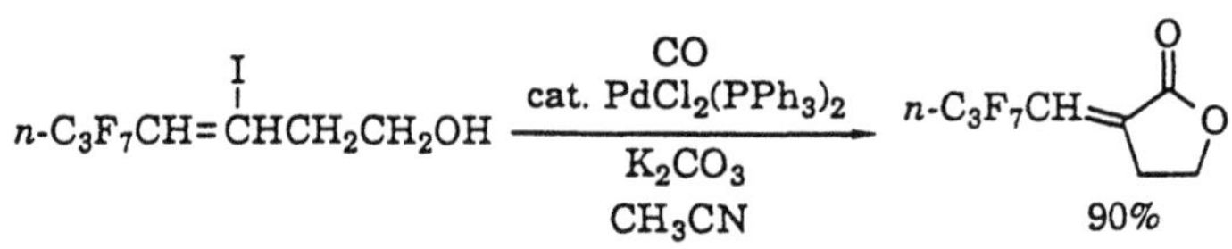

(29)[50]

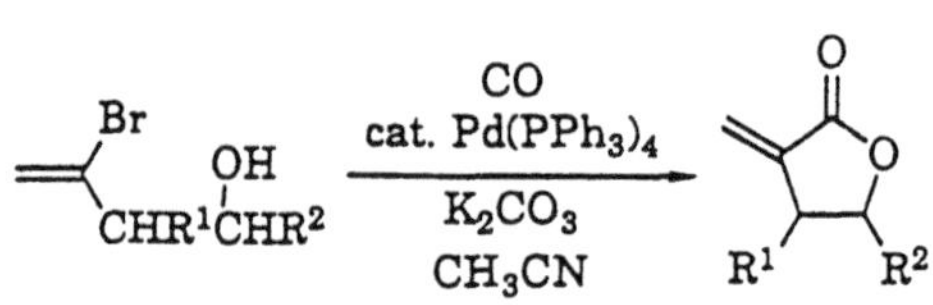

(30)[51]

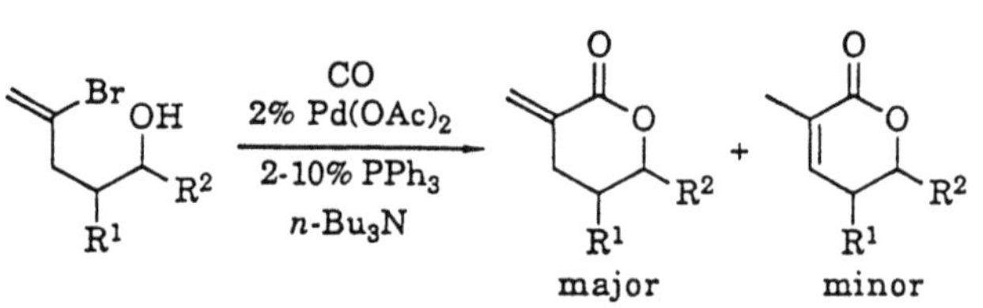

(31)[52]

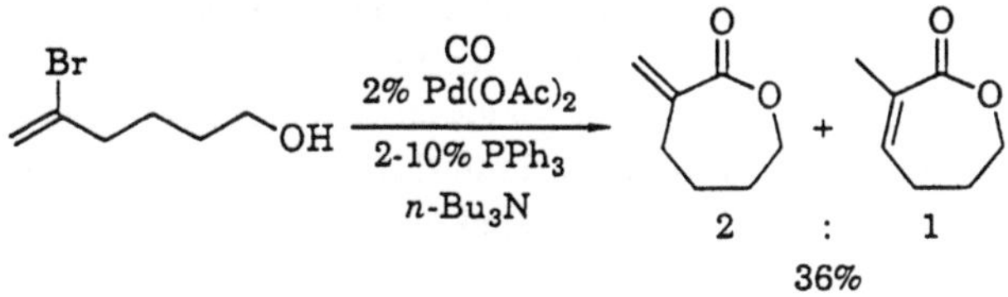

(32)[52]

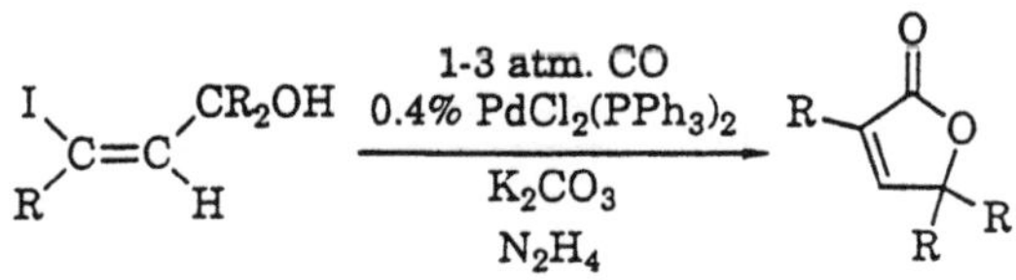

(33)[53,54]

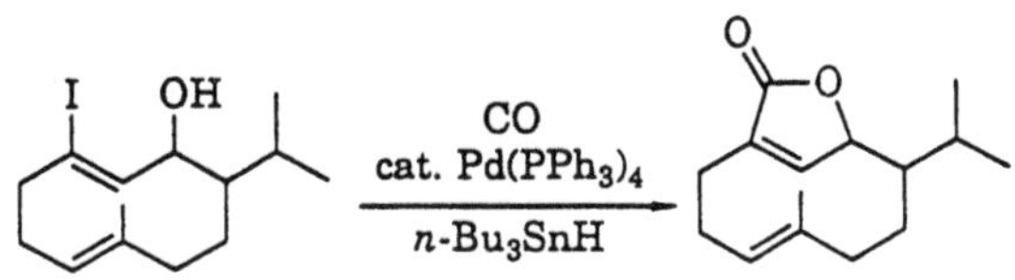

(34)

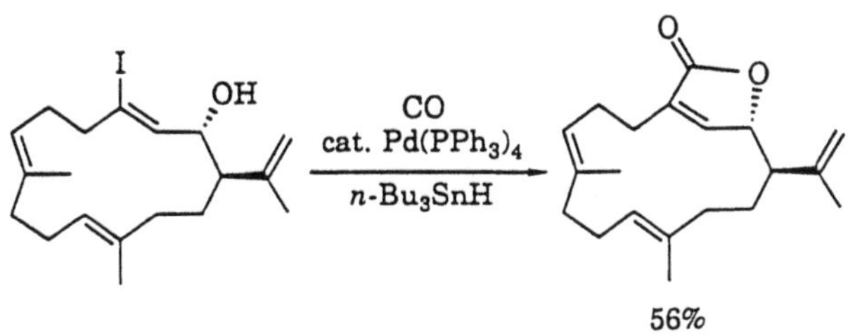

(35)

Enols can also be trapped to form lactones (Eqs. 36 and 37).[36]

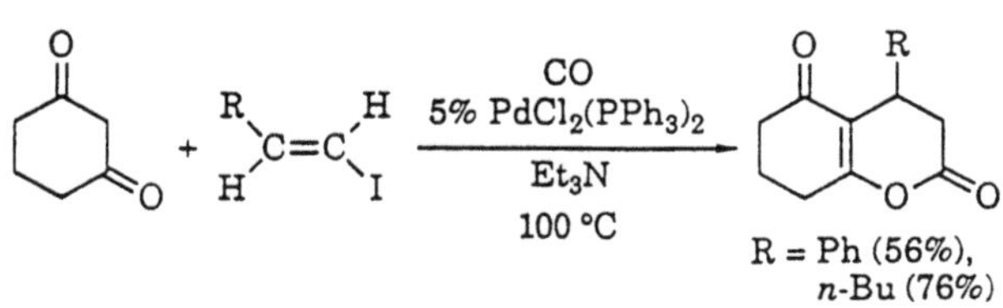

(36)

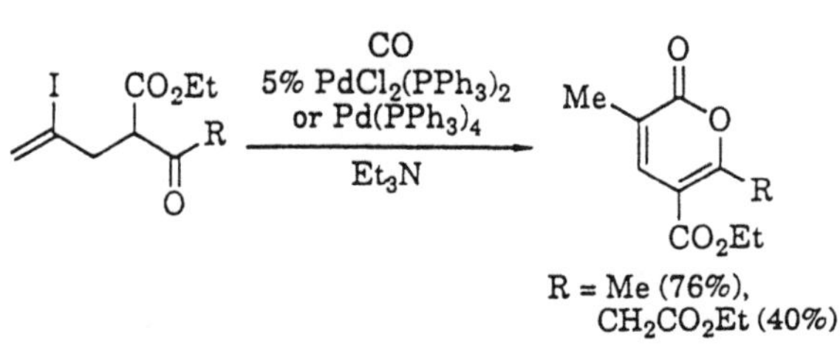

(37)

Carbonylation in the presence of 1° or 2° amines affords amides in high yields with no apparent loss of stereochemistry (Eq. 38).[58] Tertiary amines undergo C-N bond cleavage to afford amides, but the yields are low.[59] A number of α-alkylidene β-lactams have been prepared in this fashion (Eq. 39).[60–63] Five-, six- and seven-membered ring lactams can also be prepared via carbonylation, but isomerization of the double bond is a problem (Eq. 40).[52]

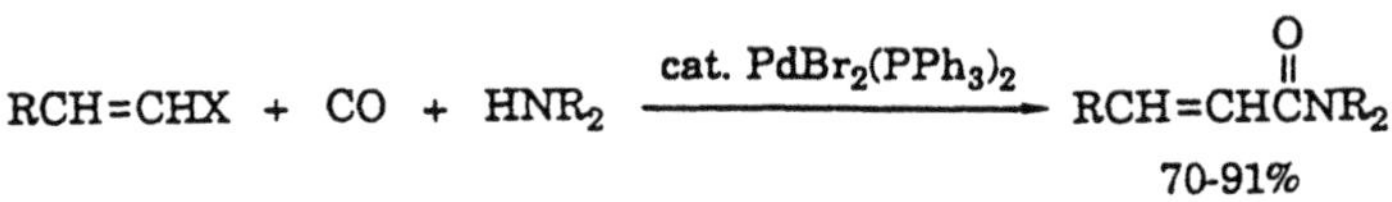

(38)

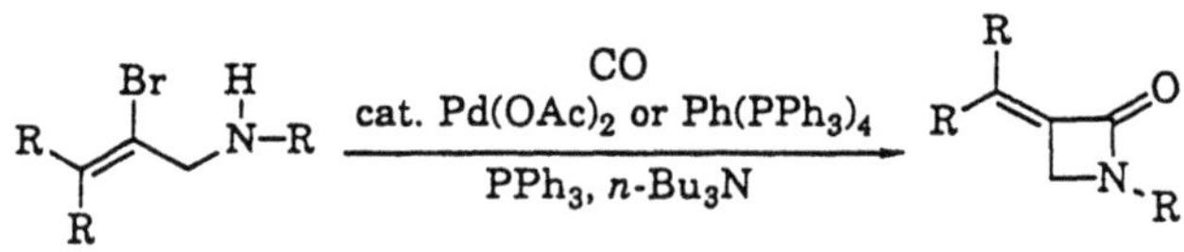

(39)

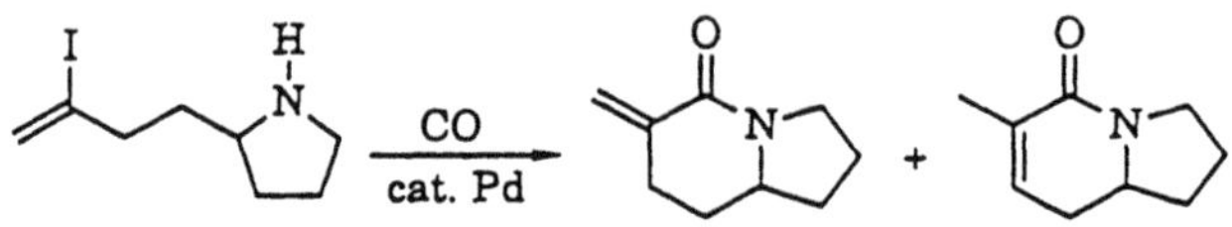

(40)

Intramolecular cyclization onto amides is also possible (Eq. 41).[64]

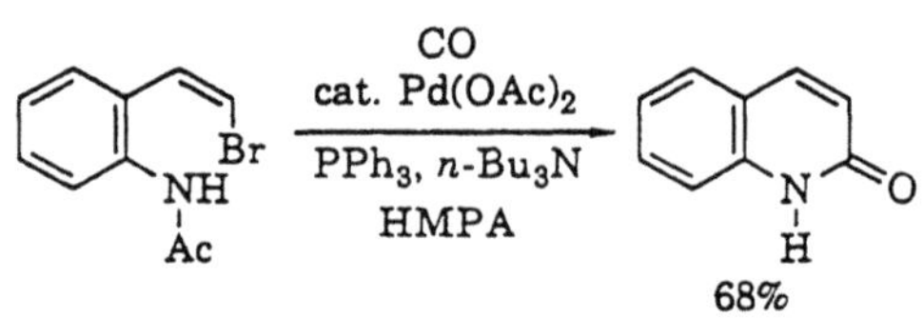

(41)

Novel approaches to imides have also been reported (Eqs. 42 and 43).

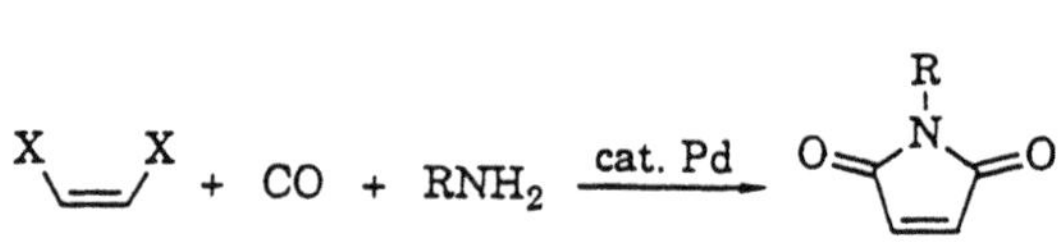

(42)[65]

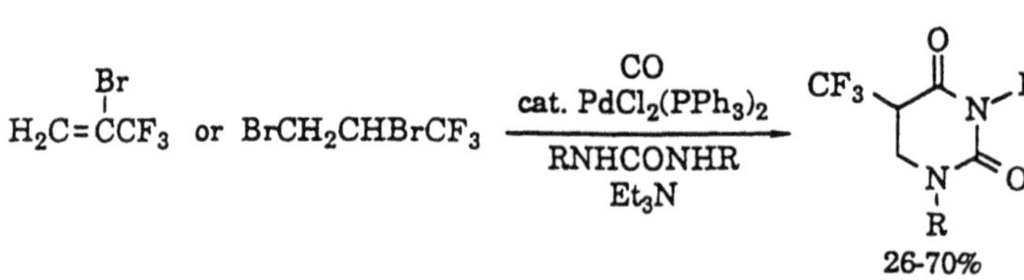

(43)[66]

While the high pressure carbonylation of vinylic halides in an alcohol solvent affords primarily the simple ester,[67] α-keto amides can be prepared in good yield when amines and appropriately substituted vinylic halides are employed (Eq. 44).[68–70] Simple amides are the major by-product.

$$\mathrm{RCH{=}CHX + 2\,CO + HNR_2 \xrightarrow{\text{cat. Pd}} RCH{=}CH\overset{O}{\overset{\|}{C}}CONR_2} \tag{44}$$

The carbonylation of a vinylic halide in the presence of an epoxide results in halohydrin ester formation (Eq. 45).[71]

$$\mathrm{PhCH{=}CHBr + CO + H_2C\overset{O}{-}CHCH_3 \xrightarrow[70\%]{\text{cat. Pd}}} \tag{45}$$

$$\mathrm{\underset{64}{PhCH{=}CHCO_2CHMeCH_2Br} + \underset{36}{PhCH{=}CHCO_2CH_2CHMeBr}} \quad (64 : 36)$$

Electroreductive carboxylation of styryl bromide using carbon dioxide affords low yields of a malonic acid (Eq. 46).[72]

$$\mathrm{PhCH{=}CHBr \xrightarrow[\text{electrolysis}]{CO_2,\ \text{cat. Pd(II)}} \underset{36\%}{PhCH_2CH(CO_2H)_2}} \tag{46}$$

The mono- and multi-insertion of isocyanides into vinylic palladium species has been studied, but has not been put to use in organic synthesis.[73]

Coupling with Acyclic Alkenes

The palladium-catalyzed cross-coupling of vinylic halides and simple alkenes has become a very useful route to 1,3-dienes (Eq. 47).[74–80] The reaction is most commonly run using catalytic amounts of $Pd(OAc)_2$ in the presence of a tertiary amine and a triarylphosphine at temperatures of 100–150 °C. Because of the elevated temperatures, Diels-Alder adducts of the anticipated diene are sometimes observed as side products.[76] The stereochemistry of the starting vinylic halide is generally retained.[76]

$$\mathrm{RCH{=}CHX + H_2C{=}CHR' \xrightarrow[\text{base}]{\text{cat. Pd}} RCH{=}CHCH{=}CHR'} \tag{47}$$

This reaction, first reported by Heck, apparently proceeds as follows (Scheme 1).

The Heck reaction can also be run using palladium-graphite as the catalyst,[81] but a recent modification introduced by Jeffery[82,83] involving the use of catalytic

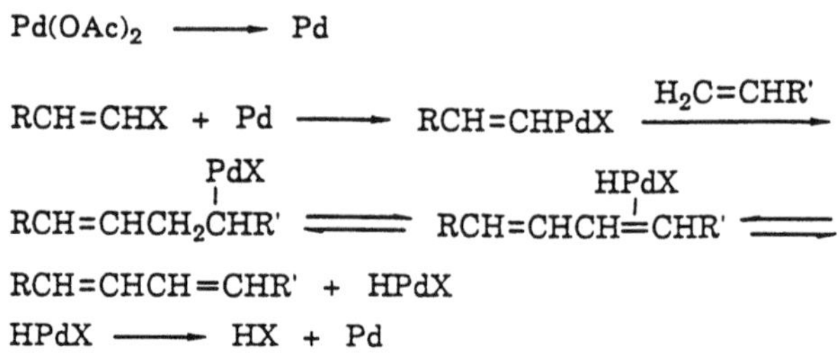

Scheme 1.

amounts of $Pd(OAc)_2$, potassium carbonate, and n-Bu_4NCl in DMF as the solvent appears to be the procedure of choice, because the reaction can often be run at room temperature and affords very good yields, although some loss of stereospecificity is observed.

The Heck reaction has been carried out using a wide variety of vinylic halides including the following (Eqs. 48–53):

$$RCF{=}CBrR' + H_2C{=}CHPh \longrightarrow \underset{71\text{-}95\%}{RCF{=}CRCH{=}CHPh} \quad (48)^{84}$$

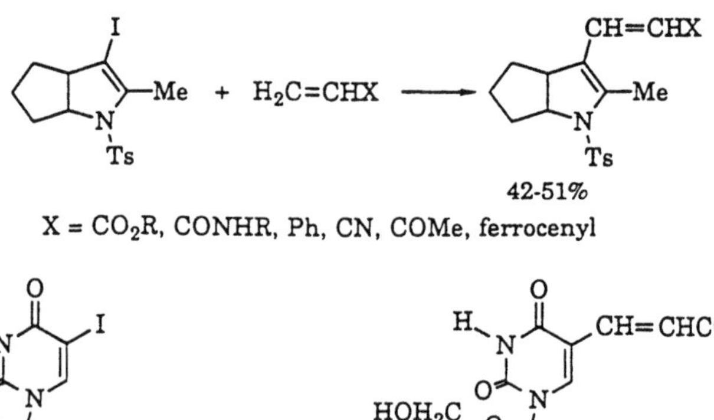

42-51%

X = CO_2R, CONHR, Ph, CN, COMe, ferrocenyl (49)[48]

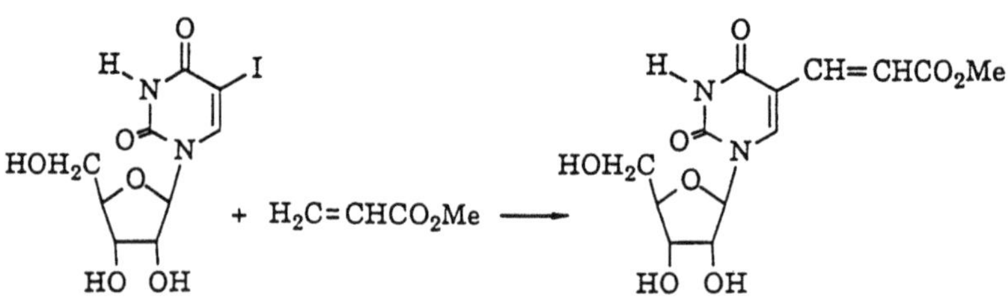

53% (50)[85,86]

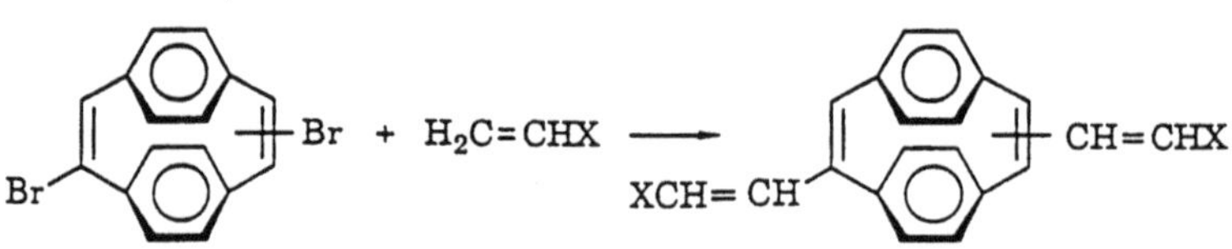

X = H, $SiMe_3$, CHO, CO_2Me, Ar 19-80% (51)[87]

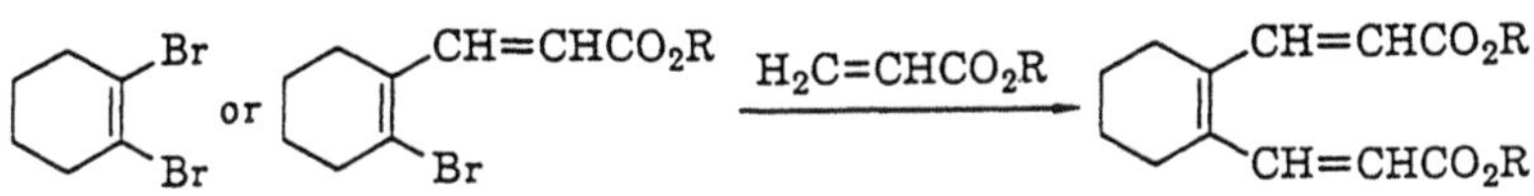

(52)[88]

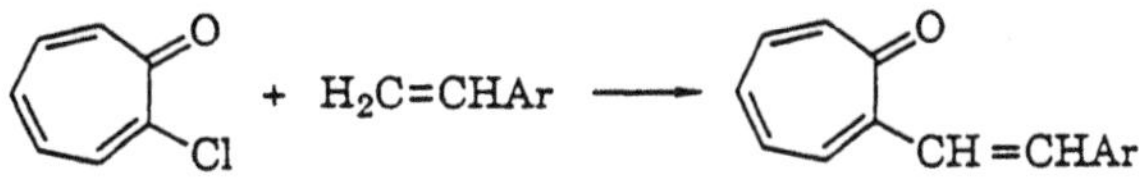

(53)[49]

As indicated in the preceding examples, the reaction has most successfully been run using terminal alkenes bearing hydrogen, aryl, carbonyl or nitrile substituents. Detailed studies have been carried out on enones,[89] carboxylic acids and derivatives,[90] and acetals.[91] Alkenes substituted by perfluoroalkyl,[92] trialkylsilyl[93] and alkoxy groups[94] may also be employed. Reactions with vinylic silanes are best run in the presence of silver salts which enhance the rate of reaction and suppress desilylation.[93]

Although most reactions proceed with good regioselectivity for terminal vinylic hydrogen substitution, vinylic ethers afford predominantly the product of internal substitution (Eq. 54).[94]

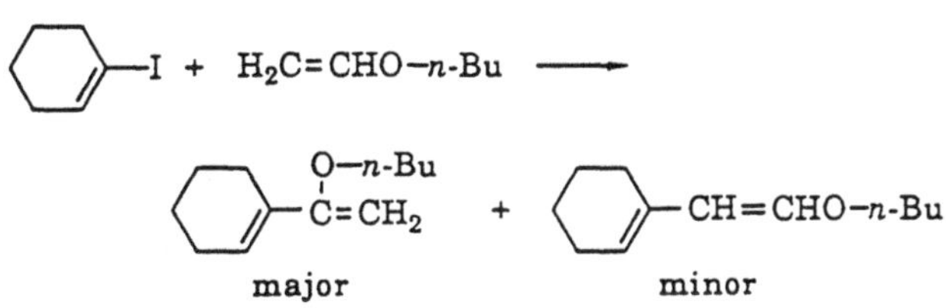

(54)

Polyenes also undergo successful substitution, although the yields are sometimes low (Eqs. 55 and 56).[80,95,96] Further examples will be discussed later (Eqs. 103–107).

$$\mathrm{PhCH{=}CHBr} + \mathrm{H_2C{=}CHCH{=}CHCO_2H} \longrightarrow \underset{57\%}{\mathrm{Ph(CH{=}CH)_3CO_2H}} \qquad (55)^{95}$$

$$2\ \mathrm{PhCH{=}CHBr} + \mathrm{H_2C{=}CHCH{=}CHCH{=}CH_2} \longrightarrow \underset{14\%}{\mathrm{Ph(CH{=}CH)_5Ph}} \qquad (56)^{96}$$

This chemistry has proven useful in the synthesis of the simple natural products pellitorine (**2**)[97] and 15-dehydroprostaglandin B_1 (**3**).[98]

There are a number of examples of the intramolecular cyclization of vinylic halides on to acyclic alkenes (cyclic alkenes will be covered later). The size of the ring depends on the precise palladium catalyst used, or the presence or absence of

E,E- n-$C_5H_{11}CH{=}CHCH{=}CHC(O)NH$-$i$-Bu

2

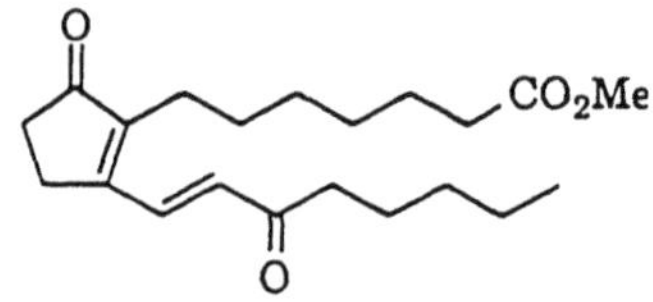

3

a base (Eq. 57).[99] A variety of carbocycles and heterocycles have been prepared in this fashion (Eqs. 58–62).

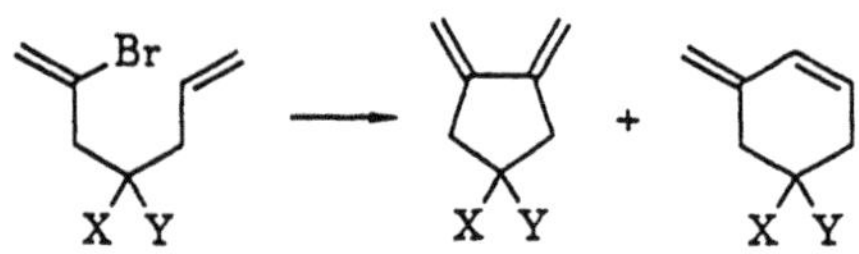

(57)

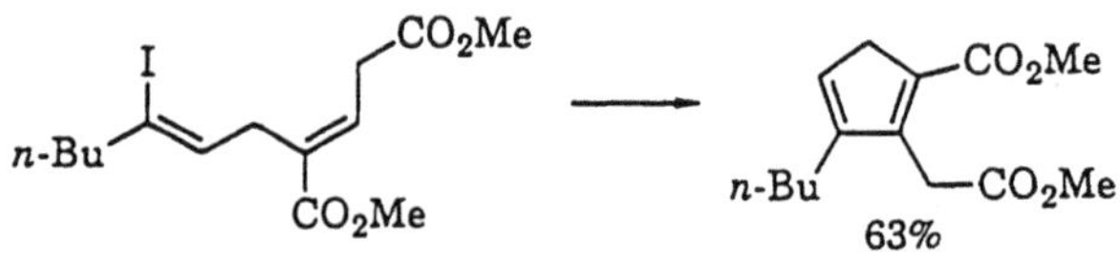

(58)[100]

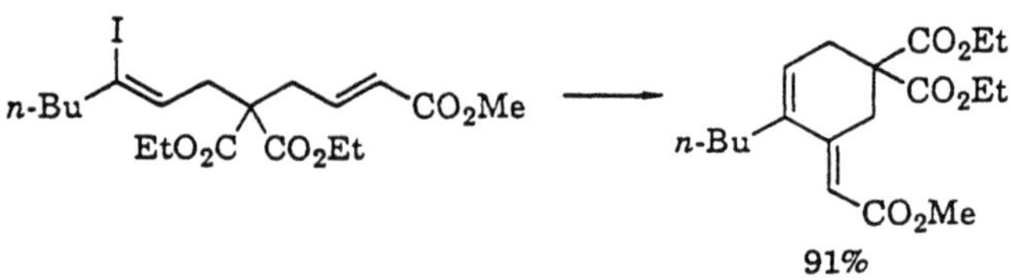

(59)[100]

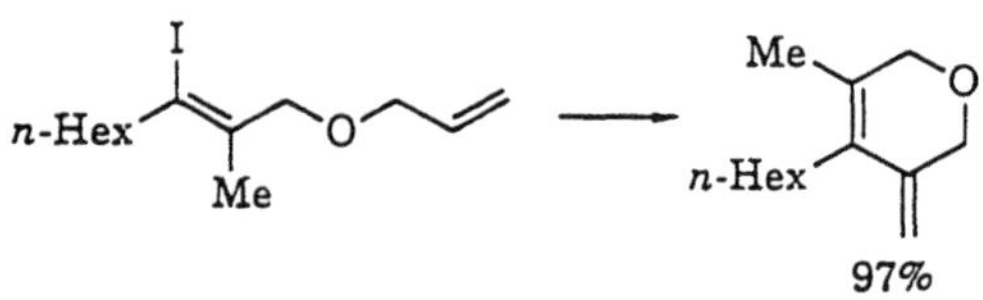

(60)[101]

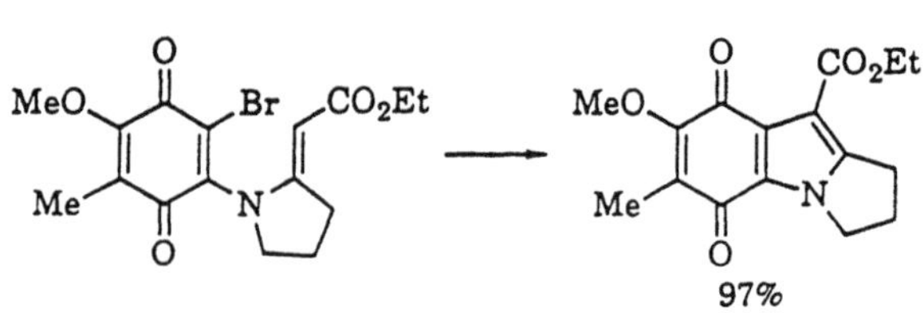

(61)[102]

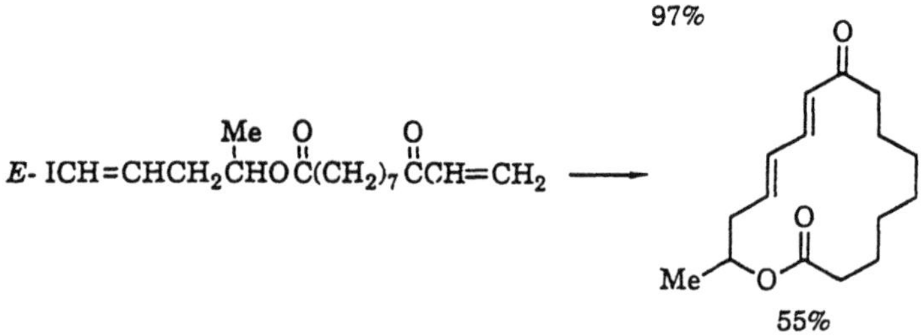

(62)[89]

When palladium hydride elimination is blocked, alternative cyclization processes have been observed (Eqs. 63 and 64).[103]

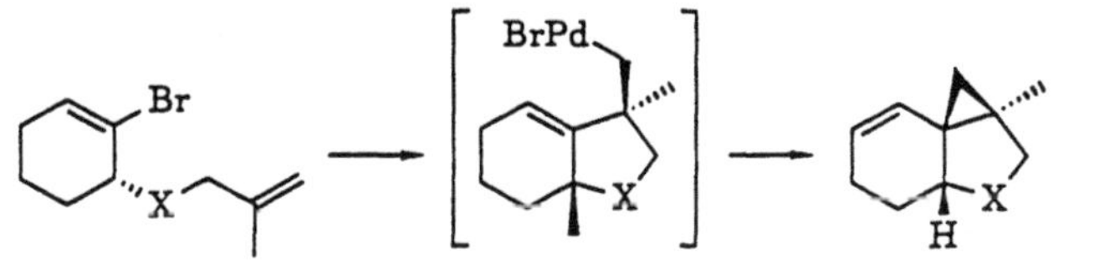

(63)

X = NSO_2Ph (78%), $C(CO_2Me)_2$ (67%)

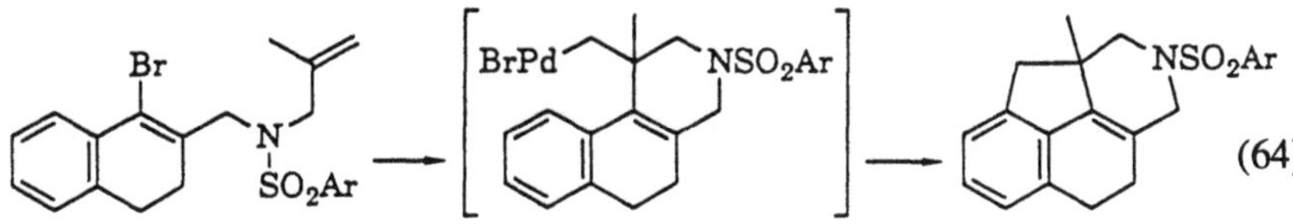

(64)

50%

The presence of certain functional groups in the alkene can profoundly change the nature of the products. For example, vinylic halides cross-couple with allylic alcohols to produce intermediates **4** which can eliminate to dienols (Scheme 2, path a) or unsaturated carbonyl compounds (path b).[104] The dienol product can be prepared with high regio- and stereoselectivity by simply running the reaction in the presence of AgOAc or Ag_2CO_3 (Eq. 65).[105] By using Ag_2CO_3 and *n*-Bu_4NHSO_4, one can obtain good yields of enals instead (Eq. 66).[106] The palladium-catalyzed coupling of aryl and vinylic halides with 2-butene-1,4-diol affords 4-substituted 2-hydroxytetrahydrofurans (Eq. 67).[107]

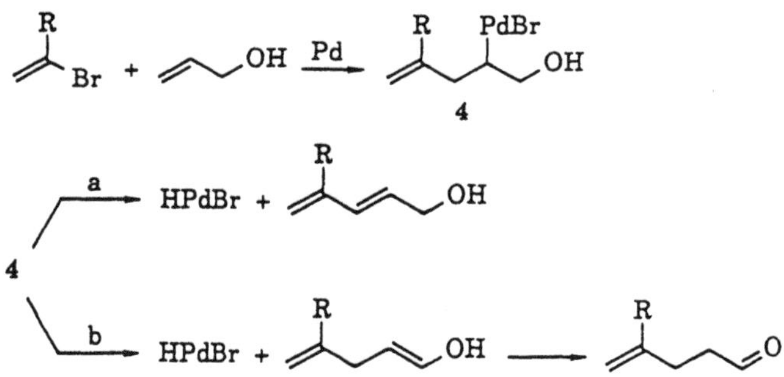

Scheme 2.

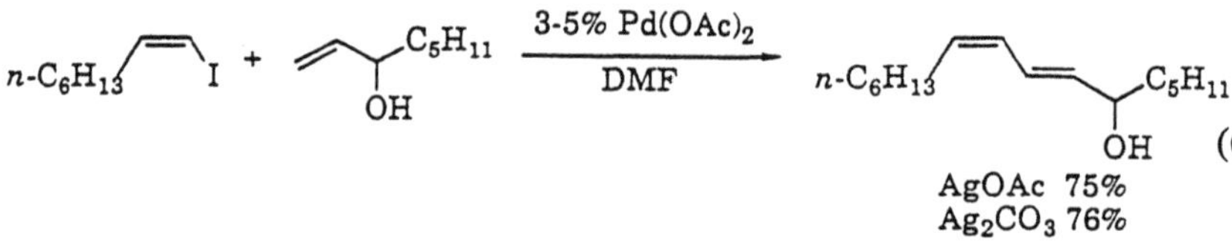

(65)

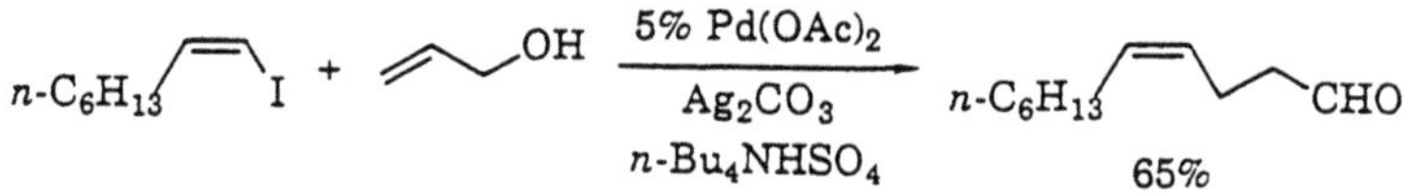

(66)

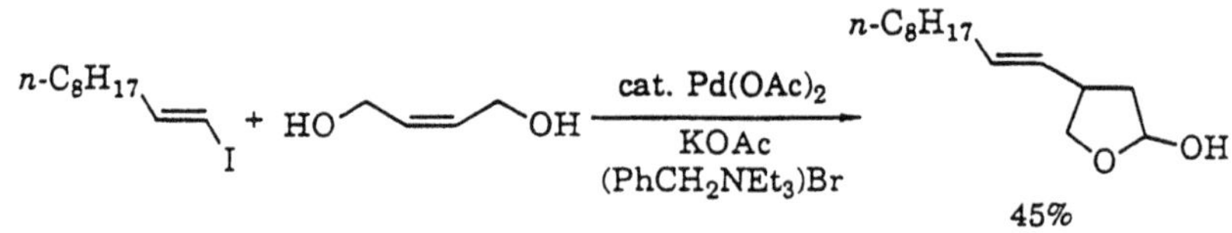

(67)

Vinylic halides and allylic esters have been intramolecularly cross-coupled to generate the corresponding 1,4-dienes by using n-$Bu_3SnAlEt_2$ in the presence of a palladium(O) catalyst (Eq. 68).[108]

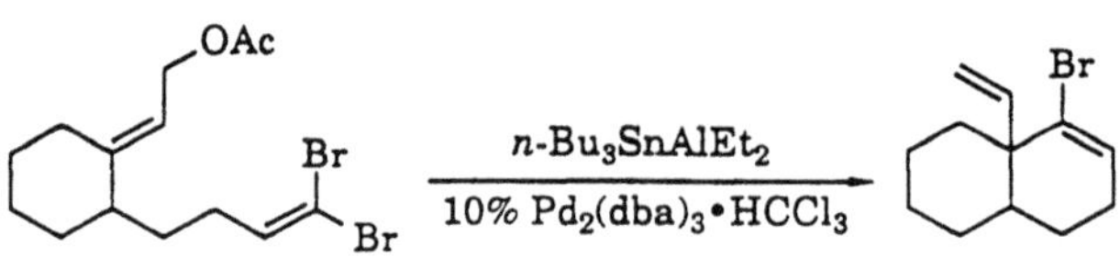

(68)

Vinylic halides will also react with vinylic esters in the presence of n-Bu_3SnOMe and a palladium catalyst to produce 3-alkenones, presumably via tin enolate formation (Eq. 69).[109]

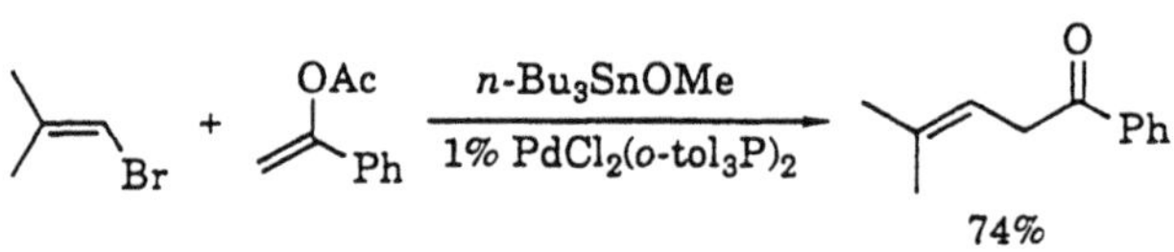

(69)

The homoallylic palladium compound generated by coupling a vinylic palladium compound and an alkene can also rearrange by β-hydride elimination and readdition to produce a π-allylpalladium complex (Eq. 70). Products of subsequent displacement by various nucleophiles have been observed. For example, the reaction of vinylic halides, secondary amines and simple alkenes (Eq. 71),[110] allylic alcohols[104] or acrolein acetals[91] afford allylic amine products. Bromo dienes have been cyclized in a similar fashion to produce various five-, six- and seven-membered carbocyclic[111] or heterocyclic[112] allylic amines (Eq. 72).

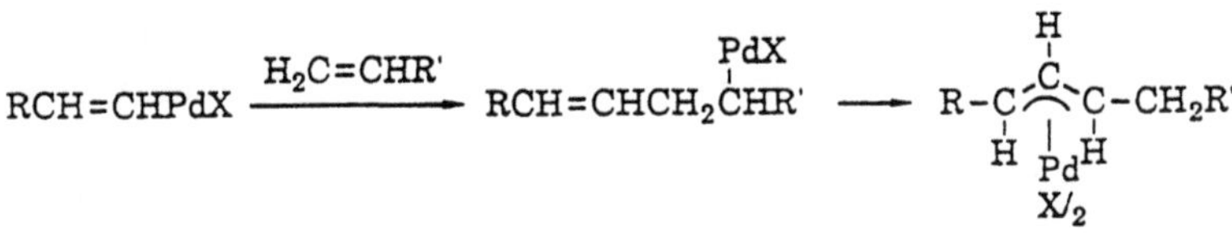

(70)

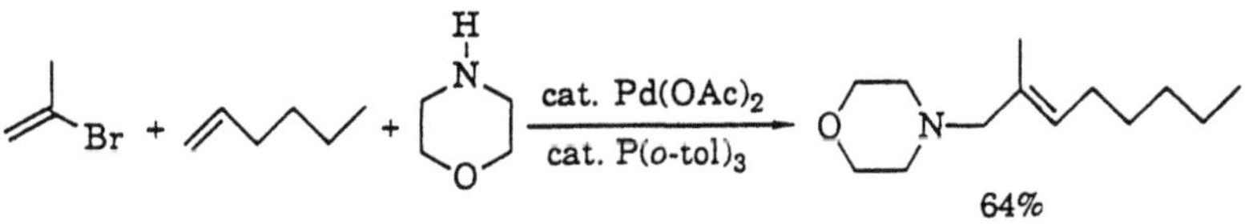

(71)

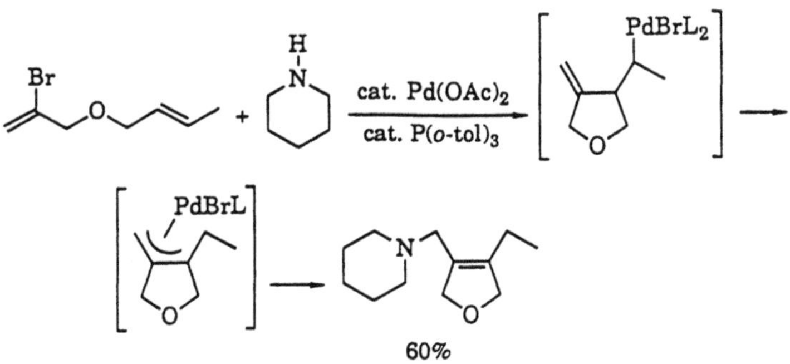

(72)

We have developed an intramolecular variation of this reaction to produce unsaturated lactones (Eq. 73).[113]

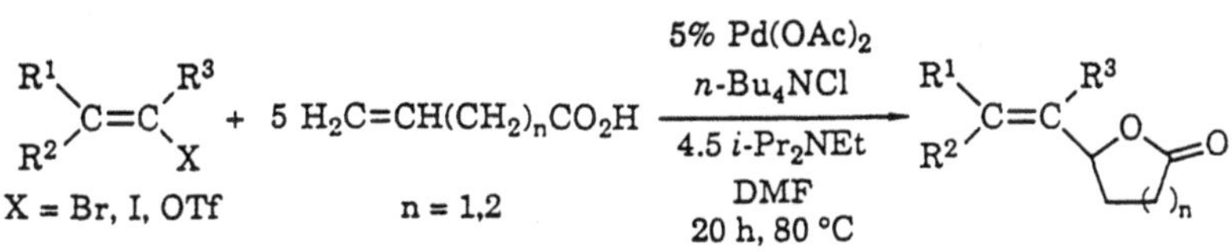

(73)

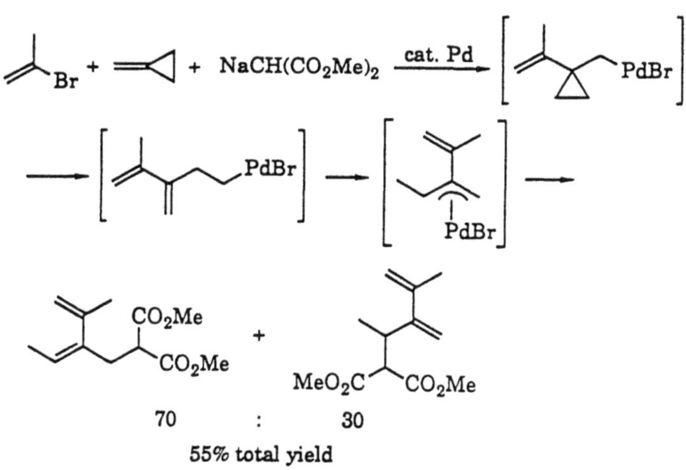

(74)

The cross-coupling of vinylic halides, alkylidene cyclopropanes and stabilized carbanions also proceeds via ring-opening through π-allylpalladium intermediates (Eq. 74).[114,115]

In a mechanistically quite different process, the reaction of a vinylic halide and unsaturated malonate anion produces a cyclopentane derivative (Eq. 75).[116]

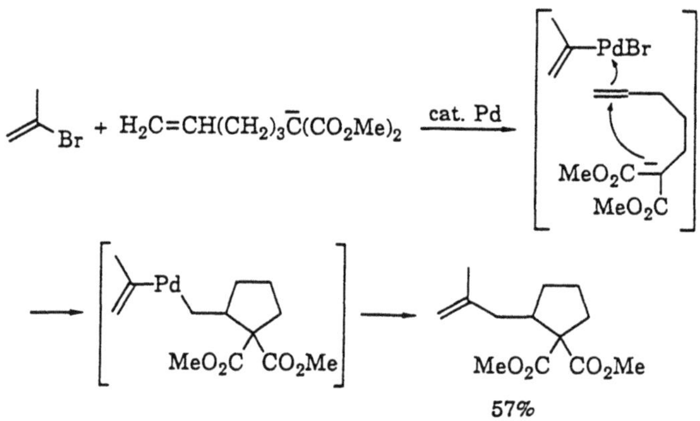

(75)

Coupling with Cyclic Alkenes

The palladium-catalyzed vinylation of cyclic alkenes requires that the subsequent *syn*-palladium hydride elimination occur away from the vinylic group, resulting in 1,4-diene formation (Eq. 76).[90] We recently reported two mild, general procedures to effect such reactions (Eqs. 77 and 78).[117] These reactions are sensitive to the structure of the cyclic alkene and the latter procedure is more effective for less reactive alkenes or where double bond isomerization is a problem.

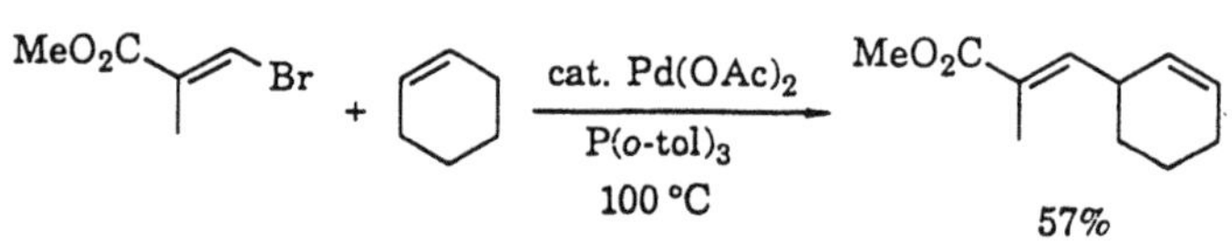

(76)

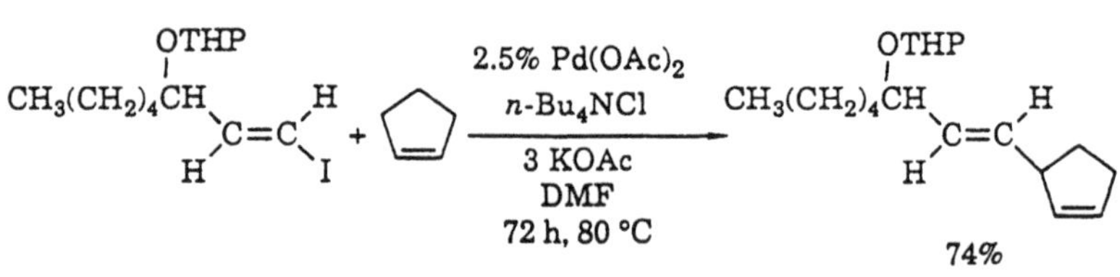

(77)

This reaction has recently been applied to 2-cyclopenten-1-ol and the corresponding *tert*-butyldimethylsilyl ether (Eq. 79),[118,119] as well as *cis*-4-cyclopentene-1,3-diol, where it affords a novel new route to prostaglandins (Eqs. 80 and 81). While

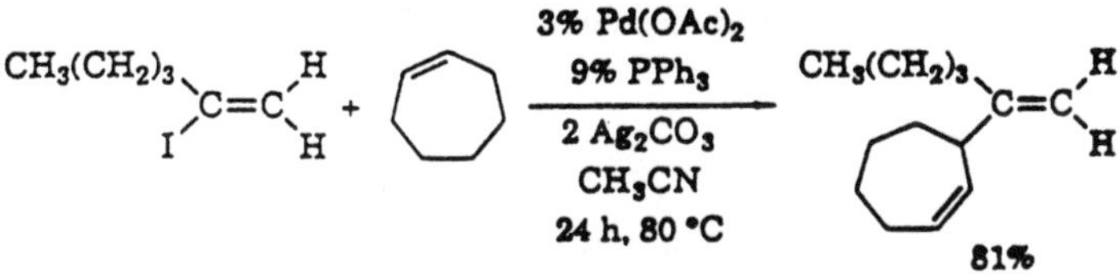

(78)

use of the *trans*-vinylic iodide (Eq. 80) exhibited no diastereoselectivity, the corresponding reaction of the *cis*-vinylic iodide was >99% diastereoselective (Eq. 81).

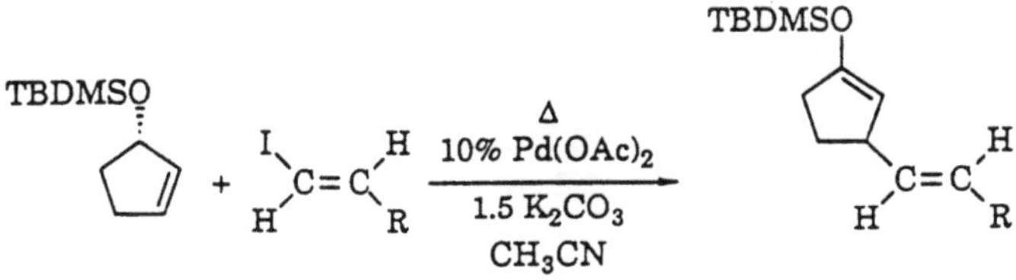

(79)

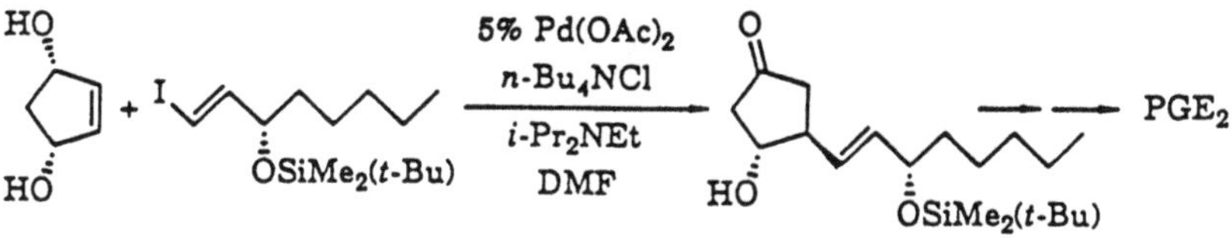

(80)[120]

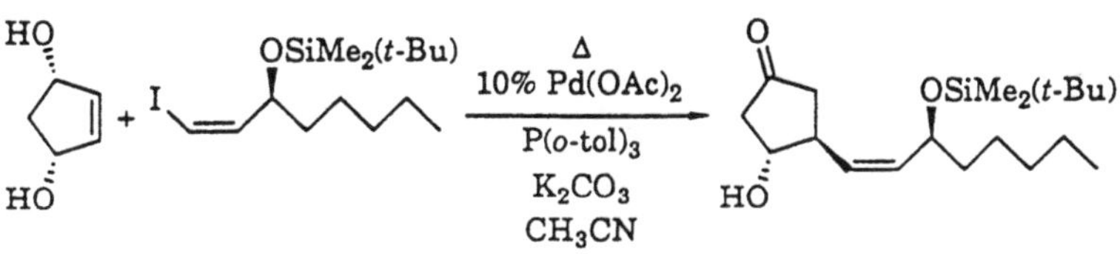

(81)[119]

There are now numerous examples of the intramolecular vinylation of cyclic alkenes. Fused, bridged and spirocyclic hydrocarbons[121–123] and heterocyclic[101,124,125] systems can be efficiently constructed in this fashion (Eqs. 82–85). Recently, an asymmetric version of this process has been reported (Eq.

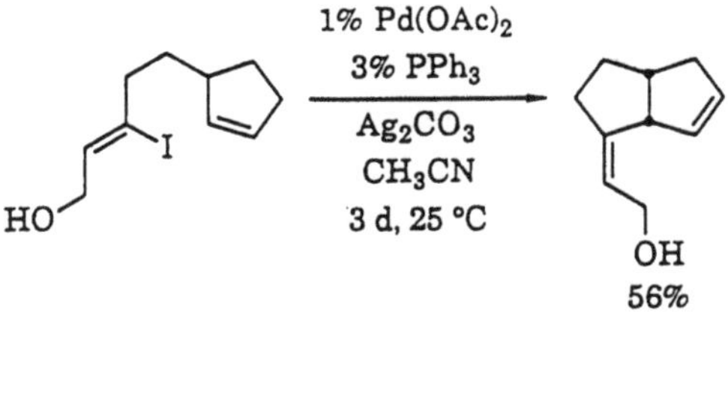

(82)[121]

86).[126,127] The reaction will even proceed when double bond transposition is impossible (Eq. 87).[123]

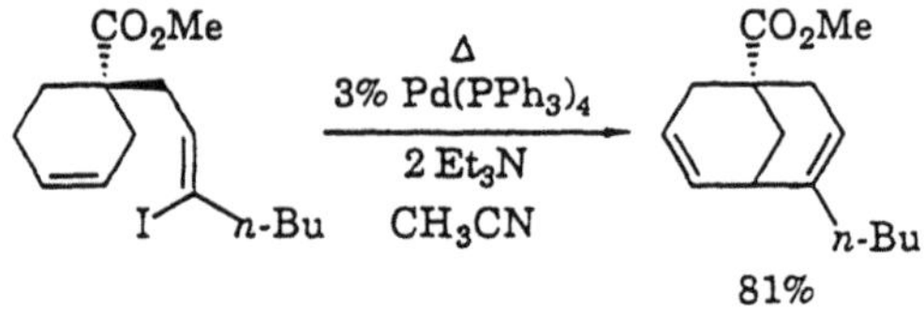

(83)[122]

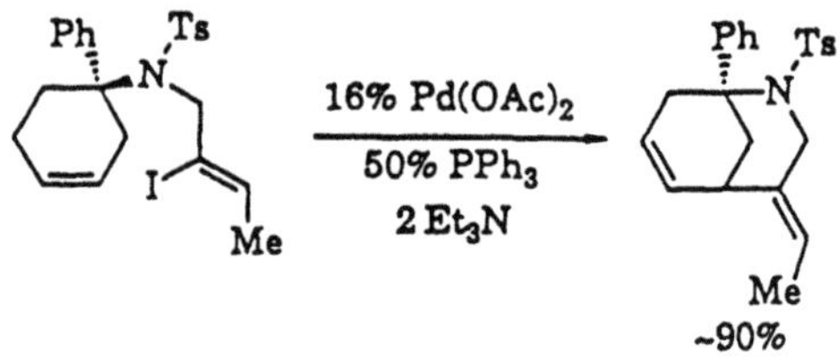

(84)[124]

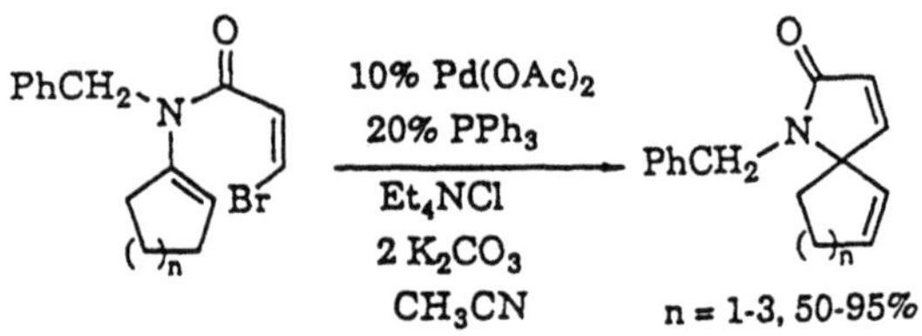

(85)[125]

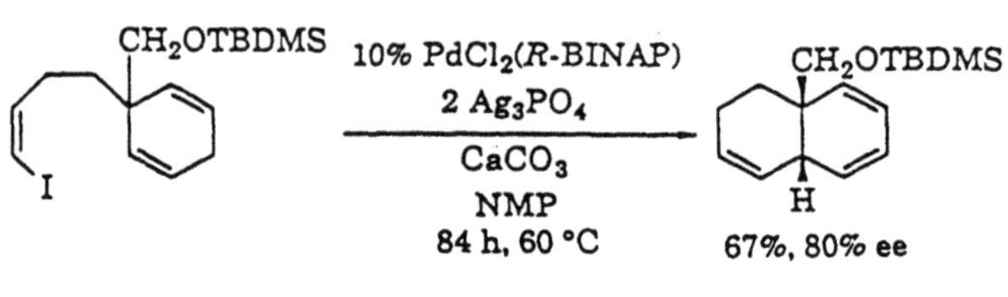

(86)

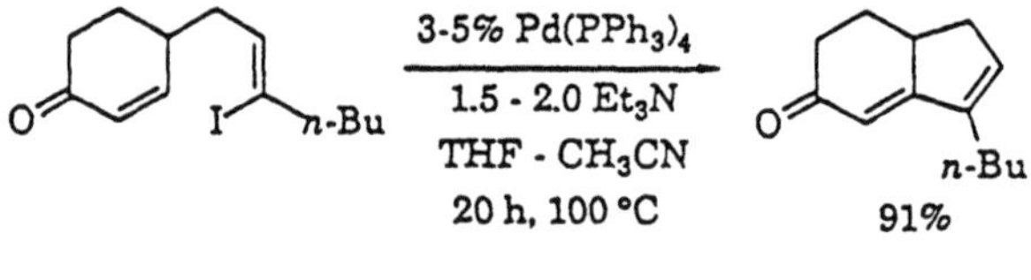

(87)

Coupling with Bicyclic Alkenes

With certain bicyclic alkenes, such as norbornene, *syn*-β-hydride elimination is impossible and other processes are observed. For example, cyclization to a cyclopropane and subsequent β-hydride elimination (Eq. 88),[128] or insertion of a second alkene, followed by cyclization and β-hydride elimination (Eq. 89)[129,130] have been reported. On the other hand, use of a dienyl halide reportedly affords a different cyclopropane, the mechanism of formation of which is unclear (Eq. 90).[128,129]

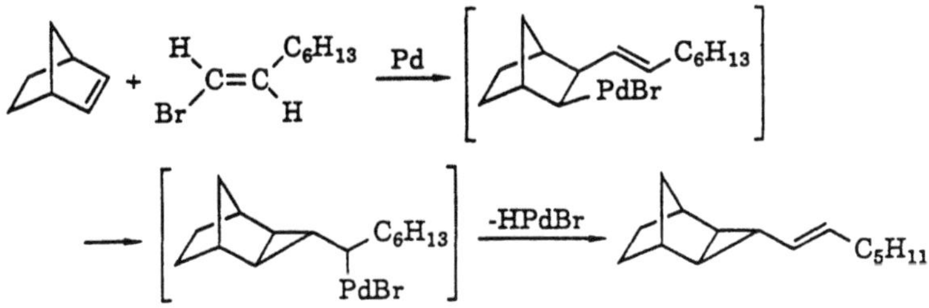

(88)

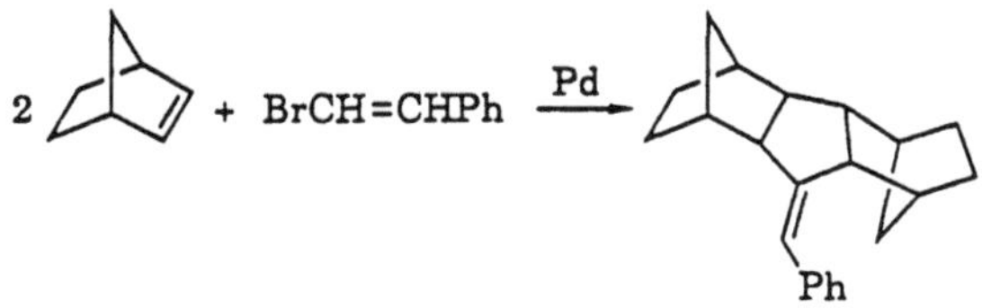

(89)

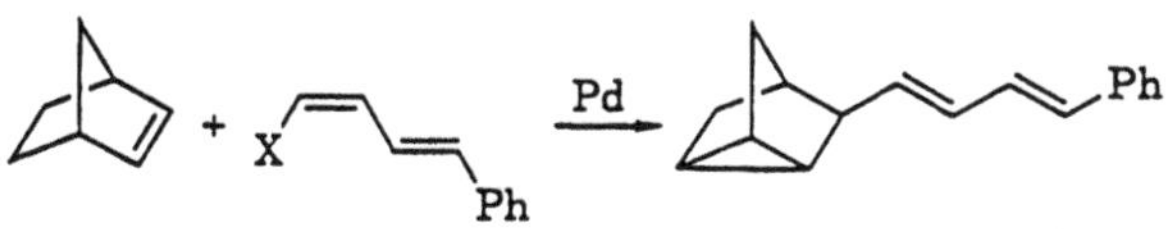

(90)

In the presence of ammonium formate, however, the latter reaction produces cyclization products (Eq. 91)[128,129] With styryl bromide, ammonium formate provides primarily the benzyl-substituted cyclopropane,[131] but use of a trialkylammonium formate affords predominantly the simple hydrovinylation product (Eq. 92).[132] The use of potassium formate also provides excellent yields of hydrovinylation products (Eq. 93).[133]

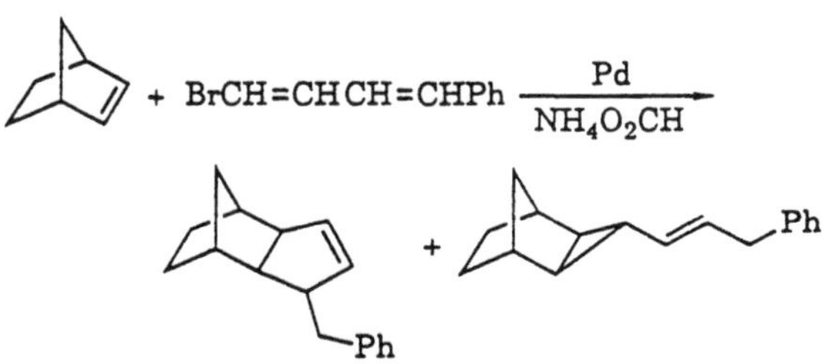

(91)

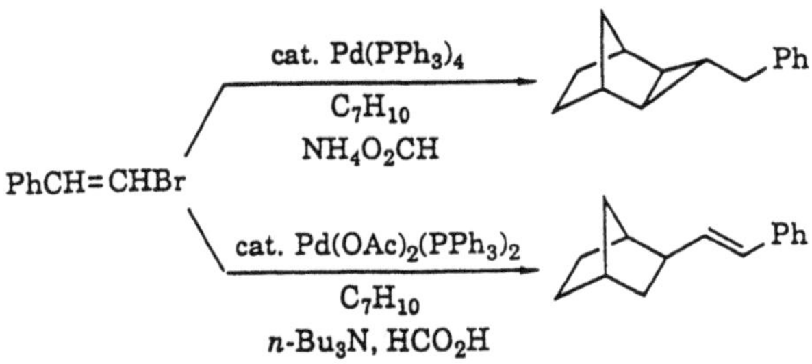

(92)

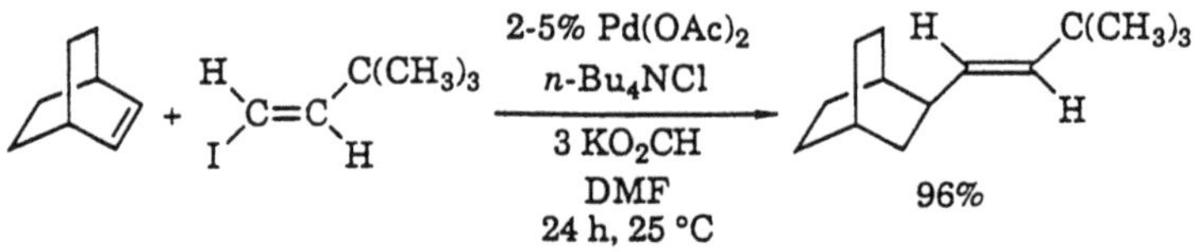

(93)

In the presence of an amine, norbornene and styryl bromide react to generate a cyclopropyl carbinyl amine (Eq. 94).[134]

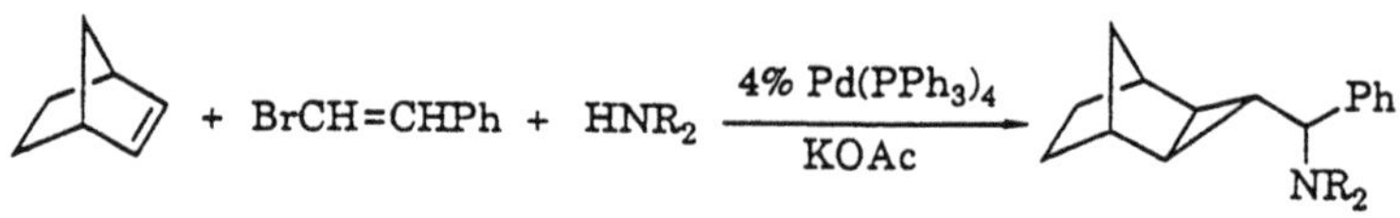

(94)

2,3-Disubstituted norbornenes can be prepared by using vinylic halides and organotin compounds (Eq. 95)[135,136] or terminal alkynes (Eq. 96).[137,138]

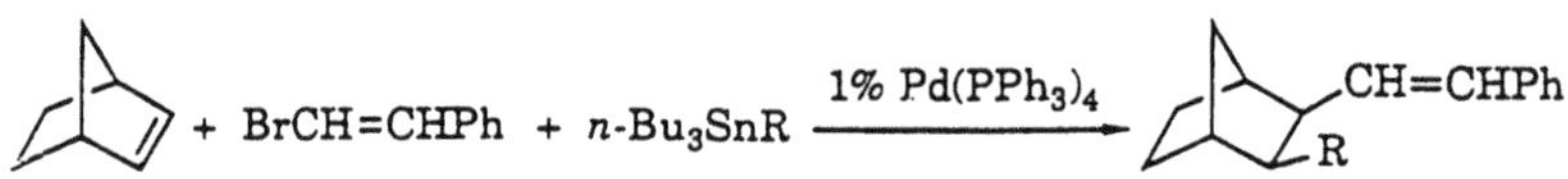

(95)

R = Ph (71%), $CH_2CH{=}CH_2$ (36%), $CH{=}CH_2$ (91%), C≡CPh (64%)

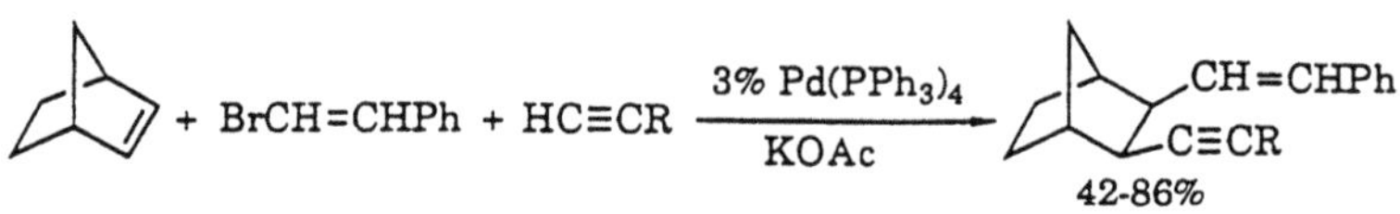

(96)

Carbonylation affords a unique route to the corresponding anhydrides (Eq. 97).[129,139]

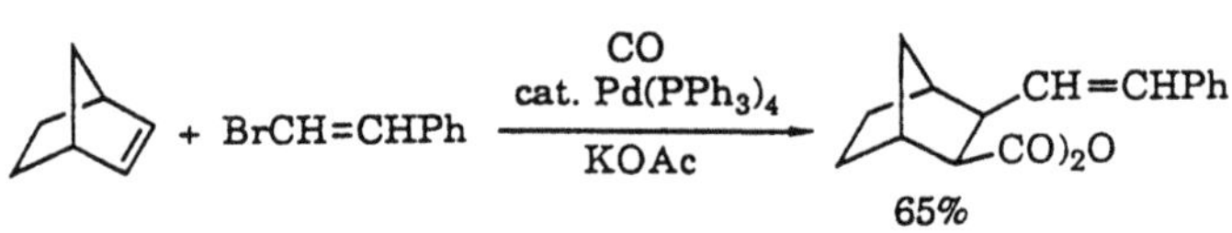

(97)

In the presence of KCN, nitriles are observed (Eq. 98).[140]

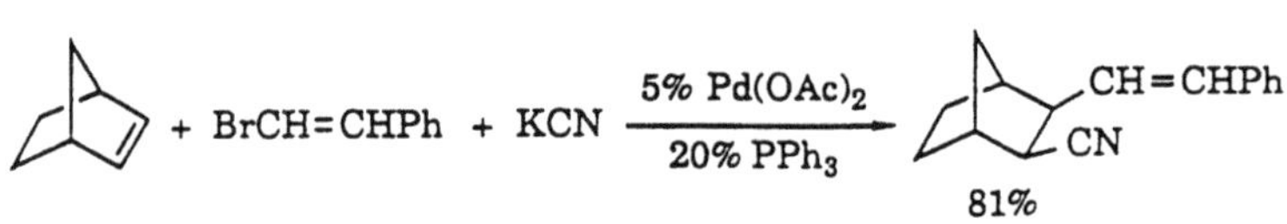

(98)

Reaction with Dienes

Vinylic palladium addition to allenes produces π-allylpalladium intermediates. Allenic alcohols undergo such additions, followed by β-hydride elimination and tautomerization to produce α,β-unsaturated carbonyl products (Eq. 99).[141]

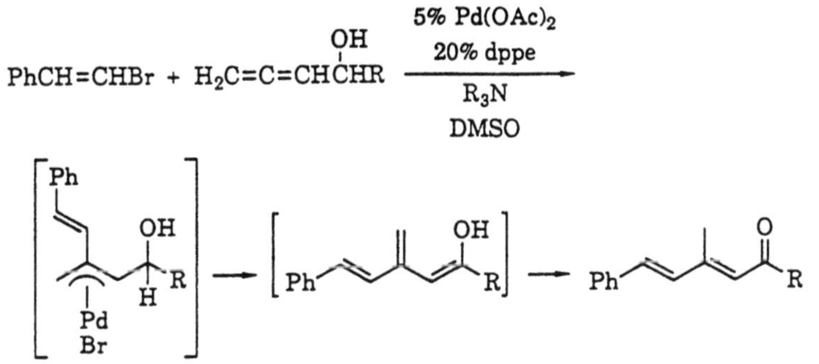

(99)

R = H (69%), Ph (94%), *i*-Bu (77%)

With simple allenes and secondary amines, allylic amines are formed (Eq. 100).[142]

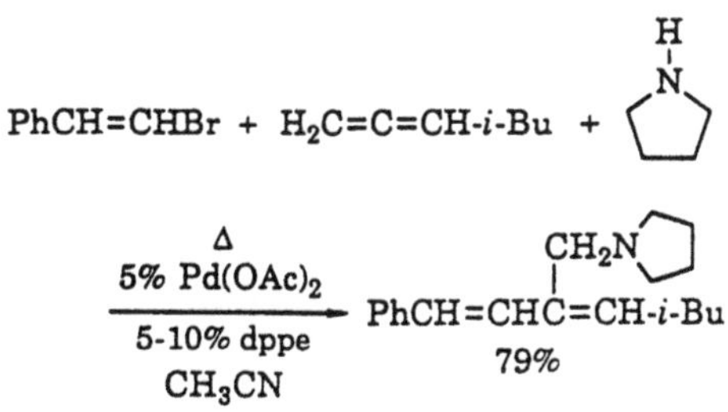

(100)

When stabilized carbanions are utilized as nucleophiles, a variety of functionalized dienes are generated often in high yield (Eq. 101).[143–146] Aryl halides, silyl-substituted[146] vinylic halides, and a variety of allenes and stabilized carbanions have been successfully employed in this reaction. Intramolecular versions of this reaction have also been reported (Eq. 102).[147,148] The ring system formed is dependent on the organic halide used. Vinyl bromide affords exclusively the cyclopropane, while the other organic halides listed yield 75–100% of the cyclopentene.

As briefly noted earlier (Eqs. 55 and 56),[80,95,96] vinylic halides can react with 1,3-dienes to produce trienes (Eq. 103).[149] Polyenes can also be generated by the mono- or divinylation of alkenes and polyenes, although the yields are usually low

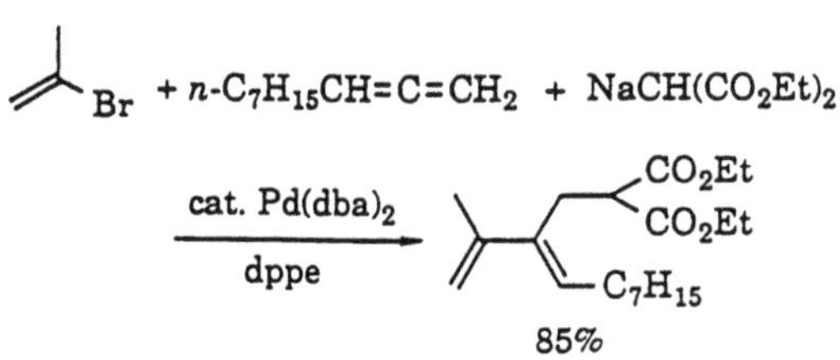

(101)

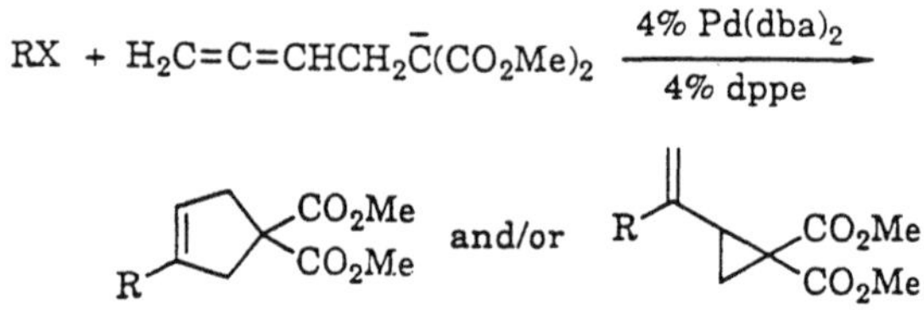

(102)

R = phenyl, vinyl, isopropenyl, cyclohexenyl

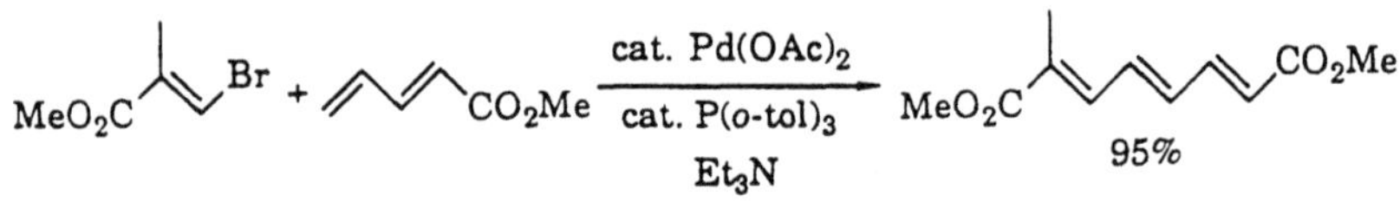

(103)

(Eqs. 104–107).[149] These reactions appear to require particularly reactive vinylic halides bearing electron-withdrawing groups.

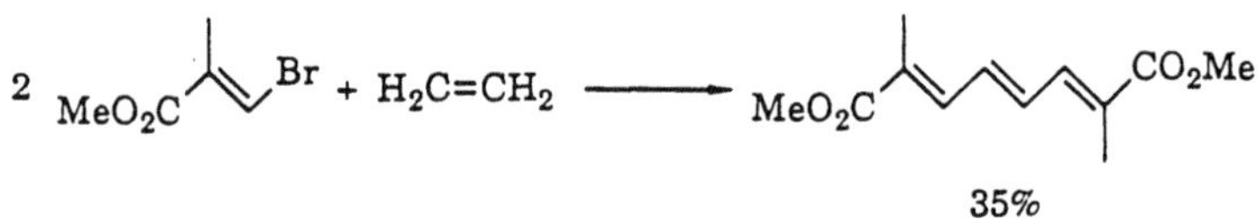

(104)

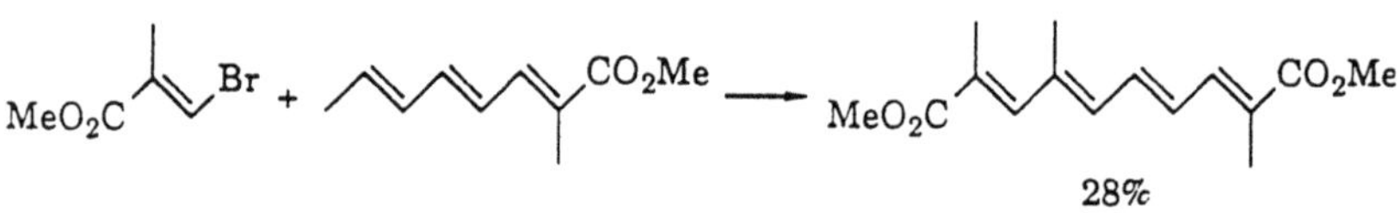

(105)

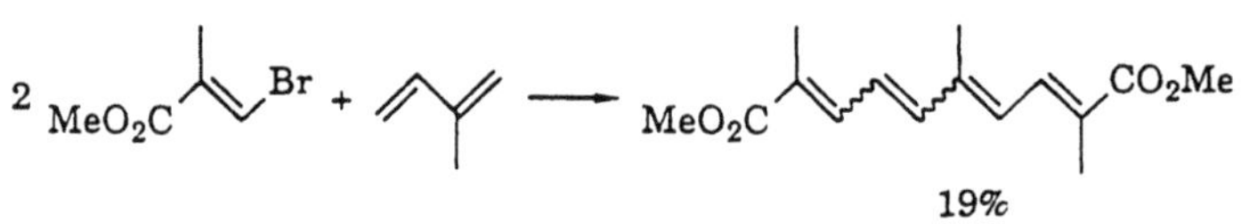

(106)

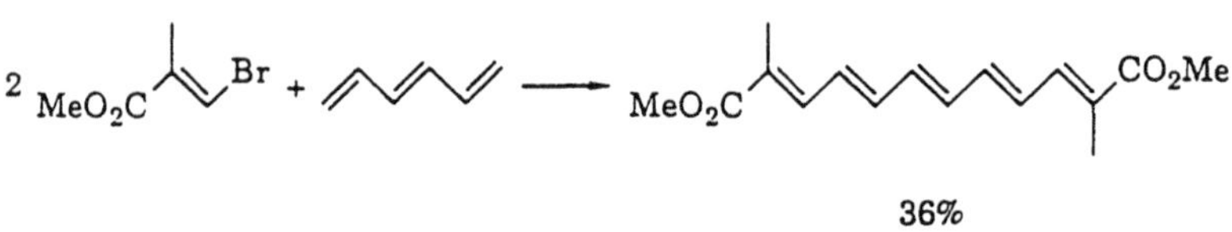

(107)

The vinylation of a diene presumably generates π-allylpalladium intermediates which one should be able to capture using appropriate nucleophiles. Indeed, intramolecular vinylation in the presence of a formate salt produces the anticipated reduction product (Eq. 108).[150] Capture by acetate anion has also been observed,

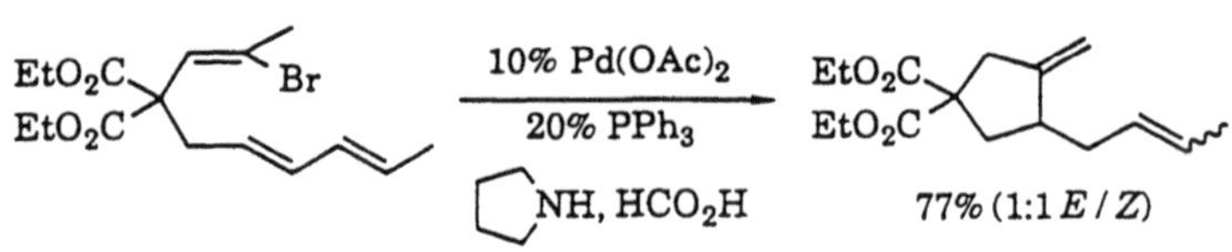

(108)

although vinylic triflates are more effective as starting materials for asymmetric versions of this process (see Eq. 317).

Amines have also been employed as nucleophiles in such reactions (Eq. 109).[151] 1,4-Dienes undergo vinylation and subsequent palladium migration via β-hydride elimination-readdition to afford dienyl amines (Eq. 110).[152] However, mixtures of isomeric products are generally produced by this process.

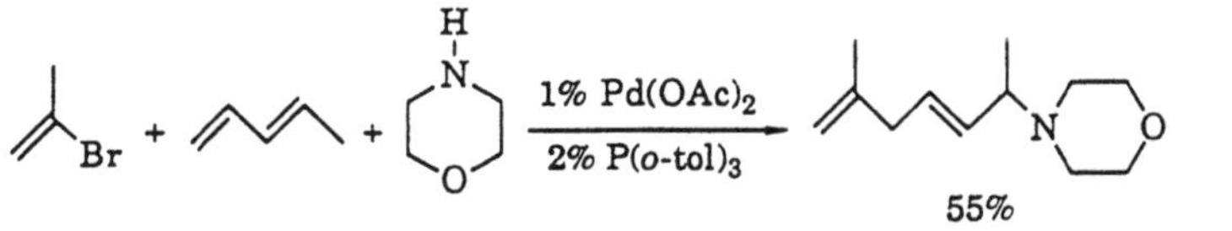

(109)

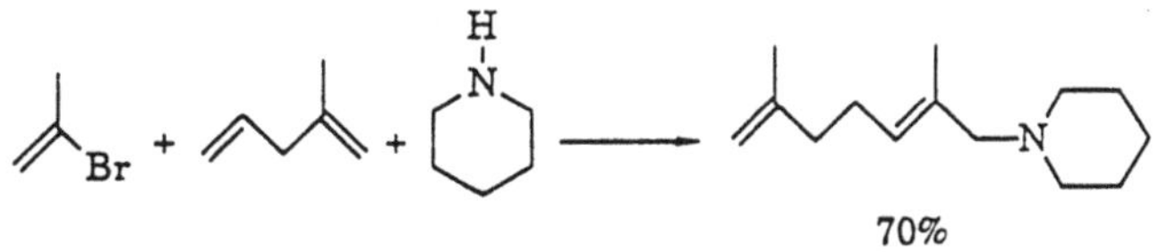

(110)

There is one example in the literature of the palladium-catalyzed addition of a vinylic group and a carbanion to a 1,3-diene (Eq. 111).[153]

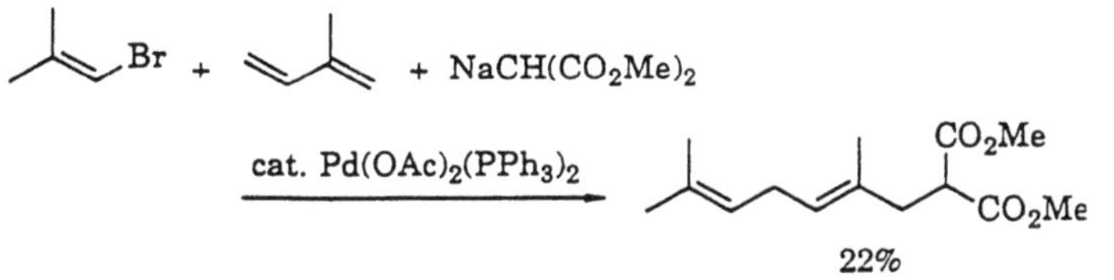

(111)

Reaction with Alkynes

Palladium catalyzes the hydrovinylation of internal alkynes to produce 1,3-dienes (Eq. 112).[154] The best results are obtained using alkynes bearing aryl or tertiary alcohol groups. Alkyl acetylenes give lower yields.

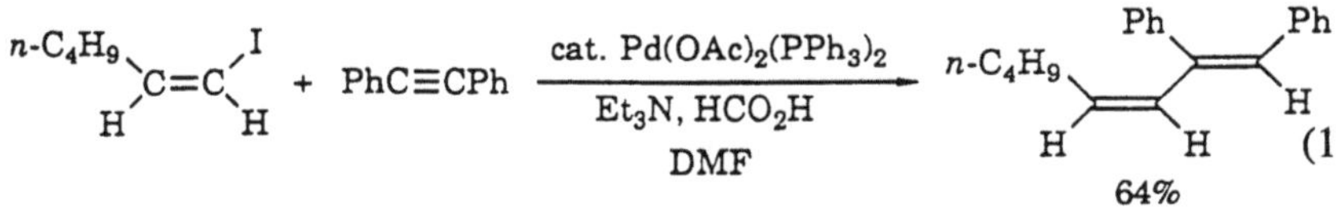

(112)

The intermediates obtained by vinylpalladium addition to alkynes may be trapped by internal alkenes (Eqs. 113–115)[155] or terminal alkynes (Eq. 116).[156] An aryl zinc reagent has also been utilized as a trap (Eq. 117).[157]

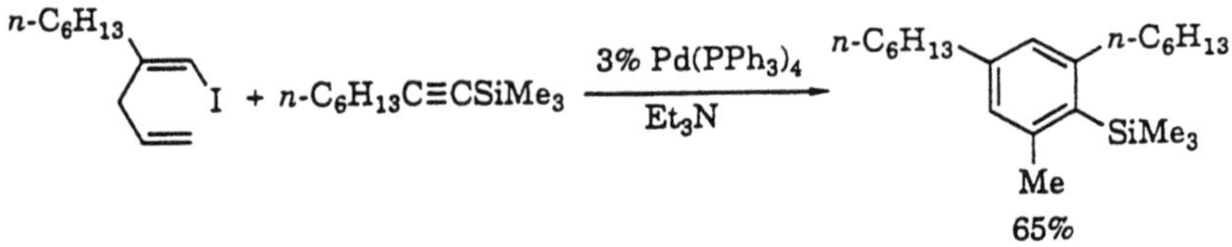

(113)

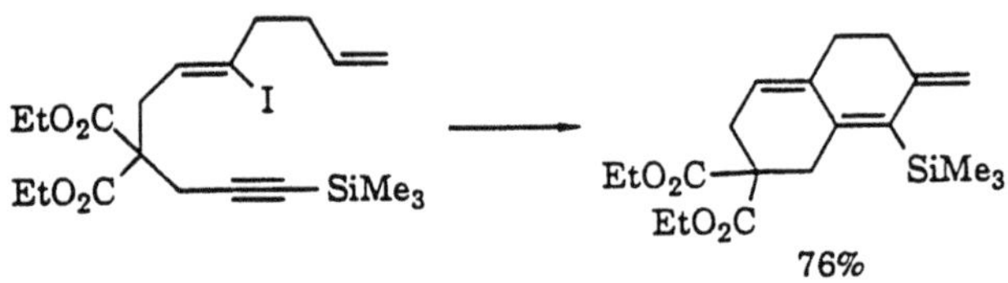

(114)

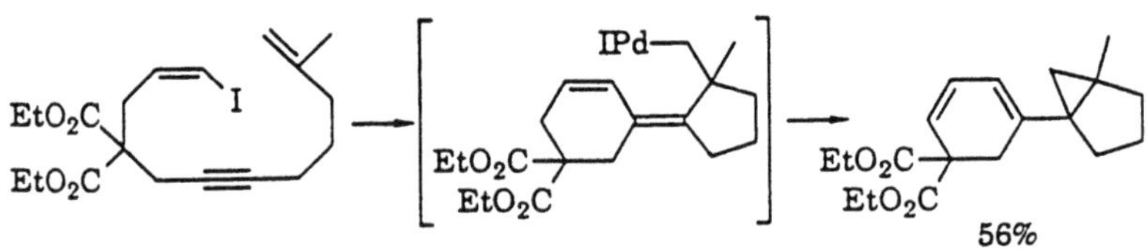

(115)

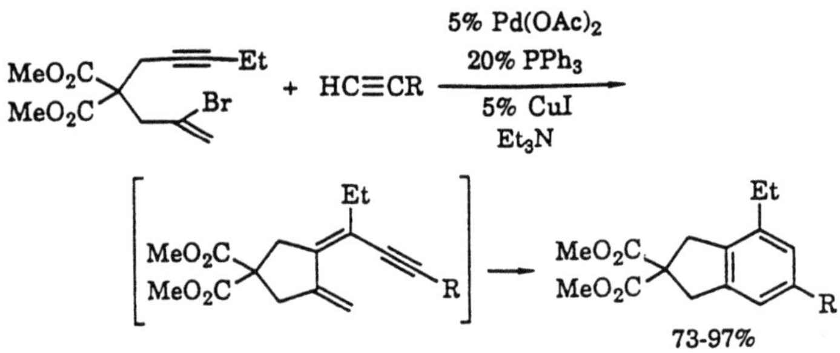

(116)

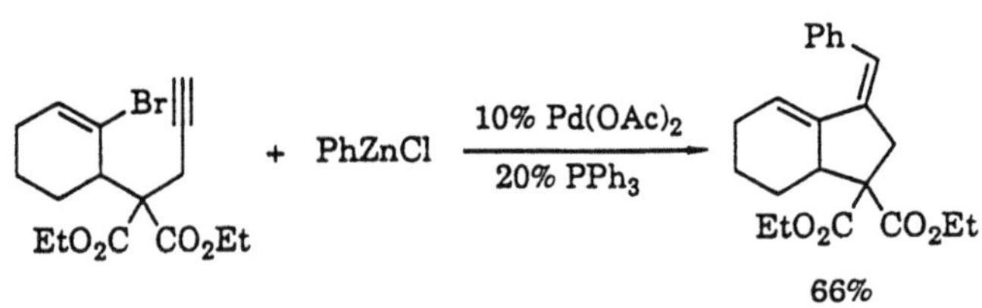

(117)

One of the most widely used palladium-catalyzed processes of vinylic halides involves their coupling with terminal alkynes to produce enynes (Eq. 118). Three different experimental procedures were reported in 1975 to effect this transforma-

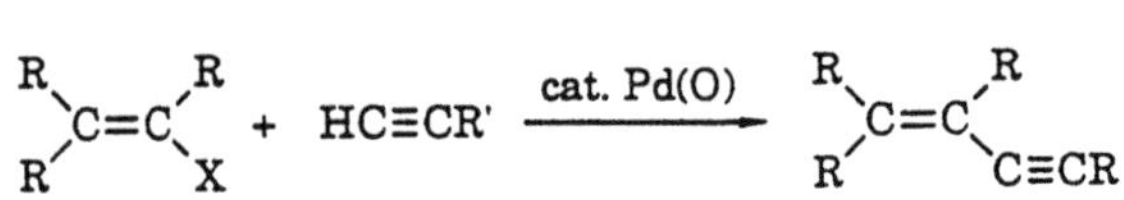

(118)

$$RCH{=}CHX + Pd \longrightarrow RCH{=}CHPdX \xrightarrow{HC\equiv CR \text{ or } CuC\equiv CR}$$

$$RCH{=}CHPdC\equiv CR \longrightarrow RCH{=}CHC\equiv CR + Pd$$

Scheme 3.

tion: (1)[158] 1% $PdCl_2(PPh_3)_2$, 0.5% CuI, excess Et_2NH; (2)[159] 6% $PdCl_2(PPh_3)_2$ or $Pd(PPh_3)_4$, 1.2 NaOMe; and (3)[160] 1–2% $Pd(OAc)_2(PPh_3)_2$, excess Et_3N or piperidine. All three procedures work with a variety of aryl and vinylic halides to afford good yields of aryl alkynes and enynes respectively. More recently procedures involving: (1)[161] 1% $PhPdI(PPh_3)_2$, 1% CuI, and 2 Et_3N; (2)[162] 1% $Pd(PPh_3)_4$ or $PdCl_2(PPh_3)_2$, 7% CuI, and excess Et_2NH; or (3)[163] 1% $Pd(PPh_3)_4$ or $PdCl_2(PPh_3)_2$, 1% CuI, 1% $(PhCH_2NEt_3)Cl$, aqueous NaOH, and benzene; plus numerous variations of Sonogashira's original 1975 procedure[158] have been reported to effect this useful transformation.

The reaction is assumed to proceed as shown in Scheme 3. It has in fact been shown that a vinylic palladium iodide bis(triphenylphosphine) complex will react with a terminal alkyne and Et_3N to give the corresponding enyne.[162]

This interesting coupling process has been reported to accommodate considerable functionality. For example, silyl enynes are easily prepared (Eq. 119).[164] Note that the use of the alkynol provides a useful method for introducing the ethynyl group. Enynoate esters can be stereospecifically prepared by this process, even when an allylic ester is present (Eqs. 120 and 121).

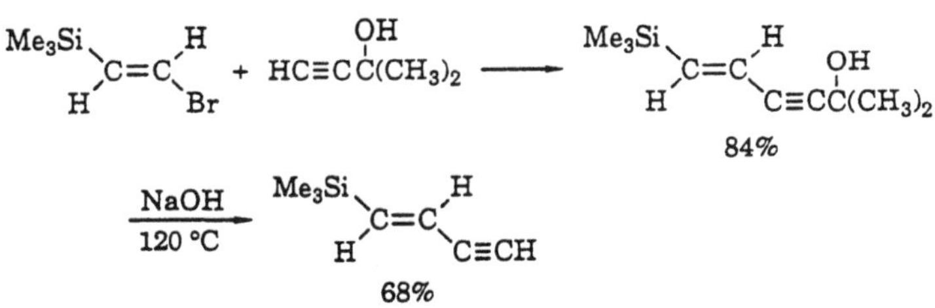

(119)

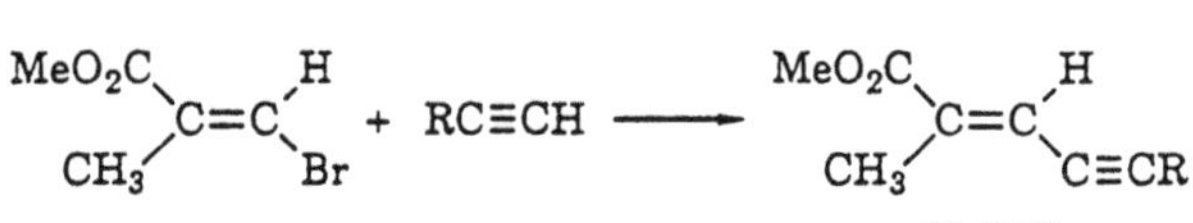

(120)[165]

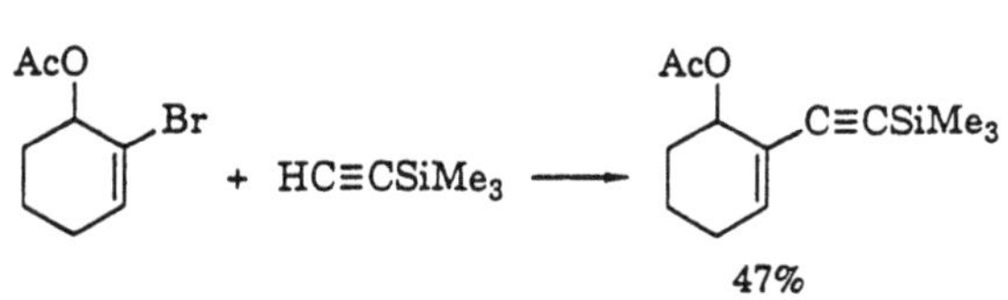

(121)[166]

Dihalo compounds bearing different halogens can be made to undergo selective substitution (Eqs. 122–124).

$$R^1CF{=}CBrR^2 + HC{\equiv}CR^3 \longrightarrow R^1CF{=}CR^2C{\equiv}CR^3 \qquad (122)^{84}$$

$$ArCH{=}CFBr + HC{\equiv}CR \longrightarrow ArCH{=}CFC{\equiv}CR \qquad (123)^{167}$$

$$R^1CF{=}CFI + HC{\equiv}CR^2 \longrightarrow R^1CF{=}CFC{\equiv}CR^2 \qquad (124)^{168}$$

With identical halogens, both mono- and disubstitution have been reported. For example, 1,1-dichloroethylene sequentially affords enynes and enediynes (Eq. 125).[169] Analogous reactions with (*E*)- and (*Z*)-1,2-dichloroethylene have proved very valuable in synthesis. Monosubstitution proceeds stereospecifically to the corresponding (*E*)- or (*Z*)-chloroenynes (Eq. 126).[170,171] The (*Z*)-chloroenynes are readily converted to 1,3-diynes (Eq. 127).[172] Stereospecific sequential or simultaneous disubstitution of 1,2-dichloroethylene (Eqs. 128 and 129), 1,2-diiodoethylene (Eq. 130),[175] and 1,2-dibromocyclopentene (Eq. 131)[176] have also been accomplished.

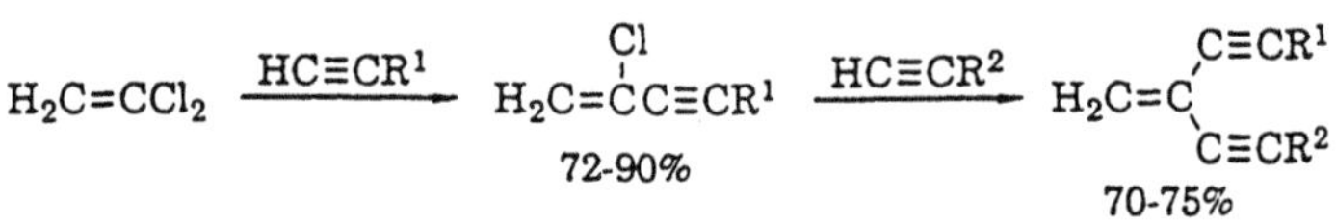

(125)

$$ClCH{=}CHCl + HC{\equiv}CR \longrightarrow \underset{65\text{-}100\%}{ClCH{=}CHC{\equiv}CR} \qquad (126)$$

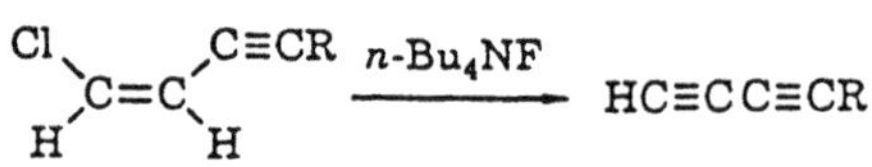

(127)

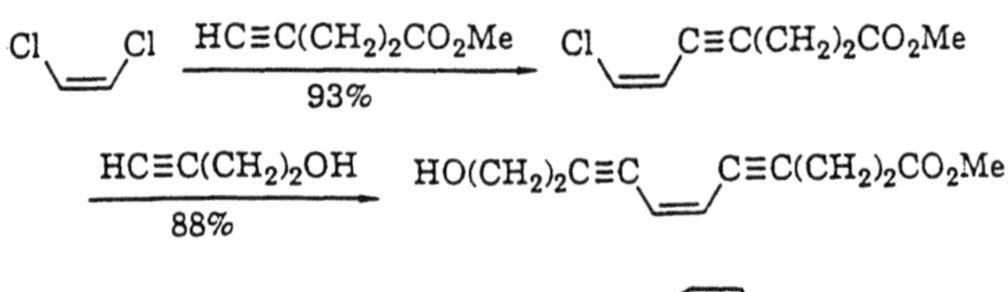

(128)[173]

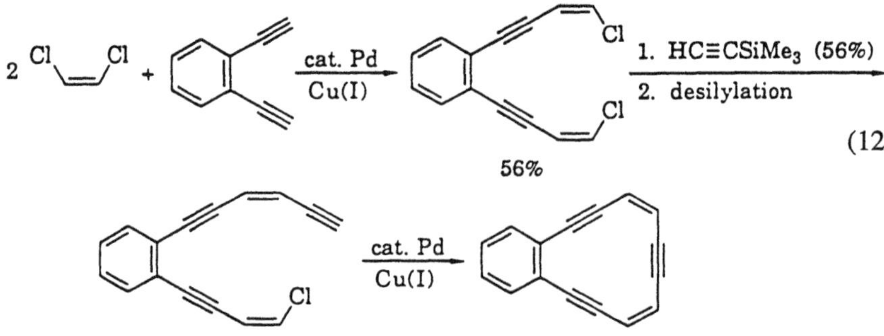

(129)[174,175]

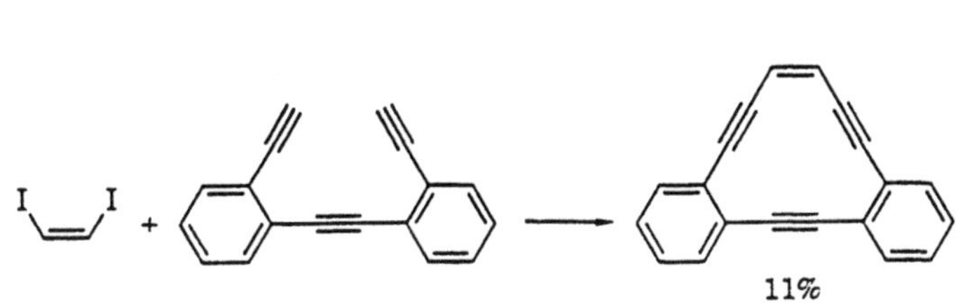

(130)

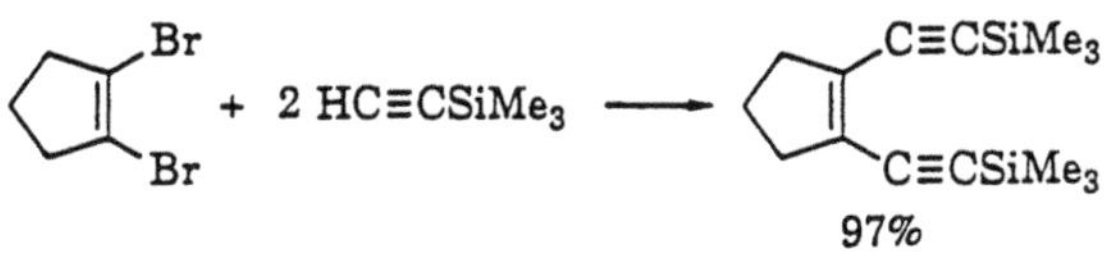

(131)

The coupling process has proven useful for the synthesis of heterocycles (Eqs. 132 and 133) and numerous natural products and compounds of biological interest.

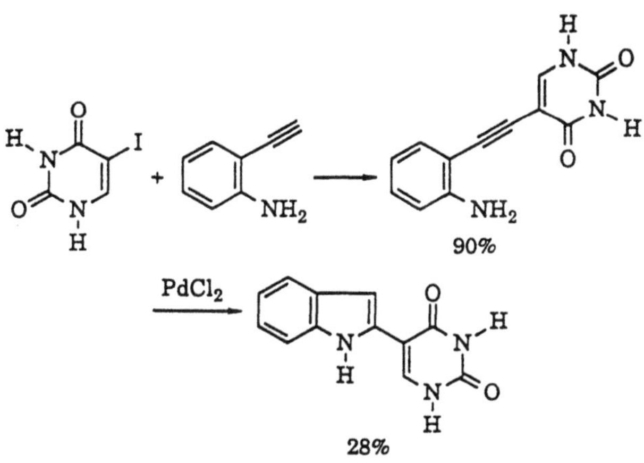

(132)[177]

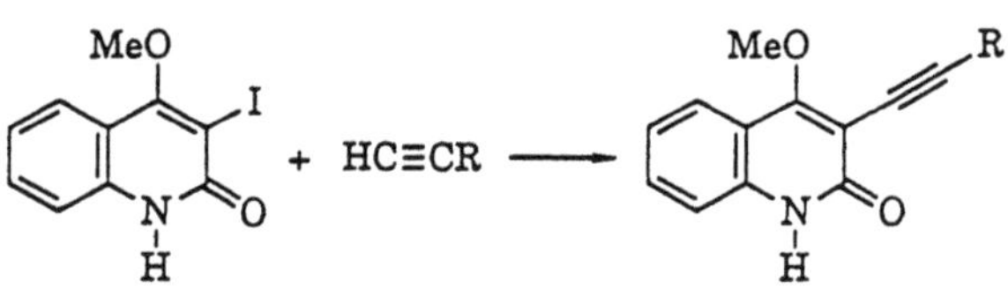

(133)[178]

Various polyunsaturated natural products have been prepared using this chemistry (Eqs. 134–137).

This chemistry is quite practical for the synthesis of polyunsaturated insecticides (Eqs. 138–141) and pyrethins (Eq. 142).

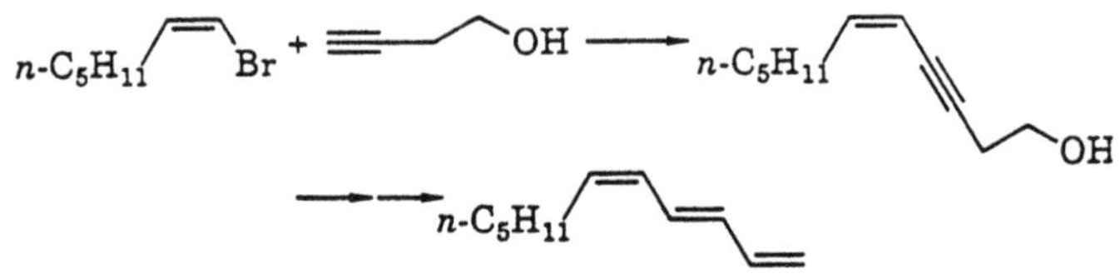

(134)[179]

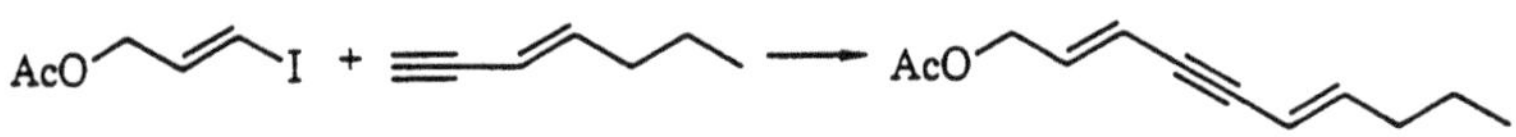

(135)[180,181]

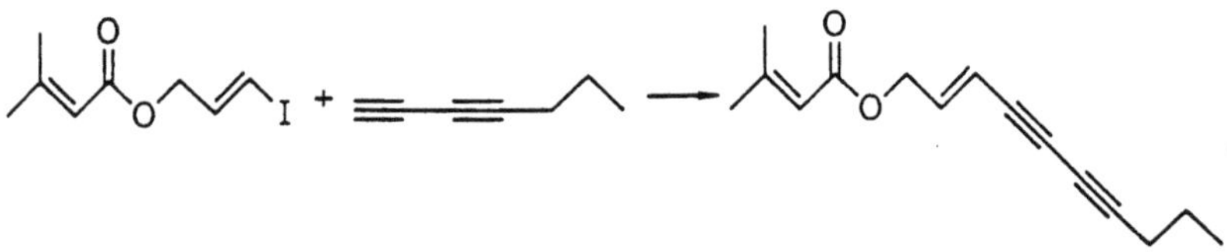

(136)[181]

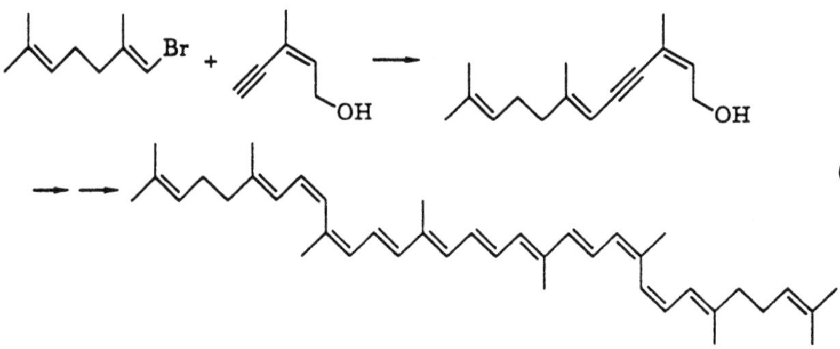

(137)[182]

$$\underset{E \text{ or } Z}{\mathrm{ArCH{=}CHBr}} + \underset{n = 3,\,5}{\mathrm{HC{\equiv}C(CH_2)_nOH}} \longrightarrow \mathrm{ArCH{=}CHC{\equiv}C(CH_2)_nOH}$$

$$\longrightarrow\longrightarrow \mathrm{Ar(CH{=}CH)_2(CH_2)_x(CH{=}CH)_yCONH\text{-}\mathit{i}\text{-}Bu} \quad x = 2, 4;\ y = 1, 2$$

(138)[183]

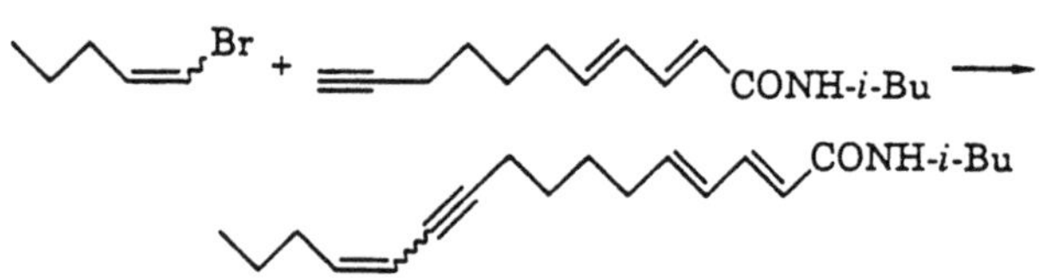

(139)[184]

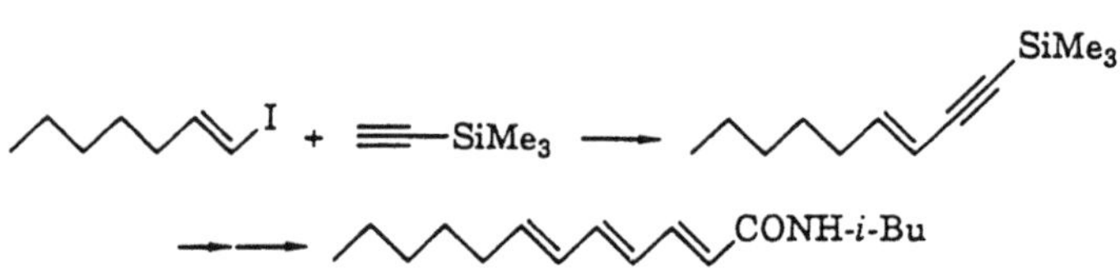

(140)[185]

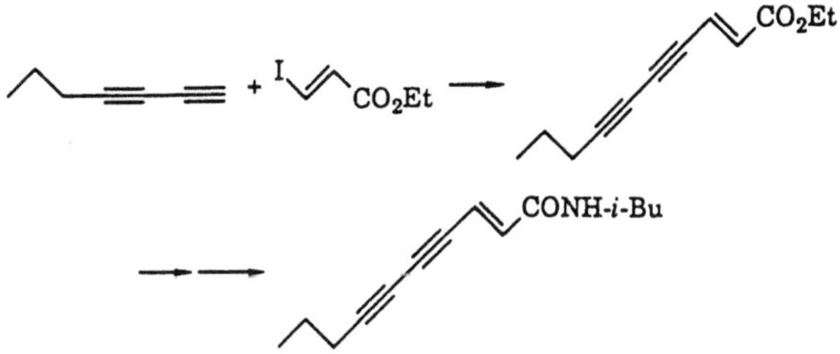

(141)[181]

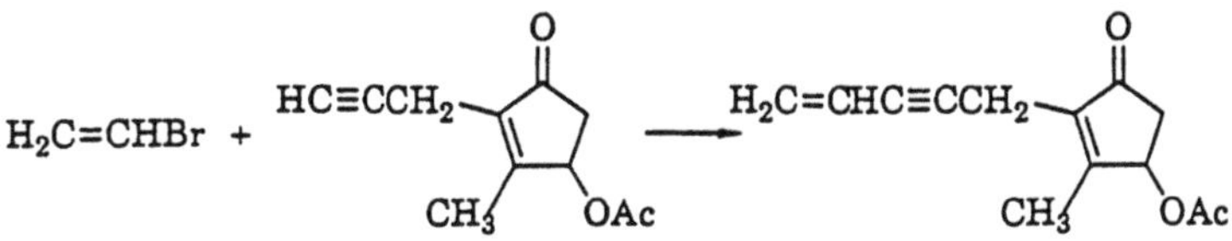

(142)[186]

Naturally occurring thiophene-containing enynes and analogues can be prepared in like fashion (Eqs. 143 and 144).

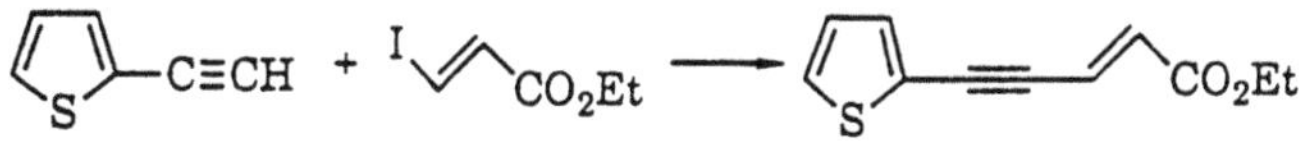

(143)[164,181]

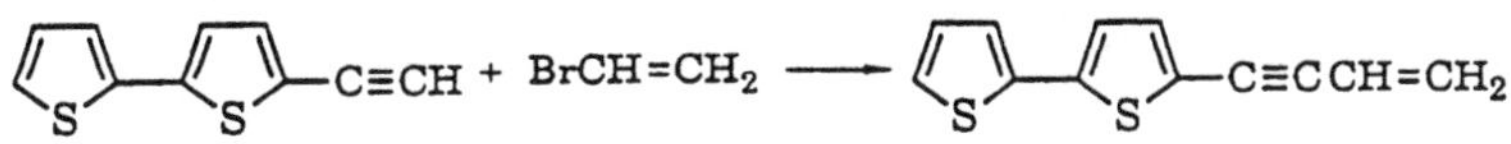

(144)[187]

Syntheses of compounds as diverse as histrionicotoxins (Eq. 145),[188] serine protease suicide substrates (Eq. 146),[189,190] vitamins,[191] and antitumor antibiotics, such as neooxazolomycin,[192] have utilized this coupling reaction.

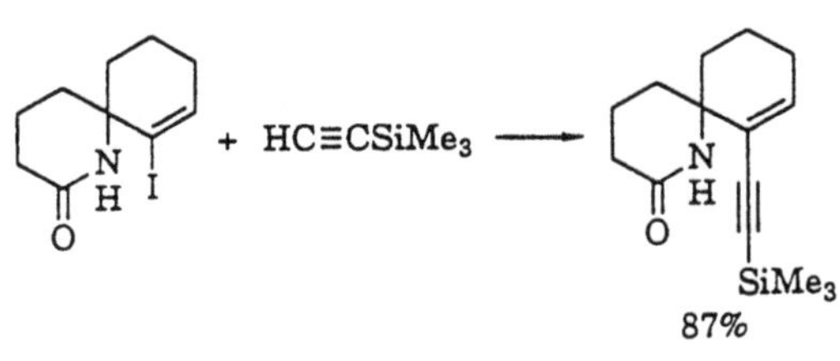

(145)

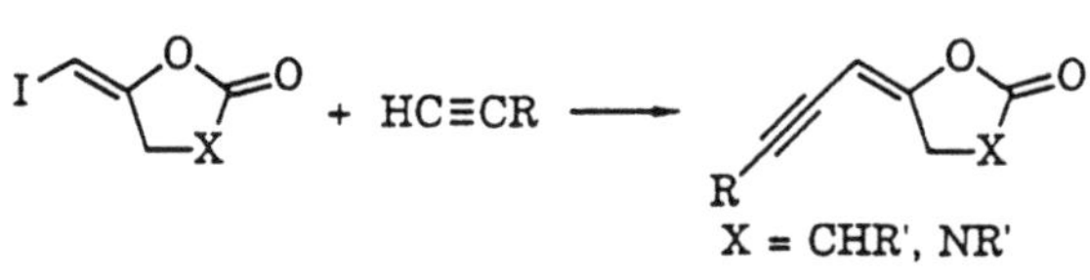

(146)

This enyne synthesis is particularly amenable to the synthesis of pheromones or reputed pheromones, as illustrated by the following synthesis of the sex pheromone

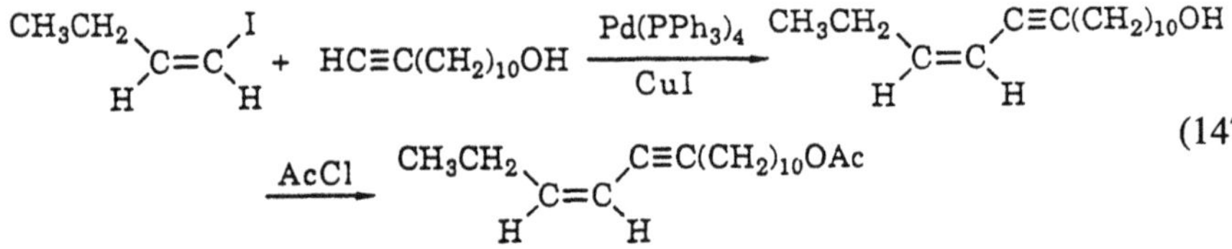

(147)

of the oak processionary moth, *Thaumetopoea pityocampa* (Eq. 147).[193–196] Subsequent reduction of the carbon-carbon triple bond affords a very general route to pheromones containing conjugated double bonds.[163,193,195,197–199] Naturally occurring trienes can be prepared in a similar fashion (Eq. 148).[200]

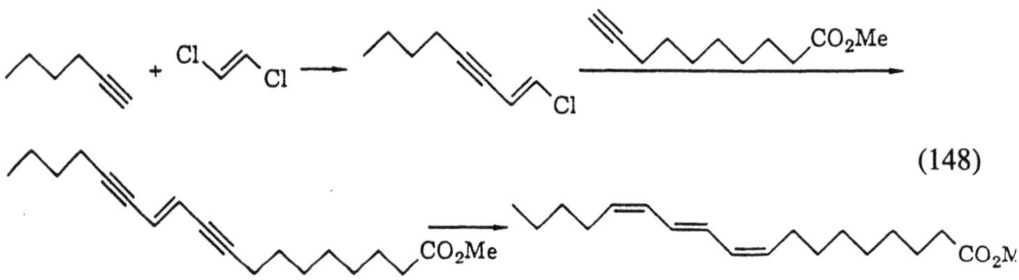

(148)

Leukotrienes are also readily prepared by palladium-catalyzed enyne formation and subsequent reduction.[181,201–209] The following example illustrates the versatility of this approach (Eq. 149).[209] The related natural products coriolic acid,[210] dimorphecolic acid[211] and an irreversible inhibitor of soybean lipoxygenase[212] have been similarly prepared.

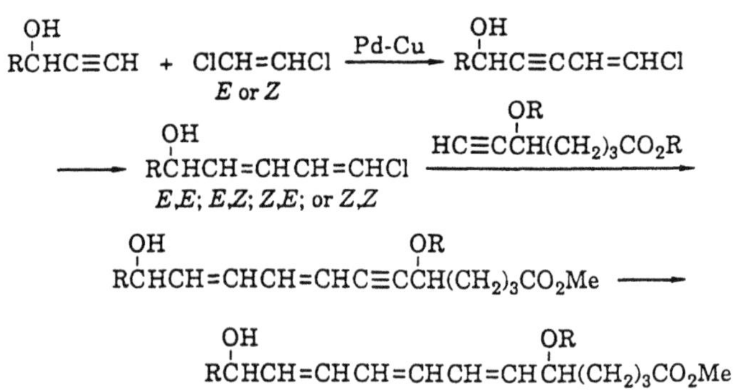

(149)

Palladium-catalyzed alkynylation has been used to prepare nucleosides substituted in various positions (Eqs. 150–152).

Recently, this chemistry has found extensive use in the synthesis of the enediyne core present in the antitumor antibiotics calichemicin, esperamicin, dynenaicin and neocarzinostatin (Eqs. 153 and 154).[225–245]

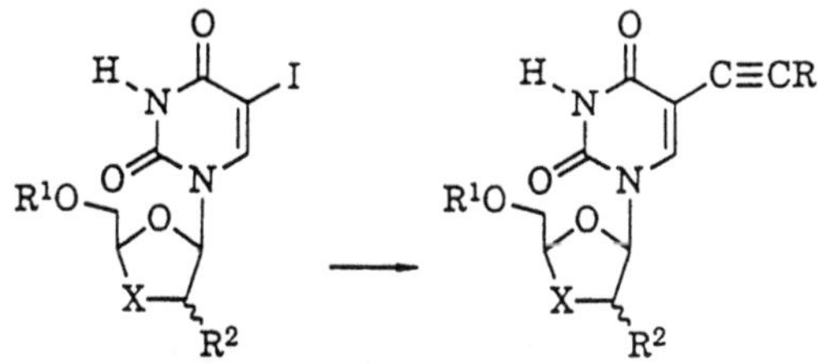

(150)[213–222]

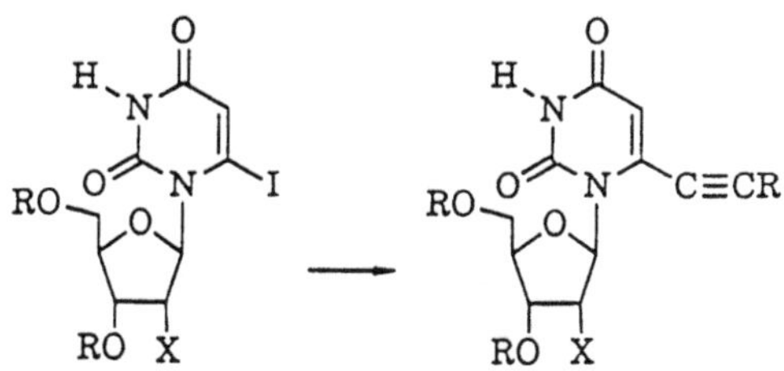

(151)[223]

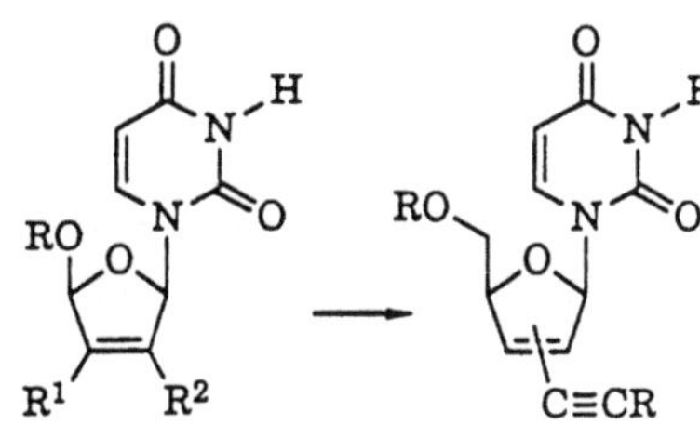

R^1, R^2 = H, Br or Br, H

(152)[224]

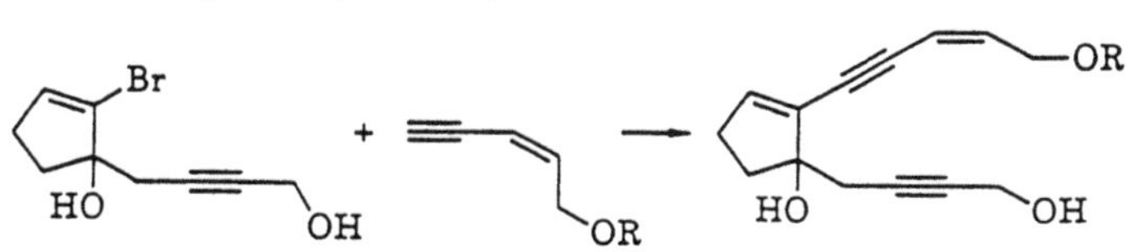

(153)[225]

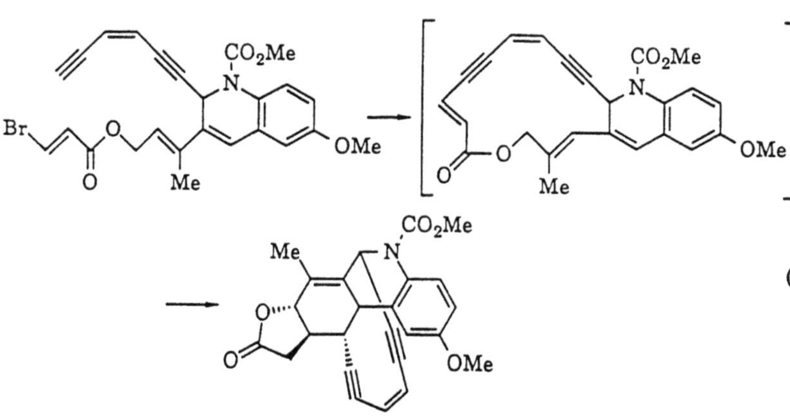

(154)[226]

Finally, the alkynylation of an iodocubane has been reported to effect a novel rearrangement to an enyne (Eq. 155).[246]

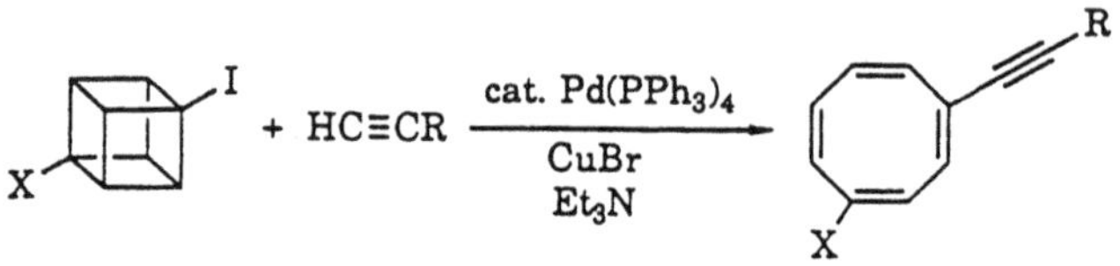

(155)

Upon only a slight variation in the experimental procedure, vinylic halides react with terminal alkynes to give low yields of fulvenes by a sequence involving two sequential alkyne insertions, followed by cyclization (Eq. 156).[247] Internal alkynes react similarly.[248]

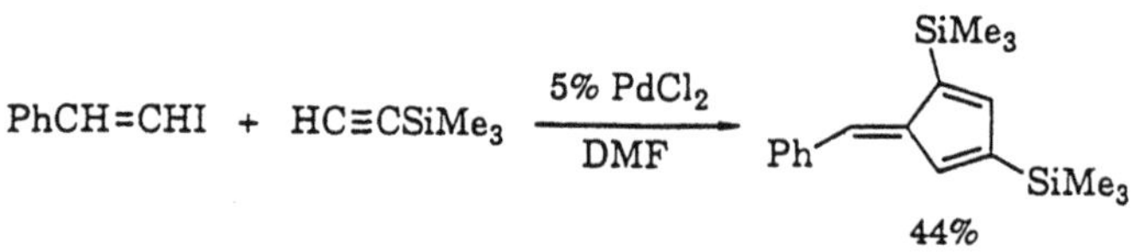

(156)

Reaction with Arenes

While vinylic palladium complexes can apparently undergo intermolecular[249] and intramolecular[250] reaction with arenes (Eqs. 157 and 158), this reaction has thus far seen very limited use.

$$Cl_2C{=}CHPdCl(PPh_3)_2 + C_6H_6 \xrightarrow[HOAc]{AgOAc} \underset{80\%}{Cl_2C{=}CHC_6H_5} \qquad (157)$$

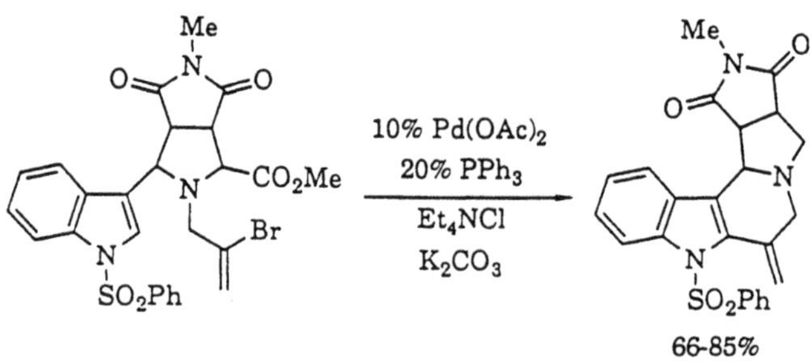

(158)

Coupling with Vinylic Halides and Triflates

There are few examples also of vinylic halides being cross-coupling with organic halides using palladium. Both the inter- and intramolecular coupling of vinylic halides have been reported (Eqs. 159 and 160).[251,252] Vinylic halides and vinylic triflates may also be intramolecularly coupled by n-$Bu_3SnSiMe_3$ and a palladium catalyst (Eq. 161).[253]

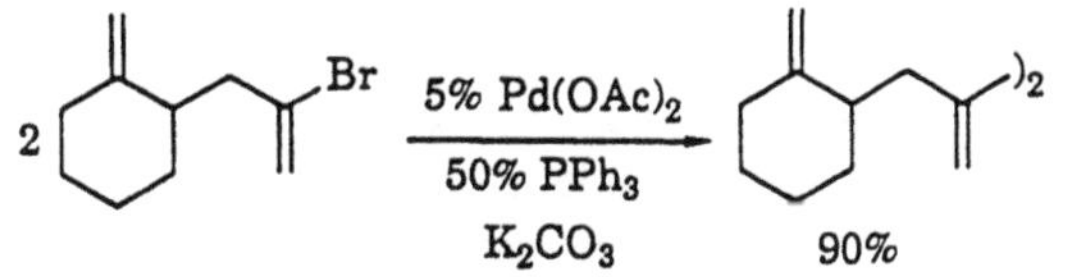

(159)

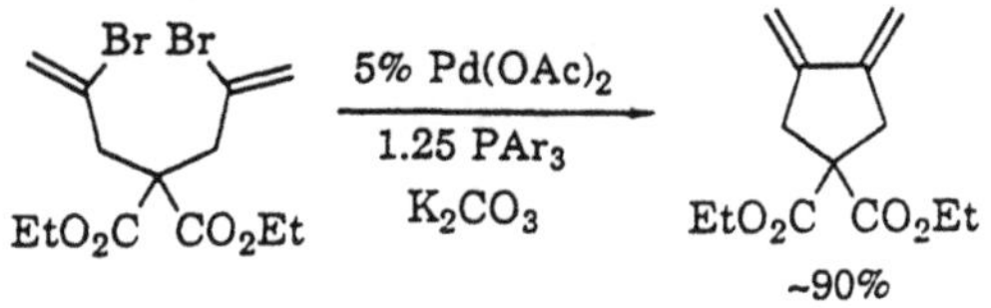

(160)

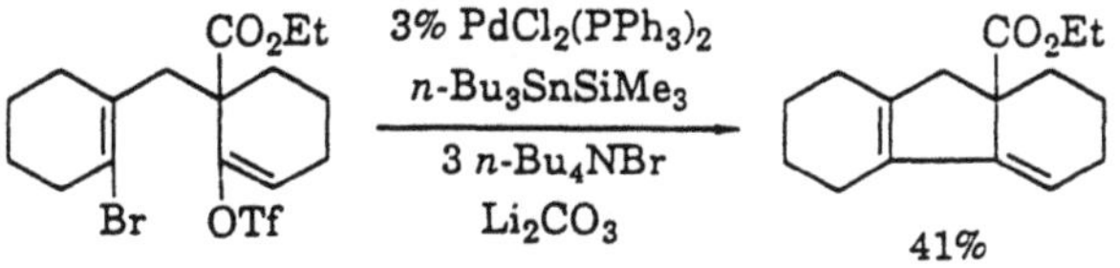

(161)

Vinylation of Enolates

There appears to be only one report of the reaction of relatively stabilized carbanions reacting with vinylpalladium intermediates. Ketones bearing vinylic halides can be cyclized using palladium (Eq. 162).[254] The use of tin enolates allows intermolecular cross-coupling, as noted earlier (Eq. 69).[26,109]

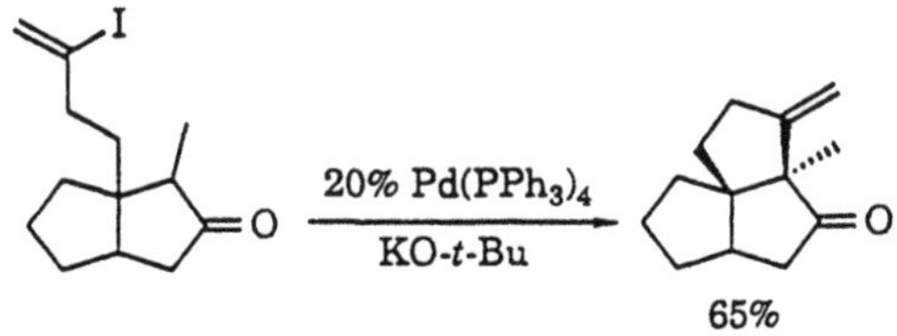

(162)

Alkylation by Organometallics

The palladium-catalyzed coupling of vinylic halides and organometallics has been extensively explored. A number of alkyl organometallics can be employed in this process. Alkyllithium reagents can effect this transformation using stoichiometric amounts of palladium (Eq. 163).[255–258] The reaction proceeds with nearly complete retention of stereochemistry. When alkyllithium reagents containing β-hydrogens, such as n-BuLi, are utilized, the yields drop substantially. The mechanism appears to involve alkyl vinylic palladium(II) and -(IV) intermediates.[258]

$$PhCH{=}CHBr + LiCH_3 \xrightarrow{Pd(acac)_2 \text{ or } Pd(PPh_3)_4} PhCH{=}CHCH_3 \quad (163)$$

The use of alkylmagnesium reagents allows one to run the reaction using only catalytic amounts of palladium. Methyl-,[255,257] primary alkyl-,[259–264] secondary alkyl-,[262,263,265,266] and even tertiary alkylmagnesium[267] reagents can be effectively employed. Carbon-carbon double bonds can be accommodated in the Grignard reagent.[263,268] While $Pd(PPh_3)_4$ is suitable as the catalyst when methyl[255,257] and primary alkylmagnesium[259,263] reagents are employed, palladium complexes bearing ferrocenylphosphine ligands are more effective, especially for secondary and tertiary alkyl Grignard reagents.[262,266,267] A polymer bound palladium catalyst has also proven effective with MeMgI.[269] Chiral ferrocenylphosphines are useful for asymmetric catalysis.[265]

This chemistry can be quite valuable in pheromone synthesis (Eq. 164).[260,261,268]

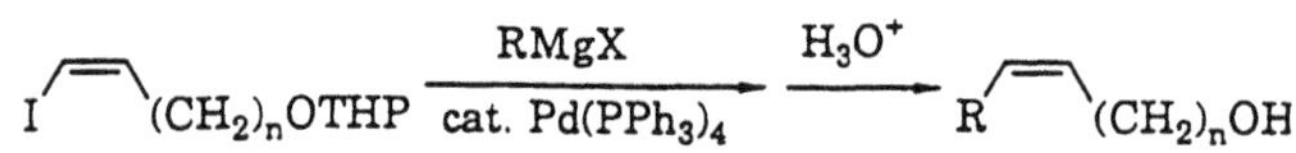

260
(164)

Allylic (Eq. 165)[163,270–273] and vinylic (eqs. 166 and 167)[274] silanes have been prepared in this fashion.

$$RCH{=}CHX + R\overset{\displaystyle MgX}{\overset{|}{C}}HSiMe_3 \xrightarrow{\text{cat. Pd}} RCH{=}CHCHRSiMe_3 \quad (165)$$

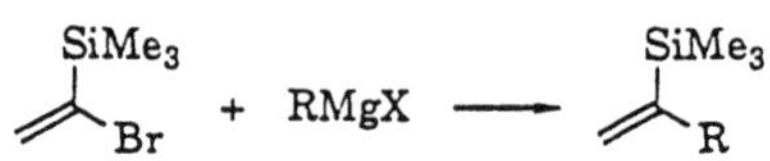

(166)

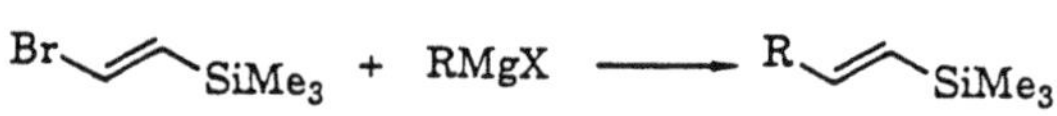

(167)

The stereospecific alkylation of sulfur-containing vinylic halides is also feasible (Eq. 168).[275–277]

$$RMgX + BrCH{=}CHSPh \xrightarrow{\text{cat. Pd}} RCH{=}CHSPh \quad (168)$$

1,1-Dichloro-1-alkenes can be dialkylated in the presence of a third chloride (Eq. 169).[278] Trichloroethylene, on the other hand, affords 1,1-dichloro-1-alkenes (Eq. 170).[279]

$$2\ RMgCl + Cl_2C{=}CH(CH_2)_nCl \xrightarrow{\text{cat. } PdCl_2(PPh_3)_2} R_2C{=}CH(CH_2)_nCl \quad (169)^{278}$$

$$RMgCl + ClCH{=}CCl_2 \xrightarrow{\text{cat. } Pd(PPh_3)_4} RCH{=}CCl_2 \qquad (170)$$

Alkylzinc reagents appear to give even better results than the alkyl Grignard reagents. Simple alkyl-,[280,281] cyclopropyl-,[282,283] β-phenethyl-,[284] alkenyl-,[263,285–287] and alkynylzinc[287,288] compounds can be employed. These reagents can in some cases be prepared by unique approaches (Eqs. 171 and 172).

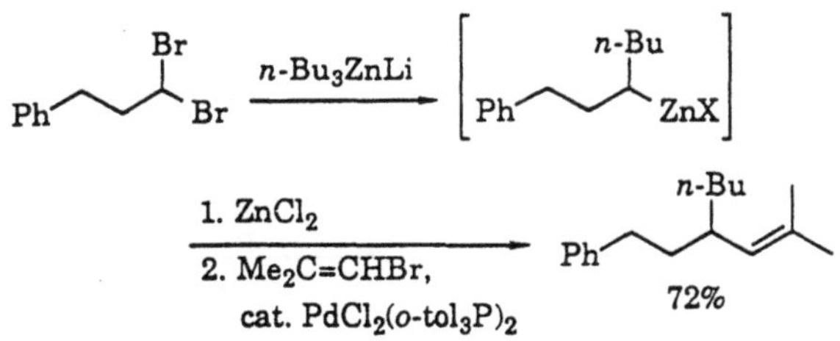

(171)[281,282]

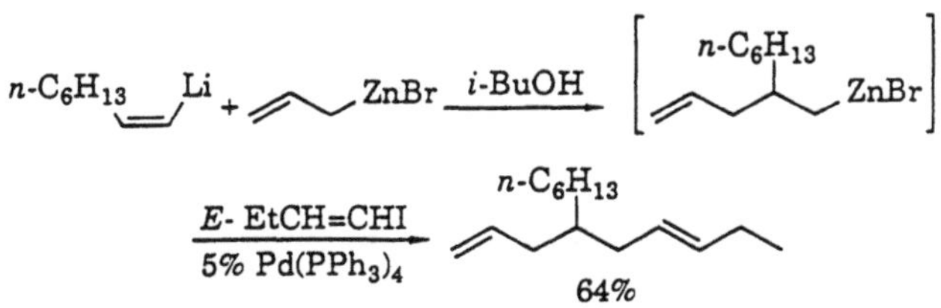

(172)[285]

Perhaps the greatest advantage of this process stems from the fact that various functionally-substituted alkylzinc reagents are readily prepared and utilized (Eqs. 173–175). Particularly important are ketone-[294] and ester-containing[295–298] organozinc reagents (Eqs. 176–178).

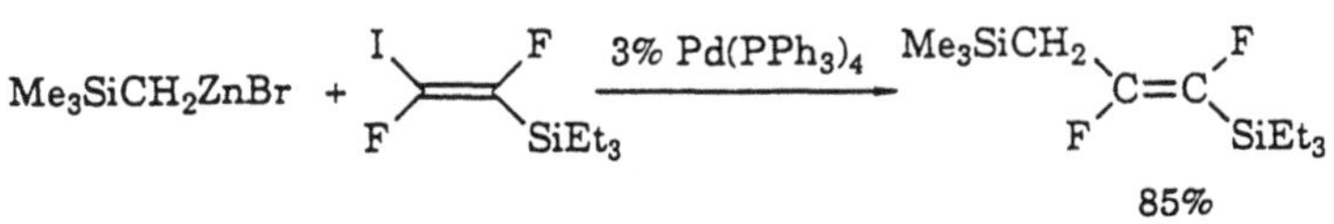

(173)[289,290]

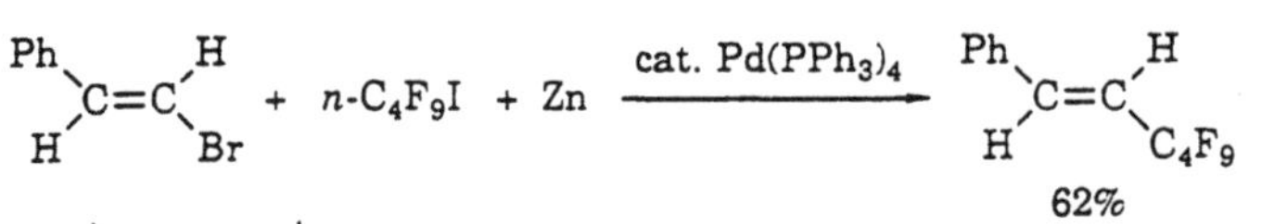

(174)[291,292]

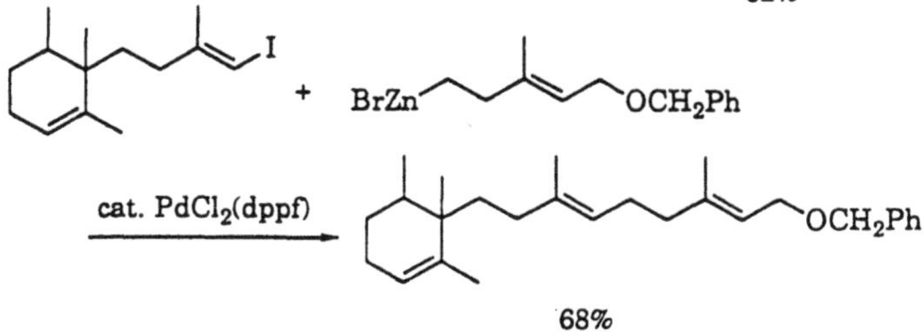

(175)[293]

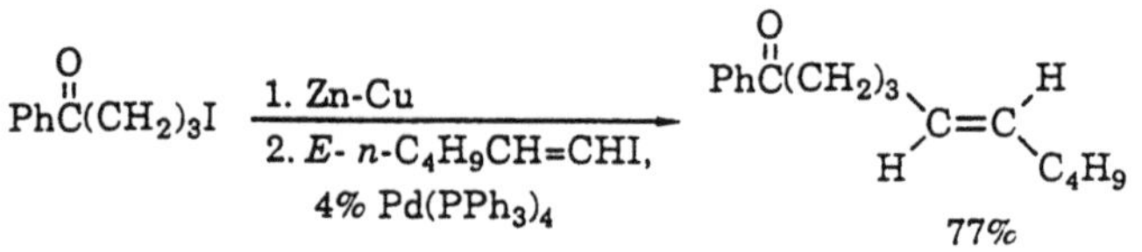

(176)

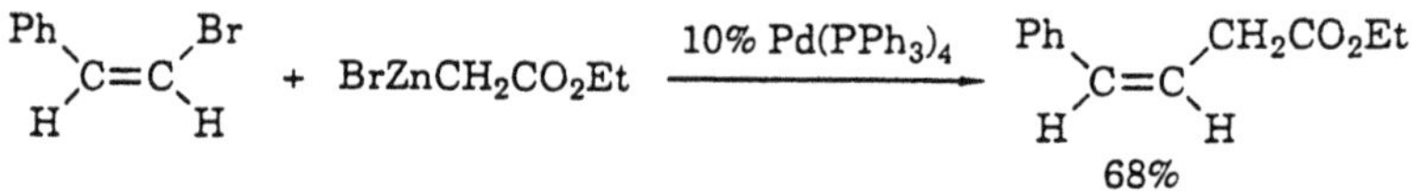

(177)[295]

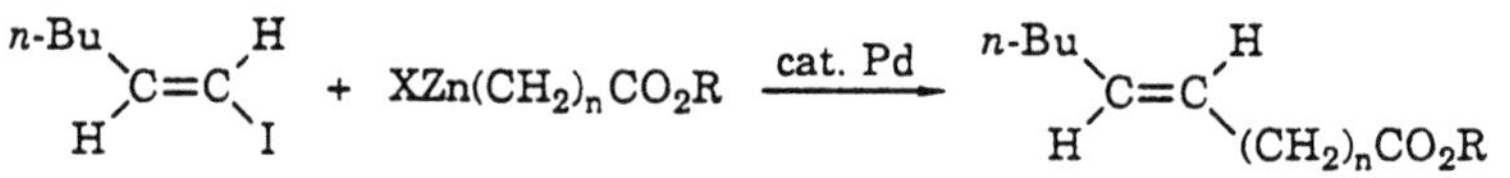

(178)

n	R	%	ref.
2	i-Pr	90	296, 298
3	Et	71	297

The vinylic halide may also accommodate considerable functionality. Ester groups are compatible with this chemistry, as illustrated in the following synthesis of mokupalide (Eq. 179).[299]

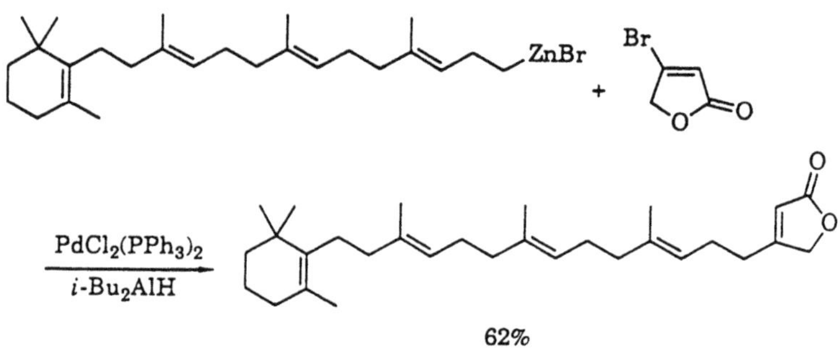

(179)

β-Halo vinylic boranes prepared by haloboration of alkynes undergo palladium-catalyzed alkylation by organozinc reagents followed by subsequent alkylation[300–302] or carbonylation[303] (Eq. 180).

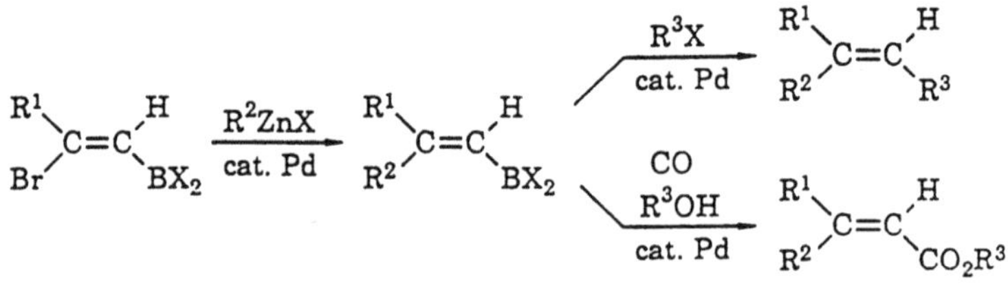

(180)

The palladium-catalyzed coupling of alkylzinc reagents and vinylic halides has been employed in the synthesis of several insect sex pheromones (Eqs. 181 and

182).[287,304–306] This chemistry has also proven valuable in the synthesis of the natural products casbene,[286] ageline A,[293] and prezizanol and prezizaene.[283]

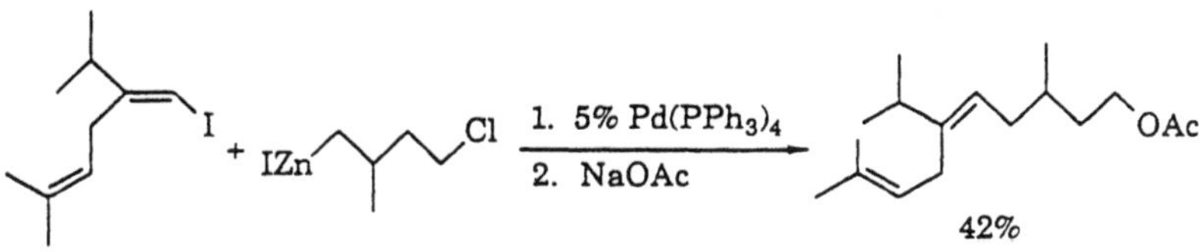

(181)[306]

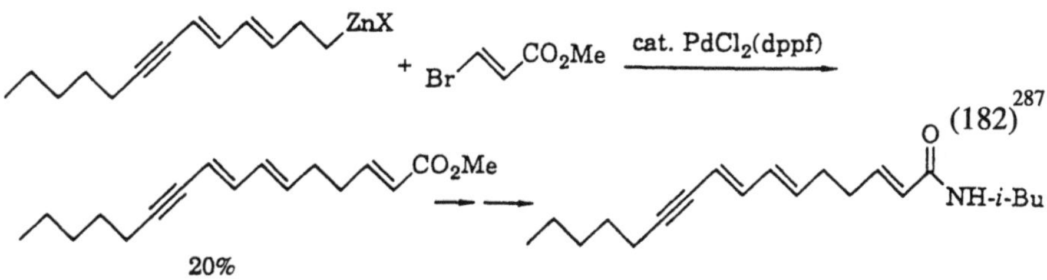

(182)[287]

Alkylboron compounds of 9-BBN[307] and derivatives,[308] as well as alkylboronate esters,[309] also alkylate vinylic halides (Eqs. 183 and 184). Intramolecular versions of this reaction are also known (Eq. 185).[307b] Vinylic sulfides have also been prepared using 9-alkyl-9-BBN compounds (Eq. 186).[310,311]

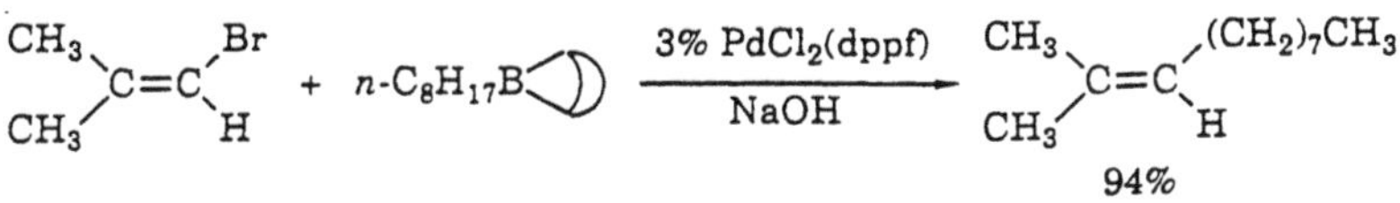

(183)

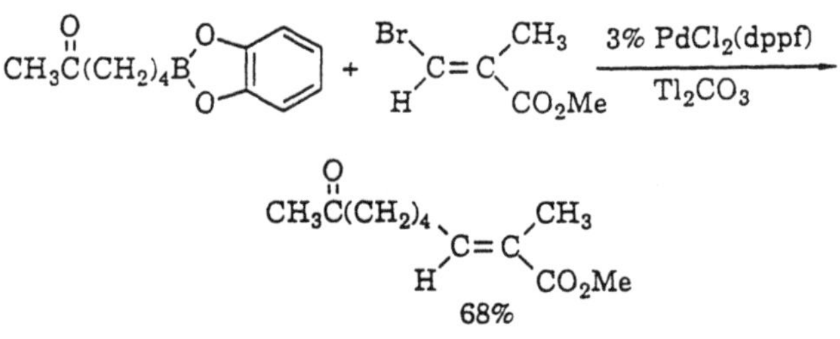

(184)

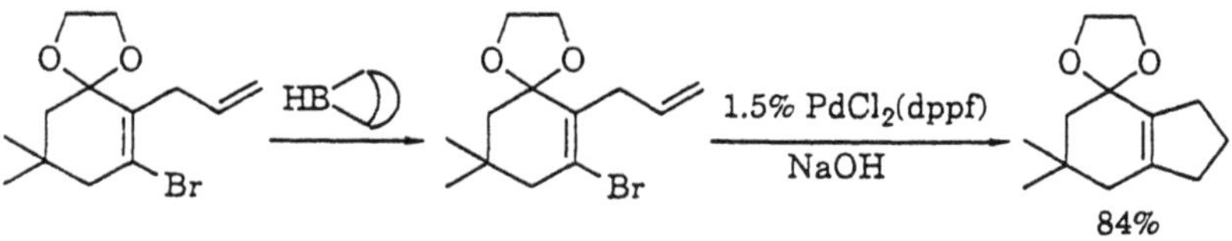

(185)

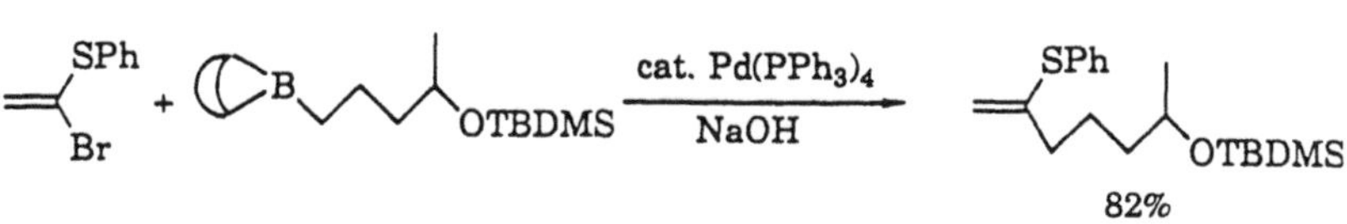

(186)[310]

Tetramethyltin and an alkoxymethyltin compound have been used to alkylate[29] vinylic halides of tosyl enamines,[48] α,β-unsaturated ketones,[312] nucleosides,[224,313] tropolones[314] and avermectins[315] (Eq. 187).

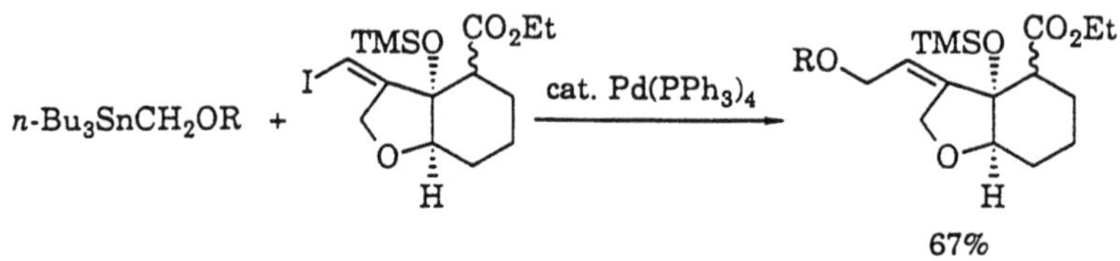
(187)

Coupling with Allylic Organometallics

Relatively few allylic organometallics have been cross-coupled with vinylic halides in the presence of a palladium catalyst. Although allylzinc chloride reacts with a bromovinylboronate ester[302] in good yield (Eq. 188), allyltrimethylsilane[316] and allylic stannanes[317,318] give low yields in this type of reaction (Eq. 189).

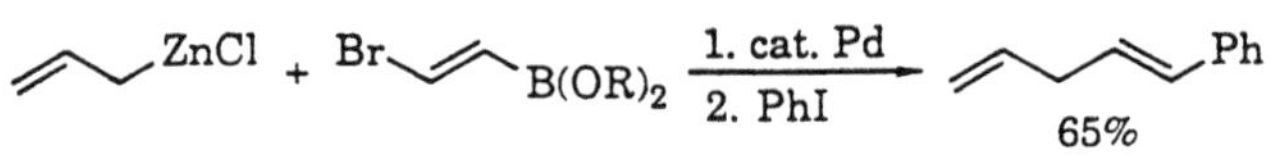
(188)

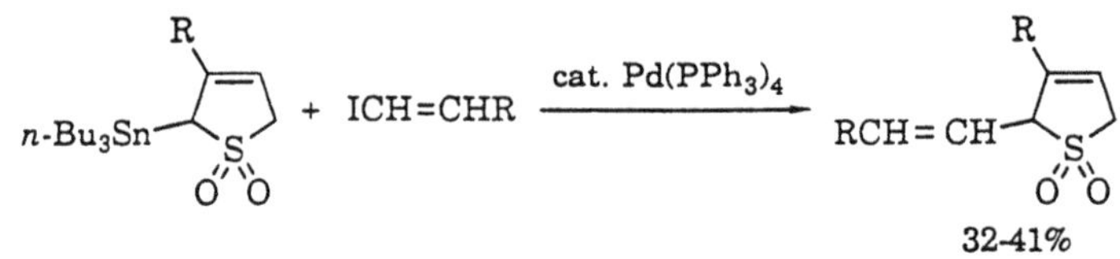
(189)

Reaction with Benzylic Organometallics

The asymmetric palladium-catalyzed alkylation of vinylic halides via benzylic magnesium halides has been extensively studied using a variety of chiral phosphines (Eq. 190).[265,319–322] Enantioselectivity of 65% has been achieved.[322] This

$$ArCHR^1MgX + XCH{=}CHR^2 \xrightarrow{\text{cat. Pd}} ArCHR^1CH{=}CHR^2 \qquad (190)$$

reaction has been used to prepare vinylic sulfides[275,277] and chiral allylic silanes[273,277,323] (Eq. 191).

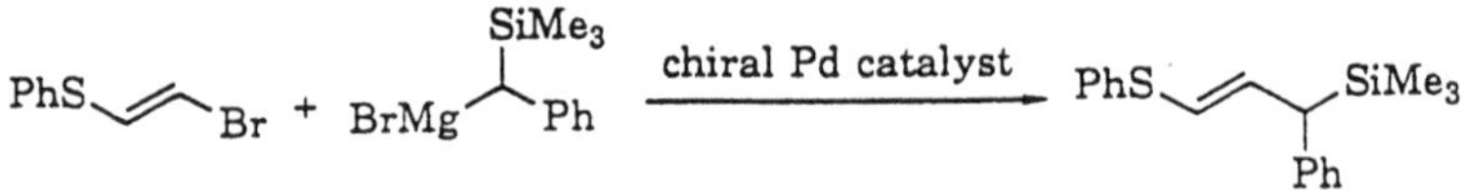
[277] (191)

Benzylic zinc reagents are perhaps even more effective for cross-coupling[324,325] and chiral phosphines give enantiomeric excesses up to 86%.[321,322]

Cross-Coupling with Vinylic Organometallics

A large number of different vinylic metallics react with vinylic halides to afford 1,3-dienes. While vinylic lithium reagents do not give good yields of 1,3-diene,[326,327] vinylic magnesium reagents do.[255,257,259,327] Retention of stereochemistry is observed. The reaction has been used to prepare silyl dienes (Eq. 192)[328] and pheromones[170,260] (Eq. 193).

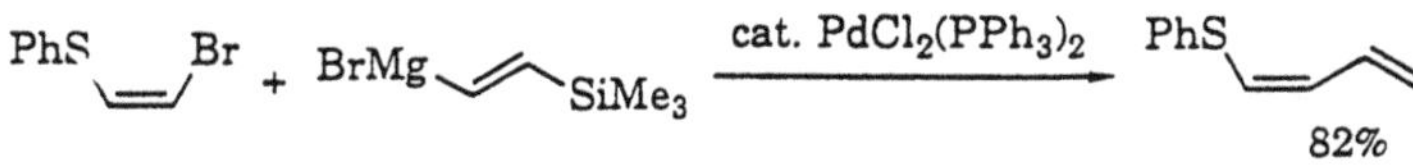

(192)

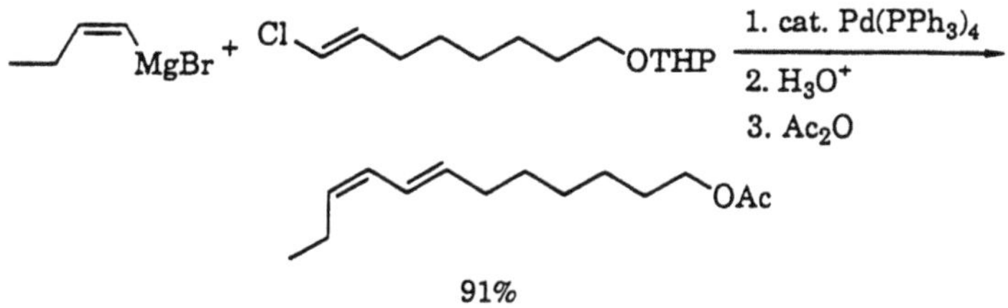

(193)[170]

Vinylic zirconium reagents, readily available by alkyne addition reactions, also cross-couple in good yield.[325] For example, the hydrozirconation (Eq. 194)[327,329,330] or zirconium-promoted carboalumination of alkynes (Eq. 195)[331,332] provides vinylmetallic intermediates which react with vinylic halides in the presence of a palladium catalyst to afford high yields of 1,3-dienes.

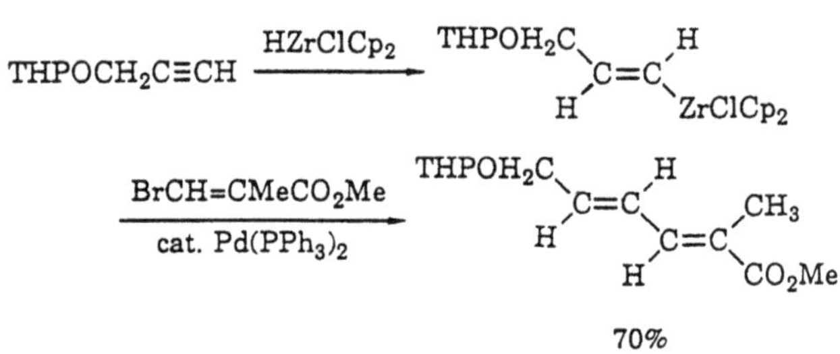

(194)[330]

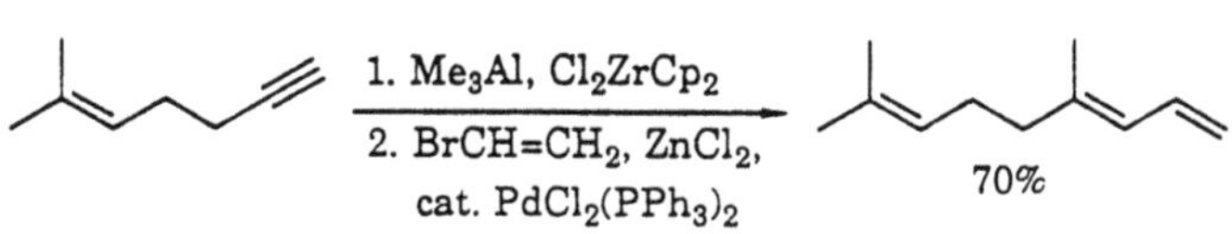

(195)[331]

This chemistry has been applied to the synthesis of pheromone analogs,[198,199] naturally-occurring and synthetic analogs of insecticidal lipid isobutyl amides (Eq. 196),[184,185] and nucleosides (Eq. 197).[333]

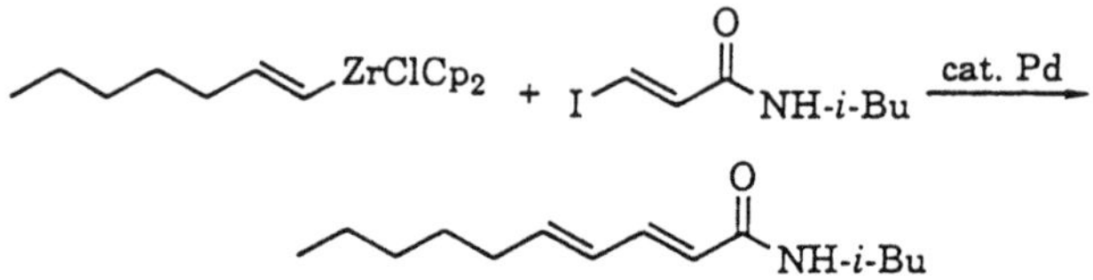

(196)

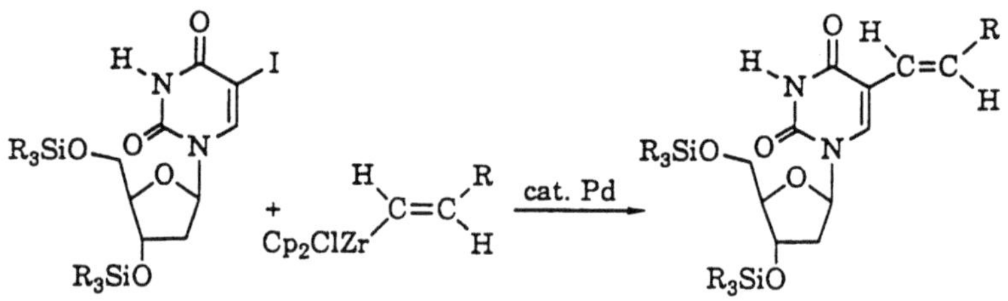

(197)

Vinylic copper reagents prepared via carbocupration of alkynes react stereospecifically with vinylic halides in the presence of a palladium catalyst to afford 1,3-dienes in high yields. Vinylic cuprates derived from organolithium reagents require the presence of $ZnBr_2$ to produce high yields (Eq. 198).[334,335] Only one of the two vinylic groups on copper is transferred. Carbonyl-containing vinylic iodides can be employed (Eq. 199),[336] and a number of pheromones have been prepared in this manner (Eq. 200).[337,338]

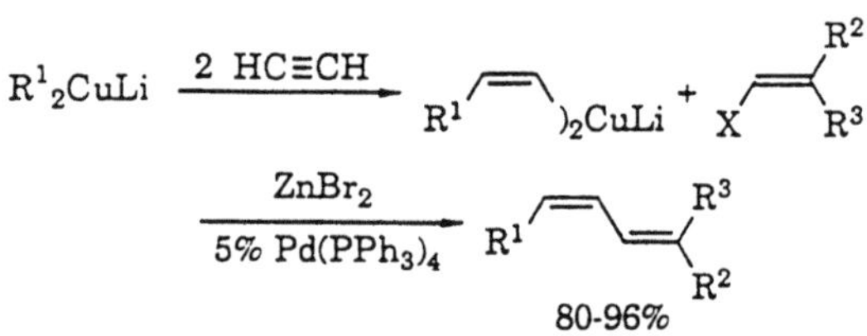

(198)

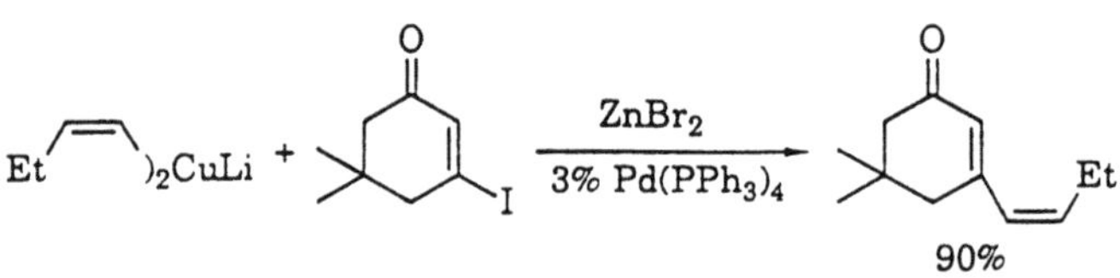

(199)

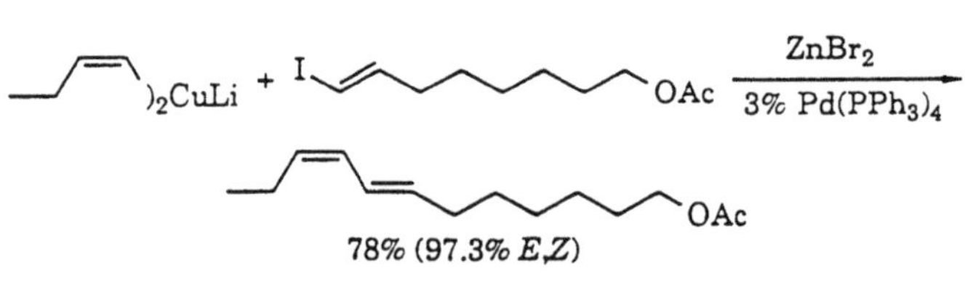

(200)[337]

The cross-coupling of vinylic copper reagents prepared by carbocupration using alkylmagnesium reagents gives more highly substituted products, makes more efficient use of the vinylic group, and no longer requires zinc bromide (Eq. 201).[339,340] Once again β-halo enones can be utilized in this process.[336]

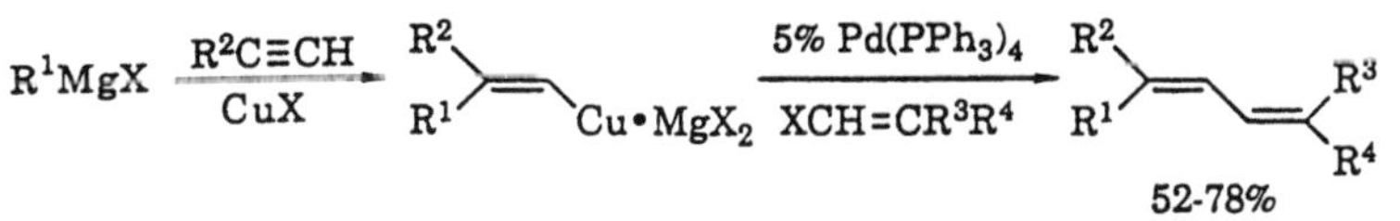
(201)

As indicated earlier, vinylic zinc reagents can be quite effective for the conversion of vinylic halides to 1,3-dienes (Eq. 202).[327,341,342] This methodology has been applied to the synthesis of the natural products (3*E*,5*E*)-1,3,5-undecatriene,[343] four isomers of methyl dimorphecolate (Eq. 203),[344] and several pheromones.[337,345–347]

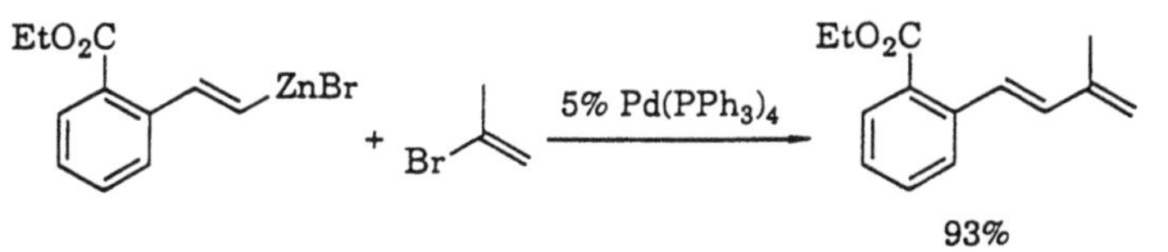
(202)[342]

$$CH_3(CH_2)_4CH{=}CHZnBr + ICH{=}CHCH(OH)(CH_2)_7CO_2Me \xrightarrow{3\%\ Pd(PPh_3)_4} CH_3(CH_2)_4CH{=}CHCH{=}CHCH(OH)(CH_2)_7CO_2Me \quad (203)$$

(first reactant: *E* or *Z*; second reactant: *E* or *Z*)

As noted above, the vinylic halide may bear certain functionality. Thus, α-iodo enones smoothly react with vinylic zinc reagents (Eq. 204).[348] Subsequent reduction affords α-alkenyl ketones. This process complements an earlier approach which also used vinylic zinc reagents (Eq. 205).[349]

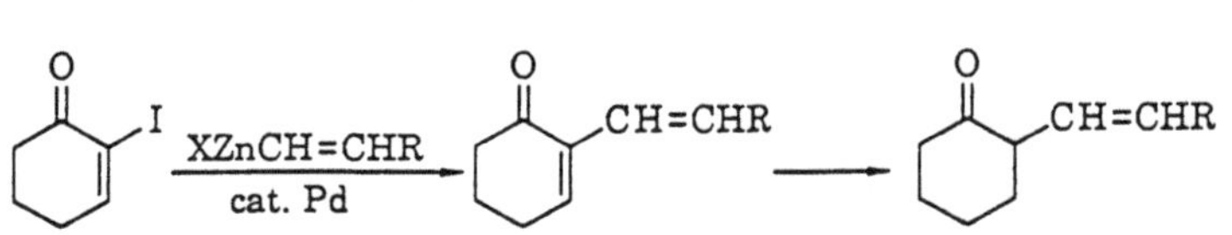
(204)

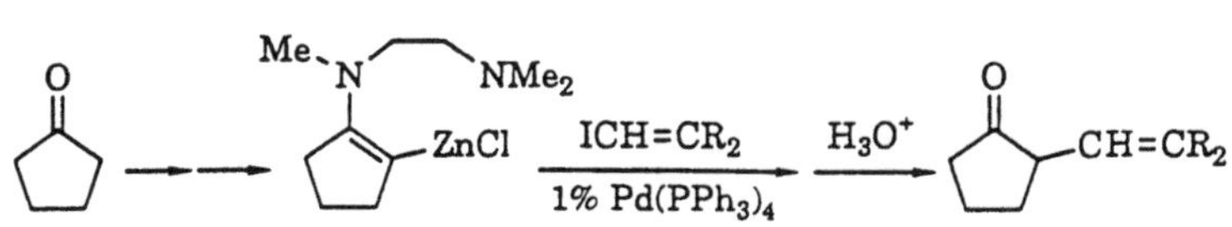
(205)

A wide variety of dienes and trienes are easily prepared by the cross-coupling of vinylic zinc reagents and bromovinylic boranes, followed by subsequent further alkylation[300–302] or carbonylation[303] (Eq. 206). Myrcene (Eq. 207)[350] and various enones, such as pseudoionone (Eq. 208),[351] have been prepared in this fashion.

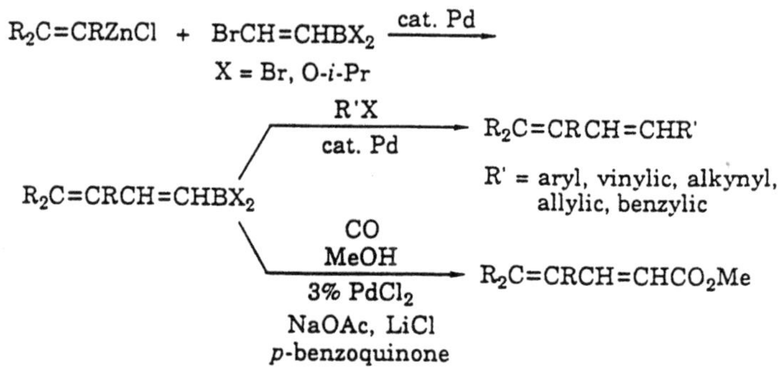

(206)

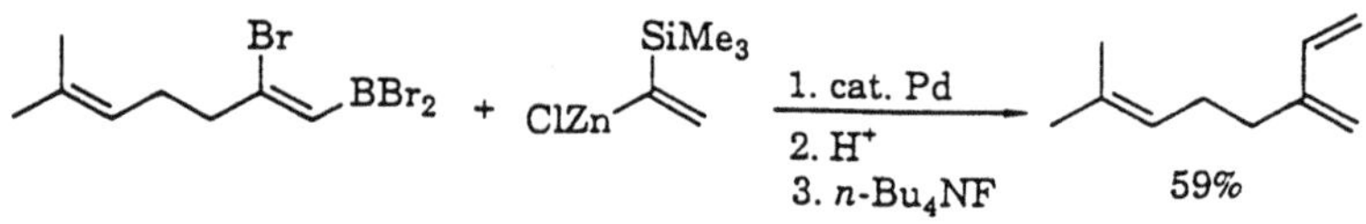

(207)

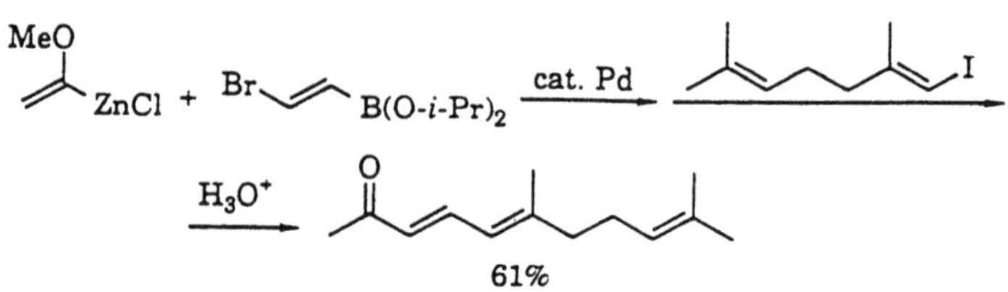

(208)

A number of fluorinated dienes have also been synthesized using vinylic zinc reagents (Eq. 209).[289,326,352–355]

$$CF_3CF{=}CFZnX + ICF{=}CF_2 \xrightarrow{2.3\%\ Pd(PPh_3)_4} CF_3CF{=}CFCF{=}CF_2 \quad \begin{matrix} Z\text{- }50\% \\ E\text{- }80\% \end{matrix} \qquad (209)^{352}$$

Cyclopropenone ketals can be metallated and reacted with vinylic halides in the presence of a palladium catalyst (Eq. 210).[356]

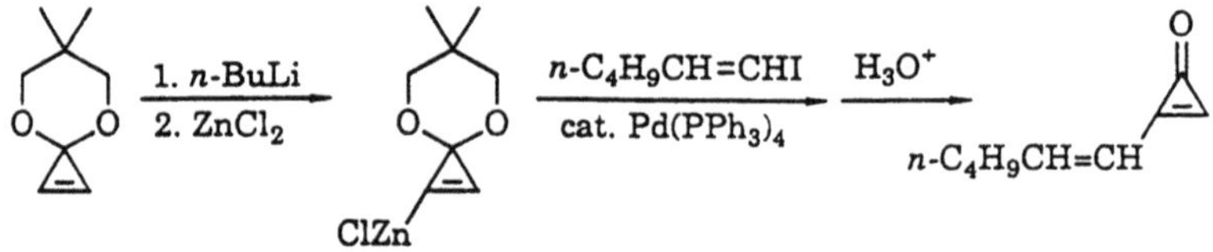

(210)

As noted in equations 207 and 208, vinylic zinc reagents bearing α-silyl,[350] -alkoxy,[351,357,358] -alkylthio,[358] and -carbamate[359] groups have been cross-coupled effectively (Eq. 211).

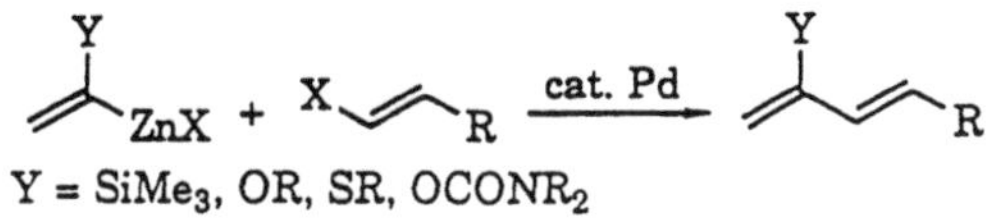

(211)

Vinylic zinc reagents derived from hydrazones have also been utilized (Eq. 212).[360]

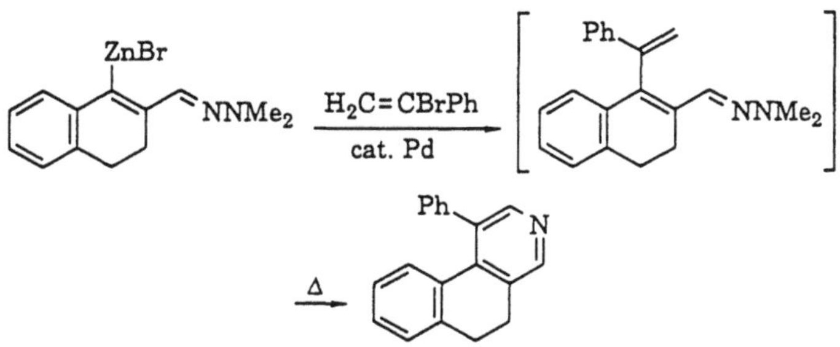

(212)

Steroids have been prepared in like manner (Eq. 213).[361–363]

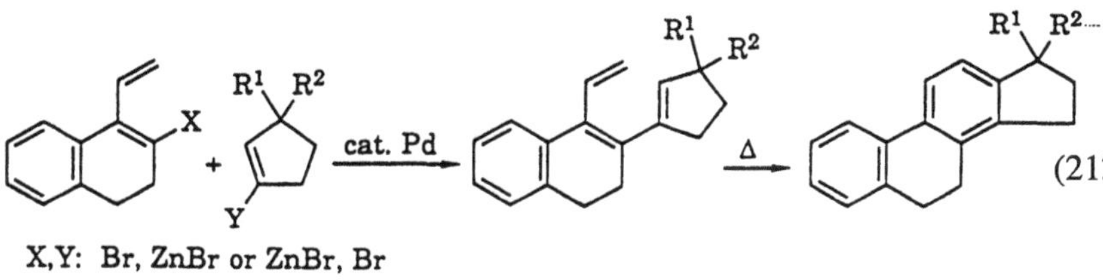

(213)

Vinylic cadmium and mercury compounds will react with vinylic halides in the presence of a palladium catalyst, but the reactions have not proven very useful.[327]

On the other hand, the cross-coupling of vinylic boranes and vinylic halides provides a very valuable approach to 1,3-dienes which has been extensively studied.[364,365] (*E*)-1-Alkenylboranes derived by the hydroboration of terminal alkynes react with (*E*)- or (*Z*)-vinylic halides stereospecifically (Eq. 214).[327,366,367] The corresponding (*Z*)-1-alkenylboranes, prepared via hydroboration of 1-halo-1-alkynes, also cross-couple stereospecifically (Eq. 215).[367–369] Other substituted vinylic boranes also undergo this process as well (Eqs. 216 and 217).

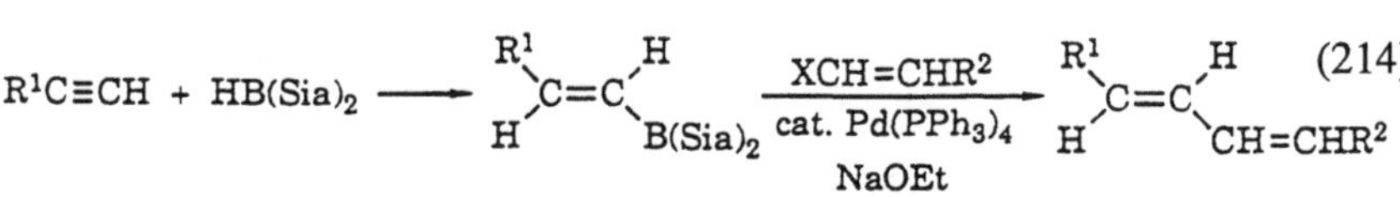

(214)

Corresponding boronate complexes also couple efficiently when $ZnCl_2$ is added (Eq. 218).[327,358]

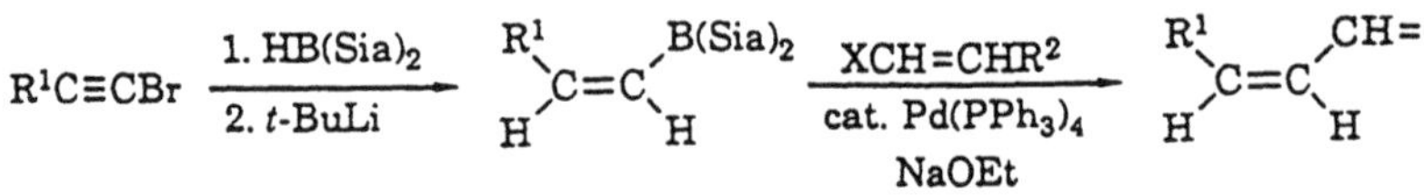

(215)

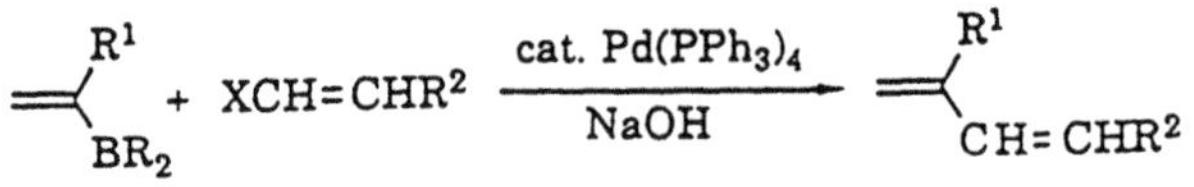

[370] (216)

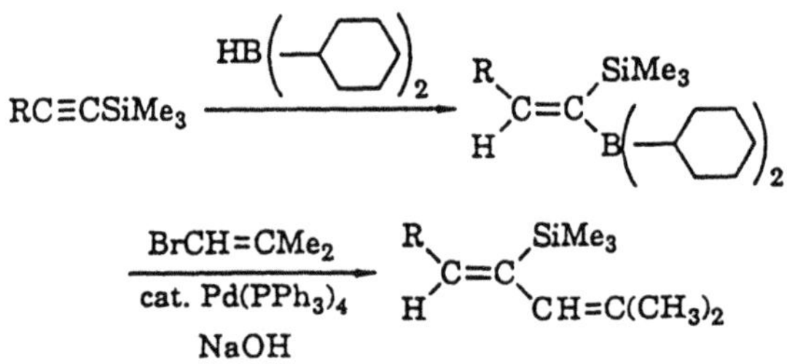

[371] (217)

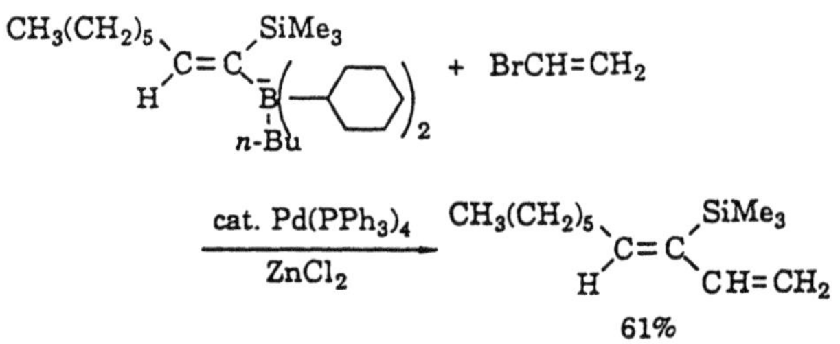

(218)

This chemistry has been applied to the synthesis of olefinic fatty acid amides (Eq. 219),[287] simple pheromones,[372–374] coriolic acid,[210] leukotrienes (Eq. 220)[201,375] and the fungal prohormone trisporol B.[376]

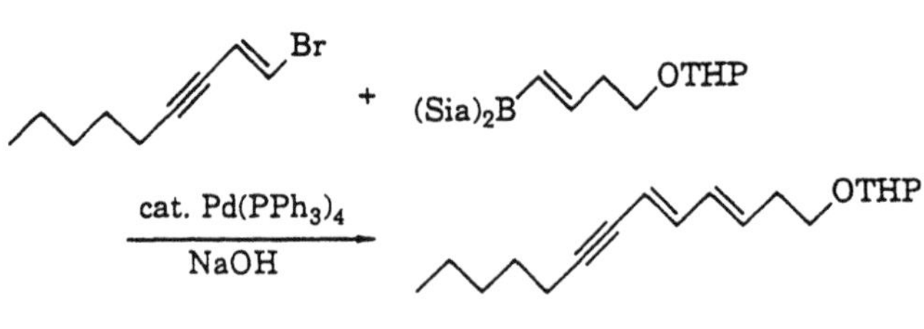

(219)

As noted earlier, vinylic boranes prepared via bromoboration of terminal alkynes can be sequentially alkylated first at the halogen and then at boron providing a versatile approach to all sorts of alkenes (Eq. 221).[300–302,350,351]

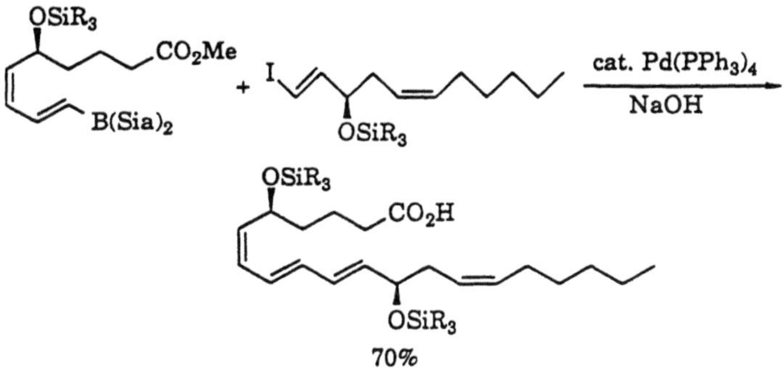

(220)

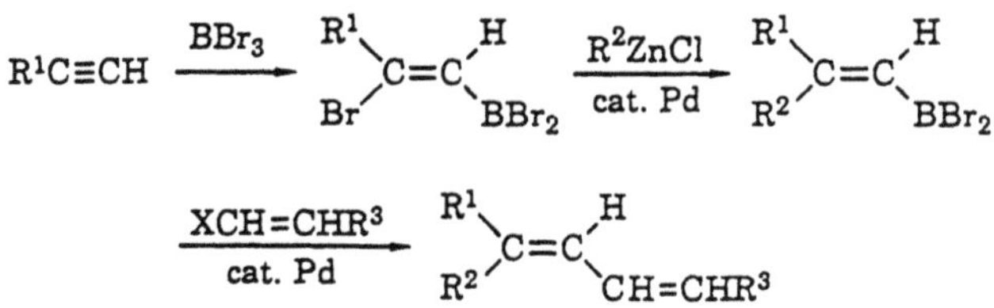

(221)

Boronate esters prepared from 1-halo-1-alkynes undergo analogous coupling processes (Eq. 222).[377–379] This reaction has been utilized to synthesize the pheromone bombykol (Eq. 223)[369] and dienyl sulfides (Eq. 224).[380]

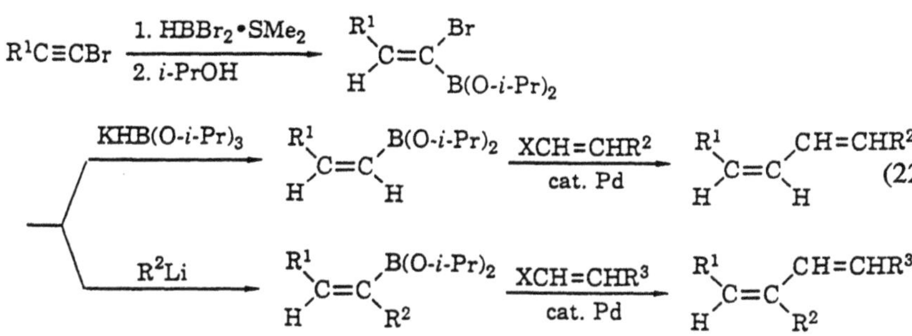

(222)

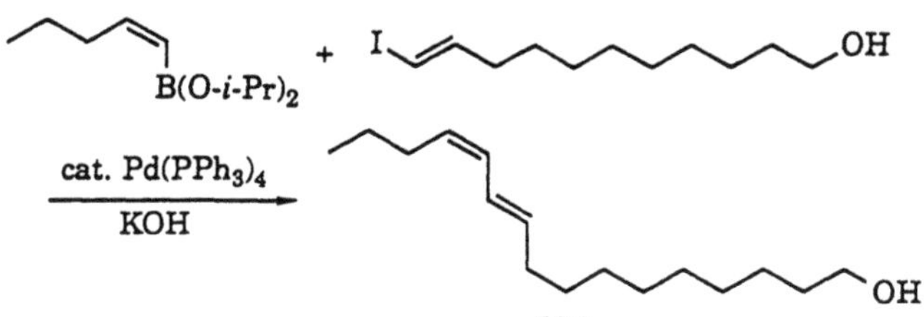

(223)

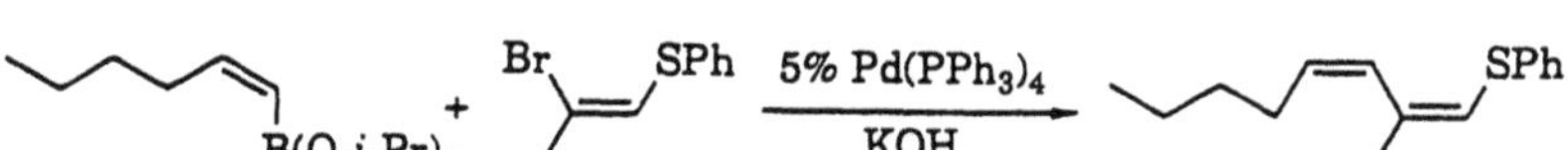

(224)

Boronate esters derived from alkynes via hydroboration using catecholborane are also effectively cross-coupled using hydroxide or alkoxide in the presence of catalytic amounts of $Pd(PPh_3)_4$ (Eq. 225).[367,368,381] This reaction has been applied to β-halovinylic sulfides,[277,380] 3-halo-2-alken-1-ones,[378] and 3-halo-2-alkenoates.[379] *trans*-(C_{10})Allofarnesene (Eq. 226)[382] and leukotrienes (Eq. 227)[206,383] have thus been prepared.

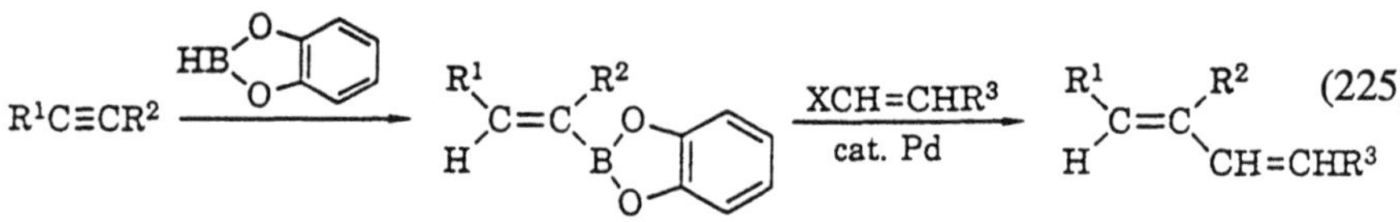

(225)

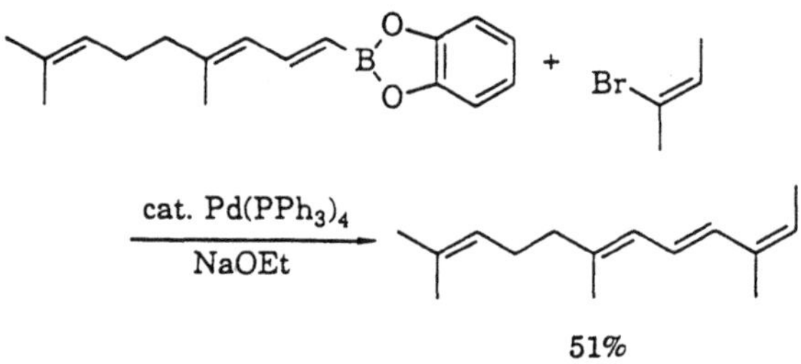

(226)

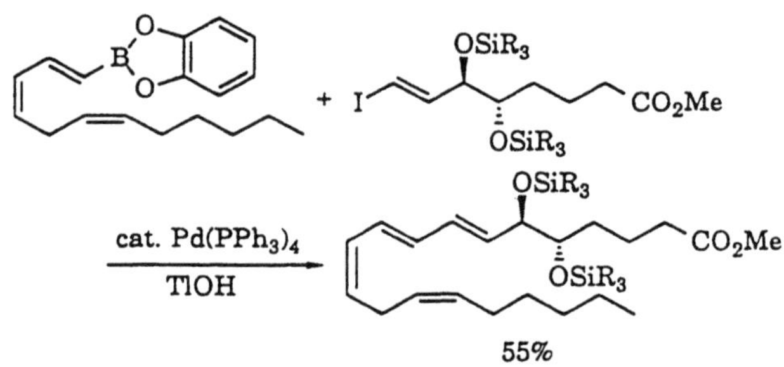

(227)

Note that thallium hydroxide has been utilized in this last process. This base has also been effective in the synthesis of chlorothricolide (Eq. 228).[384,385] The dramatic effect of TlOH in these coupling reactions was first reported in the synthesis of palytoxin using vinylboronic acids (Eq. 229).[386] It has since been used in the

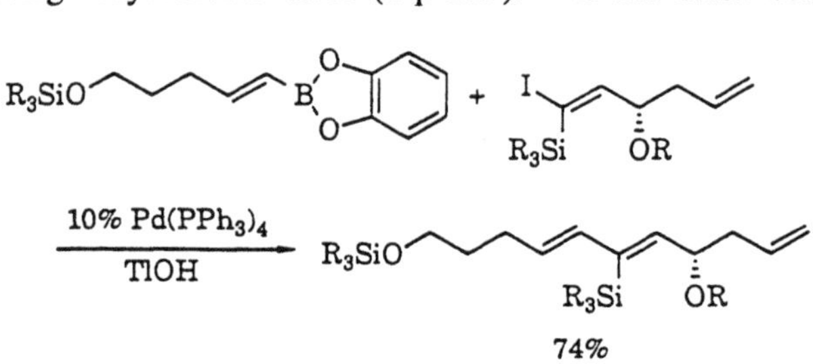

(228)

reaction of vinylboronic acids and 1,1-dibromoalkenes, where a highly stereoselective coupling is observed (Eq. 230).[387] The boronic acids are more effective in this last process than the corresponding catechol esters. Boronic acids have also been used in the synthesis of hydroprene,[388] a pheromone[389] and nucleosides.[221]

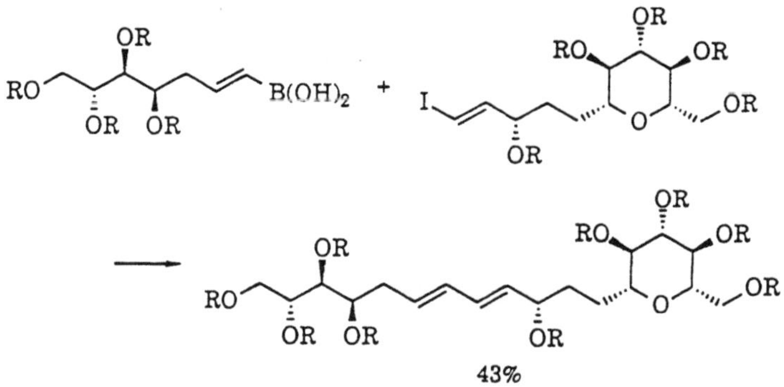
(229)

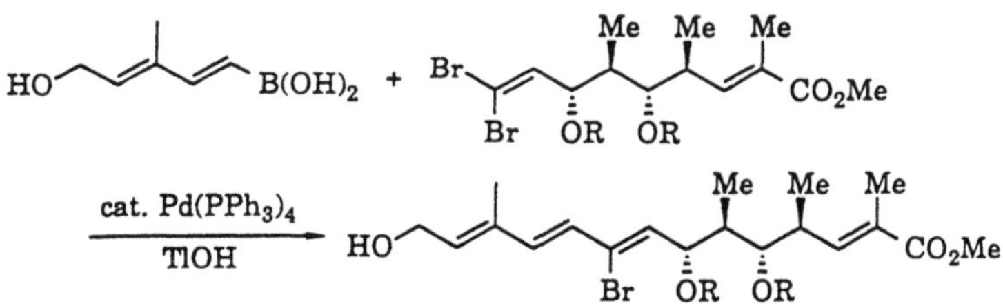
(230)

It is quite curious that the reaction of vinylic catechol boronate esters and vinylic halides in the presence of palladium black and triethylamine gives a "head-to-tail" coupling product (Eq. 231).[390] This process probably involves vinylpalladium addition to the borane carbon-carbon double bond, followed by palladium migration and palladium-boron beta elimination.

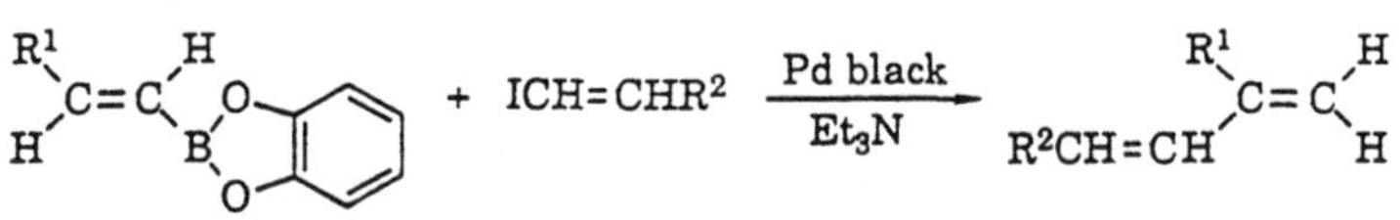
(231)

Vinylic aluminum reagents have proven nearly as effective as their zirconium and boron counterparts for the synthesis of 1,3-dienes. The hydroalumination of terminal alkynes affords (*E*)-vinylic alanes, which couple stereospecifically with vinylic halides in the presence of catalytic amounts of either $Pd(PPh_3)_2$, $Pd(PPh_3)_4$ or $Ni(PPh_3)_4$ (Eq. 232).[327,391] The palladium processes exhibit better stereospeci-

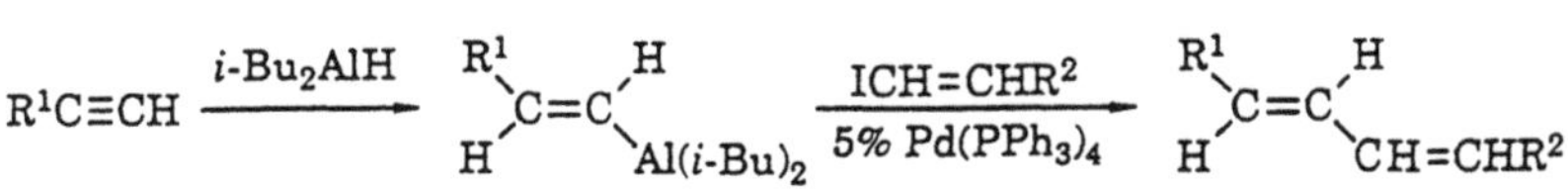
(232)

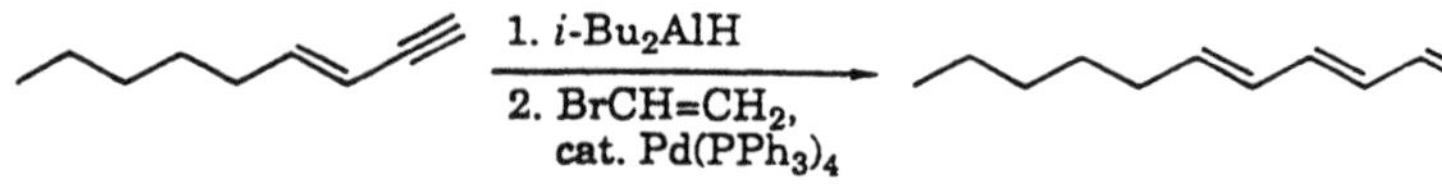

(233)

ficity. More recently, it has been found that the addition of $ZnCl_2$ accelerates the reaction and only 1% $Pd(PPh_3)_4$ is necessary.[335,392] The reaction has found utility for the synthesis of trienes (Eq. 233),[343] as well as dienyl silanes (Eqs. 234 and 235).

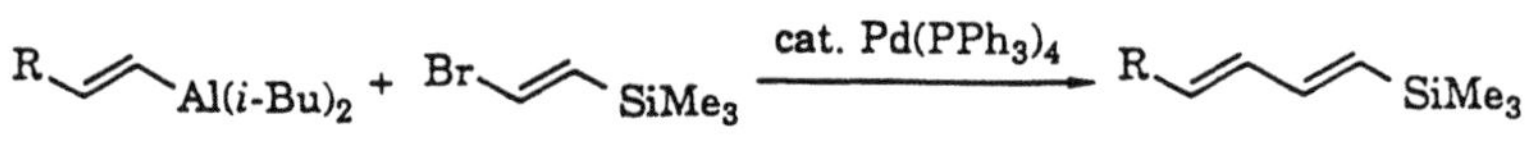

(234)[164,393]

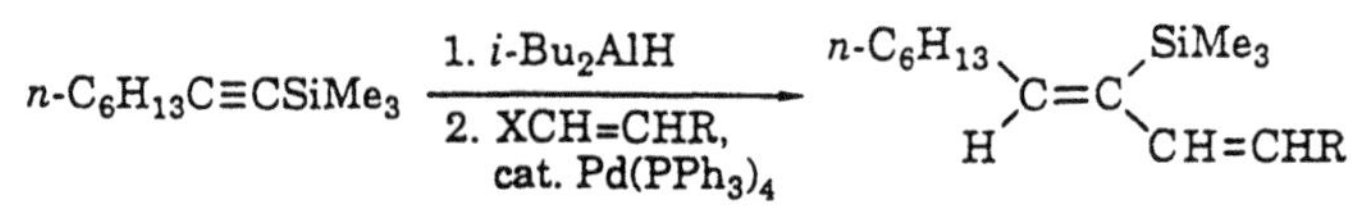

(235)[358]

The selective monosubstitution of dichloroethylenes can also be effected (Eqs. 236 and 237).

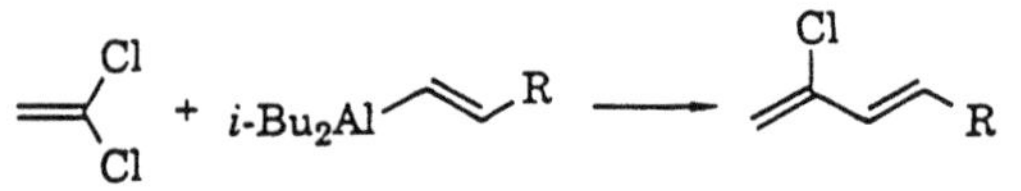

(236)[169]

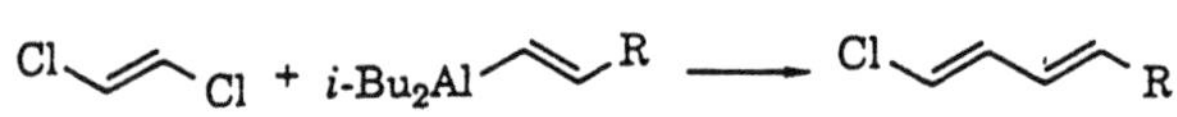

(237)[200]

As noted earlier (Eq. 195), the zirconium-catalyzed carboalumination of alkynes yields trisubstituted vinylic alanes which undergo facile palladium-catalyzed cross-coupling (Eq. 238).[325,331,332]

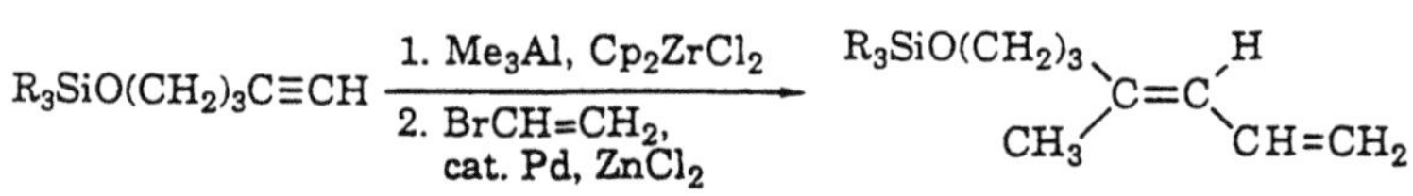

(238)[332]

The cross-coupling of vinylic silanes and vinylic halides has been far less studied. Vinylic trimethylsilanes will undergo this process in the presence of tris(diethylamino)sulfonium difluorotrimethylsilicate (TASF) with π-allylpalladium chloride dimer as the catalyst (Eq. 239).[316] Substitution of a methyl group on silicon by a fluorine affords much more reactive vinylic silanes (Eq. 240).[394] Surprisingly, the corresponding difluoromethylsilyl- and trifluorosilyl compounds are less reactive. Related alkoxy-substituted silanes may also be cross-coupled using n-Bu_4NF (Eq. 241).[395] Triethylammonium bis(catecholato)alkenylsiliconates have also been employed for cross-coupling, but the yields are only 21–29%.[396]

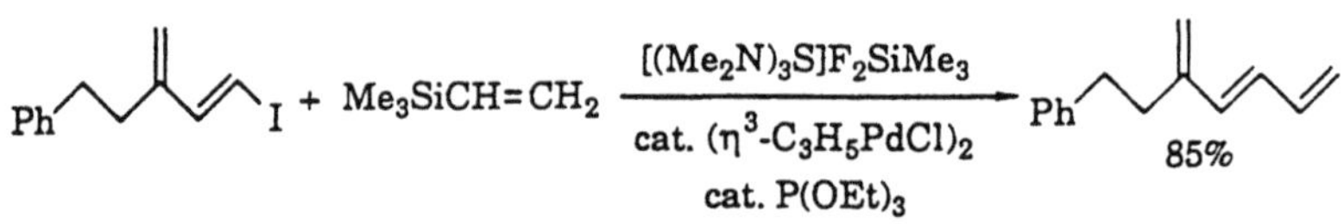

(239)

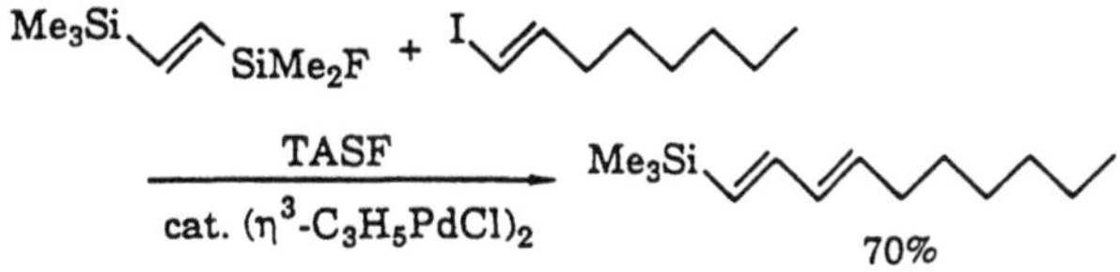

(240)

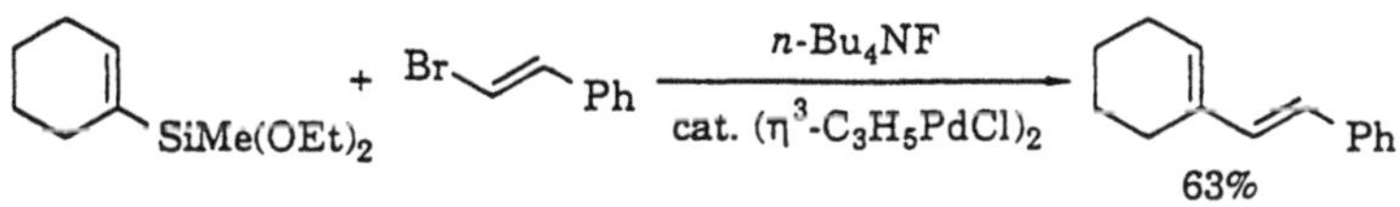

(241)

Vinylic tin compounds are among the most useful vinylic metallics for converting vinylic halides to 1,3-dienes.[26] The higher reactivity of the tin moiety has been utilized to prepare dienyl silanes (Eq. 242).[397] Numerous, stereoisomeric, functionally-substituted vinylic stannanes have been cross-coupled using $PdCl_2(CH_3CN)_2$ as the catalyst (Eq. 243).[398] Besides several simple examples of this reaction,[48,399] this process has been used to prepare tropolones,[314] abietanoid quinones,[400] plant toxins,[329] neooxazolomycin,[192] cephalosporins,[21] didemnenones,[401] avermectins,[315,402] pheromones,[398,403] coriolic acid[312] and an isomer,[404] prostaglandin B_1,[312] lipoxin B,[202] glucals,[405,406] various pyrimidines,[407,408] uracils,[409] uridine nucleosides[224,313,409,410] and related heterocycles.[411]

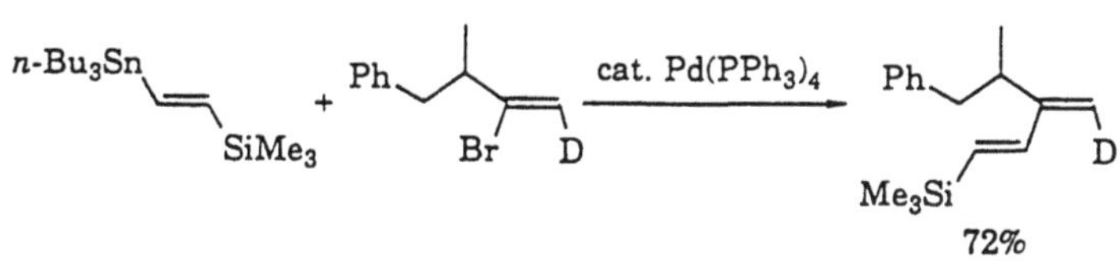

(242)

$$n\text{-}C_4H_9CH{=}CHI + n\text{-}Bu_3SnCH{=}CHCO_2Et \xrightarrow{\text{1-2\% } PdCl_2(CH_3CN)_2}$$

$$n\text{-}C_4H_9CH{=}CHCH{=}CHCO_2Et \qquad (243)$$

2-*E*, 4-*E*	83%
2-*Z*, 4-*E*	78%
2-*E*, 4-*Z*	65%
2-*Z*, 4-*Z*	62%

α-Alkoxy vinylic tin reagents serve as useful nucleophilic acylating agents (Eq. 244).[412–414]

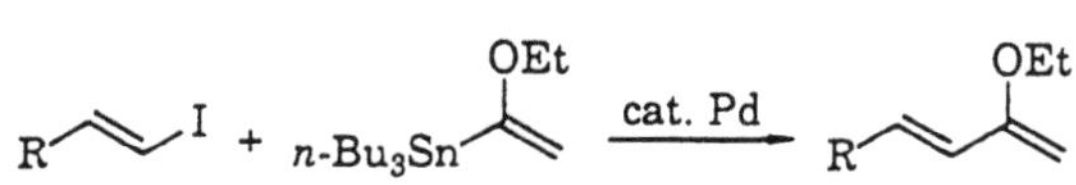

(244)

Cross-coupling can often be effected in more than one way (Eqs. 245 and 246).

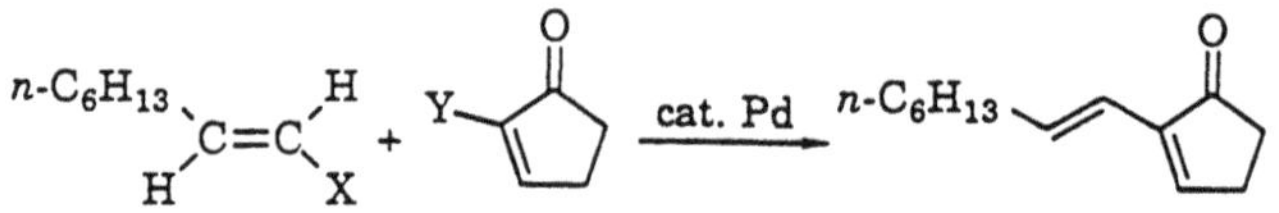

(245)[348]

X	Y	%
I	Me_3Sn	80
n-Bu_3Sn	I	64

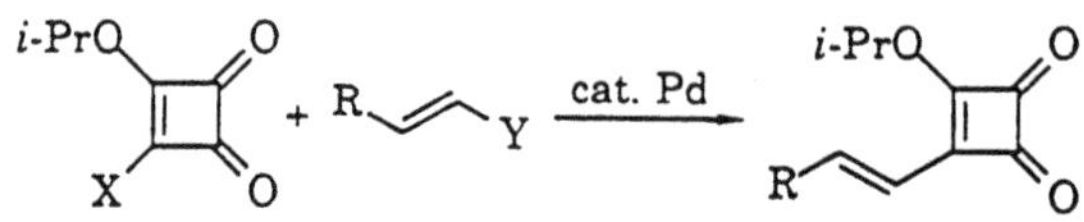

(246)[415,416]

X,Y = Cl, n-Bu_3Sn or n-Bu_3Sn, I

Coupling with Allenic Organometallics

There are only a few examples of the cross-coupling of vinylic halides and allenic organometallics by palladium catalysis. Allenic lithium,[417] magnesium,[418] copper,[418] and zinc[357,418] reagents all couple regioselectively in high yield (Eq. 247). Alkoxy-substituted allenic lithium[417] and zinc[357] reagents afford allenes readily hydrolyzed to dienones (Eq. 248).

$$\mathrm{RCH{=}C{=}CHM} + \mathrm{XCH{=}CHR'} \xrightarrow{\text{cat. Pd}} \mathrm{RCH{=}C{=}CHCH{=}CHR'} \qquad (247)$$

$$\mathrm{H_2C{=}C{=}C(OMe)M} + \mathrm{ICH{=}CHR} \xrightarrow{\text{cat. Pd}} \mathrm{H_2C{=}C{=}C(OMe)CH{=}CHR} \xrightarrow{\mathrm{H_3O^+}} \mathrm{H_2C{=}CHC(O)CH{=}CHR} \qquad (248)$$

M = Li, ZnX

Reaction with Alkynyl Organometallics

Alkynyl metallics of magnesium, copper, zinc, silicon and tin also cleanly react with vinylic halides to form enynes (Eq. 249). In view of the ease with which enynes can be prepared directly from terminal alkynes and vinylic halides, it is questionable whether any of these processes offers any synthetic advantages. Nevertheless, the magnesium process proceeds stereospecifically in high yield,[259] and has been used to prepare a pheromone,[374] a vitamin A analogue (Eq. 250),[419] and a potentially fungicidal, thiophene-containing acetylene.[187]

$$\mathrm{RC{\equiv}CM} + \mathrm{XCH{=}CHR'} \xrightarrow{\text{cat. Pd}} \mathrm{RC{\equiv}CCH{=}CHR'} \qquad (249)$$

M = MgX, Cu, ZnX, SiR_3, SnR_3

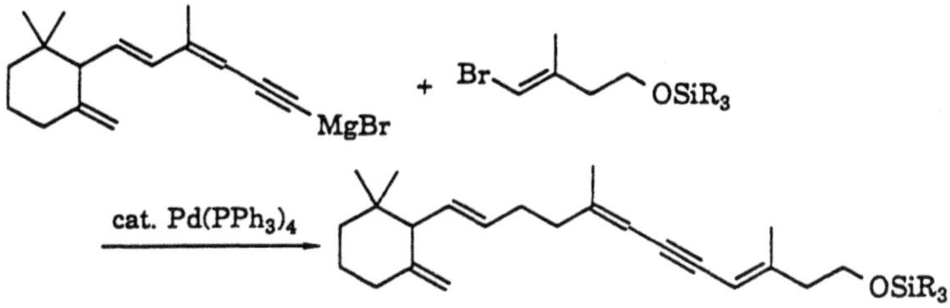

(250)

Alkynyl copper reagents undergo coupling with vinylic halides in the presence of catalytic amounts of $PhPdI(PPh_3)_2$ (Eq. 251).[29,420]

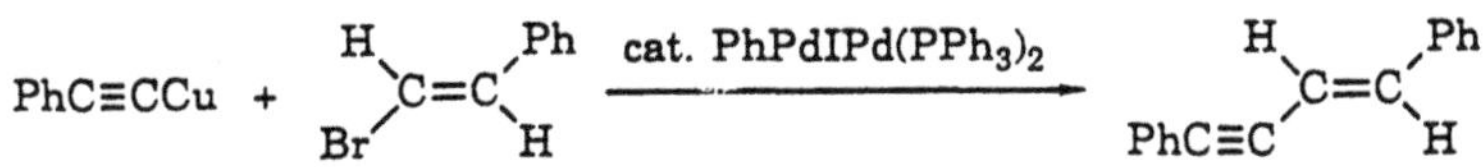

(251)

Alkynyl zinc reagents are also quite effective for stereospecific enyne formation.[421] Enediynes can also be prepared in this fashion (Eq. 252).[422] This process is highly diastereoselective for *E*-alkenyl bromides (Eq. 253).[180,181,343] Desilylation affords a terminal enyne (R = *n*-Pr) useful in the synthesis of a naturally occurring dienyne (see Eq. 135). Similar high diastereoselectivity has been observed in the reactions of (*E*)- and (*Z*)-1,2-dibromoethylene (Eq. 254).[287,343,423] Dialkynylation can be a problem here, but is easily circumvented by using 1-chloro-2-iodoethylene[424,425] or related alkenes[426] (Eq. 255). This chemistry has been applied to the synthesis of pyrethroids, where dialkynylation predominates (Eq. 256).[427]

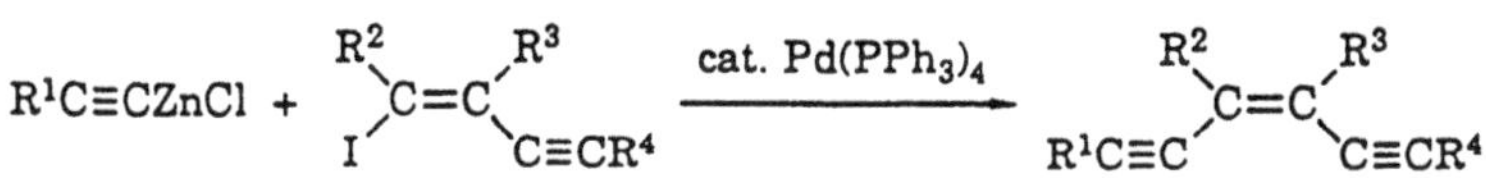

(252)

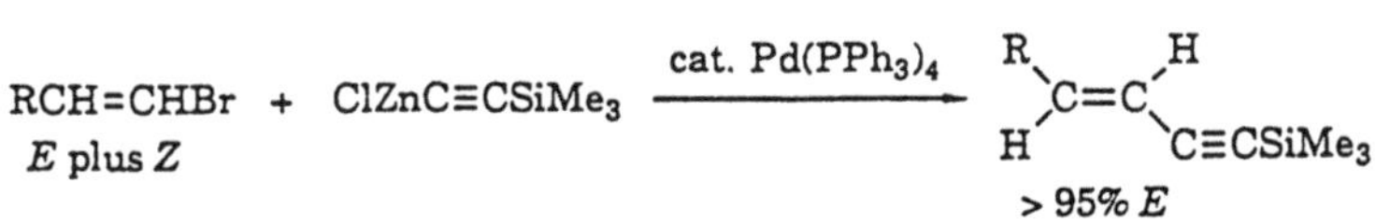

(253)

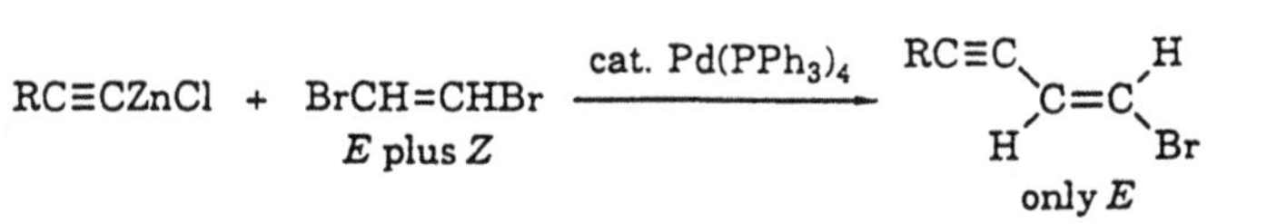

(254)

$$R^1C{\equiv}CZnCl + ICH{=}CClR^2 \xrightarrow{\text{cat. Pd}} R^1C{\equiv}CCH{=}CClR^2 \qquad (255)$$

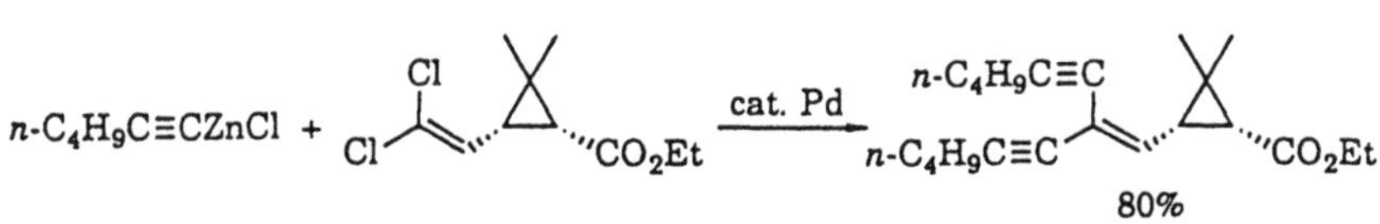

(256)

The coupling of alkynyl zinc reagents and vinylic halides has been applied to the synthesis of pheromones,[345,346,428] prostaglandins (Eq. 257),[429] and uridine nucleosides.[430]

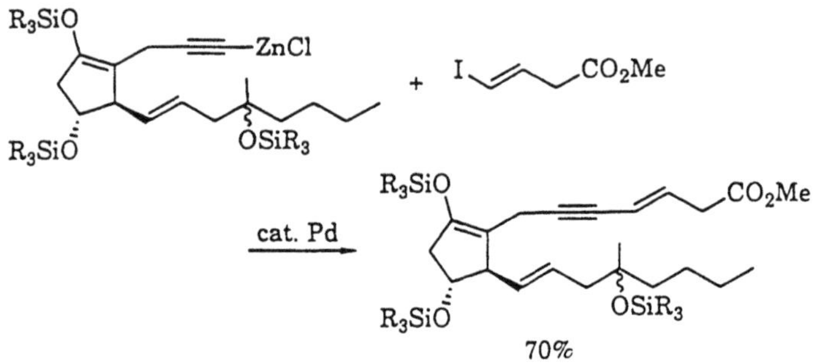

(257)

By employing β-bromovinyl metallics, enynyl silicon[164,393] and boron[300–303,350] compounds can be prepared (Eq. 258). As noted previously (Eq. 180), the vinylic boron compounds thus obtained can be further coupled with a variety of organic halides or carbonylated to esters.

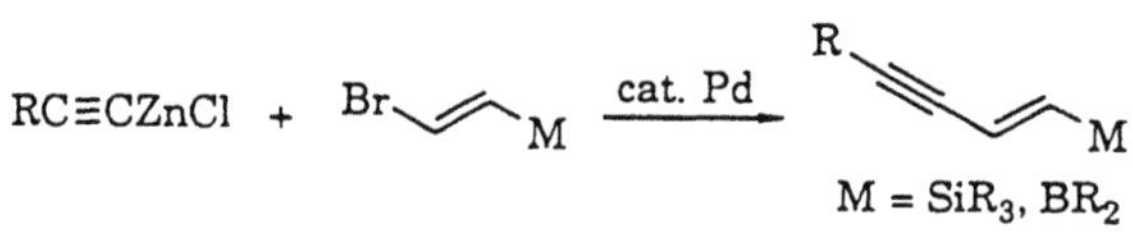

(258)

Finally, fluorinated enynes,[431,432] even containing silicon,[289] can be prepared via alkynyl zinc reagents (Eq. 259).

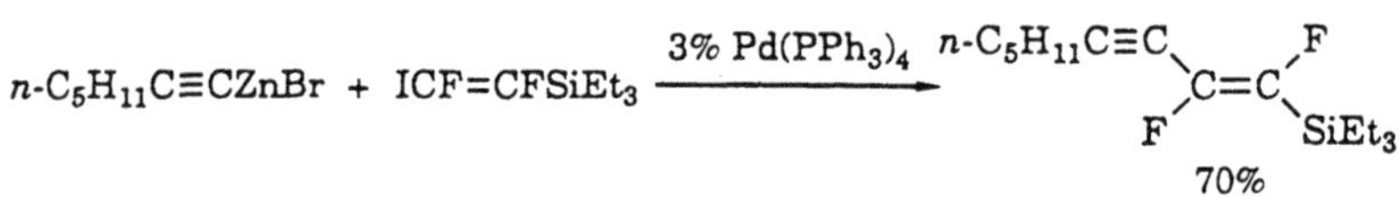

(259)

Alkynyl silanes react with vinylic halides in the presence of tris(diethylamino)sulfonium difluorotrimethylsilicate, when π-allylpalladium chloride dimer is used as a catalyst (Eq. 260).[316] Dienynes can also be prepared in this fashion (Eq. 261).[433]

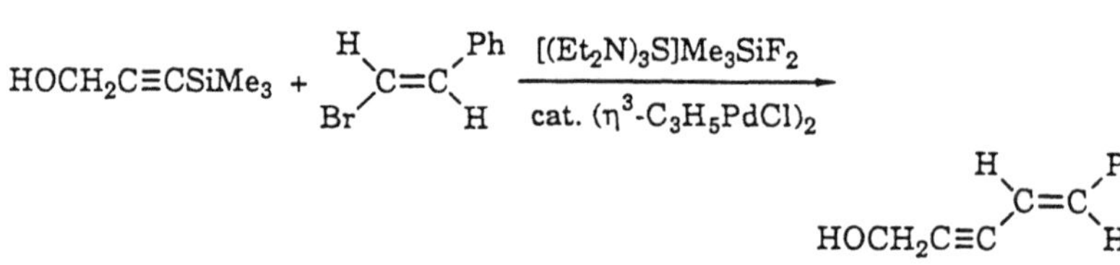

(260)

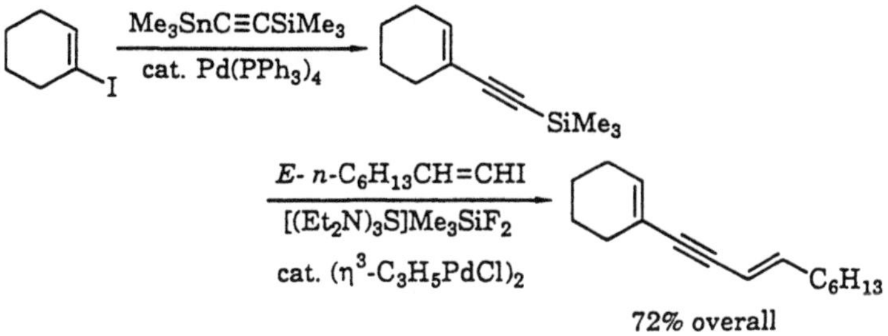

(261)

As noted above, alkynyl stannanes can be used for the preparation of enynes from vinylic halides. A general procedure using $PdCl_2(CH_3CN)_2$ as catalyst has been reported.[434] However, $Pd(PPh_3)_4$ and $Pd_2(dba)_3$ have also been employed as catalysts.

A wide variety of enynes have been prepared by this process, including enediynes (Eq. 262),[422] silyl enynes,[164,433] alkynyl cyclobutenediones (Eq. 263),[415,435] numerous pheromones,[434] the antimycotic agent terbinafine,[436] the neocarzinostatin chromaphore (Eq. 264),[437–439] five-substituted uracils and uridines,[409] and four-substituted 1,2,6-thiadiazine 1,1-dioxides.[411]

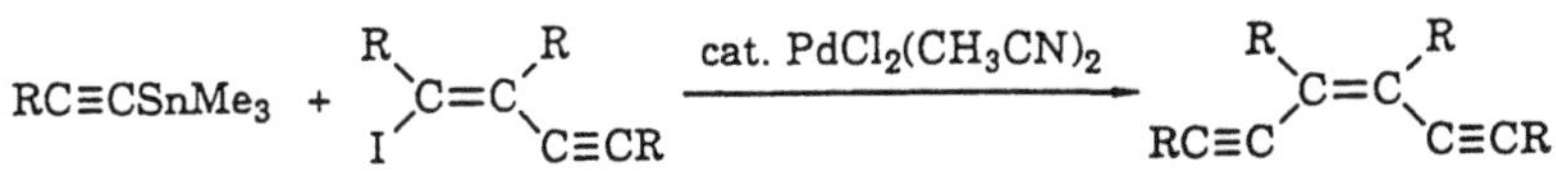

(262)

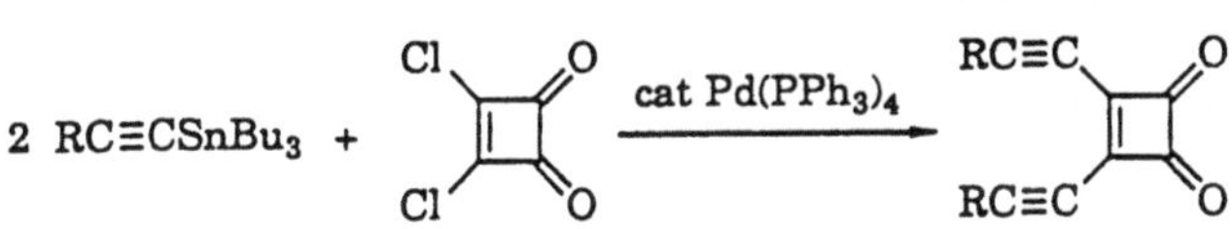

(263)[435]

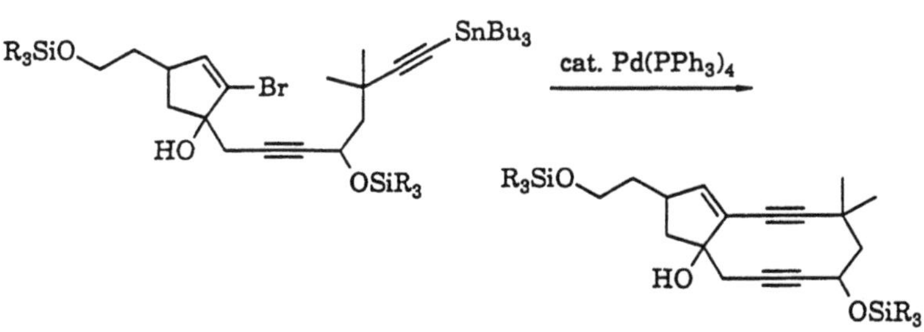

(264)[437]

Cross-Coupling with Aryl Metallics

A wide variety of aryl metallic reagents have been cross-coupled with vinylic halides in the presence of a palladium catalyst. Both aryl lithium[255,257,440] and magnesium[259] reagents react stereospecifically in the presence of $Pd(PPh_3)_4$. Using $PdCl_2(dppf)$ as catalyst and aryl magnesium reagents, the reaction is diastereoselective for (*E*)-1-bromo-1-alkenes.[268] Aryl magnesium reagents couple with bromovinyl sulfides[275] and silanes[441] to give the corresponding styryl sulfides and silanes. Antiestrogenic tamoxifen (X = NMe_2) has been prepared by this process (Eq. 265).[442] Aryl zinc, boron and tin reagents can also be employed here. 1,1-Dichloroalkenes can be regio- and stereoselectively mono- and diarylated in this manner (Eq. 266).[264] Note that aryl zinc reagents are also effective here.

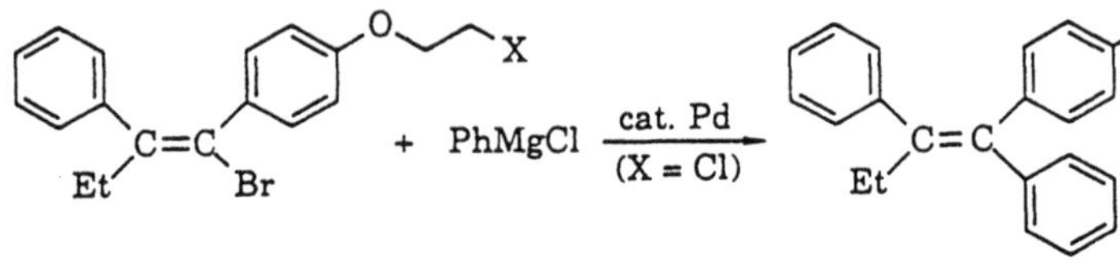

(265)

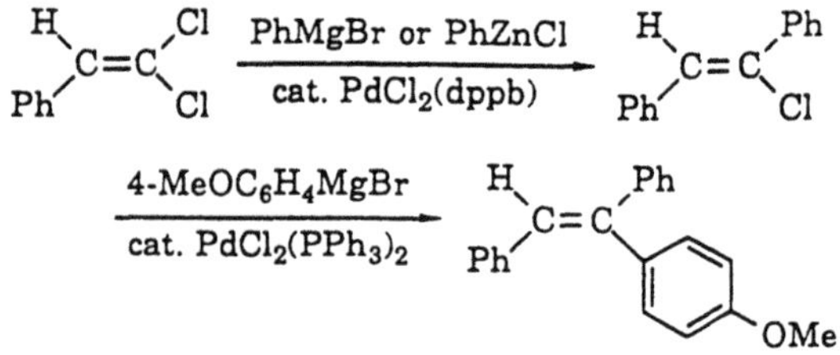

(266)

Aryl copper reagents have been cross-coupled with vinylic halides to prepare styrene complexes containing a chromium tricarbonyl moiety[443] and aryl sydnones,[444] but no other examples have been reported.

On the other hand, aryl zinc reagents, as already noted, usually cross-couple in high yield.[341] Heterocyclic zinc reagents have also been utilized,[445,446] as have halo dienes (Eq. 267).[360,362,363]

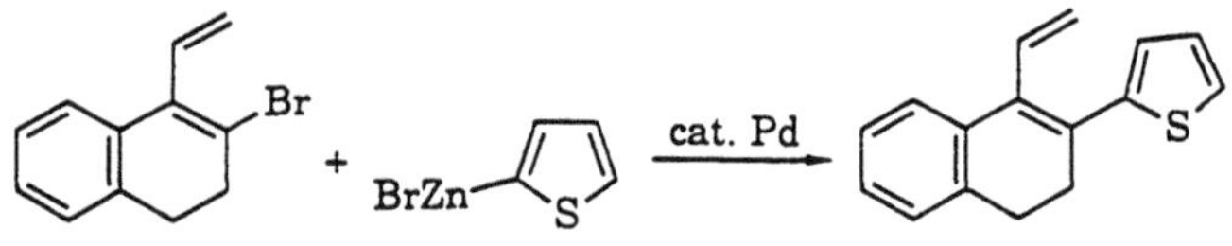

(267)

As noted in Eq. 266, the process exhibits high regio- and stereoselectivity with 1,1-dichloroalkenes. This has been utilized for the synthesis of a variety of pyrethroids (Eq. 268).[427]

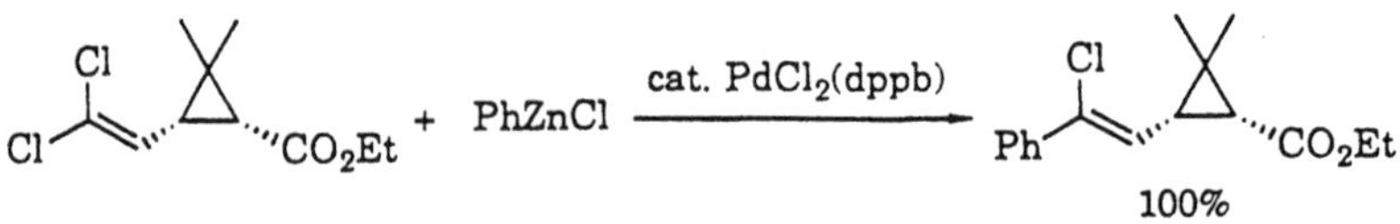

(268)

The cross-coupling of vinylic halides and aryl zinc reagents has been applied to the synthesis of tamoxifen (see Eq. 265) and derivatives,[442,447–450] broparestrol,[451]

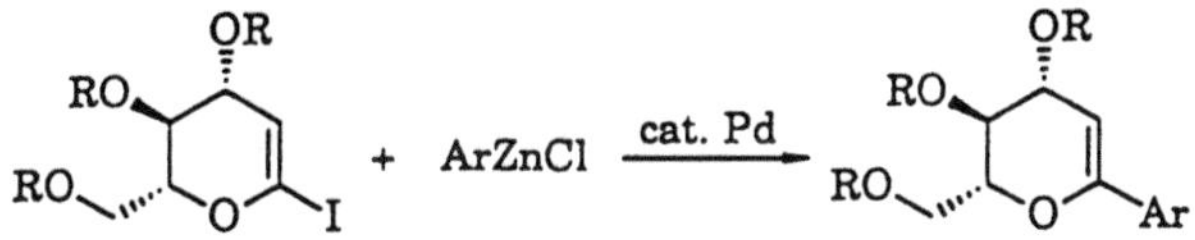
(269)

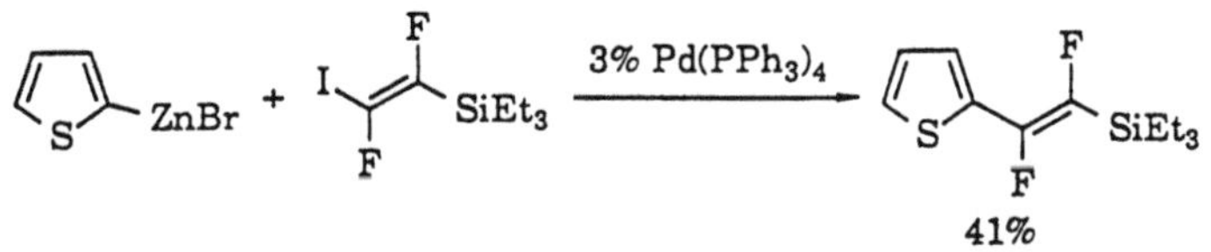
(270)

clomid,[452,453] 2,3,5-trisubstituted furan intermediates for lophotoxin and pukalide,[454] aryl tropolones,[455] aryl uridines,[456] and aryl glucals (Eq. 269).[405] Arylboronic acids can also be utilized in this last process. Arene-containing fluoroolefins can also be prepared in this manner (Eq. 270).[289,326] Attempts to cross-couple heterocyclic zinc reagents with (1-bromovinyl)trimethylsilane results in a mixture of substitution products (Eq. 271).[446,457] On the other hand, bromovinyl boron compounds couple cleanly with aryl zinc reagents (Eq. 272). The resulting vinylboranes can be further alkylated[300–302,351] or carbonylated.[303]

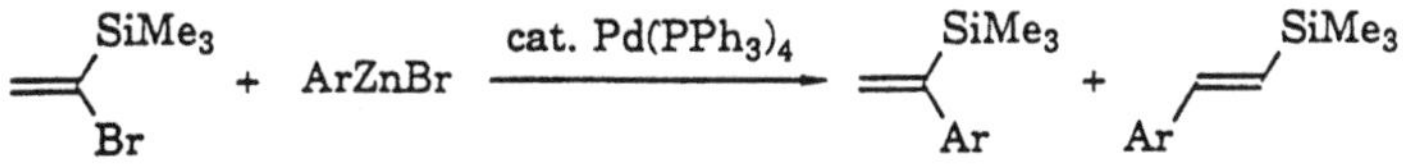
(271)

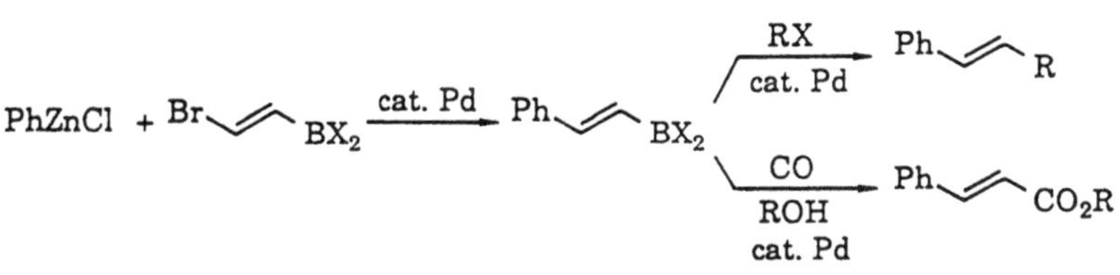
(272)

As noted above (Eq. 269), arylboronic acids have been effectively cross-coupled with vinylic halides to produce aryl glucals.[405] This process has also been utilized to arylate tropones,[458,459] chromones (Eq. 273),[460,461] and uridines.[462]

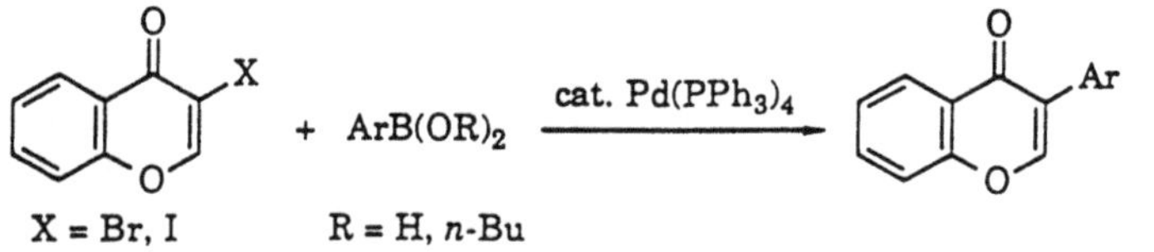
(273)

The one example in the literature of the palladium-catalyzed coupling of phenyl pentafluorosilicate with styryl bromide afforded a low yield of rearranged product (Eq. 274).[463]

$$K_2[PhSiF_5] + PhCH{=}CHBr \xrightarrow{\text{cat. Pd}} \underset{23\%}{Ph_2C{=}CH_2} \qquad (274)$$

Aryl tin reagents are among the most effective organometallics for cross-couplings with vinylic halides.[26,464] This process has been effectively used to arylate cyclobutenediones,[415] tropolones,[314] 1,2,6-thiadiazine 1,1-dioxides,[411] uracil,[409] and uridines.[224,313,462] The coupling of arylstannanes and vinylic halides has also been utilized in the synthesis of tamoxifen (Eq. 265),[442] cembranolides,[454] and macrolides, such as zearalenone (Eq. 275).[465] The greater reactivity of the tin moiety over a boronic acid group has been used to prepare boron-containing nucleosides (Eq. 276).[466] Heterocyclic tin compounds can also be employed for cross-coupling.[33,408]

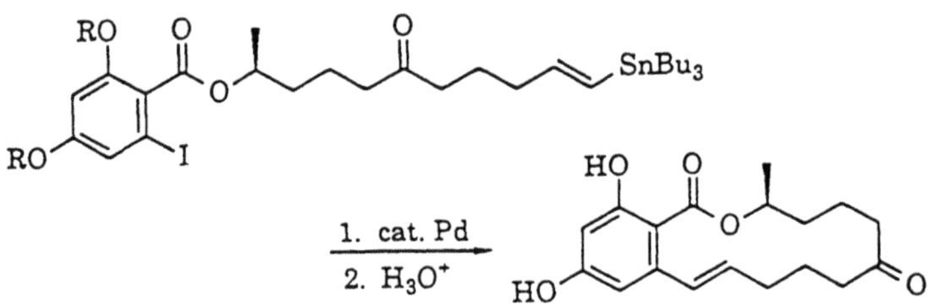

(275)

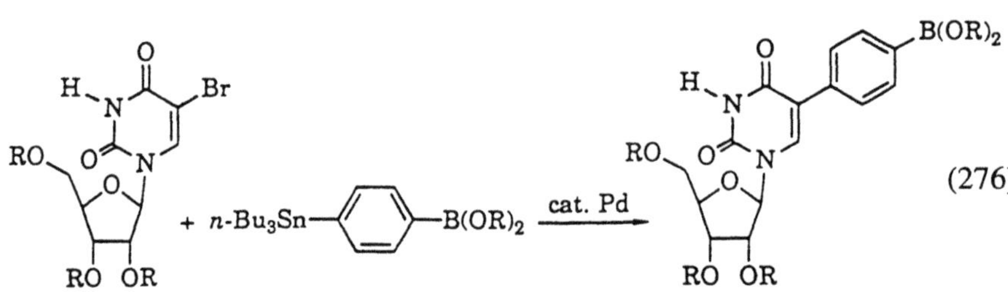

(276)

Reaction with Carbonyl Compounds

There are a number of other functional groups which can be introduced in place of the halogen of a vinylic halide through palladium chemistry. Palladium acetate plus chromium chloride catalyzes the addition of vinylic groups to aldehydes, although the preferred procedure is to use nickel chloride instead of palladium (Eq. 277).[467]

Ketone-containing vinylic halides react with a titanium isocyanate complex in the presence of a palladium catalyst to provide enamines (Eq. 278).[468]

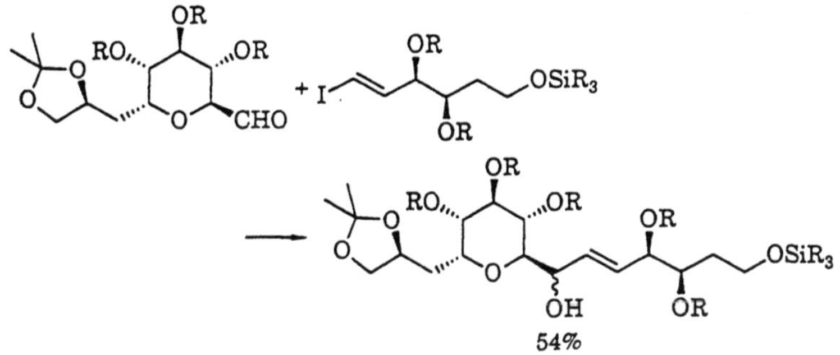

(277)

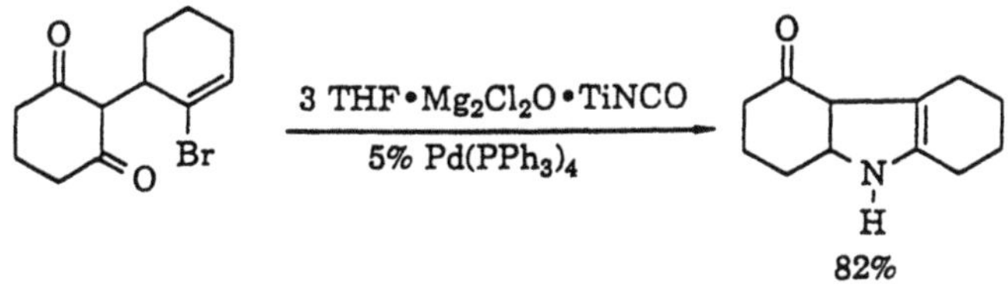

(278)

Halogenation

Vinylic palladium species react with iodine to produce vinylic iodides, but this process is of no synthetic value at presence.[162]

Substitution by Oxygen

1,1-Dibromo-1-alkenes react with hydroxide in the presence of palladium to generate carboxylic acids in good yields (Eq. 279).[469]

$$R_2C{=}CBr_2 \xrightarrow[\text{cat. Pd(dppe)}_2]{\text{NaOH or KOH}} R_2CHCO_2H \qquad (279)$$

Aryloxy groups have been introduced into cyclobutenediones through the corresponding tin reagents (Eq. 280).[415]

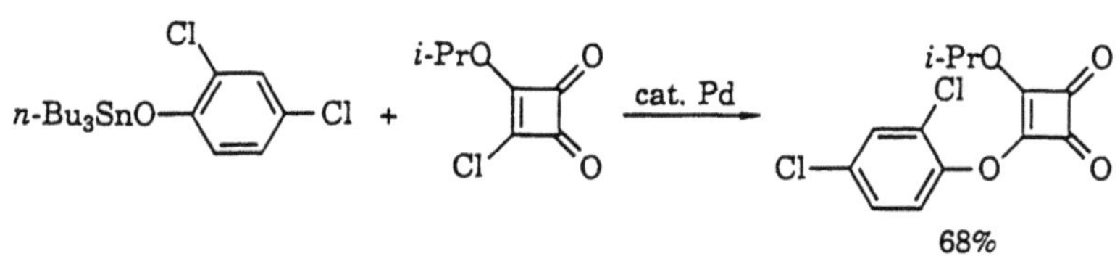

(280)

Vinylic chlorides can be transformed into enol acetates using sodium acetate with palladium chloride as catalyst (Eq. 281).[470]

$$\mathrm{PhSO_2CHPhCH_2C(Cl){=}CH_2} \xrightarrow[\text{cat. PdCl}_2,\ 65\,^\circ\text{C}]{\text{NaOAc}} \underset{71\%}{\mathrm{PhSO_2CHPhCH_2C(OAc){=}CH_2}} \qquad (281)$$

Sulfur Substitution

Palladium also catalyzes stereospecific substitution by sulfur (Eq. 282).[257,471] Tin reagents have also been utilized to prepare vinylic sulfides (Eqs. 283 and 284).

$$\mathrm{R^1CH{=}CHBr} + \underset{M\,=\,\mathrm{Li,\ Na}}{\mathrm{MSR^2}} \xrightarrow{\text{cat. Pd}} \mathrm{R^1CH{=}CHSR^2} \qquad (282)$$

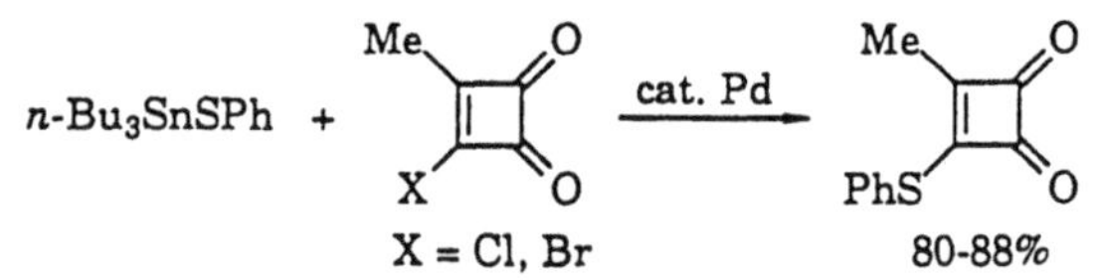

(283)[415]

$$2\ \mathrm{PhCH{=}CHBr} + \mathrm{(Et_3Sn)_2S} \xrightarrow[\text{cat. PhPdI(PPh}_3)_2]{80\text{-}88\%} \underset{100\%}{\mathrm{(PhCH{=}CH)_2S}} \qquad (284)$$[29]

Substitution by Phosphorus

Vinylic phosphorus compounds can be prepared via palladium-catalyzed vinylic substitution processes (Eqs. 285–288).

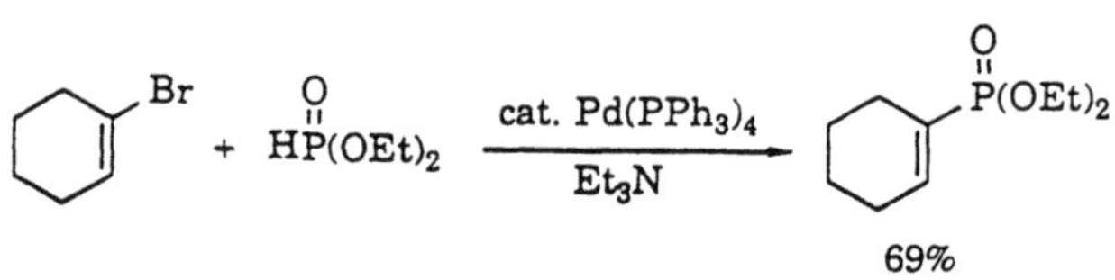

(285)[472,473]

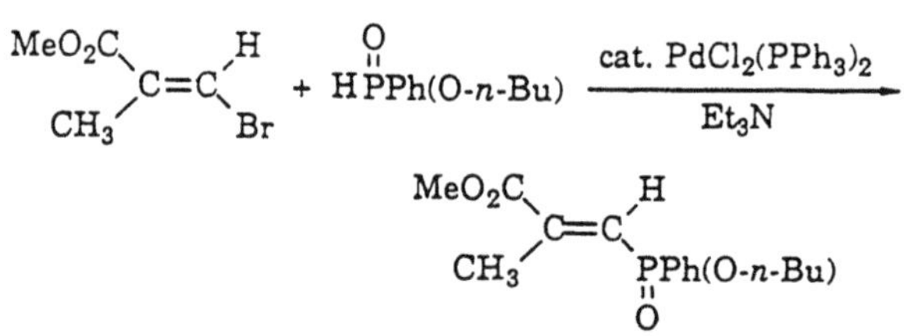

(286)[474,475]

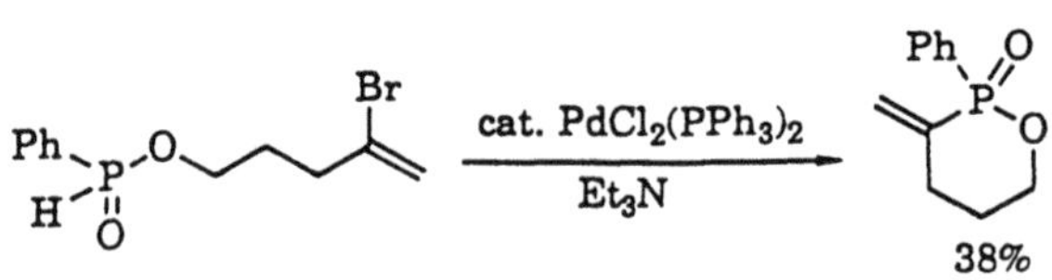

(287)[476]

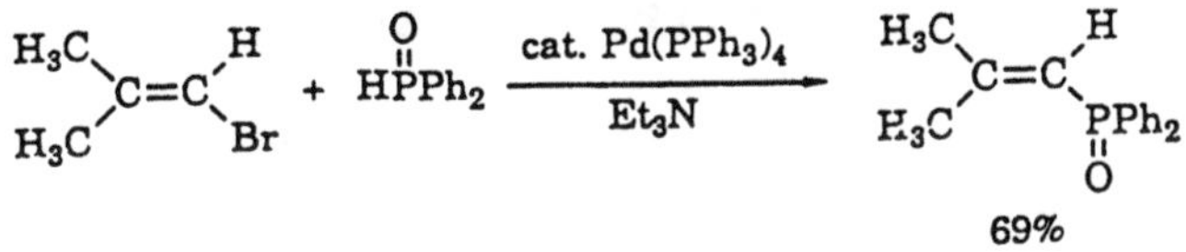

(288)[477]

Silicon Substitution

Vinylic silanes can also be prepared stereospecifically directly from vinylic halides (Eqs. 289–291).

$$n\text{-}C_5H_{11}CH{=}CHI + Al(SiMe_3)_3 \xrightarrow{\text{cat. Pd(PPh}_3)_4} n\text{-}C_5H_{11}CH{=}CHSiMe_3 \quad (289)$$ [478]

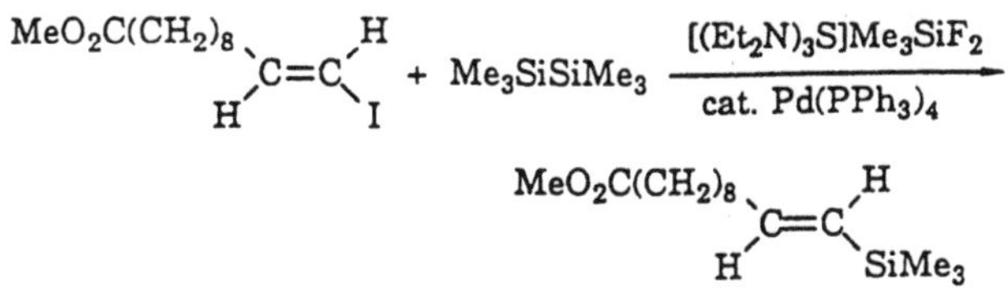

(290)[479]

$$PhCH{=}CHBr + Cl_{3\text{-}n}Me_nSiSiMe_nCl_{3\text{-}n} \xrightarrow{\text{cat. Pd(PPh}_3)_4} PhCH{=}CHSiMe_nCl_{3\text{-}n} \quad (291)$$ [480]

n = 1 (85%), 2 (83%)

Substitution by Tin

Vinylic stannanes can be prepared in like manner from sulfoxide-, sulfone-, and ester-containing vinylic chlorides (Eq. 292).[481]

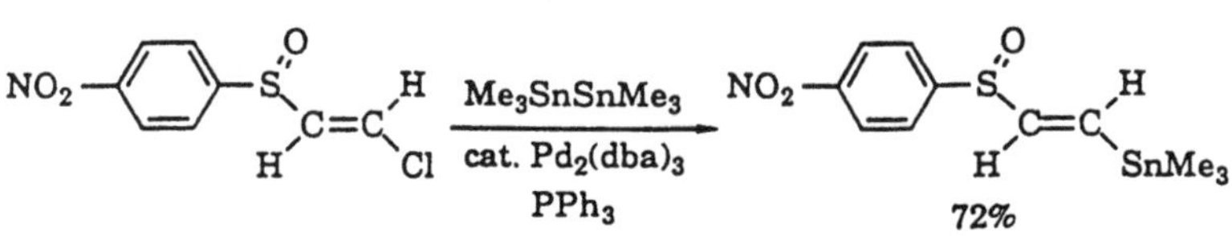

(292)

III. VINYLIC IODONIUM SALTS

Vinylic iodonium salts, though not readily available, can be used to generate vinylic palladium intermediates which undergo alkene substitution chemistry (Eq. 293).[482]

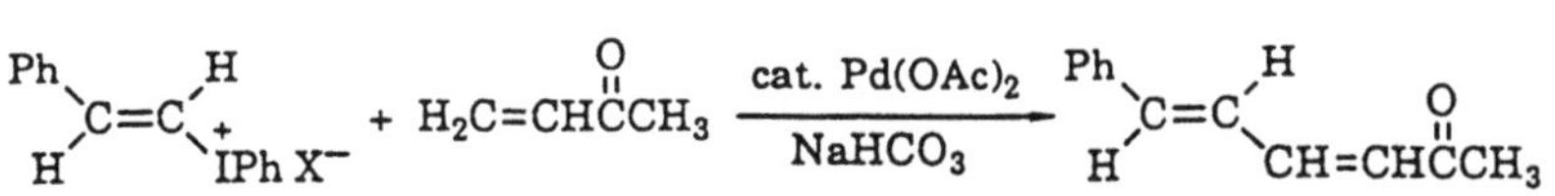

(293)

$X = BF_4$ (73%), OTs (72%)

IV. VINYLIC TRIFLATES

In recent years vinylic triflates have been shown to undergo many of the same palladium-catalyzed processes as vinylic halides.[483] Since triflates are often prepared from different starting materials than halides, their use greatly expands the synthetic utility of this chemistry.

Reduction

Several different reagents are effective for the reduction of enol triflates to alkenes. Excess formic acid and tri-*n*-butylamine in the presence of 2% $Pd(OAc)_2(PPh_3)_2$ affords alkenes in excellent yield (Eq. 294).[484] Ketones, esters, alcohols, and presumably a variety of other functional groups are unaffected by this process. The use of DCO_2D allows deuterium substitution.

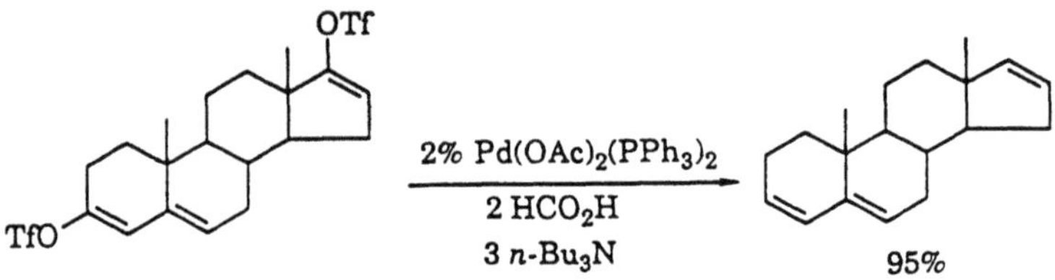

(294)

Silyl hydrides in the presence of lithium chloride and 2% $Pd(PPh_3)_4$ also reduce vinylic triflates in high yields (Eq. 295).[485]

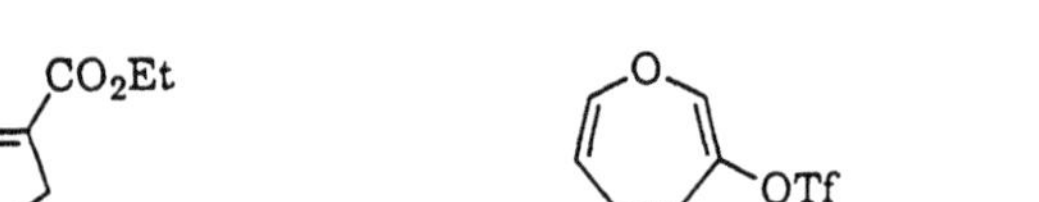

(295)

Under identical conditions, tri-*n*-butyltin hydride is also quite effective for such reductions.[486] The reaction accommodates considerable functionality as illustrated by the successful reduction of substrates **5–8**.

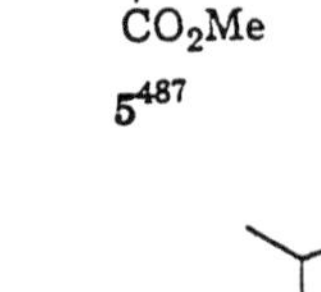

5[487]

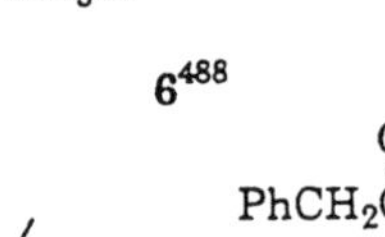

6[488]

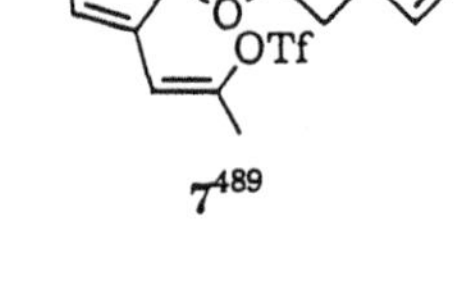

7[489]

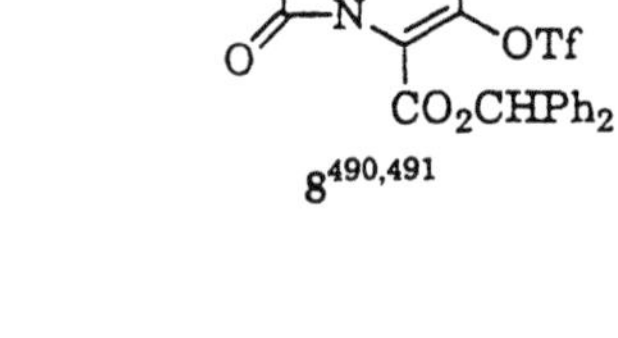

8[490,491]

Nitrile Formation

α,β-Unsaturated nitriles can be prepared from vinylic triflates using palladium catalysis (Eq. 296).[492]

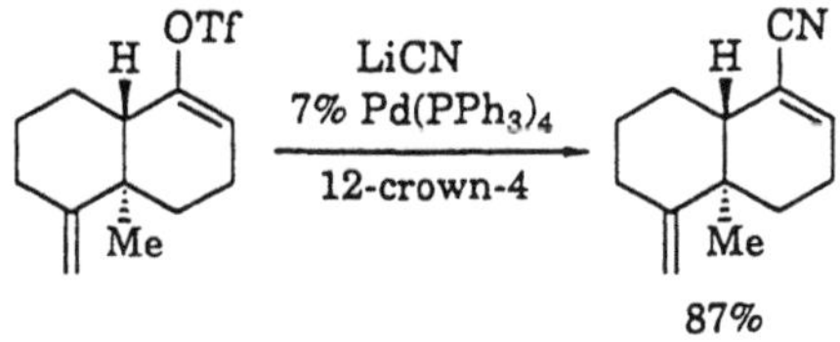

(296)

Carbonylation

The palladium-catalyzed carbonylation of vinylic triflates in the presence of a tin hydride affords unsaturated aldehydes (Eq. 297).[25,483] The presence of LiCl is critical to the success of many palladium-catalyzed reactions of vinylic triflates.

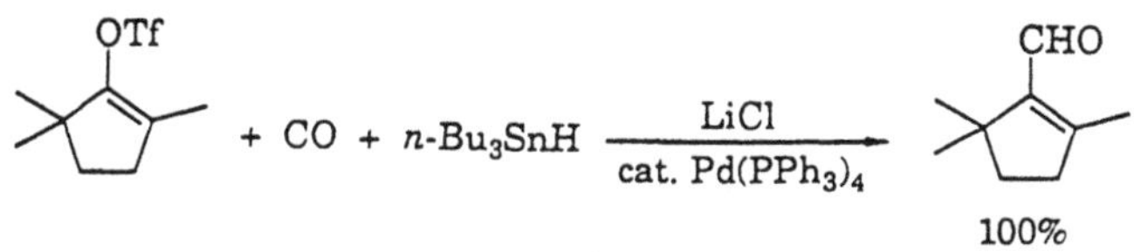

(297)

Carbonylation in the presence of organotin compounds produces enones (Eq. 298).[25,493] As indicated, alkyl, vinylic, allylic, alkynyl and aryl groups can be transferred from tin. (α-Ethoxyvinyl)trimethylstannane has also been utilized, in which case the resulting divinyl ketone can be hydrolyzed to an α-diketone (Eq. 299).[412]

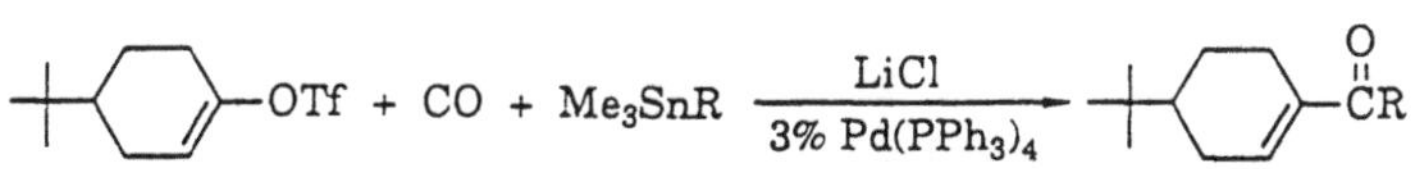

(298)

R = Me (73%), $CH{=}CH_2$ (76%), $CH_2CH{=}CH_2$ (95%), $C{\equiv}CSiMe_3$ (95%), Ph (95%)

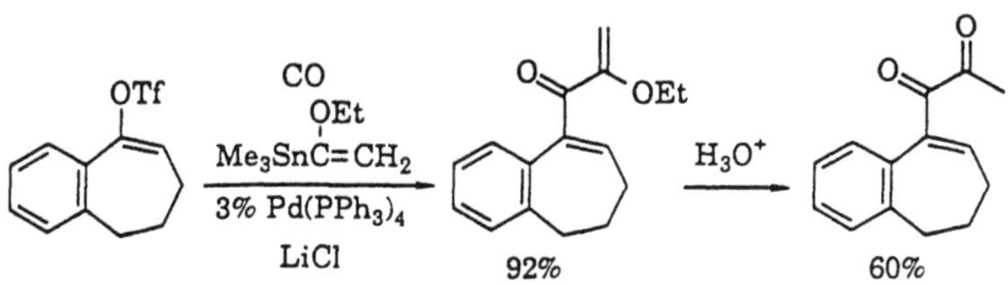

(299)

This carbonylative coupling process has been applied to the synthesis of tropolones (Eq. 300),[494] jatrophone (Eq. 301),[489] and in a novel annulation approach to capnellene (Eq. 302).[493]

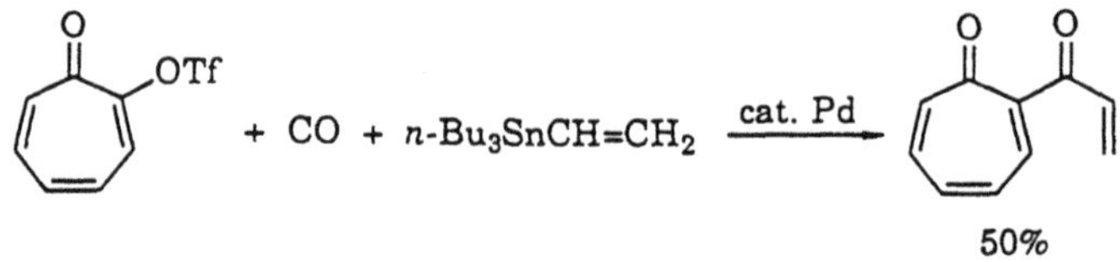

(300)

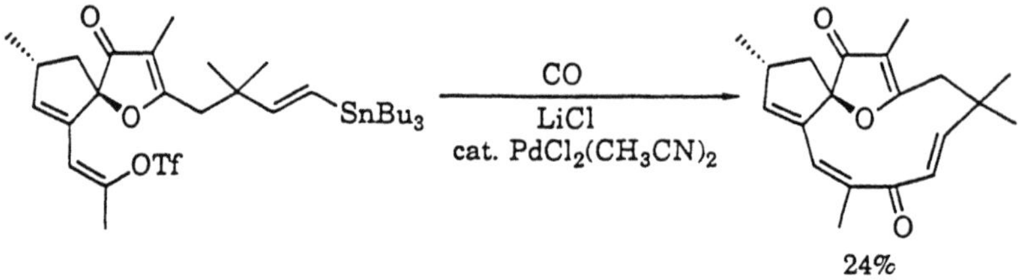

(301)

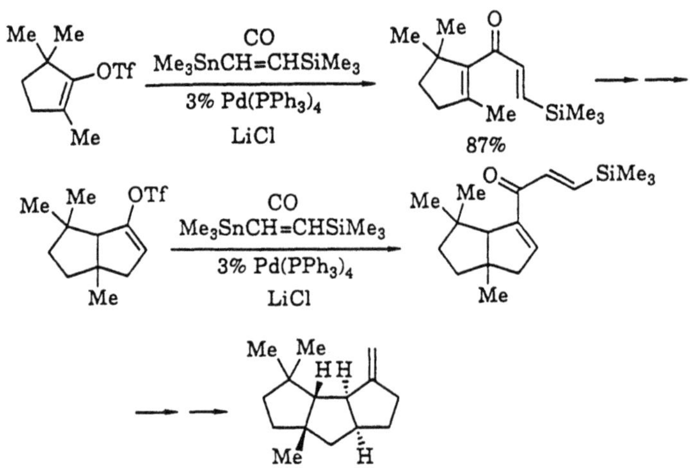

(302)

α,β-Unsaturated carboxylic acids can be prepared directly from vinylic triflates via carbonylation in water containing triethylamine and formic acid (Eq. 303).[495–497]

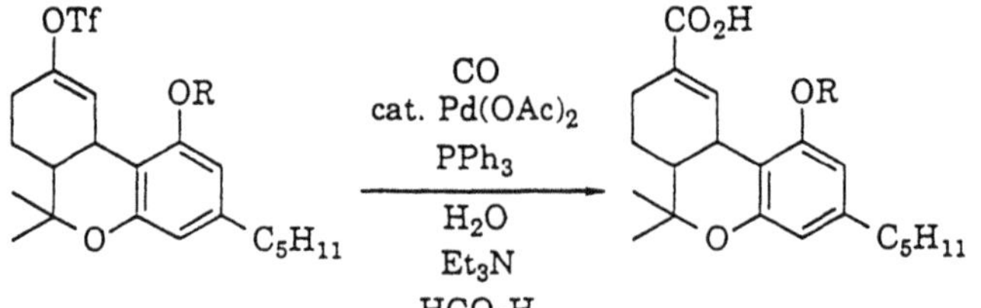

(303)

The carbonylation of vinylic triflates in the presence of an alcohol provides a useful route to α,β-unsaturated esters (Eq. 304).[498] This reaction has been used to prepare oxepin esters,[488] pentalenic acid,[499] 1-α-O-methyl-loganin,[500] aphidi-

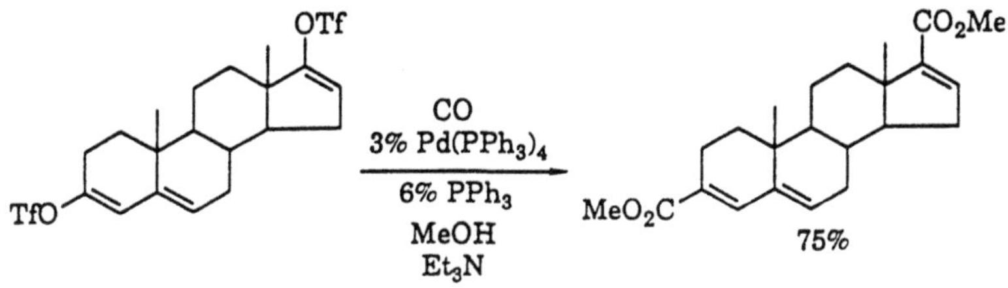
(304)

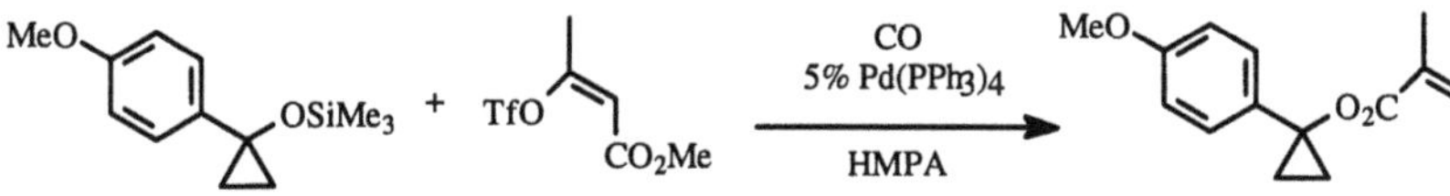
(305)

colin,[501] tetrahydrocannabinoids,[497] and steroidal 5 α-reductase inhibitors.[502] There is one example in the literature of the use of a silyl ether, instead of an alcohol (Eq. 305).[503]

The use of amines provides the corresponding amides (Eq. 306).[498,504]

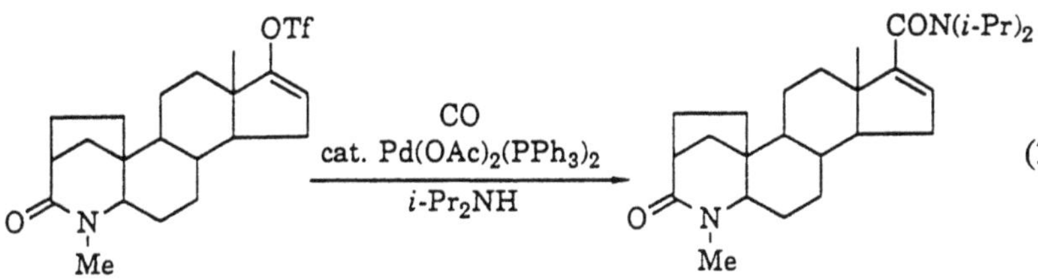
(306)

Alkene Coupling

Vinylic triflates readily undergo the Heck reaction with acyclic alkenes (Eq. 307).[505,506] Some loss of triflate stereochemistry is observed.[505] This reaction has been used to prepare a variety of dienones,[507] including β-ionone (Eq. 308).[508] It has also found utility for the synthesis of dienyl amines,[509] oxepins,[488] cardenolides,[510] lysergic acid,[511] anthramycin,[512,513] and uracils.[514] Coupling with vinylic

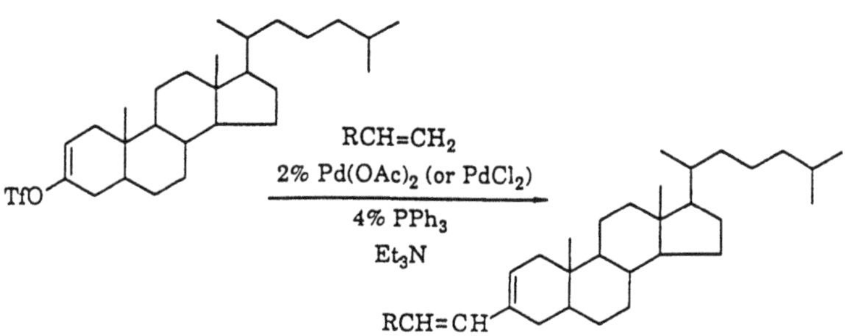
(307)

R = CO_2Me (86%), COMe (82%), CH_2OH (50%), Ph (77%), n-C_8H_{17} (60%)

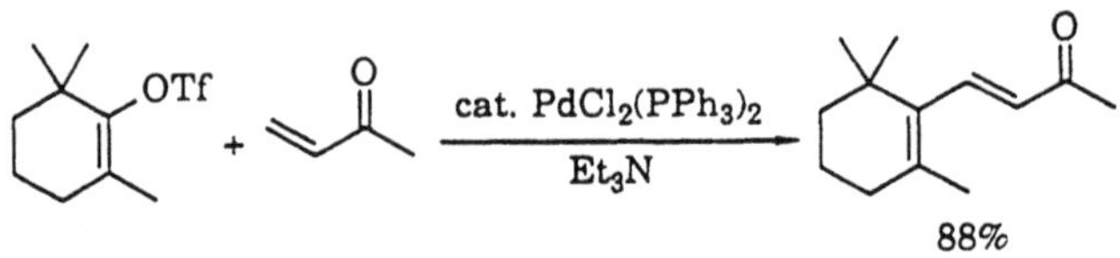

(308)

silanes,[93] vinyl acetate,[515] and vinylic ethers[94,516,517] provides silyl-, acetoxy- and alkoxy-dienes respectively. While substitution occurs β to the silyl or acetoxy group, vinylic ethers undergo substitution in the α position (Eq. 309).

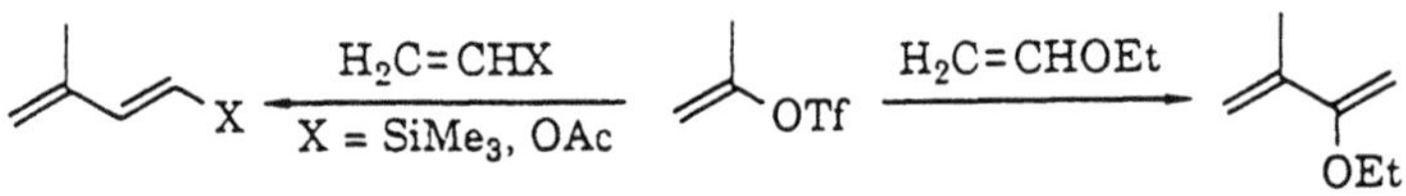

(309)

The cyclization of unsaturated vinylic triflates provides a useful approach to carbocycles (Eq. 310).[518]

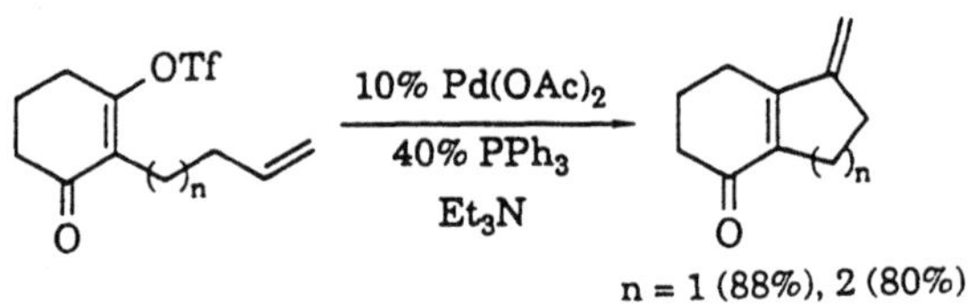

(310)

The vinylation of allylic alcohols proceeds via enol formation. By employing (*Z*)-2-butene-1,4-diol, lactols are generated (Eq. 311).[519]

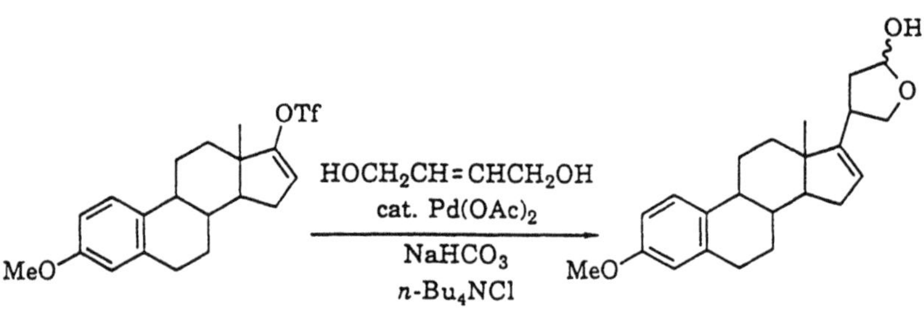

(311)

The palladium-catalyzed cross-coupling of vinylic triflates and alkenoic acids proceeds via π-allylpalladium intermediates to unsaturated lactones (Eq. 312).[113]

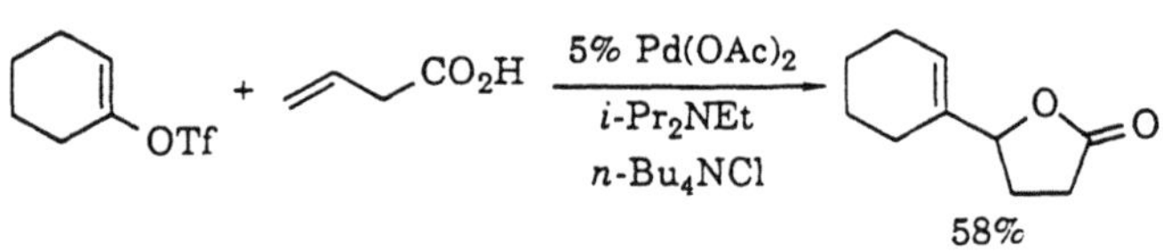

(312)

As noted earlier, vinylpalladium addition to cyclic alkenes generates intermediates which β-hydride eliminate to allylic substrates (Eq. 313).[117] With cyclic ethers, however, double bond isomerization has been observed (Eq. 314).[506] Where β-hydride elimination is blocked, conjugated dienes have been reported (Eq. 315).[520]

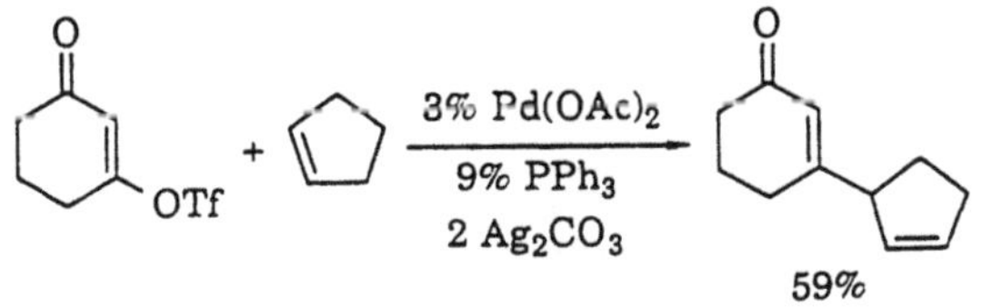

(313)

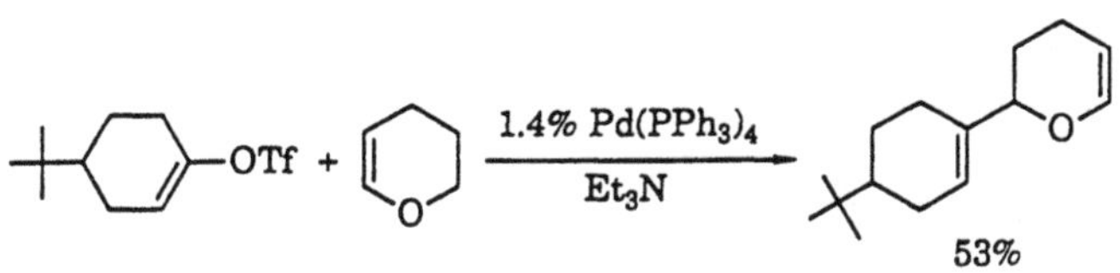

(314)

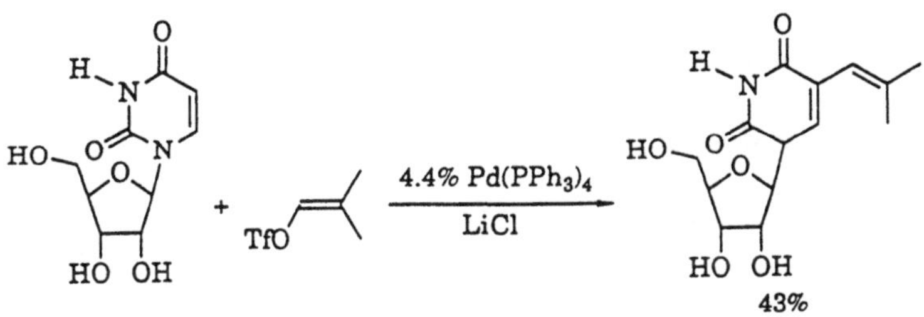

(315)

Reaction with Dienes

Allenes react with vinylic triflates and palladium to produce π-allylpalladium intermediates which undergo facile displacement by stabilized carbanions (Eq. 316).[521,522]

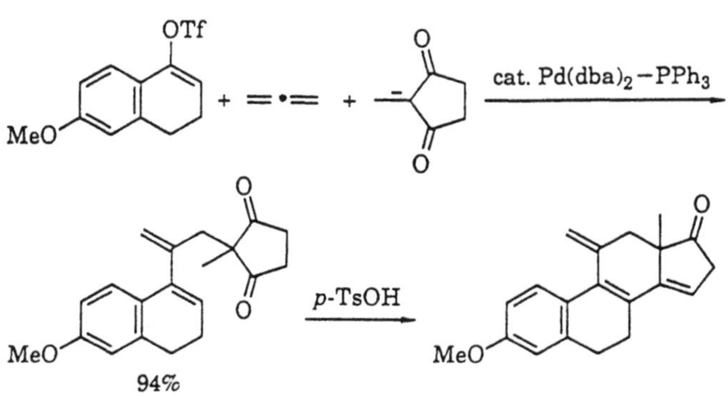

(316)

Intramolecular vinylation of 1,3-dienes provides a unique, enantioselective approach to carbocycles. By employing a chiral phosphine and trapping the intermediate π-allylpalladium species with acetate, capnellenols can be prepared in 80% enantiomeric excess (Eq. 317).[523] In a similar fashion, nonconjugated dienes undergo enantioselective multiple insertion processes (Eq. 318).[518]

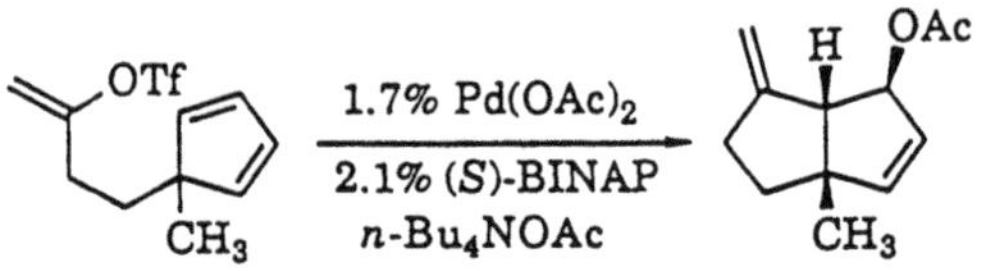

(317)

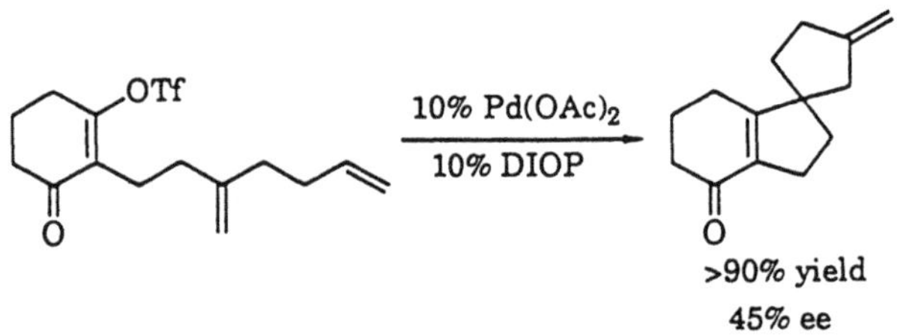

(318)

Coupling with Alkynes

Like vinylic halides, vinylic triflates react with terminal alkynes to produce conjugated enynes. Several different procedures have been reported utilizing catalytic amounts of either $Pd(OAc)_2(PPh_3)_2$ in the presence of (1) n-Bu_3N, (2) sodium acetate, or (3) diethylamine (plus CuI);[524] or 2.4% $Pd(PPh_3)_4$, plus LiCl and Et_3N.[506] 6-Alken-4-ynoic acids,[525] *o*-(3-alken-1-ynyl)anilines,[177] dienynes related to vitamin D,[526–528] 5-alkynyluracils,[514] and dienediynes related to the potent antitumor agent neocarzinostatin (Eq. 319)[529–531] have been prepared in this manner.

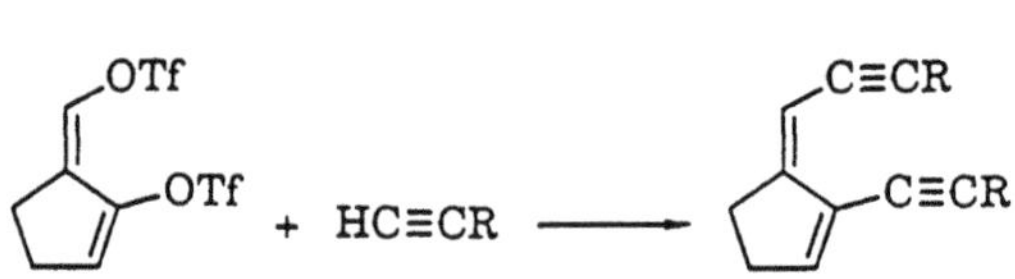

(319)

A slight variation of the above procedures affords a spirocycle (Eq. 320).[247]

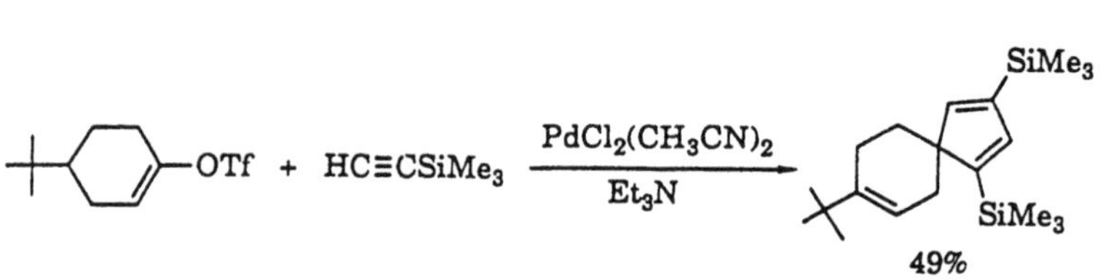

(320)

Coupling with Organic Halides

As noted earlier (Eq. 161), vinylic triflates react intramolecularly with vinylic halides, n-$Bu_3SnSiMe_3$, and a palladium catalyst to produce carbocycles. Aryl halides undergo similar processes (Eq. 321).[253]

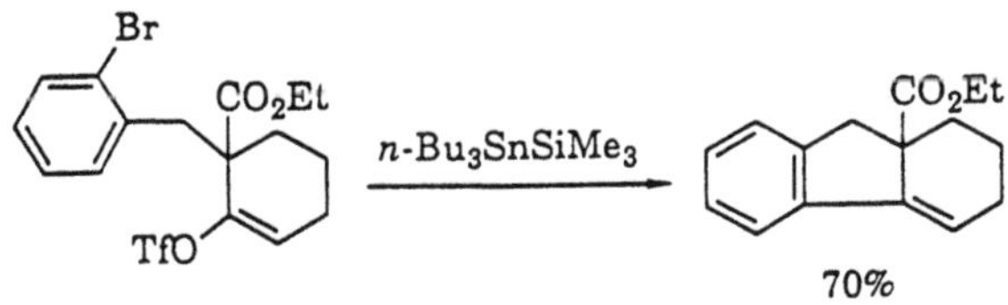

(321)

Reaction with Alkyl Metallics

In the presence of a palladium catalyst, a number of different alkyl metallic reagents cross-couple to give alkenes. Alkyl zinc reagents bearing ketone[294] and ester[297,298,532–534] functionality are particularly effective (Eqs. 322 and 323).

$$PhC(=O)(CH_2)_nZnI + H_2C{=}C(OTf)(CH_2)_3CH_3 \xrightarrow{4\%\ Pd(PPh_3)_4} PhC(=O)(CH_2)_nC(=CH_2)(CH_2)_3CH_3 \quad (322)$$

n = 3 (67%), 6 (74%)

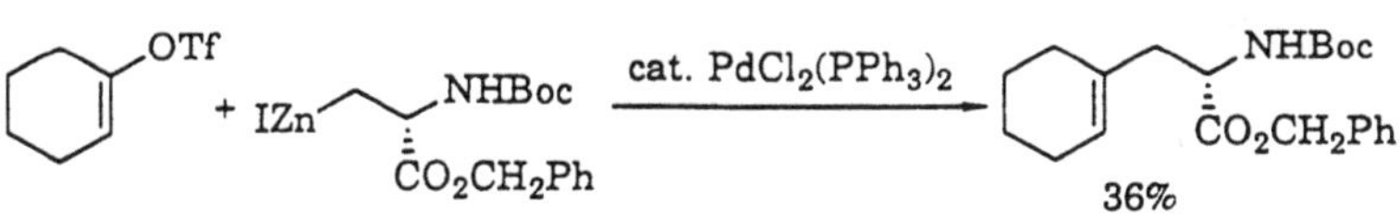

(323)[534]

Alkyl 9-BBN reagents also cross-couple with triflates in high yield (Eq. 324).[535]

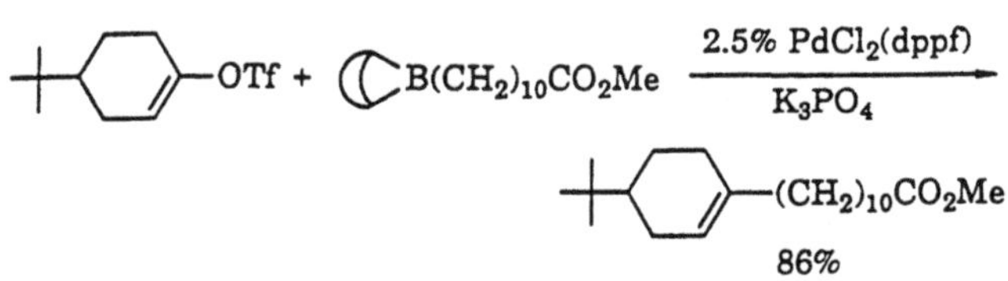

(324)

Allylic silanes have been prepared by reacting vinylic triflates with silylmethyl aluminum reagents (Eq. 325).[536]

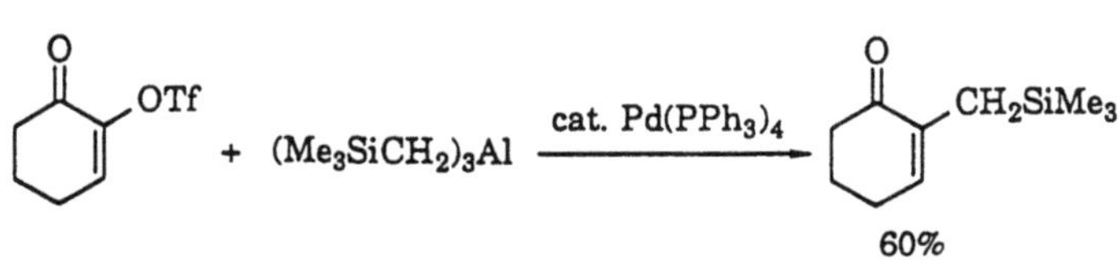

(325)

Alkyl tin reagents will also alkylate vinylic triflates (Eq. 326).[485,486] This process has been used to prepare various substituted cephalosporins (Eq. 327).[21,491]

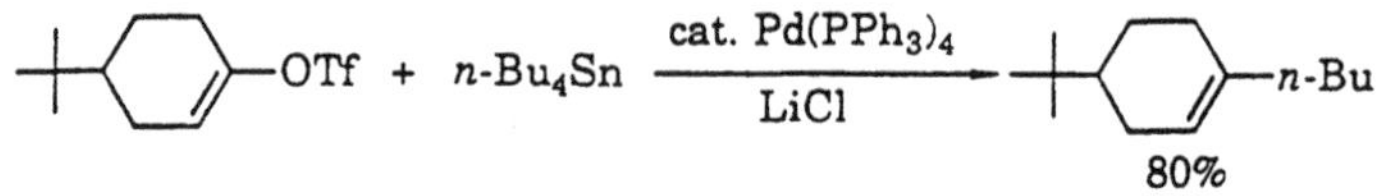

(326)

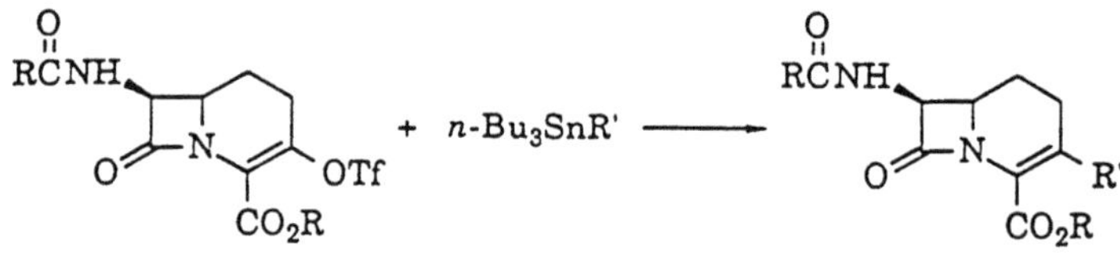

(327)

R' = Me (70%), CH_2OCH_3 (58%), CH_2O_2CPh (39%)

Allylic Cross-Coupling

Allylic silicon and tin reagents react with vinylic triflates to produce the corresponding 1,4-dienes. Allylic trifluorosilanes couple regioselectively at the more substituted end of the allylic system (Eq. 328).[537]

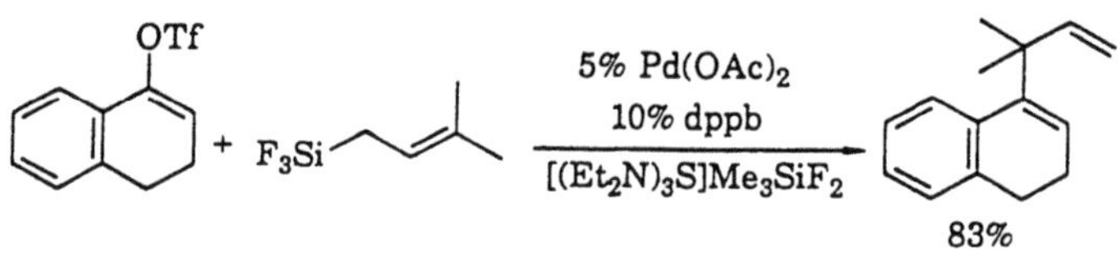

(328)

Allylic tin reagents can also be employed effectively.[485,486] The regioselectivity of this process has not been examined, but it has been applied to oxepins[488] and cephalosporins.[21,491]

Cross-Coupling with Vinylic Metallics

Vinylic organometallic reagents of zinc, boron, silicon and tin react with vinylic triflates in the presence of a palladium catalyst to produce 1,3-dienes. Simple vinylic zinc reagents cross-couple with alpha triflyloxy enones to produce the corresponding dienones (Eq. 329).[348] Zinc-containing cyclopropene ketals can also be employed.[356] Coupling with α-*n*-butoxyvinyl zinc chloride has been used to

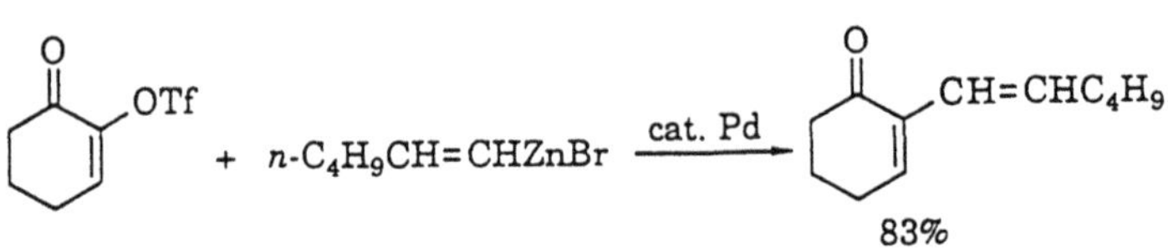

(329)

prepare the corticosteroid side chain, when ethyl vinyl ether or an α-ethoxyvinyl tin reagent proved unsatisfactory.[538]

Vinylic catechol borane reagents also cross-couple efficiently with vinylic triflates (Eq. 330).[535]

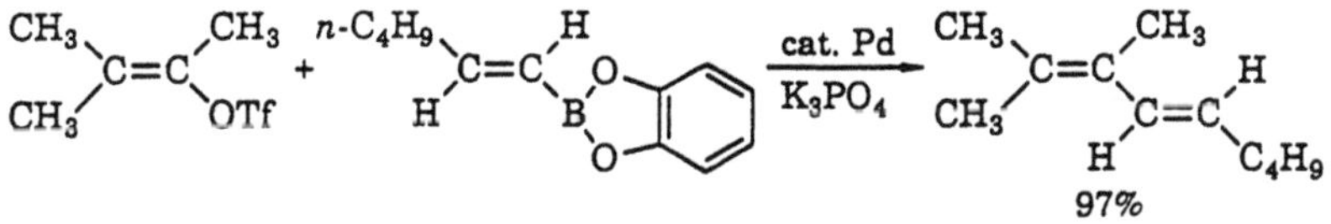

(330)

Vinylic fluorosilicon derivatives react with vinylic and aryl triflates in the presence of n-Bu_4NF and catalytic amounts of $Pd(PPh_3)_4$ (Eq. 331).[539]

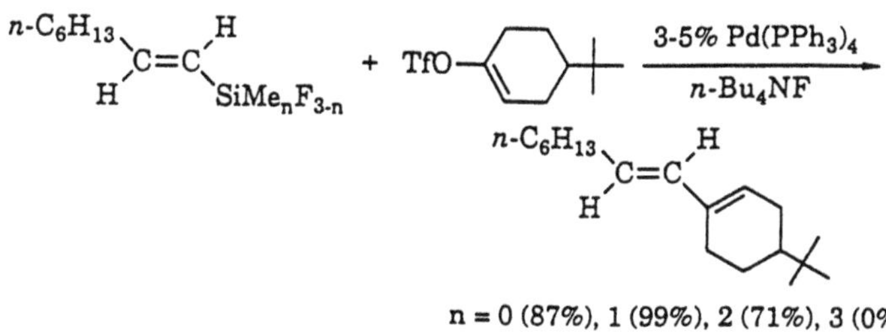

(331)

Tin reagents have been the most widely used vinylic metallics for the conversion of vinylic triflates to 1,3-dienes. Their increased reactivity relative to silicon is seen in the following useful synthesis of dienyl silanes (Eq. 332).[485,486] This approach has been

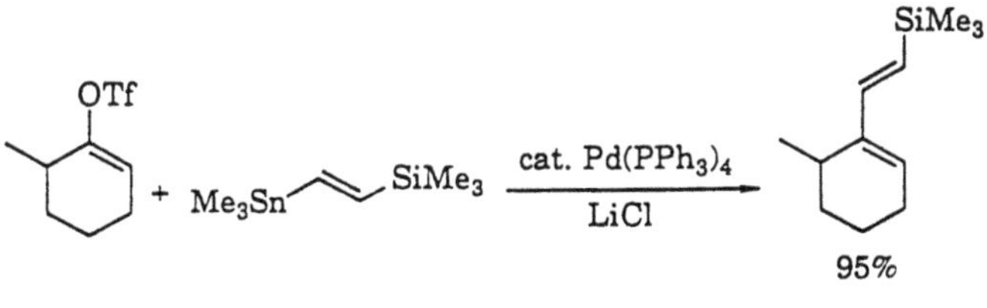

(332)

used to prepare dienyl sulfoxides,[481] 2-alkoxy-1,3-dienes,[412,517,538] and alkenyl cyclobutenedione monoacetals[416] from the corresponding functionally-substituted vinylic stannanes. It has also been applied to the synthesis of bicyclo-annulated cyclopentadienes,[540] 5-alkenyl uridines,[541] abietanoid quinones,[400,542] pleraplysillin-1,[485,486] cephalosporins,[21,490,491] anthramycin,[512,513] and glycinoeclepin A (Eq. 333).[543] Although palladium(O) reagents are usually used to catalyze this process, in this last example they apparently proved ineffective.

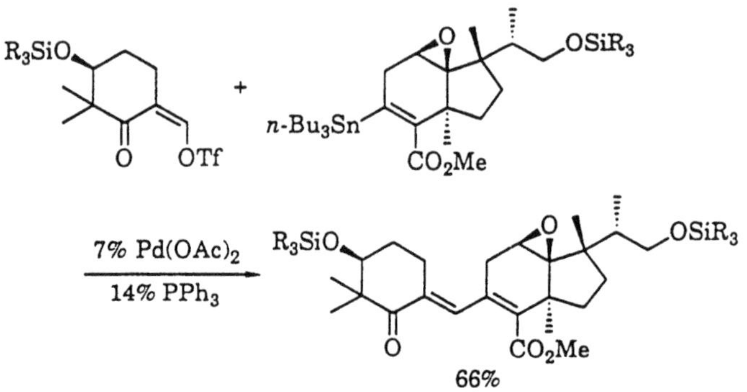

(333)

Intramolecular variants of this tin chemistry have proven useful for the synthesis of cyclic dienes,[544,545] including large ring lactones (Eq. 334),[546] and dolastane (Eq. 335)[547,548] and amphilectane[549] diterpenoids.

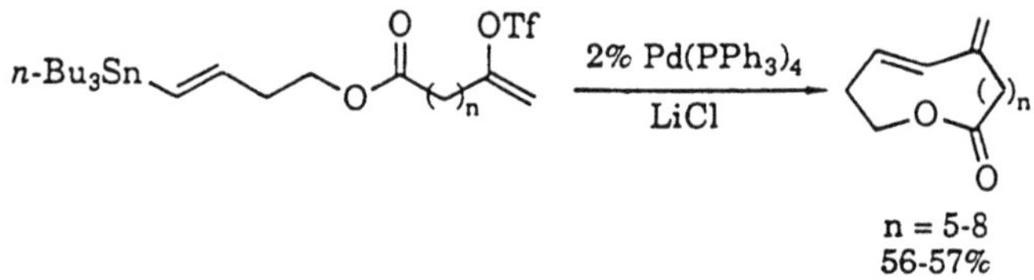

(334)

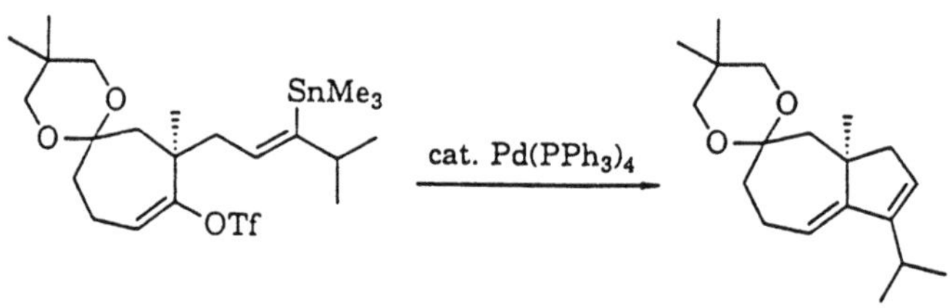

(335)

Coupling with Alkynyl Metallics

Alkynyl silicon and tin reagents readily convert vinylic triflates to enynes. Only one example of the silane reaction has been reported (Eq. 336).[539]

$$n\text{-}C_4H_9C{\equiv}CSiMe_3 + n\text{-}C_6H_{13}\overset{\overset{\large OTf}{|}}{C}{=}CH_2 \xrightarrow[n\text{-}Bu_4NF]{3\text{-}5\%\ Pd(PPh_3)_4} n\text{-}C_6H_{13}\overset{\overset{\large C\equiv CC_4H_9}{|}}{C}{=}CH_2 \quad (336)$$

81%

Alkynylstannanes are more reactive (Eq. 337).[485,486] Applications to vitamin D[526] and cephalosporin[21,490,491] synthesis have been reported.

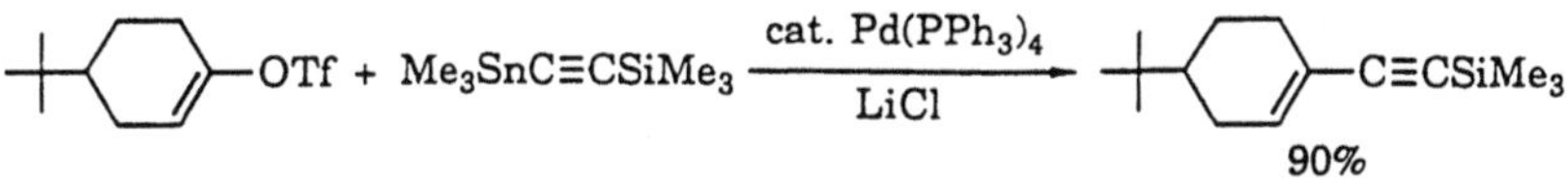

(337)

Since the direct coupling with terminal alkynes is feasible, it is questionable whether these organometallic procedures offer any advantages.

Reaction with Aryl Metallics

Aryl metallic reagents of zinc, boron and tin have been cross-coupled effectively with vinylic triflates. Aryl zinc reagents have been used to prepare arylated bicycloalkenes,[550] antiestrogenic benzocycloheptenes[551] and aryl tropolones.[552] 2-Furylzinc chloride has been utilized in steroidal furan synthesis.[553]

Arylboronate esters also arylate vinylic triflates (Eq. 338).[535]

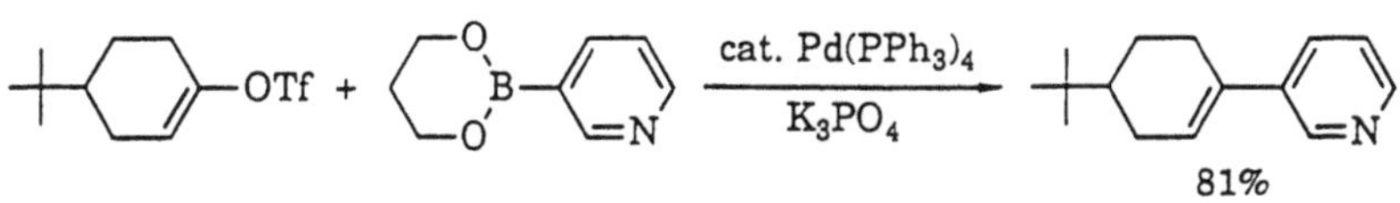

(338)

Aryl tin reagents have been more widely used to prepare conjugated arenes (Eq. 339). Although $Pd(PPh_3)_4$ is commonly used to catalyze this process, it has been found to inhibit the process.[554] The use of $Pd_2(dba)_3$ (dba = dibenzylideneacetone), with or without LiCl or $AsPh_3$, has been recommended (Eq. 339).[554] The coupling of vinylic triflates and arylstannanes has been used to prepare 4-arylcoumarins,[555] 5-aryluridines,[541] 3-arylcephems[490,491] and 2-arylcarbapenems (Eq. 340).[556] Note the variation in procedure used in this last example.

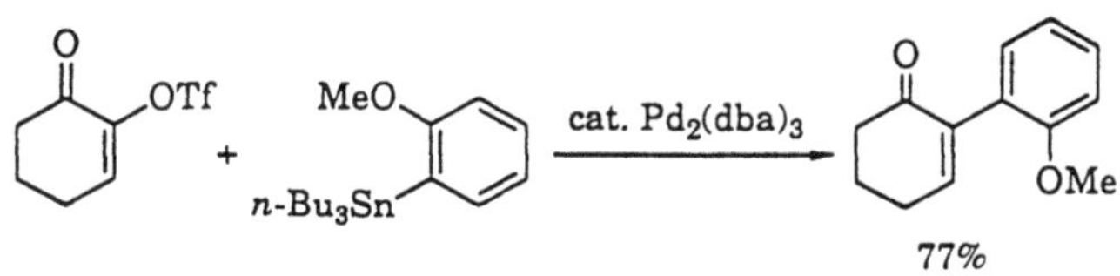

(339)

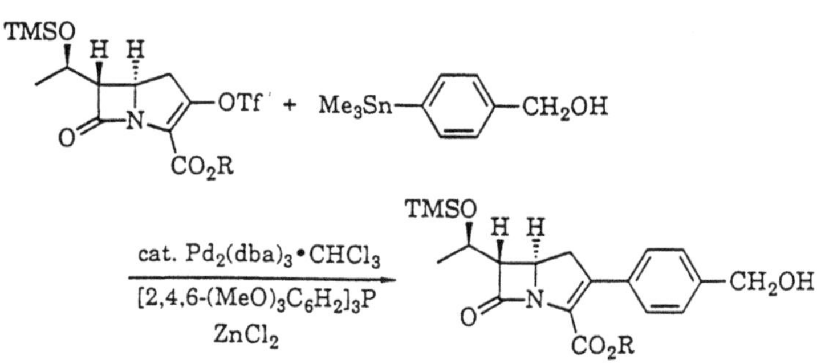

(340)

Introduction of Phosphorus

Vinylic triflates are readily converted to vinylic phosphonates,[557] phosphinates[557] and triphenylphosphonium[558,559] salts via palladium catalysis (Eqs. 341–343). This last process at least is highly stereospecific.

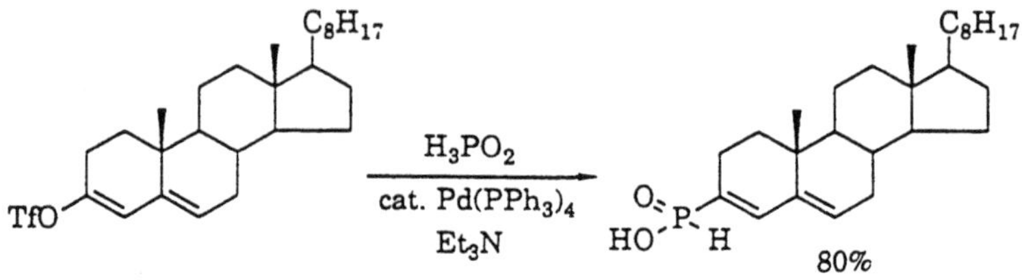

(341)

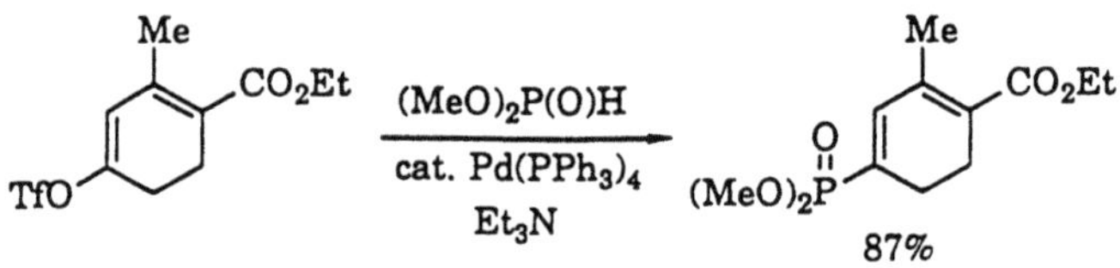

(342)

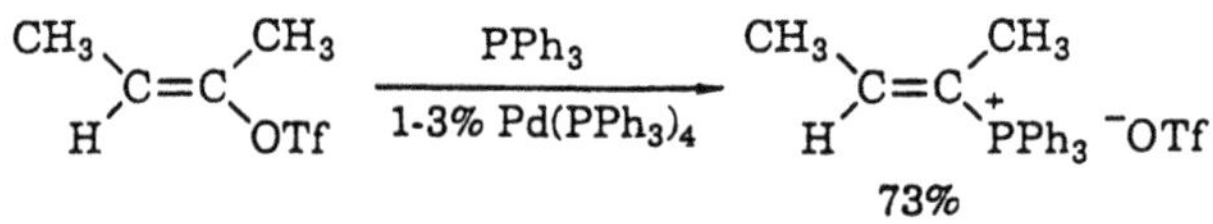

(343)

Vinylic Stannane Formation

The reaction of vinylic triflates with distannanes provides a convenient route to the corresponding vinylic stannanes.[485,550,560,561] Lactones have been transformed to stannyl pyrans in this fashion (Eq. 344).[562,563]

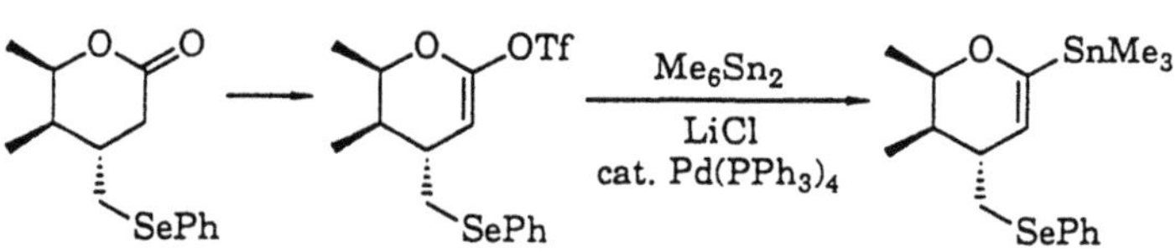

(344)

V. OTHER VINYLIC SULFONATES

A recent publication suggests that some of the chemistry carried out previously using vinylic triflates can also be carried out using the corresponding fluorosulfonate esters, which are much cheaper to prepare. Thus, the fluorosulfonate esters cross-couple readily with organostannanes (Eq. 345).[564]

Vinylic mesylates activated by a neighboring carbonyl group also undergo coupling with organotin compounds in the presence of lithium bromide (Eq. 346).[565] The reaction is believed to proceed via the corresponding vinylic bromide.

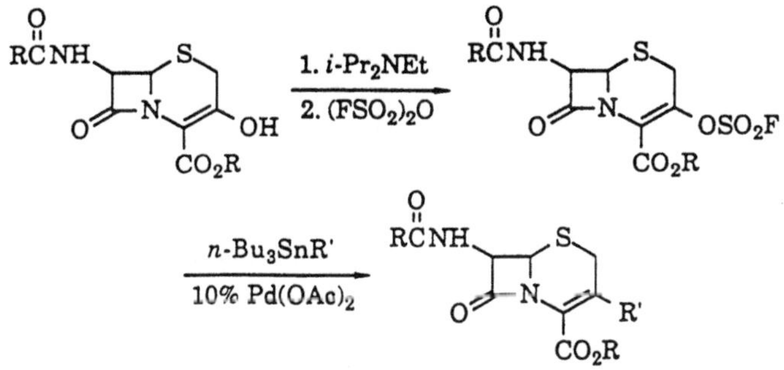

R' = aryl, vinylic

(345)

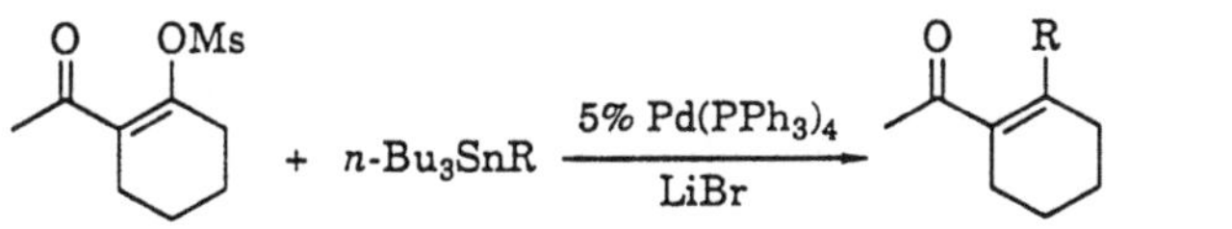

R = CH=CHR (57-74%),
Ph (33%)

(346)

Some yields are low and hydrido, alkyl, allyl and alkynyl tin reagents do not couple under the above reaction conditions.

In a similar manner, vinylic tosylates activated by a sulfoximine group react with methyl, vinylic and alkynyl aluminum reagents to produce substitution products (Eq. 347).[566]

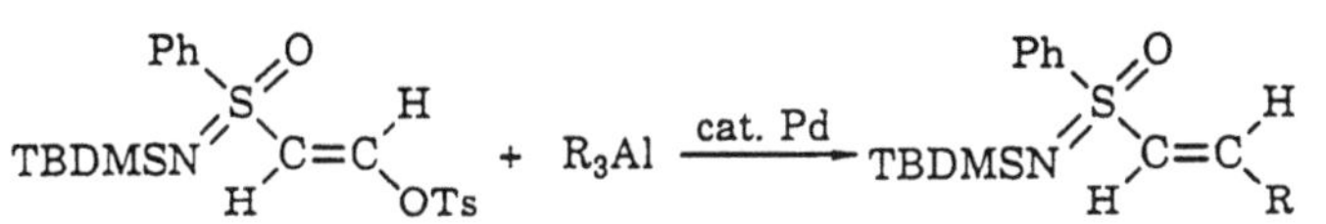

(347)

R_3Al = Me_3Al (87%), $(H_2C{=}CH)_2AlEt$ (85%), n-$BuC{\equiv}CAlMe_2$ (52%)

VI. VINYLIC PHOSPHATES

Vinylic phosphates undergo some useful palladium-catalyzed substitution processes. For example, enol phosphates derived from esters undergo palladium-catalyzed reduction by triethylaluminum (Eq. 348).[567] On the other hand, phosphates derived from thiocarboxylic S-esters react with triethylaluminum to give reduction or ethylation products, the ratio depending on the solvent employed (Eq. 349).[568,569] However, methyl-, vinylic- and alkynylaluminum reagents afford alkylated vinylic

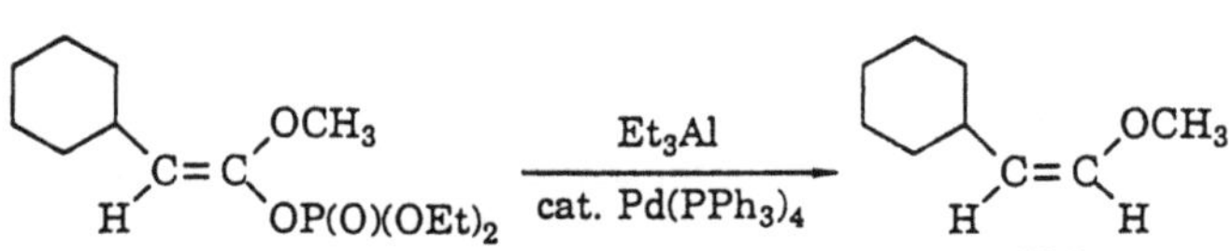

52%

(348)

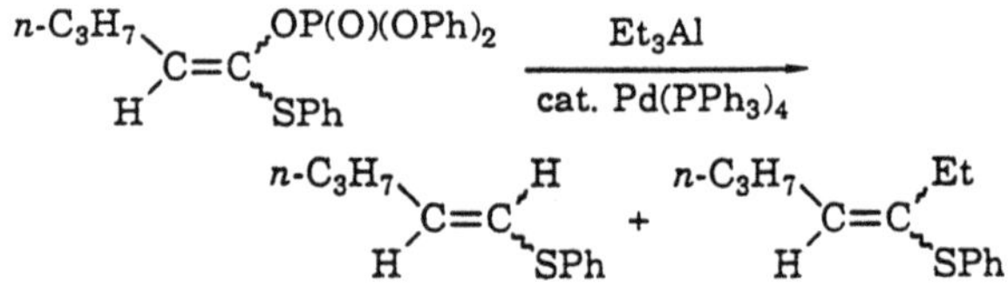

(349)

sulfides in good yields. In the absence of the neighboring oxygen or sulfur moiety, the alkylation product is produced in good yield (Eq. 350).[569,570] This reaction has been employed in a synthesis of 6,7-dideoxyforskolin (Eq. 351).[571]

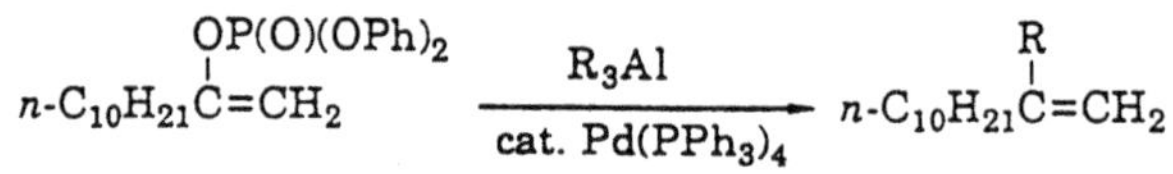

R_3Al: Me_3Al (91%), Et_3Al (71%), **n-$C_5H_{11}CH{=}CHAl(i\text{-}Bu)_2$** (66%) **$n$-$C_4H_9C{\equiv}CAlEt_2$** (59%)

(350)

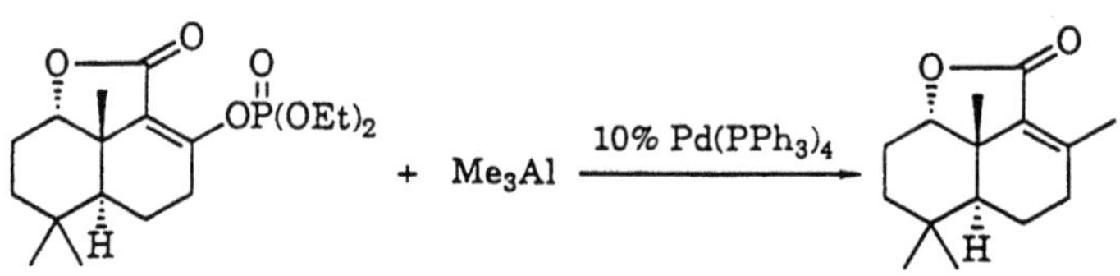

86%

(351)

Alkyl and aryl manganese reagents also effect this transformation (Eq. 352).[572]

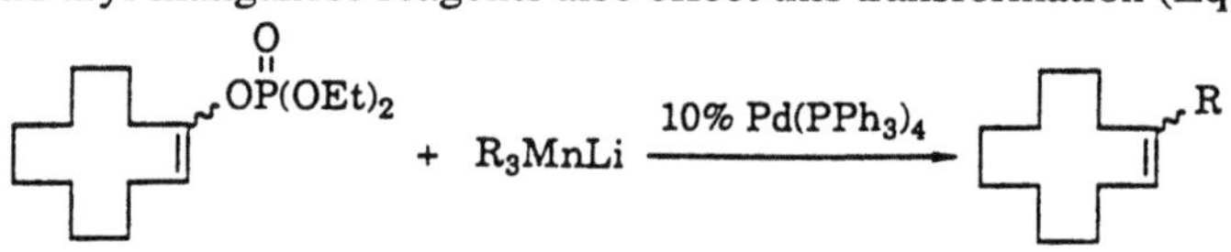

(352)

R = Me (67%), n-Bu (50%), Ph (82%)

Allylic silanes have been prepared from vinylic phosphates by employing silylmethylmagnesium reagents (Eq. 353).[573] Either palladium or, more often, nickel catalysts have been utilized.

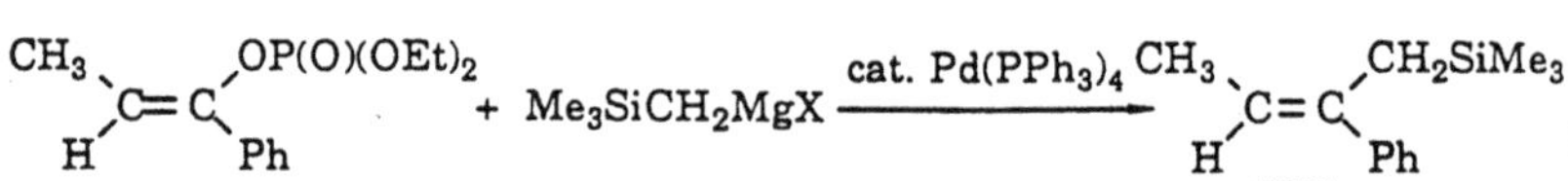

78%

(353)

Vinylic silanes can also be prepared directly from enol phosphates (Eq. 354).[574]

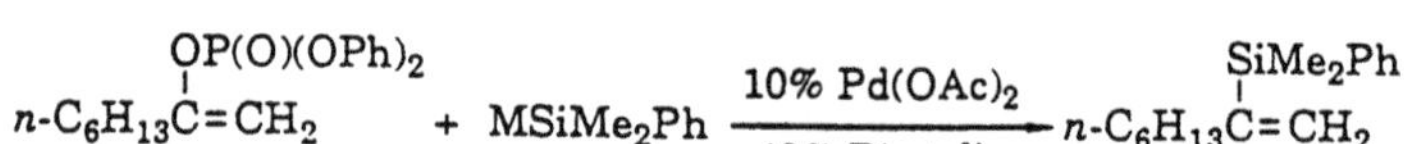

(354)

M = MgMe (74%), $AlEt_2$ (55%)

VII. VINYLIC ORGANOMETALLICS

Vinylic Lithium Reagents

Vinylic lithium reagents react with palladium(II) phosphine complexes to provide relatively stable vinylic palladium complexes.[341,575,576] These compounds have found relatively little use in organic synthesis however. The use of excess vinylic lithium reagent affords 1,3-dienes (Eq. 355).[341,577]

$$2\,(CH_3)_3CCH{=}CHLi \xrightarrow{PdCl_2(PPh_3)_2} [(CH_3)_3CCH{=}CH]_2 \qquad (355)$$

Vinylic Magnesium Compounds

α-Trimethylsilylvinylmagnesium bromide reacts with allylic acetates to produce dienyl silanes (Eq. 356).[578]

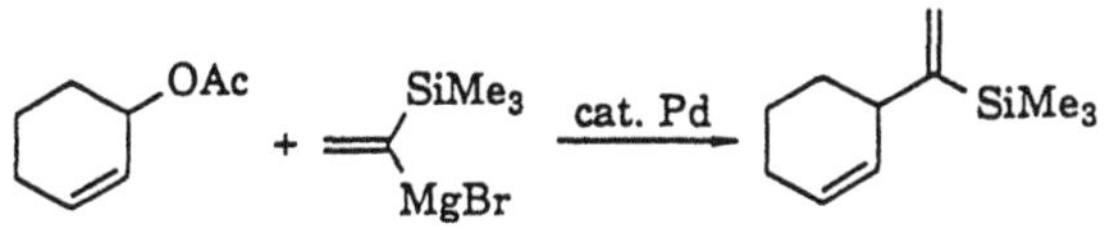

(356)

Vinylic magnesium halides also cross-couple nicely with aryl halides in the presence of a palladium catalyst (Eq. 357).[579,580] Alternatively, one may use isolated arylpalladium complexes to produce styrenes.[581,582] This reaction can be useful when cyclopalladation intermediates are employed (Eq. 358).[583] Vinylic silanes can also be prepared in this fashion (Eq. 359).[441] Aryl dienes have also been synthesized by this approach (Eq. 360).[584]

$$ArI \;+\; H_2C{=}CHMgX \xrightarrow{cat.\ Pd} ArCH{=}CH_2 \qquad (357)$$

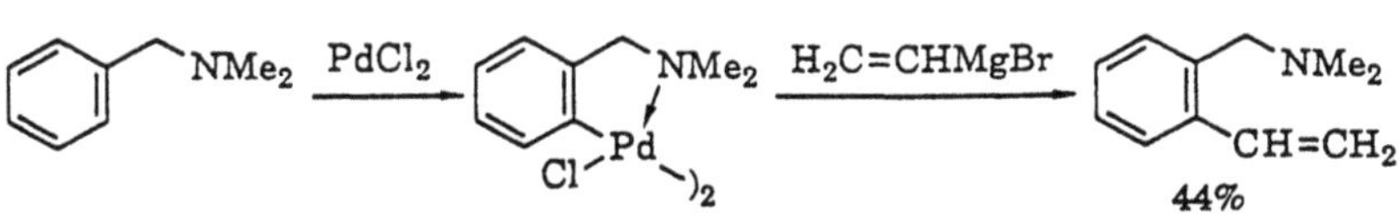

(358)

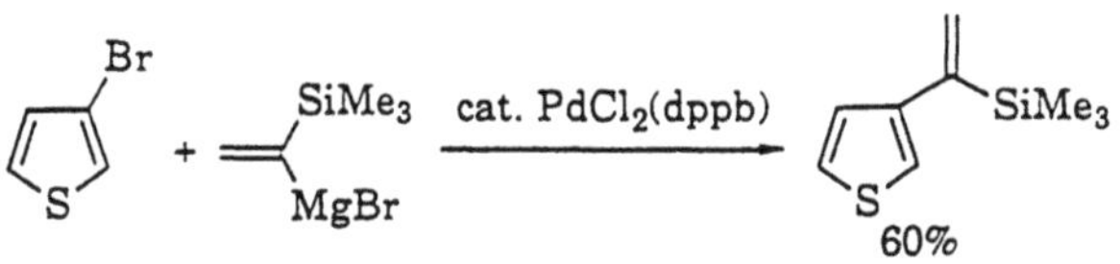

(359)

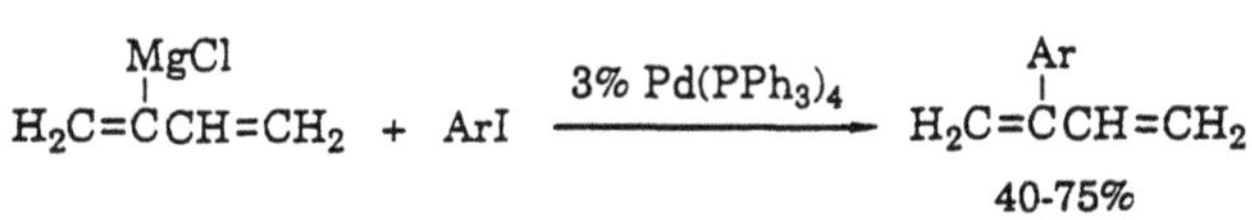

(360)

Vinylic Zirconium Reagents

As noted earlier (Eqs. 194–197), vinylic zirconium reagents can be cross-coupled with vinylic halides to produce 1,3-dienes. An analogous process using allylic halides affords 1,4-dienes (Eq. 361).[199] Using (*Z*)-crotyl bromide and crotylpalladium chloride dimer as a catalyst, retention of the allylic halide stereochemistry is observed (Eq. 362).[585,586]

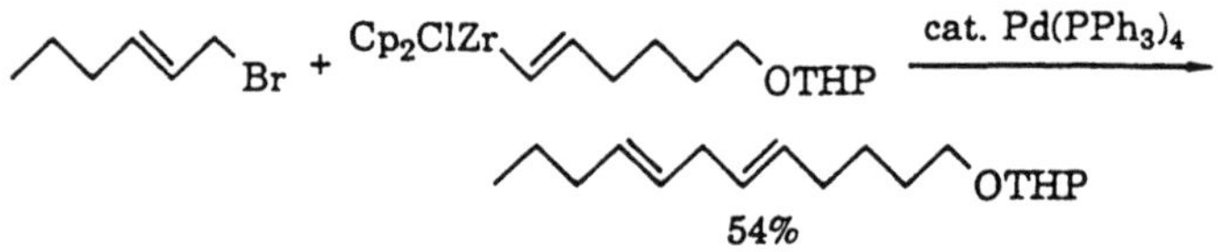

(361)

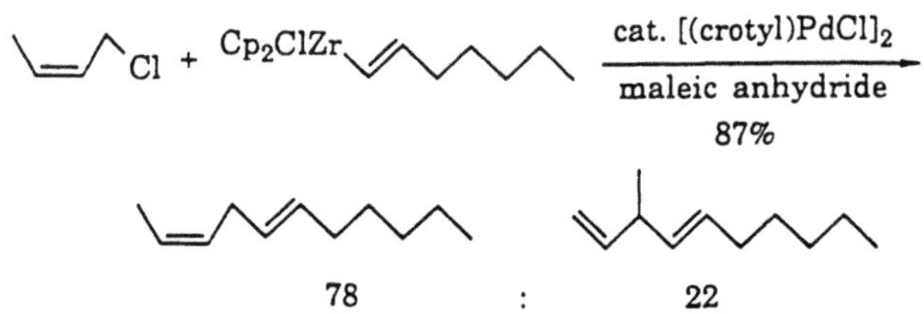

(362)

This chemistry has also been applied to allylic acetates (Eq. 363).[586] Note the inversion of stereochemistry at the allylic carbon. This is consistent with π-allylpalladium formation with inversion, transmetallation of the vinylic group from Zr to Pd, and reductive elimination with retention. Indeed, the cross-coupling of π-allylpalladium intermediates with vinylic zirconium reagents has been carefully studied with regard to its regio- and stereochemistry.[587,588] As noted below, the presence of maleic anhydride has a significant effect on the regiochemistry.

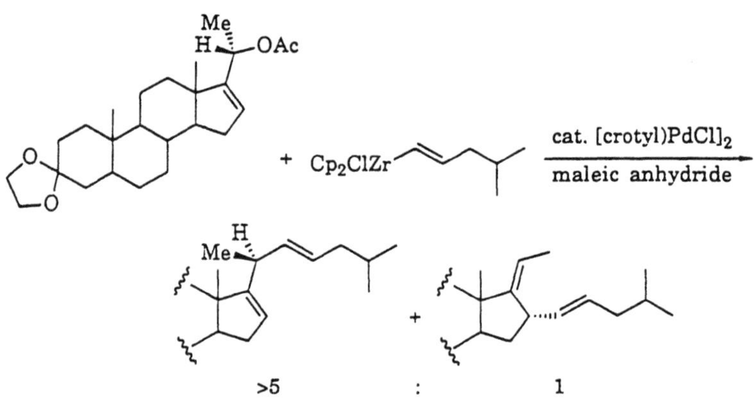

(363)

The coupling of vinylic zirconium reagents and π-allylpalladium complexes has been applied to the synthesis of 25-hydroxycholesterol (Eq. 364),[589] humulene (Eq. 365),[590] and flexibilene.[591]

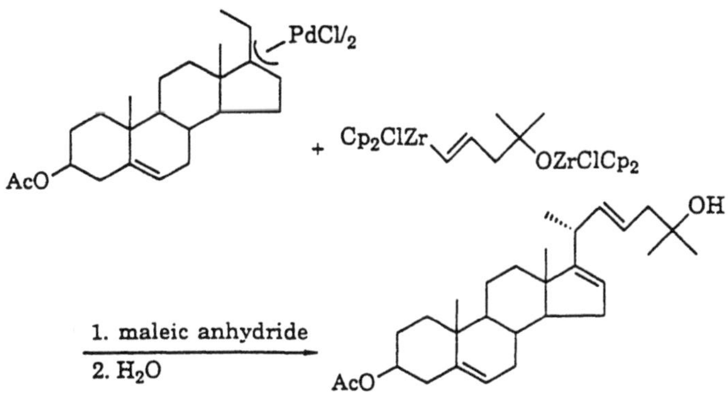

(364)

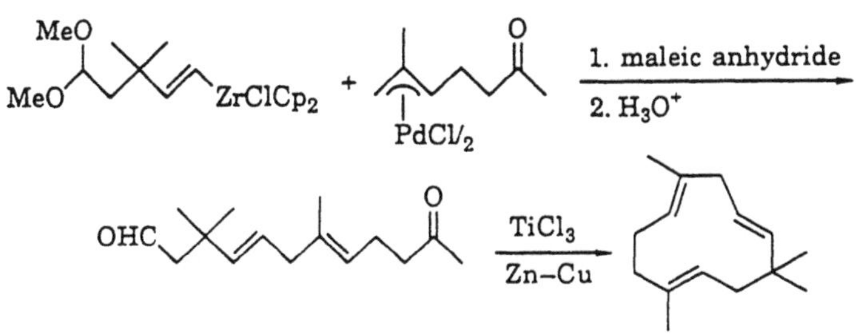

(365)

Vinylic aluminum reagents prepared by the zirconium-catalyzed carboalumination of alkynes also undergo facile coupling with organic halides. This chemistry will be discussed later under vinylic aluminum reagents.

Vinylic zirconium reagents also add to enones in a conjugate fashion when catalyzed by palladium.[592] No actual examples are reported, however, as nickel reagents seem to be the preferred catalysts.

Vinylic Copper Compounds

Vinylic copper compounds prepared by acetylene addition reactions can be coupled with aryl halides using one equivalent of zinc bromide and a palladium catalyst (Eq. 366).[335,593]

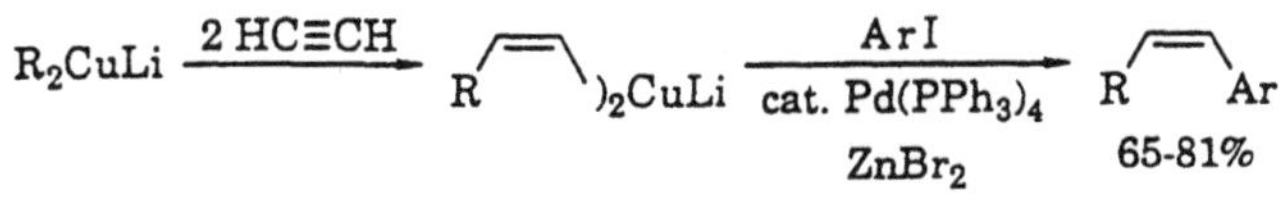

(366)

These same vinylic copper lithium reagents couple under identical conditions with acid halides or mixed anhydrides to produce enones and dienones (Eqs. 367 and 368).[336,594] Vinylic copper magnesium reagents prepared by addition to terminal alkynes undergo analogous reactions in the absence of zinc bromide (Eq. 369). α,β-Unsaturated esters can be prepared in a similar manner (Eq. 370).

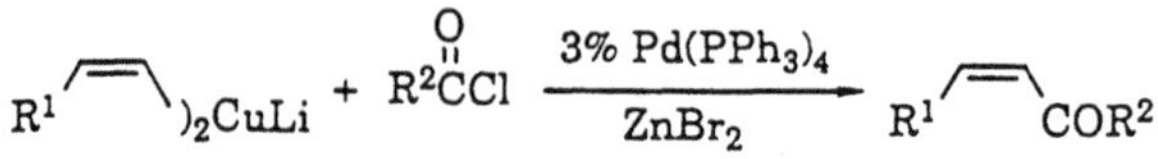

(367)

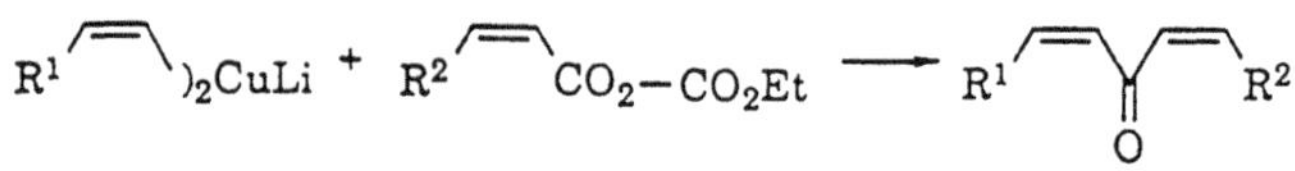

(368)

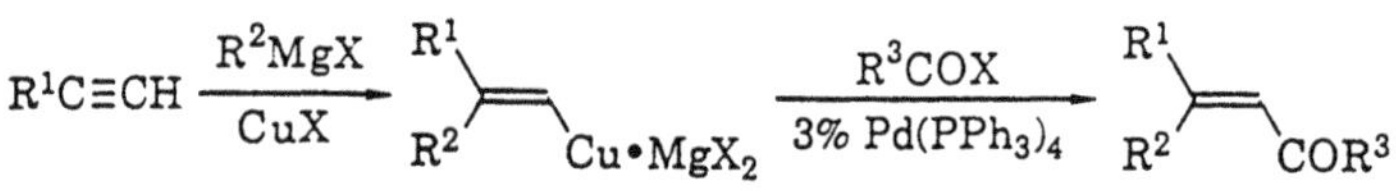

(369)

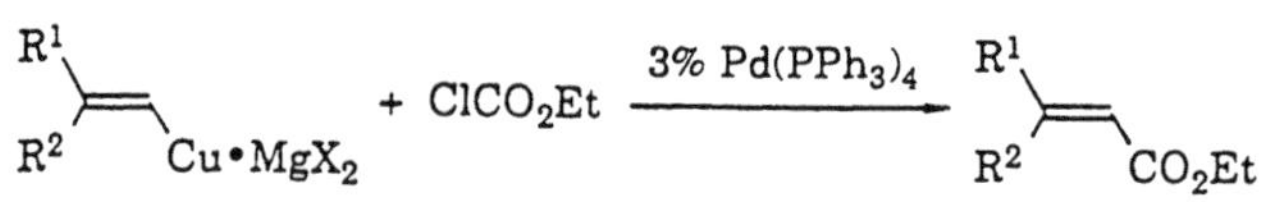

(370)

We have coupled vinylic copper reagents with stable alkylpalladium complexes to produce prostaglandin endoperoxide analogues (Eq. 371).[595,596]

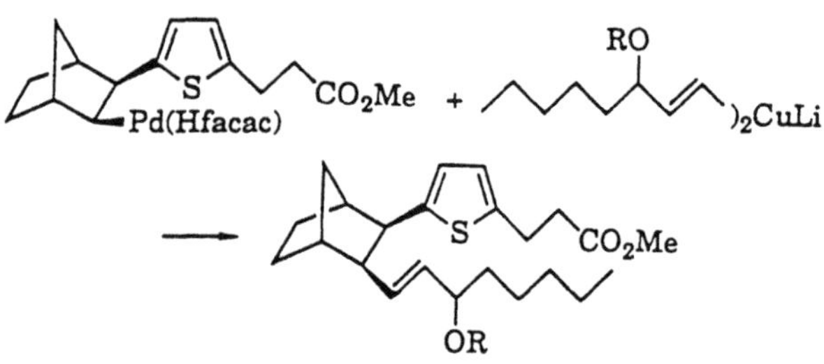

(371)

Vinylic Zinc Reagents

Vinylic zinc reagents undergo a variety of useful coupling reactions catalyzed by palladium. For example, vinylzinc bromide reacts with a dihydropyranyl acetate to produce a 1,4-diene (Eq. 372).[597] Note that this reaction proceeds with inversion of stereochemistry.

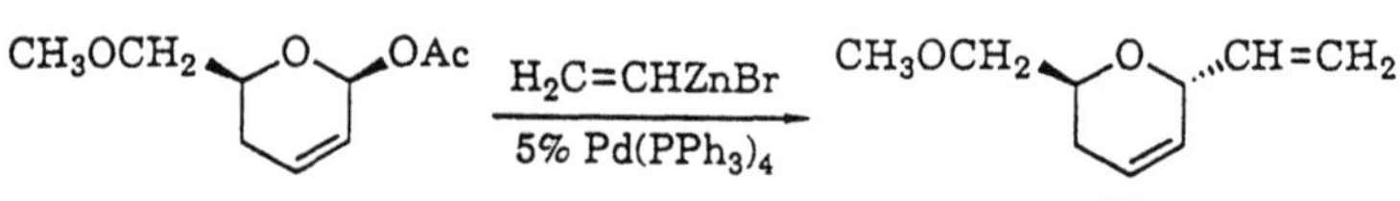

(372)

Vinylic zinc reagents generated by allylzincation of alkynes undergo intramolecular coupling with allylic ethers (Eq. 373).[598] A unimolecular version of this process has also been reported to produce bicyclic dienes.[599]

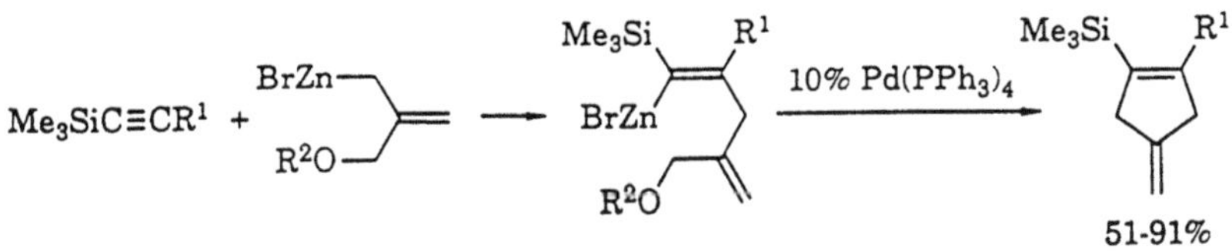

(373)

Vinylzinc bromide can be catalytically cross-coupled with allenic halides to produce 1,2,4-trienes (Eq. 374).[600] Analogous products are formed when starting with propargylic substrates (Eqs. 375 and 376).

$$\mathrm{PhCH{=}C{=}CHBr + H_2C{=}CHZnCl \xrightarrow{2\%\ Pd(PPh_3)_4} PhCH{=}C{=}CHCH{=}CH_2} \quad (374)$$

$$\mathrm{R_2\overset{X}{C}C{\equiv}CH + H_2C{=}CHZnCl \xrightarrow{2\%\ Pd(PPh_3)_4} R_2C{=}C{=}CHCH{=}CH_2} \quad (375)^{600,601}$$

X = OS(O)Me, OSO_2Me, OAc

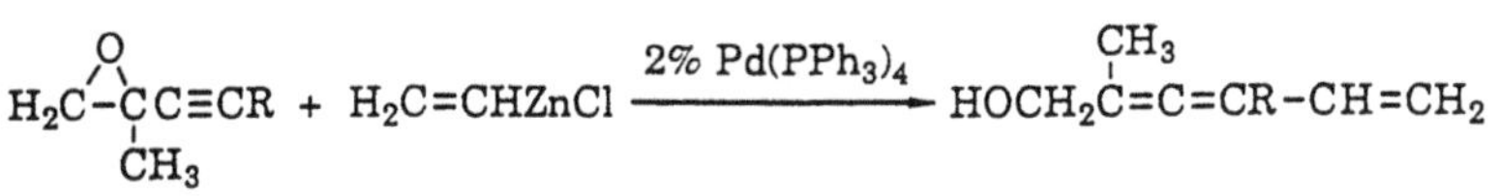

(376)[602]

R = H, CH_3

The catalytic coupling of allenic esters or sulfinates also provides trienes (Eq. 377).[603]

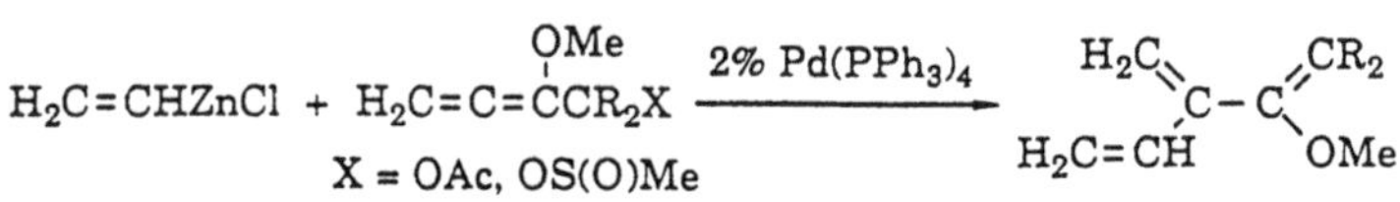

(377)

Fluorinated vinylic zinc halides can be coupled with 1-halo-1-alkynes to produce fluorinated enynes, although significant amounts of diyne are also formed (Eq. 378).[432]

$$\mathrm{RCF{=}CFZnX + XC{\equiv}CR \xrightarrow{cat.\ Pd(PPh_3)_4} RCF{=}CFC{\equiv}CR} \quad (378)$$

A number of examples of the palladium-catalyzed coupling of vinylic zinc reagents and aryl halides have been reported. Besides simple zinc reagents,[327,341,604,605] α-substituted heteroatom-containing reagents have been cross-coupled in good yield (Eq. 379).

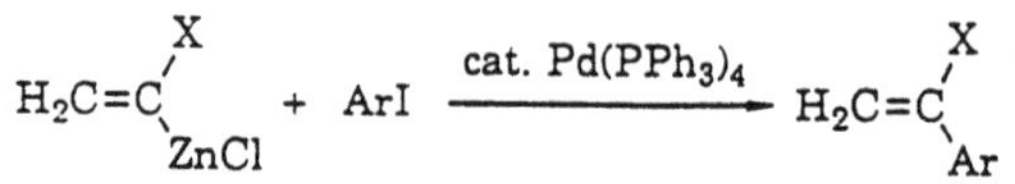

(379)

X = OEt,[358] SEt,[358] $OCONEt_2$,[359] $SiMe_3$[441,457]

This reaction has been employed for the α-arylation of ketones (Eq. 380).[349]

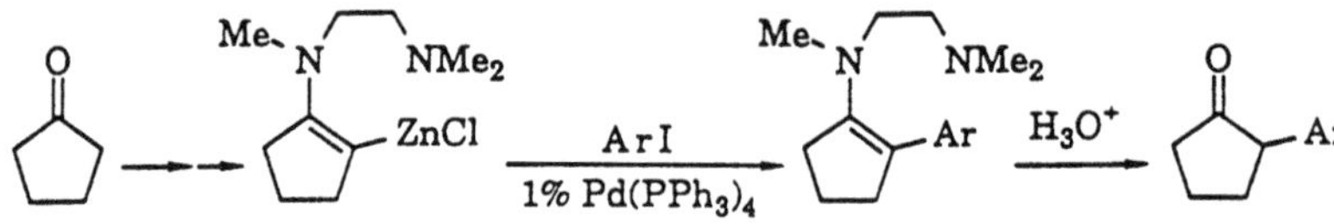

(380)

Fluorovinylic zinc reagents have been utilized to prepare a variety of fluorinated styrenes (Eq. 381).[355,606–609]

$$RCF{=}CFZnCl + ArI \xrightarrow{\text{cat. Pd}} RCF{=}CFAr \qquad (381)$$

Aryl-substituted cyclopropenone ketals,[356] polycyclic aromatic hydrocarbons,[360,362,363] and vineomycinone B_2 methyl ester (Eq. 382)[610,611] have all been prepared by the coupling of vinylic zinc reagents and aryl halides.

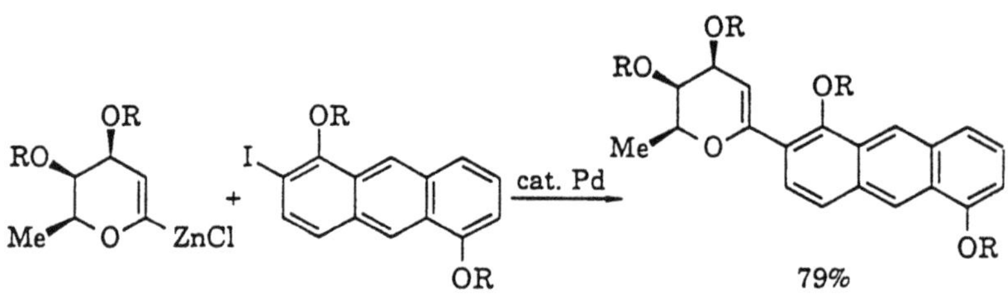

(382)

The coupling of vinylic zinc halides and acyl halides or chlorocarbonates is also catalyzed by palladium (Eqs. 383 and 384).[612] Fluorovinylic zinc reagents have also been employed in these processes.[355,606] Mixed anhydrides can also be employed in this ketone synthesis (Eq. 385).[336]

(383)

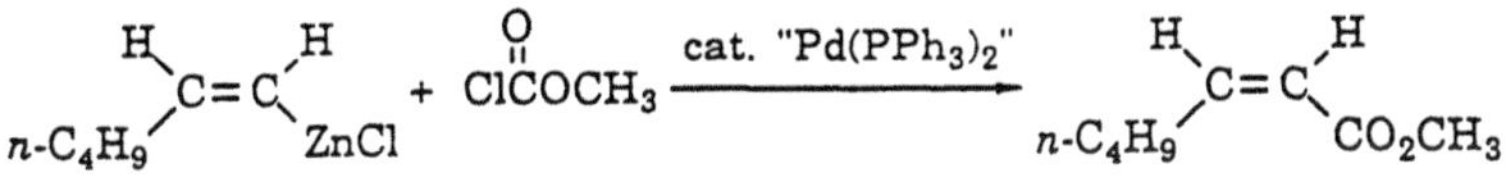

(384)

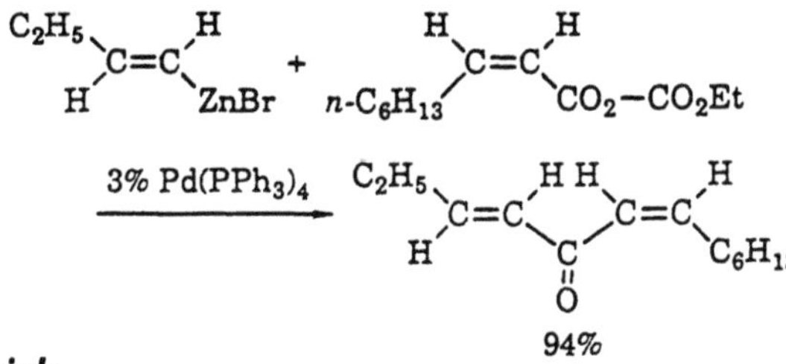

(385)

Vinylic Mercurials

Vinylic mercurials were probably the first vinylic metallics used to generated vinylic palladium species and, with the possible exception of vinylstannanes, are the most widely studied. β-Chlorovinylpalladium chloride generated in this fashion undergoes protonolysis by HCl as would be expected from work discussed earlier (Eq. 386).[613]

$$\text{ClCH=CHHgCl} \xrightarrow{\text{PdCl}_2(\text{PhCN})_2} \text{ClCH=CHPdCl} \xrightarrow{\text{HCl}} \text{ClCH=CH}_2 \qquad (386)$$

Considerable work has been reported on the reaction of vinylic mercurials and acyclic alkenes in the presence of palladium. This reaction has been reported by us to provide π-allylpalladium compounds in good yields (Eq. 387).[614,615] The reaction undoubtedly involves vinylpalladium addition to the alkene, followed by palladium hydride elimination and readdition to place the palladium in the allylic position. When alkenoic acids are employed in the reaction and π-allylpalladium formation is followed by heating with potassium carbonate, unsaturated lactones are produced (Eq. 388).[616]

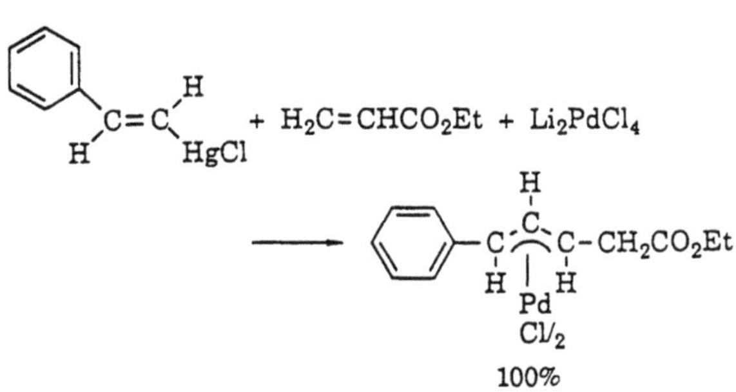

(387)

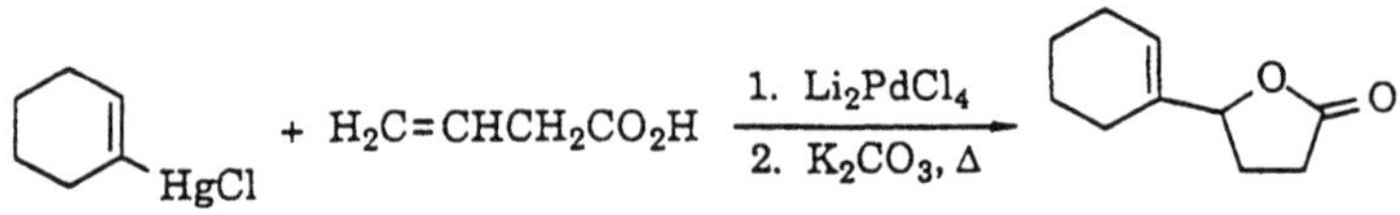

(388)

In a similar manner, the reaction of functionalized vinylic mercurials and unsaturated cyclopropanes proceeds via ring-opening and π-allylpalladium formation to afford lactones (Eq. 389).[617]

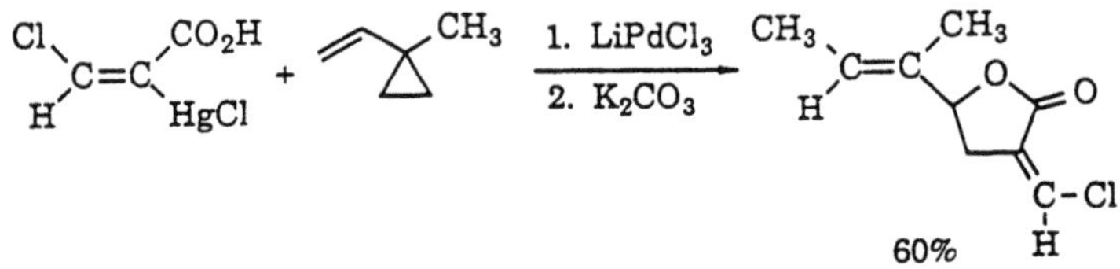

(389)

There are a number of examples of reactions of vinylic mercurials with alkenes which produce dienes. For example, β-chloro[618] and β-acetoxy[619] vinylic mercurials react with alkenes and palladium and copper salts to generate dienes (Eqs. 390 and 391).

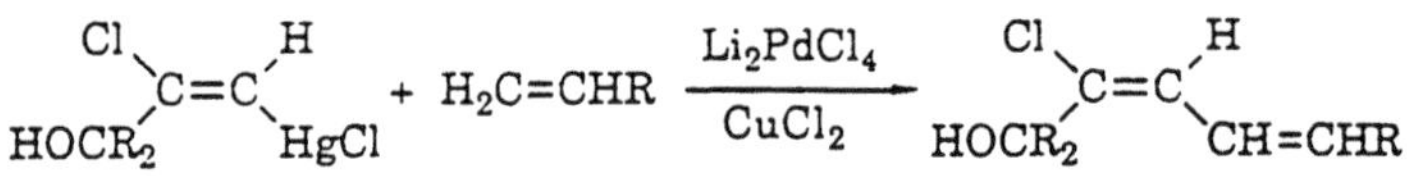

(390)

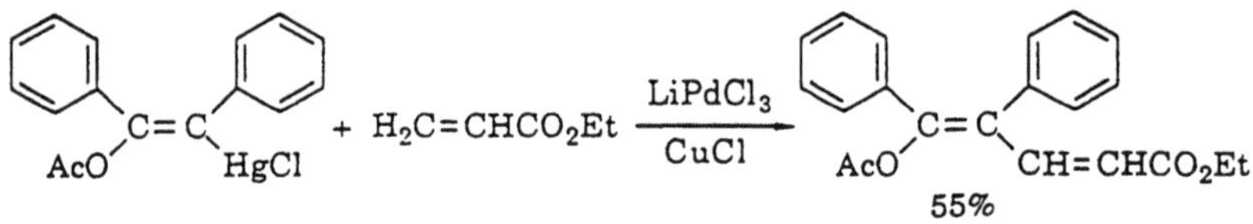

(391)

This type of reaction has been applied to the synthesis of 4-alkenyl isoquinoline derivatives (Eq. 392)[620] and numerous alkenyl nucleosides (Eq. 393).[85,214,621–628]

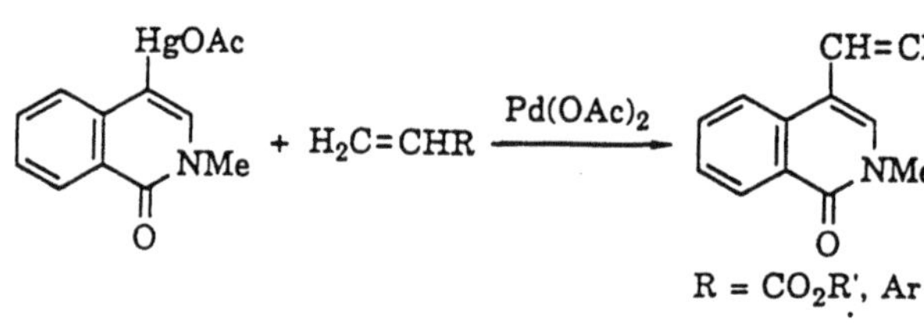

(392)

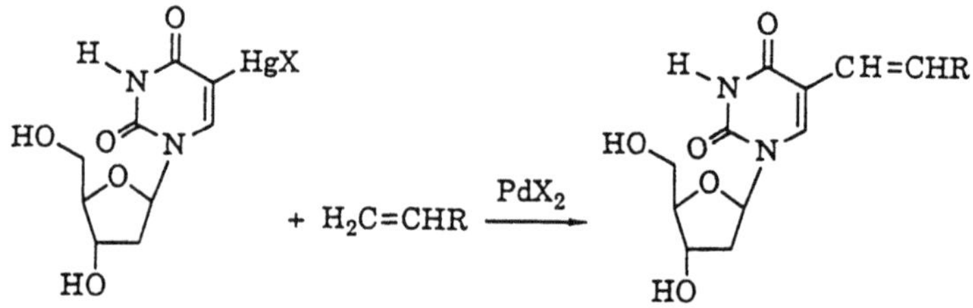

(393)

Vinylic mercurials react with allylic halides and palladium to produce 1,4-dienes (Eq. 394).[629] Several allylic nucleosides have been prepared in this manner.[214,627,630,631]

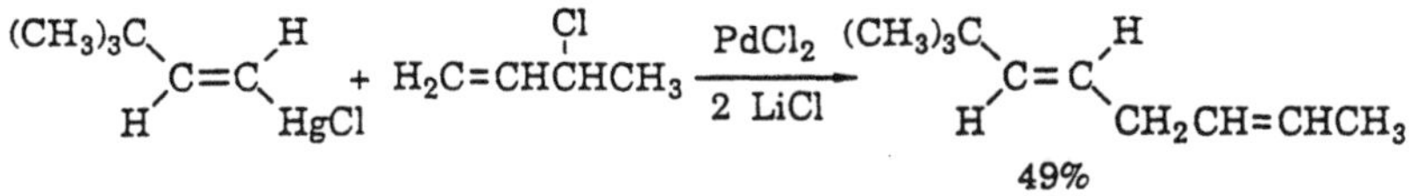

(394)

Allyl triflate in the presence of 4% $Pd(PPh_3)_4$ and 3 equivalents of LiCl has also been used to produce allyl nucleosides.[520] Allylic acetates afford only very low yields in such reactions.[630]

Relatively hindered vinylic mercurials will react with unsaturated epoxides,[632] oxetanes[633] and β-lactams[634] to afford ring-opened allylic and homoallylic alcohols, and β,γ-unsaturated amides respectively (Eqs. 395 and 396).

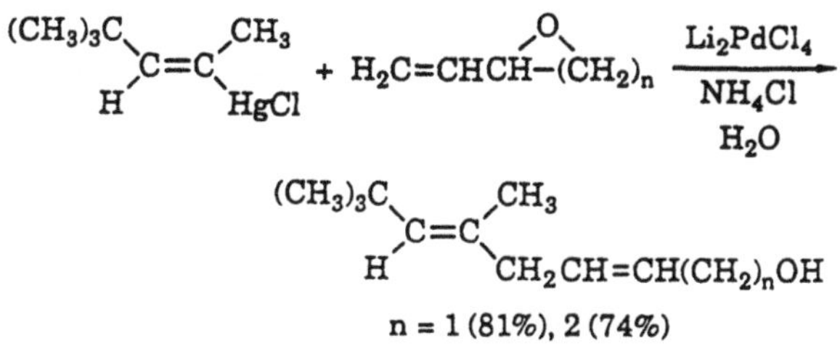

(395)

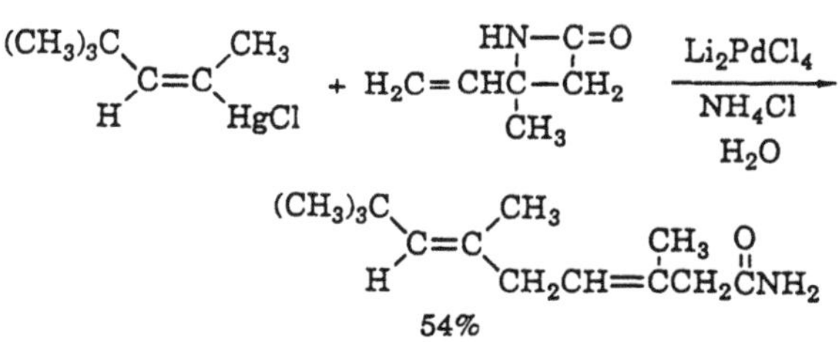

(396)

Vinylic mercurials also react with cyclic alkenes and Li_2PdCl_4 to produce π-allylpalladium compounds (Eq. 397).[635,636] In the presence of triethylamine, 1,4-dienes are produced instead (Eq. 398).

Analogous reactions with various glycals have been studied extensively as a useful route to nucleosides (Eq. 399).[214,637–651] When the glycal contains an allylic

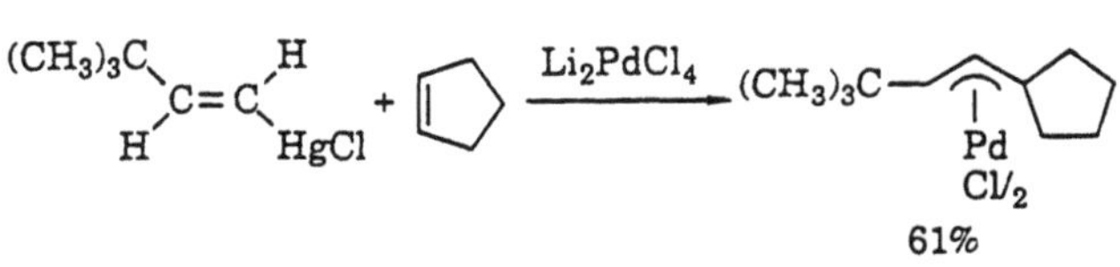

(397)

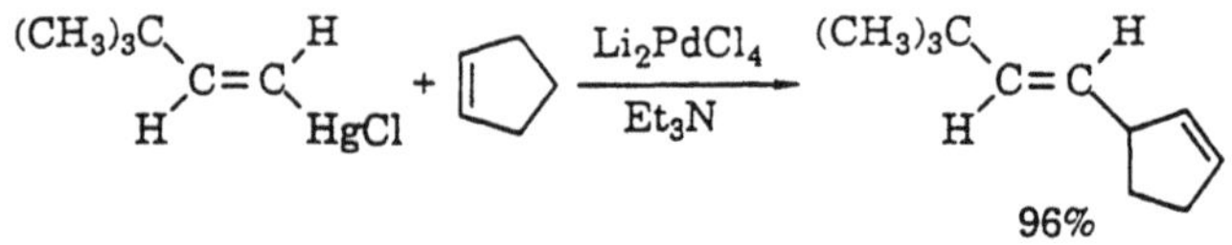

(398)

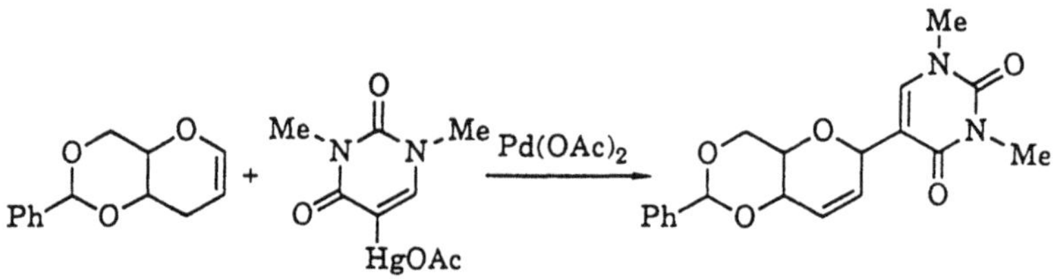

(399)[637]

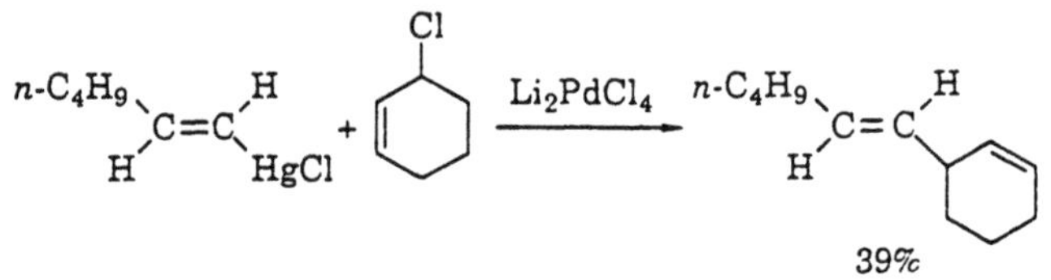

alcohol or acetate group, depending on the stereochemistry of the cyclic alkylpalladium intermediate, either β-hydride, hydroxide or acetate elimination may occur.

The reaction of simple vinylic mercurials and cyclic allylic halides affords the expected halide elimination product in modest yield (Eq. 400).[629]

(400)

The vinylation of *cis*-4-cyclopentene-1,3-diol has been employed in a very efficient approach to prostaglandins (Eq. 401).[652]

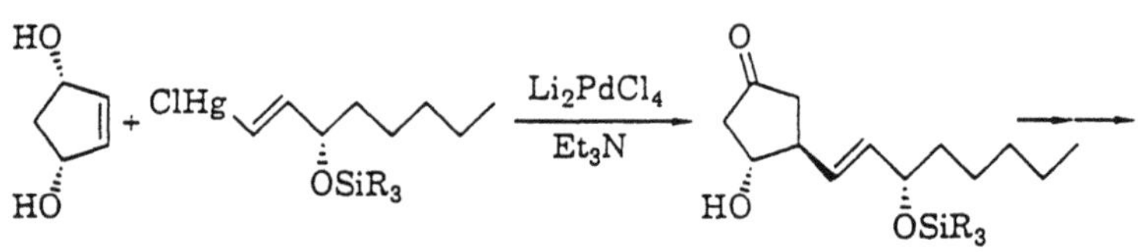

(401)

Vinylmercurials react with Li_2PdCl_4 and bicyclic alkenes to produce stable, isolable addition compounds (Eq. 402).[635,653] A number of subsequent reactions of these palladium compounds have been reported by us (Scheme 4).[635,653,654] This chemistry has proven useful in the synthesis of prostaglandin endoperoxide analogues (Eq. 403).[655]

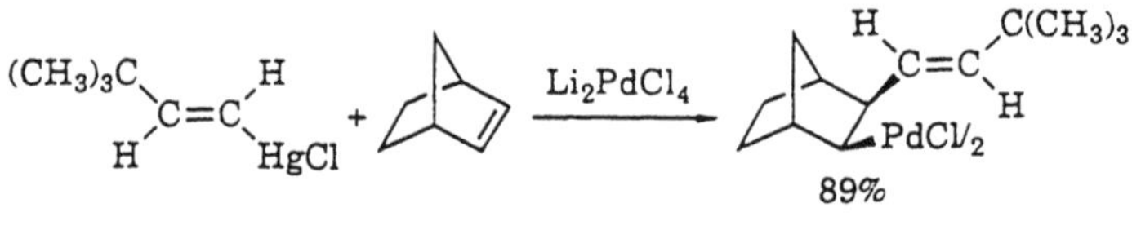

(402)

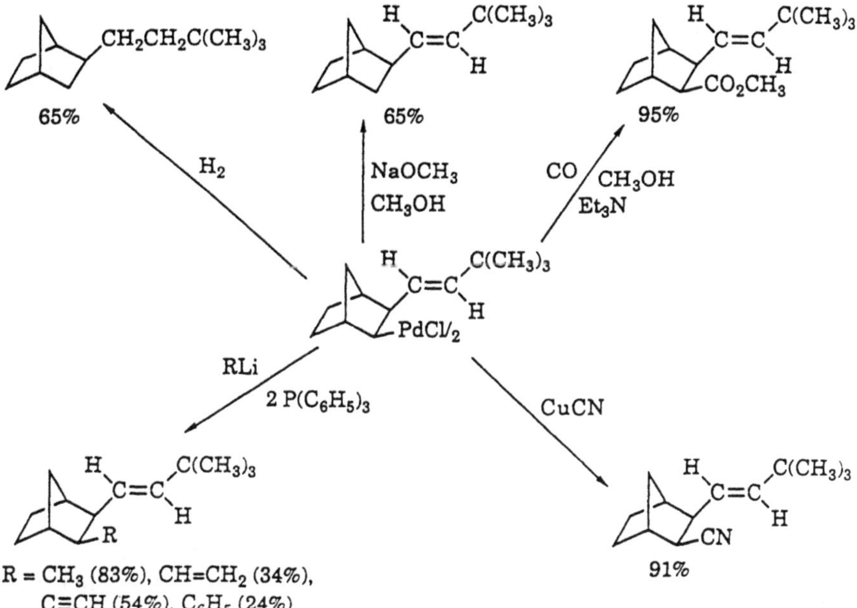

Scheme 4.

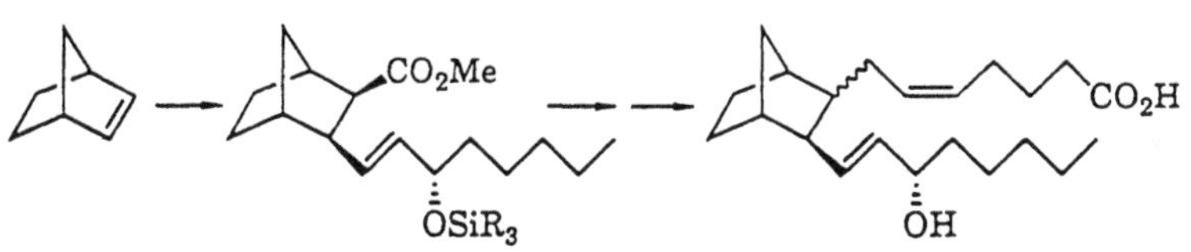

(403)

The vinylation of norbornadiene[653] or a related palladium adduct[656,657] affords nortricyclyl products (Eqs. 404 and 405).

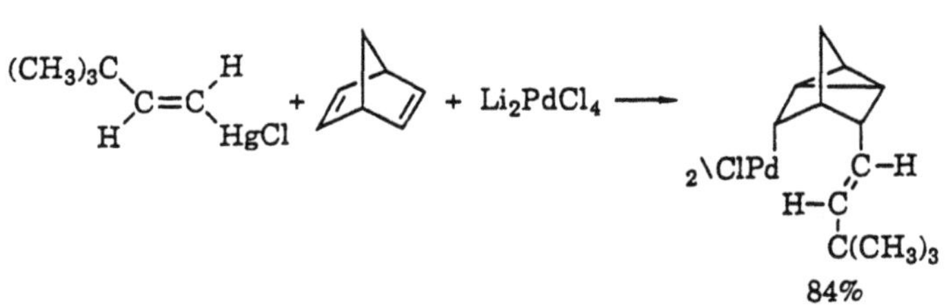

(404)

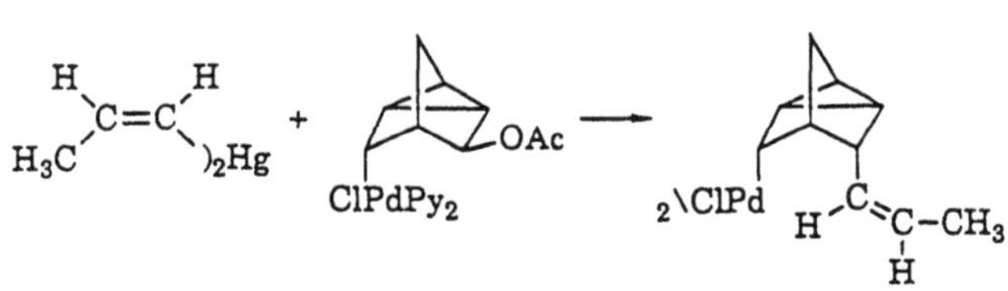

(405)

Functionalized vinylmercurials have been reacted with 1,3- and 1,4-dienes to afford annulation products via intramolecular π-allylpalladium displacement (Eqs. 406 and 407).[617,658] Palladium migration is obviously taking place in this latter process.

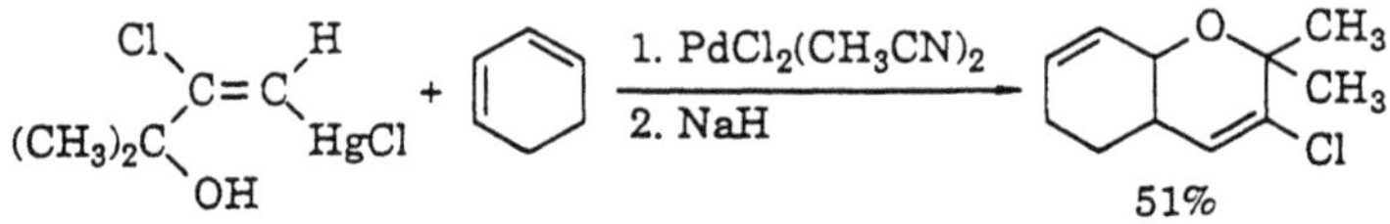

(406)

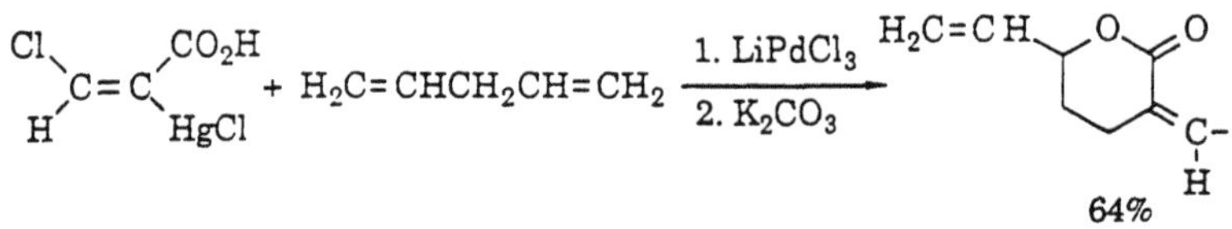

(407)

Vinylic mercurials can be acylated using acyl chlorides and a palladium catalyst, but this transformation is best effected using $AlCl_3$, instead of palladium (Eq. 408).[659]

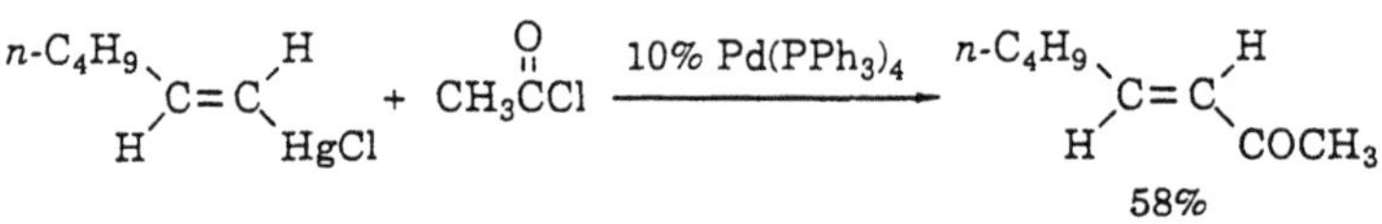

(408)

The palladium-catalyzed carbonylation of vinylic mercurials affords dienones or dienediones (Eq. 409).[660]

$$\mathrm{PhCH{=}CHHgCl} \xrightarrow[\substack{\text{pyridine} \\ \mathrm{I_2}}]{\substack{\mathrm{CO} \\ 10\%\ \mathrm{PdCl_2(PPh_3)_2}}} \mathrm{(PhCH{=}CH)_2CO} \text{ or } \mathrm{(PhCH{=}CHCO)_2} \quad (409)$$

The carbonylation of vinylmercurials in the presence of water or alcohols, catalytic amounts of palladium, and cupric chloride as a reoxidant affords α,β-unsaturated acids and esters respectively (Eq. 410).[661] The mechanism of the carbonylation of *trans*-β-chlorovinylmercuric chloride to the corresponding acrylic acid or ester has been examined.[662]

$$\mathrm{RCH{=}CHHgCl} \xrightarrow[\mathrm{R'OH}]{\substack{\mathrm{CO} \\ 10\%\ \mathrm{PdCl_2}\text{ or }\mathrm{Pd/C}}} \mathrm{RCH{=}CHCO_2R'} \quad (410)$$

The carbonylation of vinylic mercurials prepared by the intramolecular solvomercuration of allenes provides a useful route to 2-(2-tetrahydrofuranyl)-acrylates[663] and nonactate esters (Eq. 411).[664]

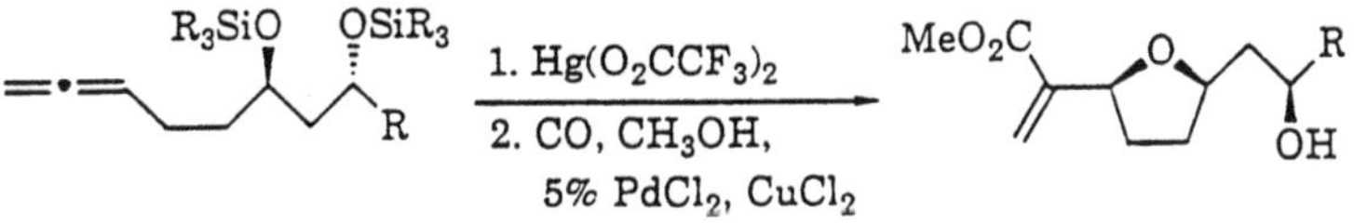

(411)

Vinylmercurials prepared by mercuric acetate addition to alkynes also undergo carbonylation to esters (Eq. 412).[665]

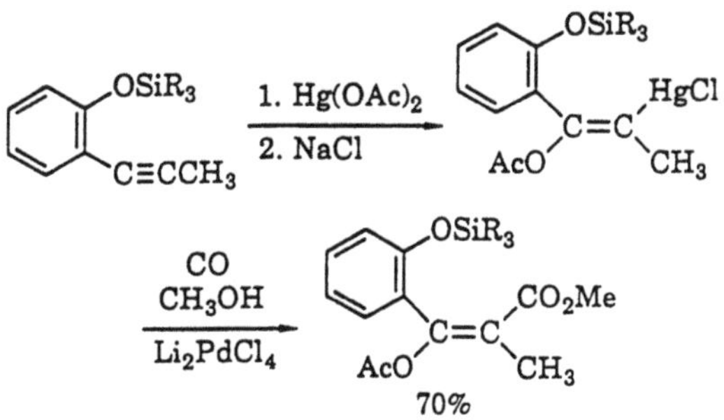

(412)

The carbonylation of vinylmercurials derived from analogous additions to propargylic alcohols provides a novel approach to butenolides (Eq. 413).[666,667] The mercuration of 4-hydroxy-2-alkyn-1-ones and subsequent carbonylation affords furan products rather than the anticipated butenolides.[668]

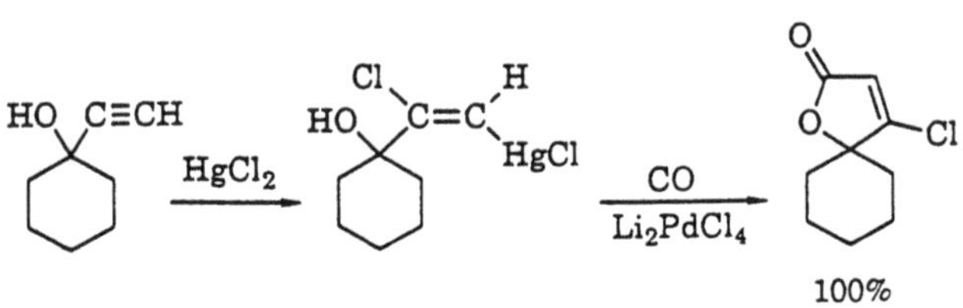

(413)

In the absence of other reagents, vinylic mercurials react with palladium reagents to afford 1,3-dienes.[669] Symmetrical dienes and polyenes are best obtained when the reaction is run in the presence of LiCl, using HMPA as the solvent (Eq. 414).[670] Divinylmercurials are catalytically coupled using $Pd(PPh_3)_4$, but some loss of

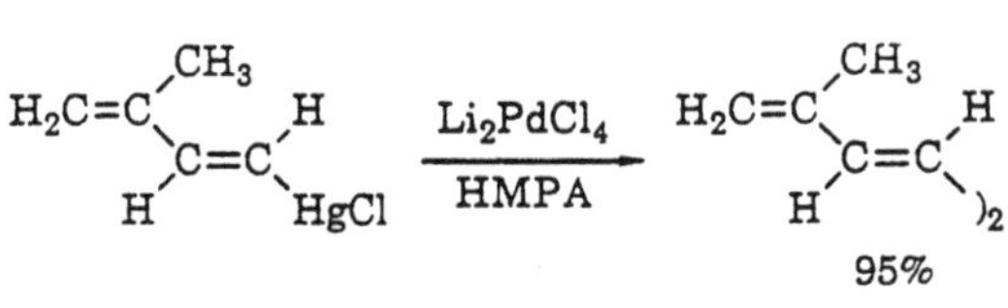

(414)

stereochemistry is observed (Eq. 415).[671] One can also employ cupric chloride and palladium chloride to effect a catalytic process, but all possible stereoisomers have been observed in the one example reported (Eq. 416).[672]

$$(CH_3CH{=}CH)_2Hg \xrightarrow{\text{cat. } Pd(PPh_3)_4} (CH_3CH{=}CH)_2 \qquad (415)$$

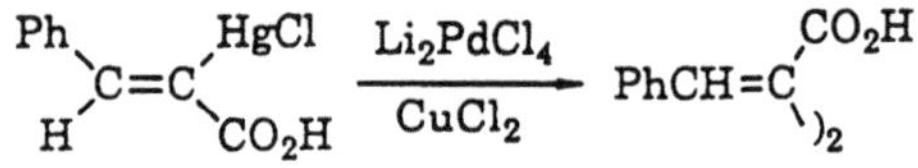
(416)

When this type of dimerization is carried out in a nonpolar solvent, such as benzene, a surprising "head-to-tail" dimerization ensues (Eq. 417).[673] Pyrimidinyl-mercury salts undergo an analogous process.[674]

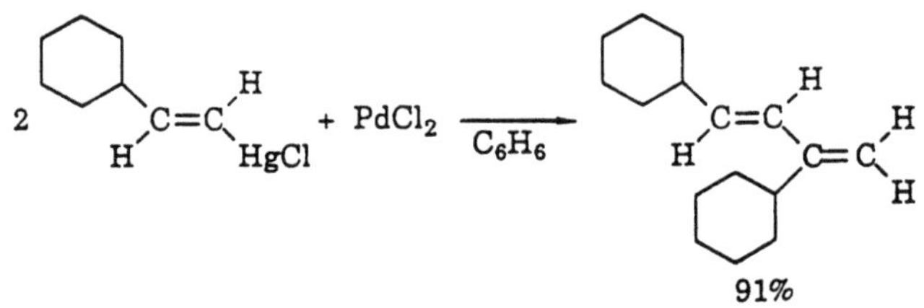
(417)

The reaction of vinylic mercurials and palladium acetate affords quite different products, enol acetates (Eq. 418).[675] This reaction which has been used to prepare simple enol esters and ene diacetates becomes catalytic in palladium when mercuric acetate is added as a reoxidant.

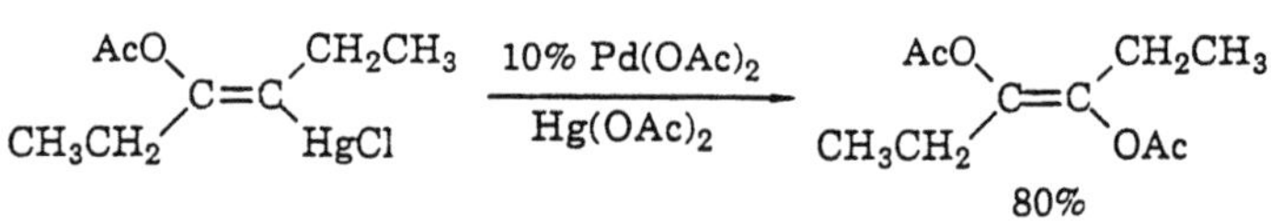
(418)

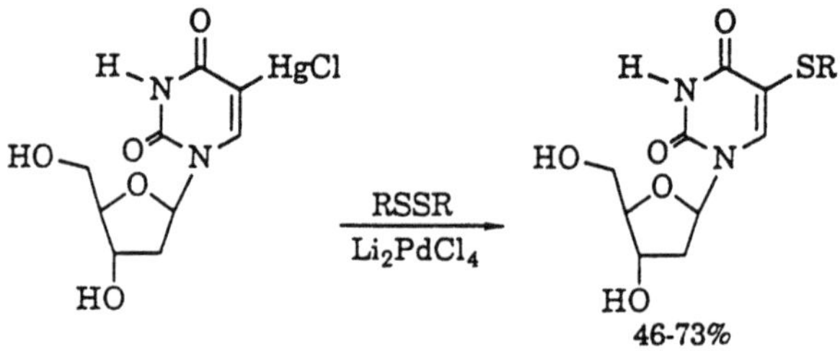
(419)

Finally, palladium can promote the substitution of mercury by sulfur.[676,677] Sulfur-containing nucleosides have been prepared in this fashion (Eq. 419). The mechanism of this reaction is unknown.

Vinylic Boranes

Organoboranes also undergo a variety of very useful palladium-catalyzed vinylic substitution reactions.[350,364,365] For example, palladium acetate catalyzes the substitution of boron by hydrogen (Eq. 420).[678,679] The analogous reaction of vinylboranes derived from terminal alkynes, when run in the presence of triethylamine and stoichiometric amounts of palladium acetate, affords rearranged alkenes instead (Eq. 421).[679,680]

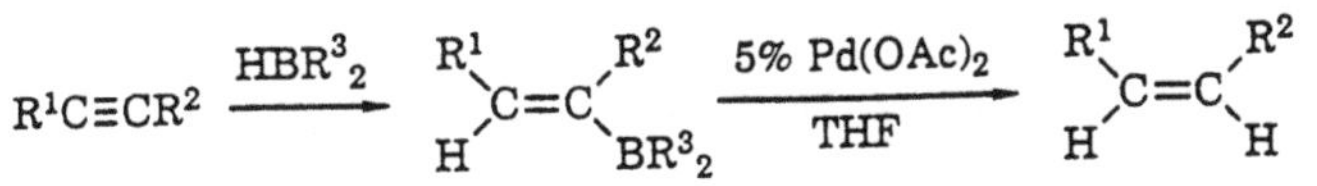
(420)

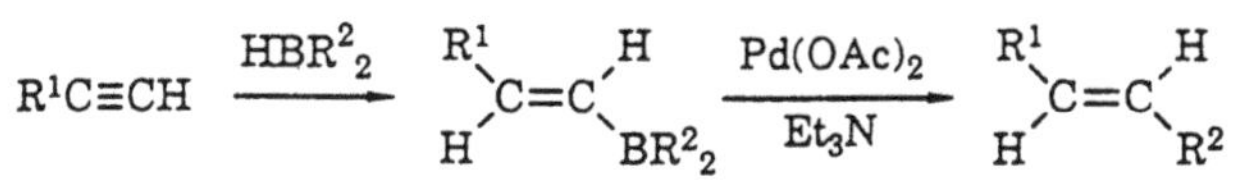
(421)

There are a few examples of the Heck reaction using vinylic boranes[679] or boronic acids[93,681] (Eq. 422). Dienyl silanes have been prepared by this process.[93]

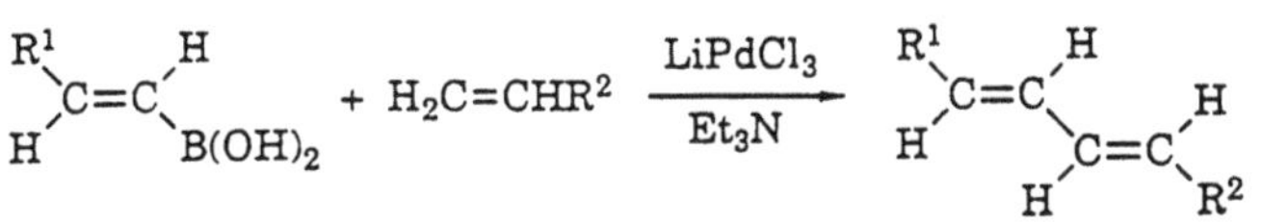
(422)[681]

R^2 = *n*-Bu, Ph, CO_2Me, CONH-*t*-Bu 48-80%

Vinylic boron compounds react with olefinic epoxides and Ni or Pd catalysts to produce regioisomeric mixtures of dienols (Eq. 423).[682]

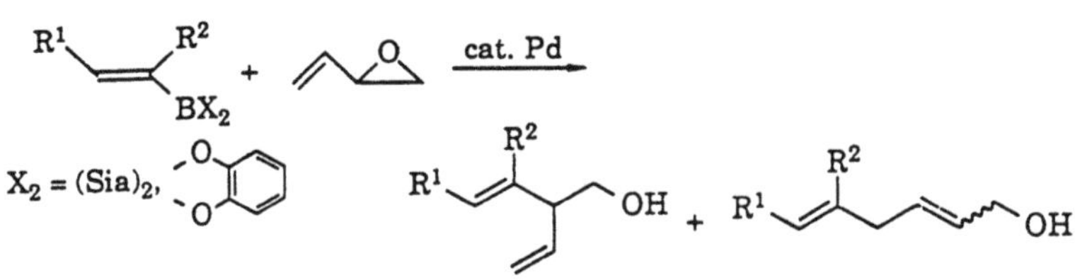
(423)

Vinylboranes can be effectively cross-coupled with a variety of organic halides through palladium catalysis.[350,365,683] Allylic halides react in the presence of catalytic amounts of palladium acetate[679] or better, $Pd(PPh_3)_4$ in the presence of hydroxide (Eq. 424).[367,684] An intramolecular version of this reaction has been utilized for the synthesis of humulene (Eq. 425).[685]

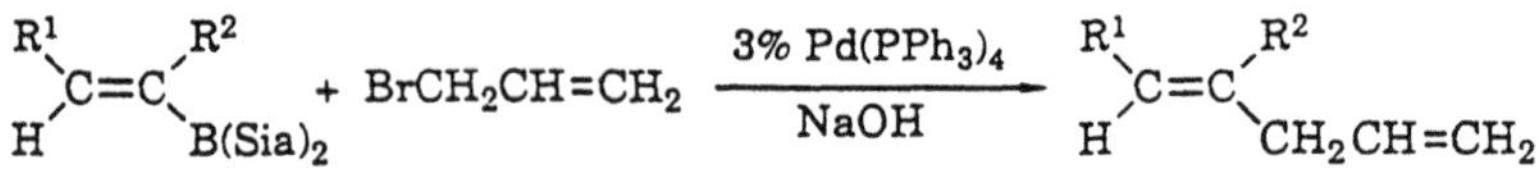

(424)

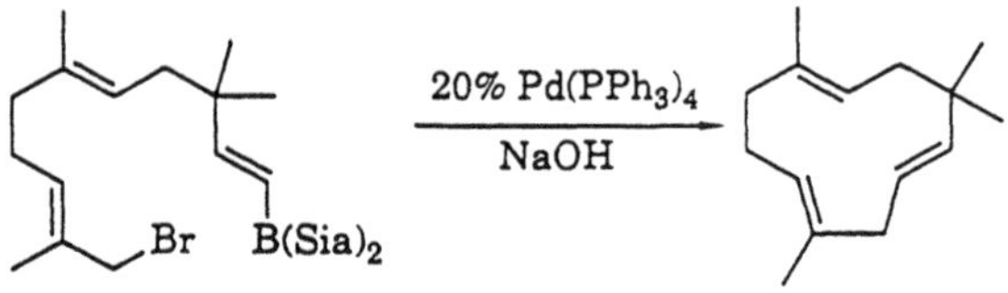

(425)

Allylation of vinylic boron compounds can also be effected in good yield by using allylic phenyl ethers (Eq. 426).[367,686]

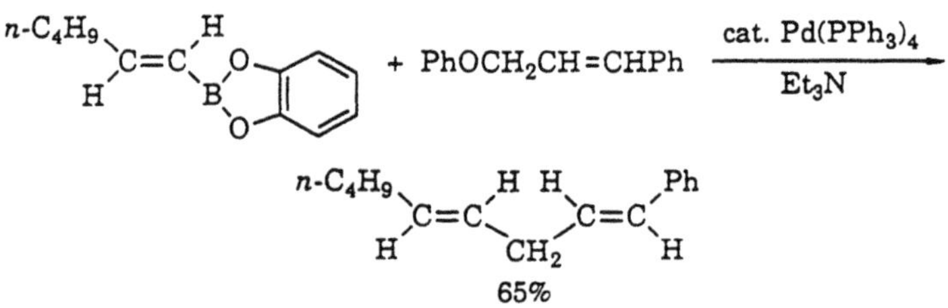

(426)

Vinylic boranes derived from terminal alkynes by copper-catalyzed stannylboration undergo allylation to stannyl dienes (Eq. 427).[687] In a similar manner, arylation can be effected.

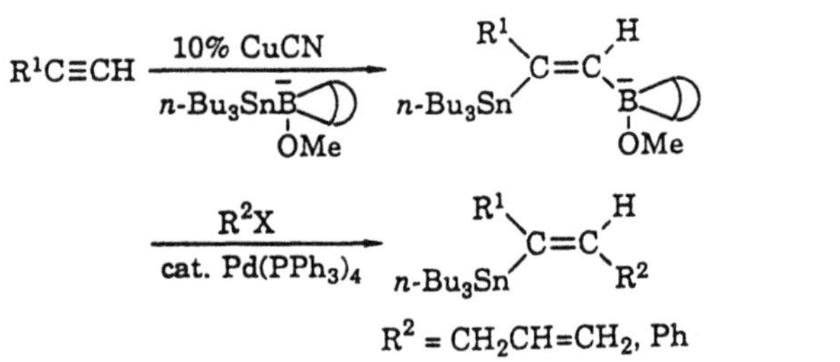

(427)

Vinylic boranes derived from alkynes via bromoboration can be alkylated sequentially at first the halogen and then the boron (Eq. 428).[301,302,351]

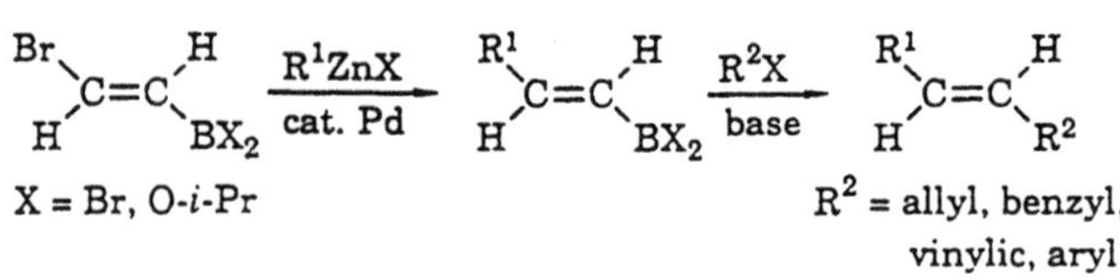

(428)

As just noted, benzylic halides take part in these palladium-catalyzed processes (Eq. 429).[684] Vinylic ethers have been prepared by such coupling processes (Eq. 430).[688] This reaction using an aryl halide has been employed in the synthesis of various styryl ethers,[689] as well as aurantioclavine.[690]

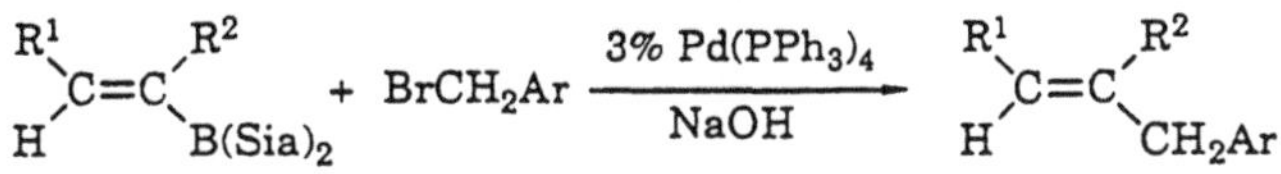
(429)

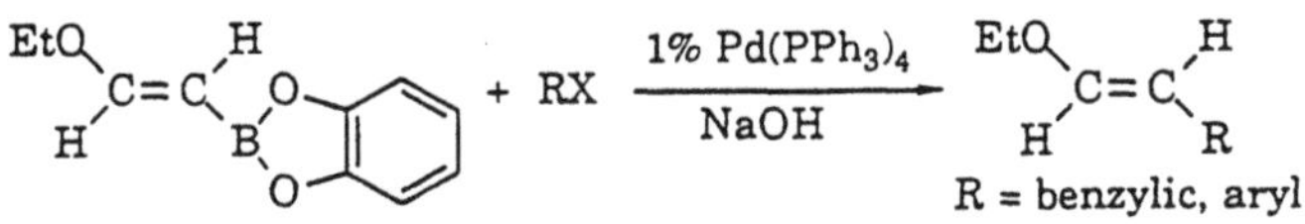
(430)

1-Halo-1-alkynes react with vinylic boron compounds to produce enynes (Eq. 431).[366–368,370] The reaction is highly stereoselective and has been employed in the synthesis of the sex pheromone of *Lobesia botrana*.[374]

$$R^1CH{=}CHB(Sia)_2 + BrC{\equiv}CR^2 \xrightarrow[NaOCH_3]{cat.\ Pd(PPh_3)_4} R^1CH{=}CHC{\equiv}CR^2 \qquad (431)$$

Besides the earlier mentioned examples of the arylation of vinylboranes, a number of other examples exist.[369–371,691,692] The reaction is usually effected using catalytic amounts of $Pd(PPh_3)_4$ in the presence of sodium ethoxide. Silyl styrenes can be prepared by this process (Eq. 432).[371] When palladium black is used as a catalyst in the presence of triethylamine, an unusual "head-to-tail" coupling is observed (Eq. 433).[390]

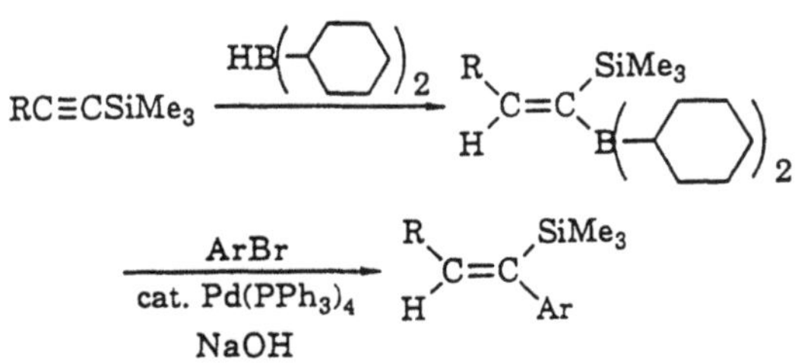
(432)

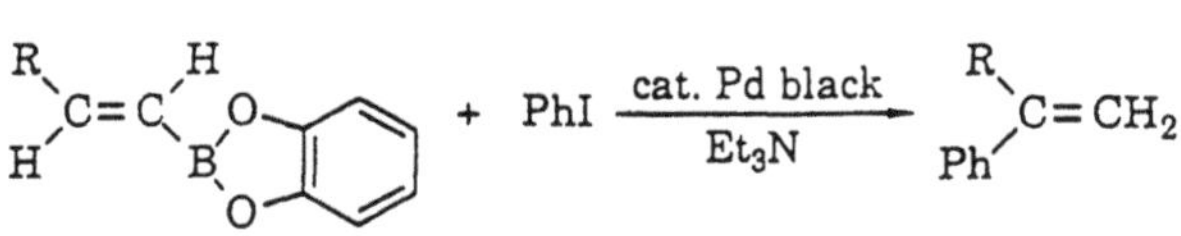
(433)

Vinylic boronate esters prepared by the hydroboration of alkynes also undergo facile palladium-catalyzed carbonylation in an alcohol solvent to produce α,β-unsaturated esters (Eq. 434).[303,693,694]

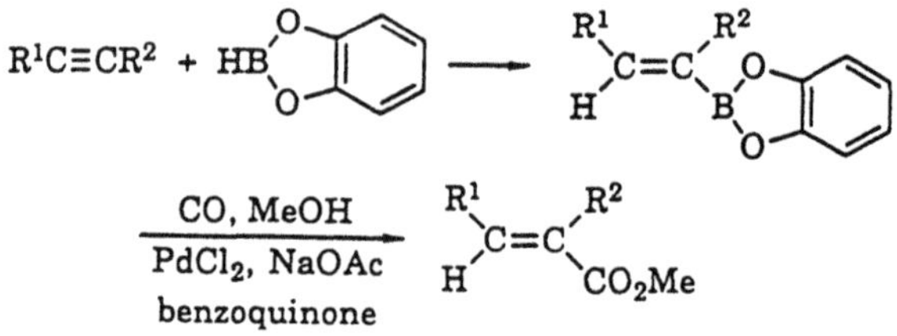

(434)

Vinylic boron reagents also couple with organopalladium compounds generated by palladium-promoted cyclization of unsaturated aryl halides (Eqs. 435 and 436).[157]

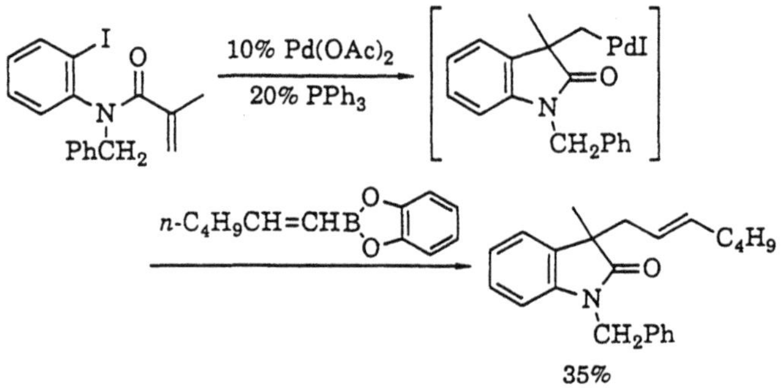

(435)

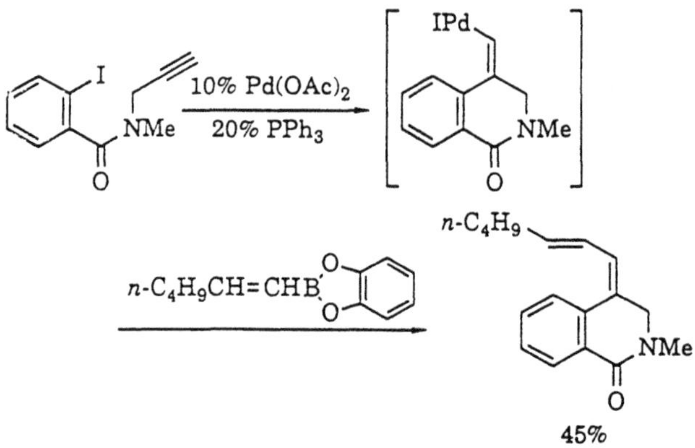

(436)

Finally, vinylic boron compounds can be dimerized to 1,3-dienes via vinylpalladium intermediates. Vinylic 9-BBN derivatives couple in modest yield using stoichiometric amounts of $Pd(OAc)_2$ plus Et_3N.[679] Vinylic catechol boronates are catalytically coupled using $Pd(OAc)_2$ with $Cu(OAc)_2$ as a reoxidant (Eq. 437).[695] Styrylboronic acids are dimerized in high yield by 10% $PdCl_2$, LiCl and Et_3N.[696]

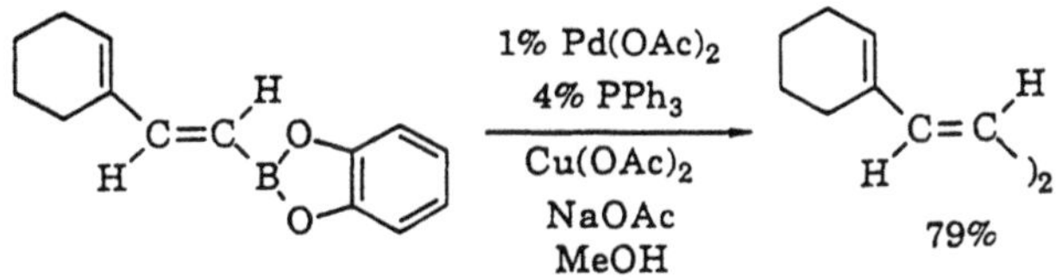

(437)

Vinylic Alanes

Vinylic aluminum reagents cross-couple with a variety of organic halides in the presence of a palladium catalyst.[325] Vinylic alanes prepared from terminal alkynes via zirconium-catalyzed carboalumination can undergo facile, highly stereospecific cross-coupling with allylic halides to form 1,4-dienes, such as α-farnesene (Eq. 438).[697–699] However, some loss of regioselectively is observed with crotyl halides.[585]

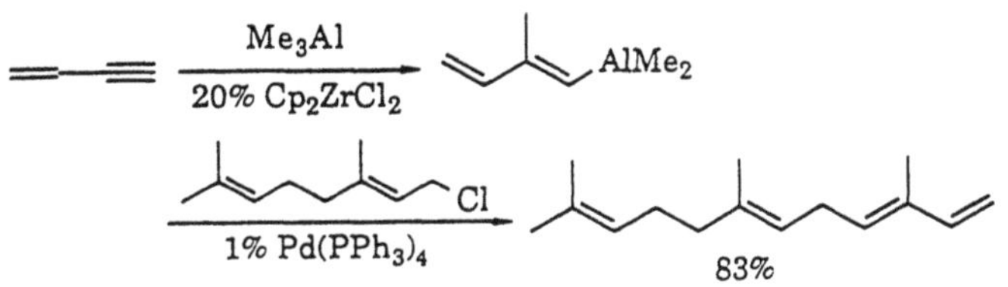

(438)

Allylic acetates undergo substitution with retention of the C–C double bond stereochemistry,[699] but inversion of the acetate stereochemistry (Eq. 439).[700]

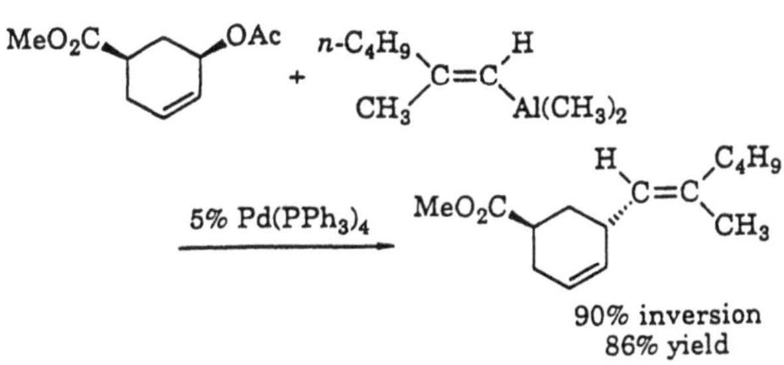

(439)

Allylic ethers of silicon and aluminum may also be utilized in this 1,4-diene synthesis, although they are less reactive than the corresponding chlorides or acetates.[699] This process has been employed in the synthesis of silyl-substituted carbocycles (Eq. 440).[701]

Allylic acetals and ortho esters can also be catalytically cross-coupled with vinylic alanes (Eq. 441).[702]

Benzylic[324] and alkynyl[331] halides also couple in good yields with vinylic alanes, although the latter process is faster using a nickel catalyst[702] (Eq. 442). Zinc

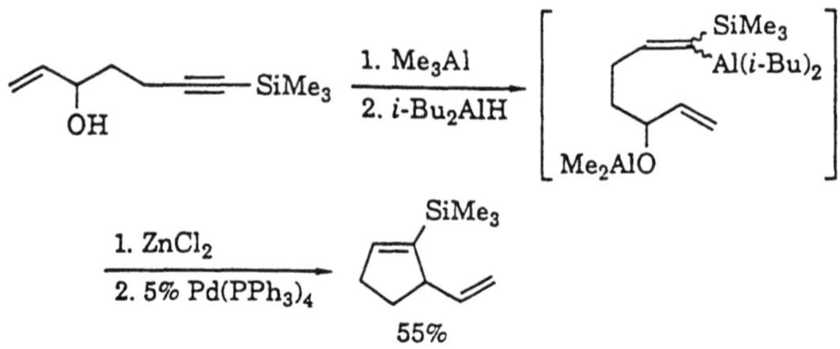

(440)

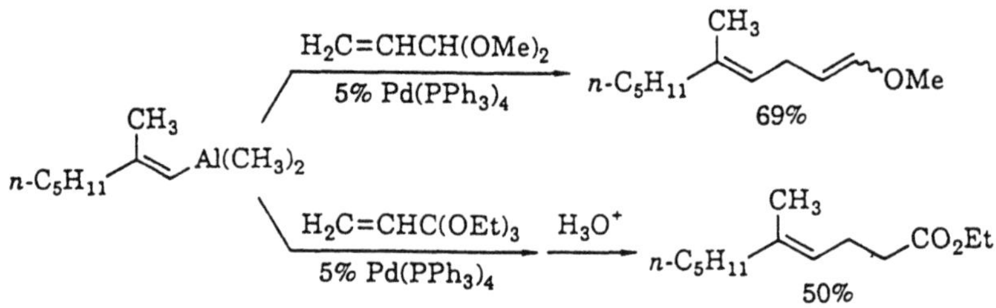

(441)

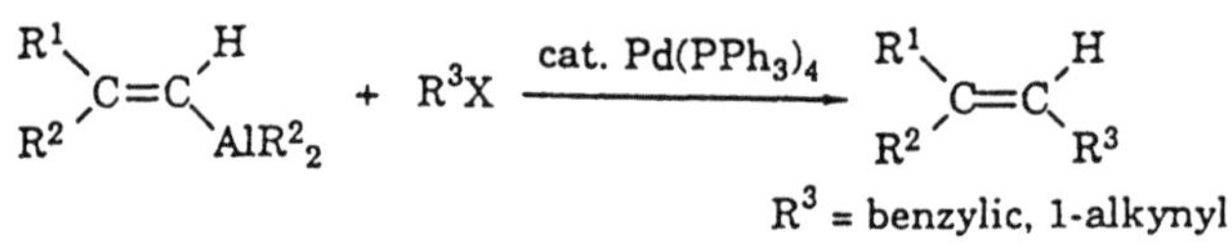

(442)

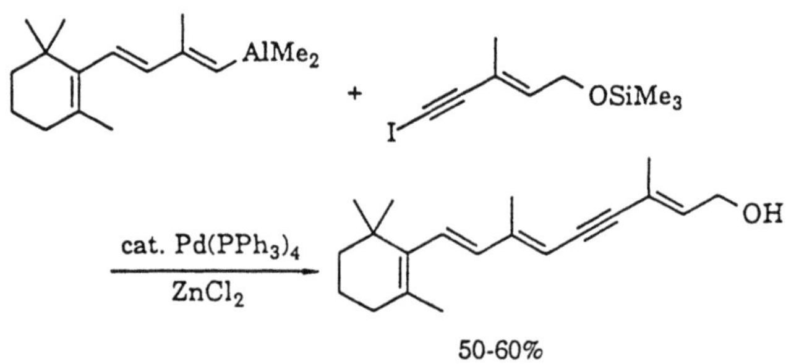

(443)

chloride has also been utilized in the synthesis of dehydrovitamin A to promote this latter process (Eq. 443).[325]

Palladium also catalyzes the arylation of vinylic alanes,[445,451,453,703–705] although $Ni(PPh_3)_4$ as catalyst affords faster coupling.[703] The arylation of hindered vinylic alanes is facilitated by $ZnCl_2$ addition. This general process has been used to vinylate heterocycles[445] and prepare triarylethylenes,[704] such as clomid,[453] tamoxifen[705] and broparestrol[451] (Eq. 444). Aryl fluorosulfonates also react in the presence of LiCl plus $Pd(PPh_3)_4$, but a mixture of products is observed (Eq. 445).[706]

Enones can be formed by the palladium-catalyzed acylation[707] or carbonylation[708] of vinylic alanes (Eqs. 446 and 447).

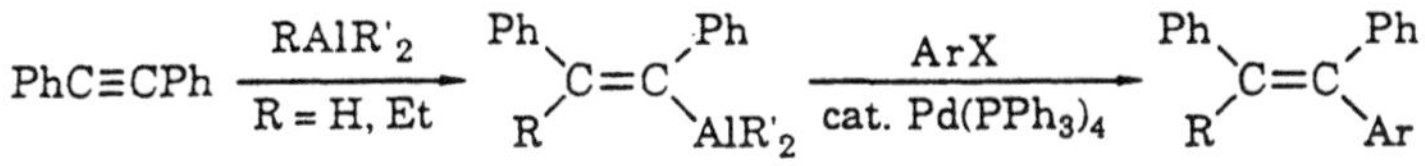
(444)

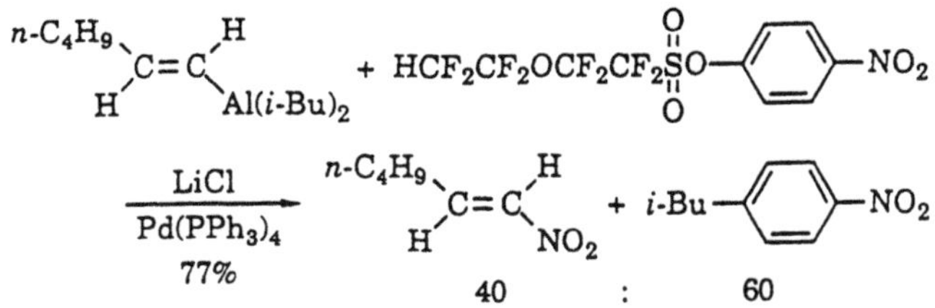
(445)

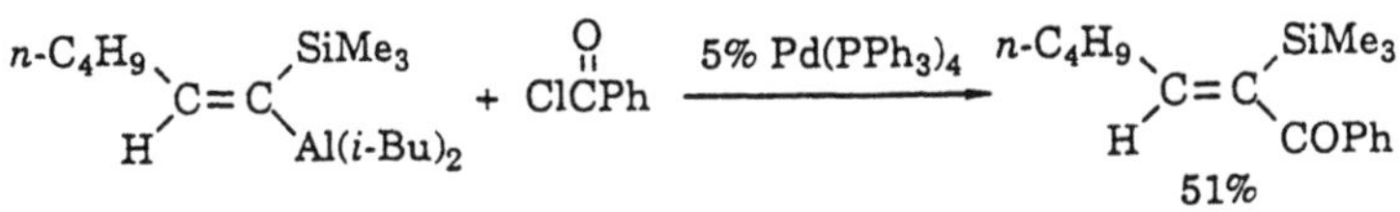
(446)

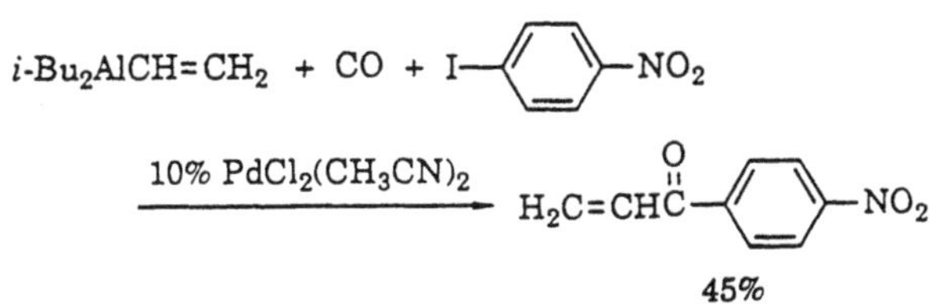
(447)

Vinylic Silanes

Vinylic silanes undergo a number of different palladium-promoted vinylic substitution reactions. The reaction of vinylic silanes and palladium chloride results in silyl-substituted π-allylpalladium compounds (Eq. 448).[709,710] Substituted vinylic silanes and trimethylvinylgermane also undergo this process.[709]

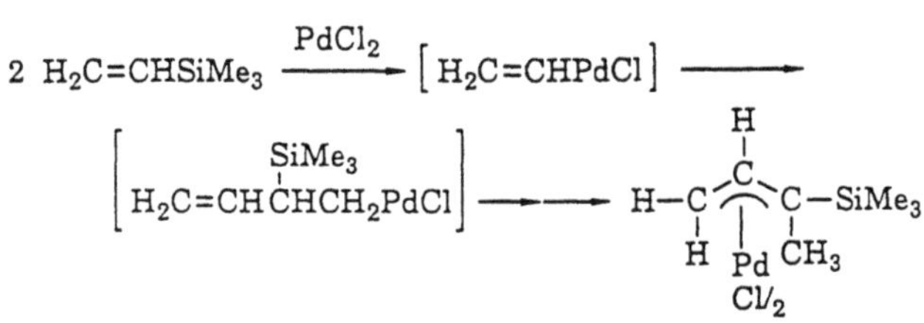
(448)

Variable yields of π-allylpalladium compounds are also observed in the palladium-promoted cross-coupling of vinylic pentafluorosilicates and alkenes (Eq. 449).[463]

Vinylic silanes also react with $PdCl_2$ in the presence of methanol[711] or cupric chloride[277] to produce 1,3-dienes and polyenes (Eq. 450). Silyl cyclopropenes can be dimerized by palladium halides (Eq. 451).[712,713] Vinylic pentafluorosilicates can also be dimerized in this fashion (Eq. 452).[463]

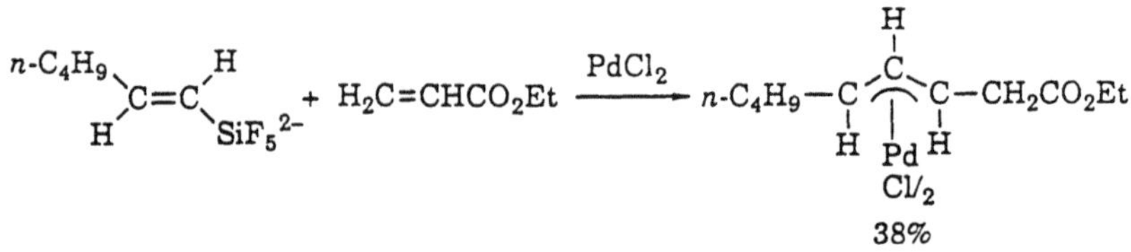

(449)

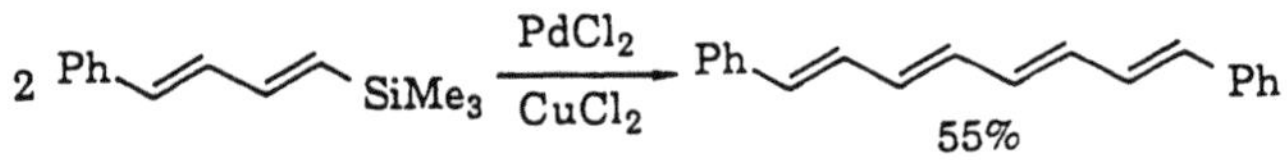

(450)

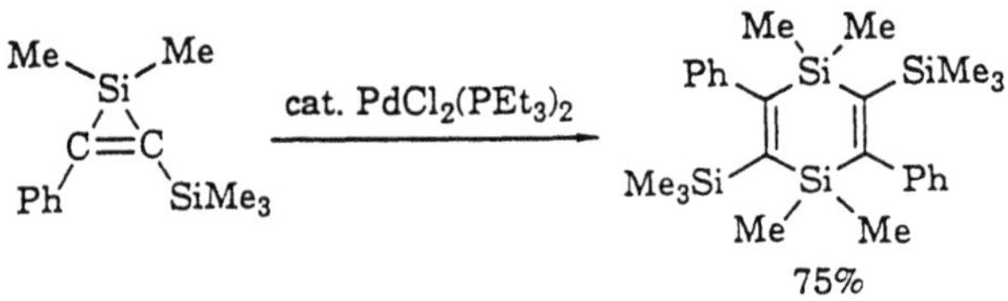

(451)

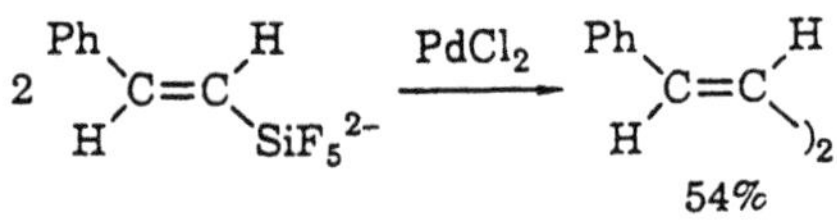

(452)

Vinylic silanes react with alkenes and palladium salts to afford a variety of products. When ethylene and $CuCl_2$ are present, a homoallylic chloride is formed in low yield (Eq. 453).[711] Alkenes bearing electron-withdrawing groups afford Heck-type products (Eq. 454). Vinylic pentafluorosilicates react similarly (Eq. 455).[463]

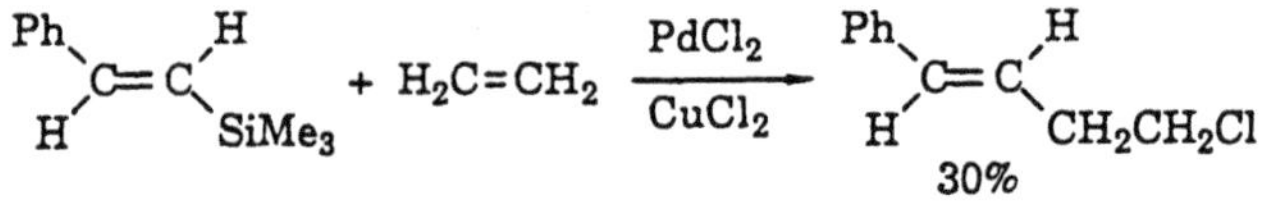

(453)

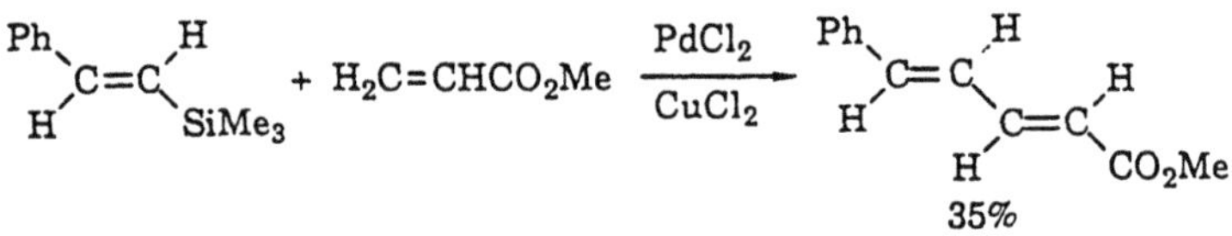

(454)

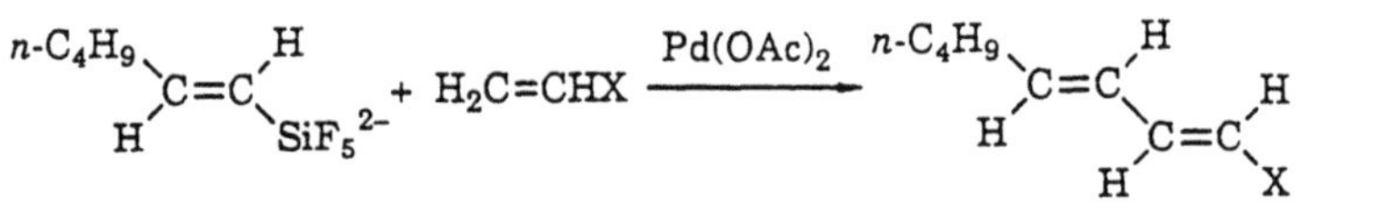

(455)

X = CO_2Me (37%), CO_2Et (48%), CHO (20%), CN (43%)

The cross-coupling of vinylic pentafluorosilicates with allylic halides provides modest yields of 1,4-dienes via double bond transposition (Eq. 456).[463,714]

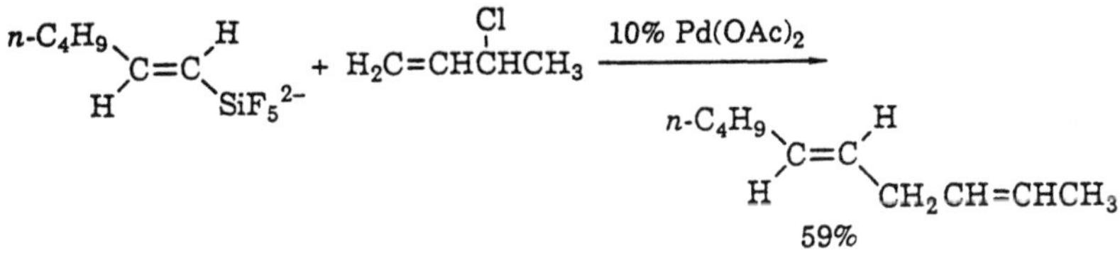
(456)

Styrenes can also be formed from vinylic silanes. For example, (*E*)-1,2-bis(trimethylsilyl)ethylene reacts with aryl halides to afford specifically (*Z*)-styrylsilanes (Eq. 457).[715] This reaction apparently proceeds by arylpalladium *syn* addition to the C–C double bond and subsequent silylpalladium *syn* elimination. The coupling of trimethylvinylsilane and aryl iodides in the presence of tris(dimethylamino)sulfonium difluorotrimethylsilicate seems to be a more general process (Eq. 458).[316]

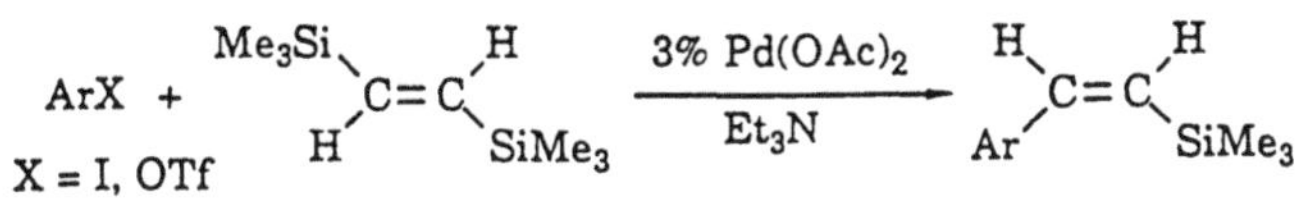
(457)

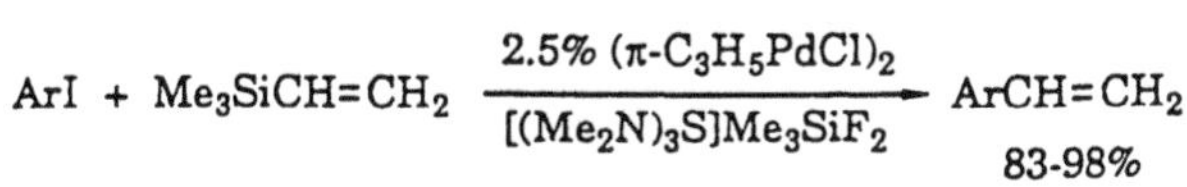
(458)

Fluoro-[394,716] and alkoxy-[395] substituted vinylic silanes can also be effectively cross-coupled with aryl and heterocyclic halides (Eqs. 459 and 460). Only low yields have so far been reported for the arylation of vinylic pentafluorosilicates[463] or bis(catecholato)alkenylsiliconates.[396]

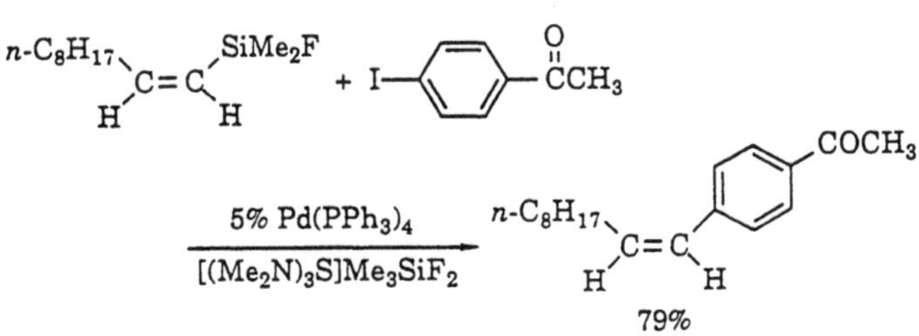
(459)

Good yields of styrenes are also provided when vinylic methyldifluorosilanes are cross-coupled with aryl triflates in the presence of tetra-*n*-butylammonium fluoride (Eq. 461).[539] The catechol-substituted vinylic silanes couple with aryl triflates to give only low yields.[396]

Aryl thallium compounds react with vinyltrimethylsilane or -pentafluorosilicate to afford styrenes, but the yields are quite low.[717]

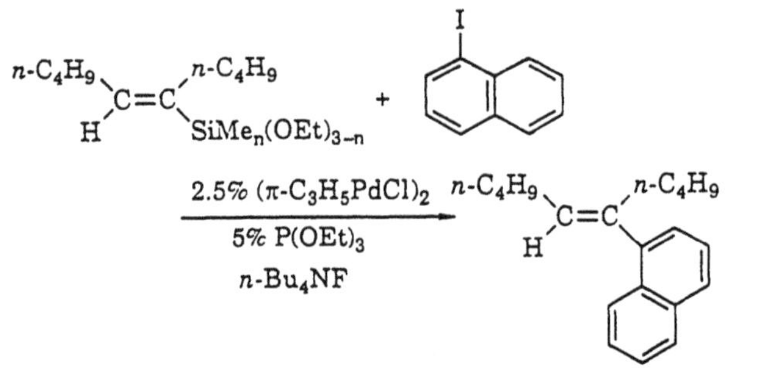

n = 0 (54%), 1 (96%), 2 (95%) (460)

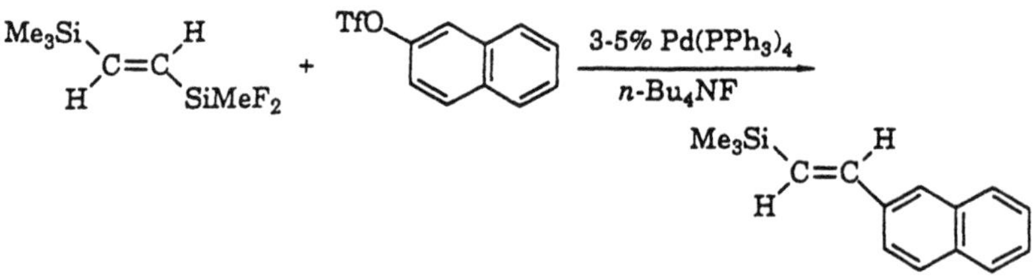

74% (461)

The carbonylation of vinylic pentafluorosilicates in methanol provides a useful route to α,β-unsaturated esters (Eq. 462).[463,718]

$$\underset{H}{CH_3O_2C(CH_2)_8}C{=}C\underset{SiF_5^{2-}}{H} \xrightarrow[\text{PdCl}_2,\ \text{NaOAc}]{\text{CO},\ \text{CH}_3\text{OH}} \underset{H}{CH_3O_2C(CH_2)_8}C{=}C\underset{CO_2CH_3}{H} \quad 72\% \qquad (462)$$

Vinylic Stannanes

Vinylic stannanes undergo a wide variety of useful palladium-promoted substitution processes.[26,30] They couple with relatively stable alkylpalladium intermediates prepared by alkene insertion (Eqs. 463 and 464). The latter type process has been used to prepare prostaglandin endoperoxide analogues (Eq. 465).[595]

A wide variety of organic halides can be cross-coupled with vinylic stannanes.[26,29] For example, perfluoroalkyl iodides react in the presence of catalytic amounts of $Pd(PPh_3)_4$ (Eq. 466).[720]

$$H_2C{=}CHNHCO_2R \xrightarrow[\text{2. NaC(CH}_3\text{)(CO}_2\text{Et)}_2]{\text{1. PdCl}_2} \left[ClPdCH_2\overset{NHCO_2R}{\overset{|}{C}H}C(CH_3)(CO_2Et)_2 \right] \qquad (463)^{719}$$

$$\xrightarrow{RCH{=}CHSnMe_3} RCH{=}CHCH_2\overset{NHCO_2R}{\overset{|}{C}H}C(CH_3)(CO_2Et)_2$$

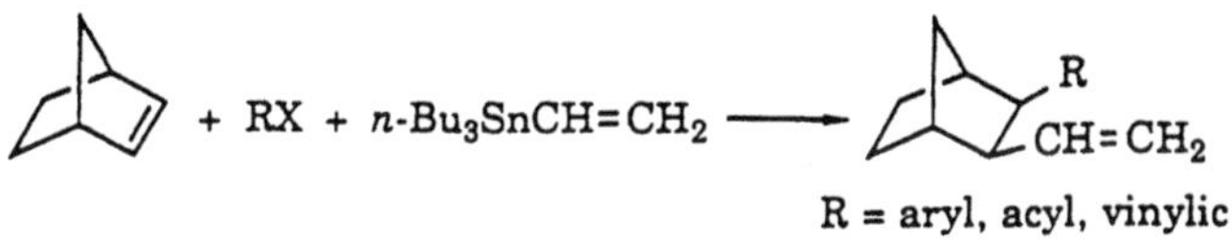

(464)[135]

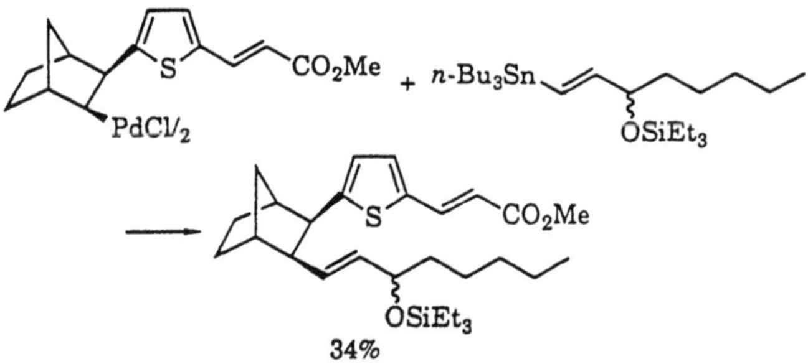

(465)

$$n\text{-}C_4F_9I + n\text{-}Bu_3SnCH{=}CHPh \xrightarrow{10\%\ Pd(PPh_3)_4} n\text{-}C_4F_9CH{=}CHPh \qquad (466)$$

E or *Z* 70%

A number of vinylic stannanes and allylic halides have been cross-coupled using palladium catalysts (Eq. 467).[721,722] Note that the vinylic stereochemistry is retained, but the halide undergoes inversion. A variety of functionally-substituted vinylic tin reagents have been employed in this process (Eqs. 468 and 469).

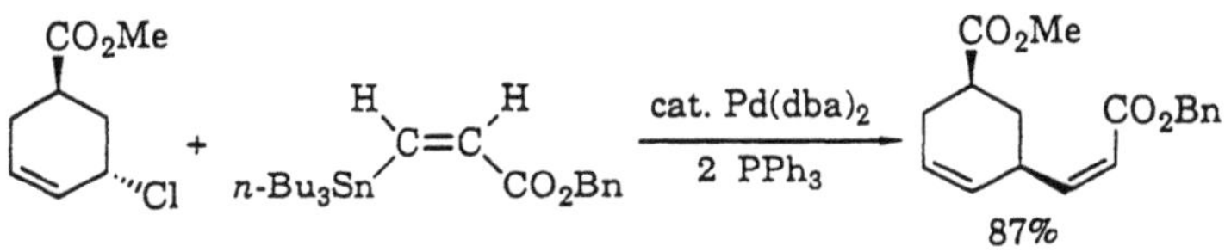

(467)

$$H_2C{=}CHCH_2Cl + n\text{-}Bu_3Sn\overset{OEt}{C}{=}CH_2 \xrightarrow{cat.\ PdCl_2(PPh_3)_2} H_2C{=}CHCH_2\overset{OEt}{C}{=}CH_2 \qquad (468)$$

(468)[413] 43%

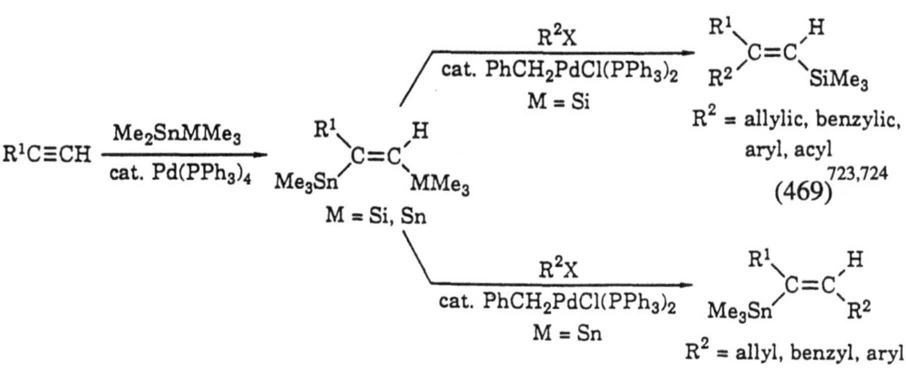

(469)[723,724]

This chemistry has been applied to the synthesis of glycals,[406] cephalosporins,[725] and prostaglandins[726–728] (Eqs. 470–472).

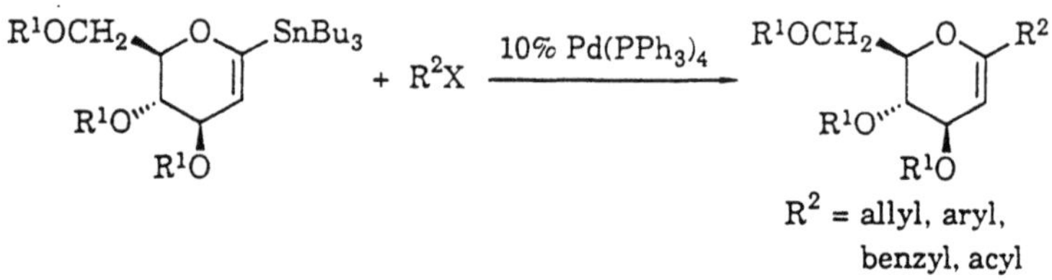

(470)

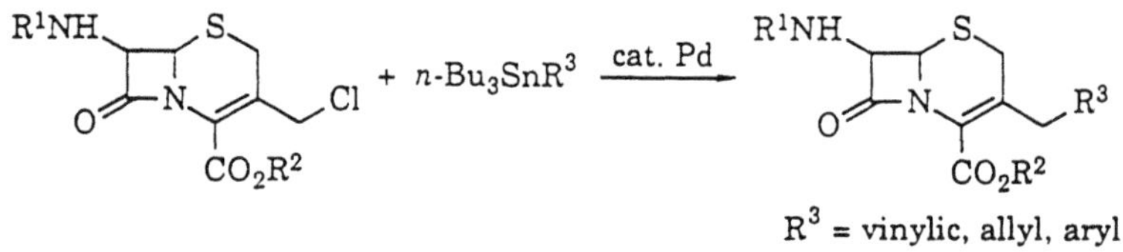

(471)

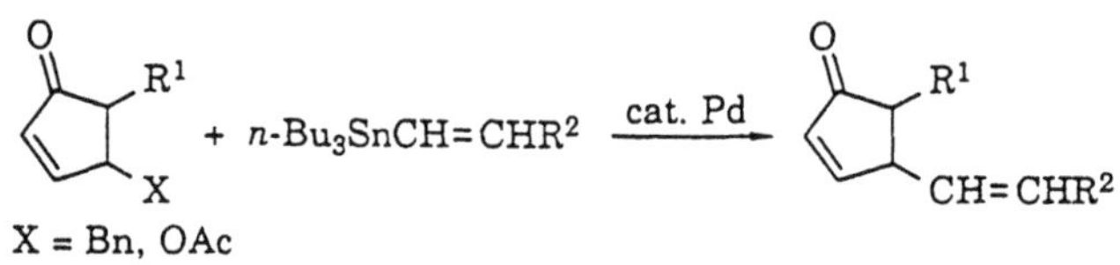

(472)

Note in this last example that allylic esters can also be utilized.[727] Besides reactions using simple allyl acetate,[729,730] more complicated allylic acetates have been studied (Eq. 473).[731] Substitution is observed to occur at the less-substituted end of the allylic system with retention of the allylic double bond stereochemistry. Allylic carbonates and phosphates also cross-couple effectively with vinylic tin reagents (Eq. 474).[732]

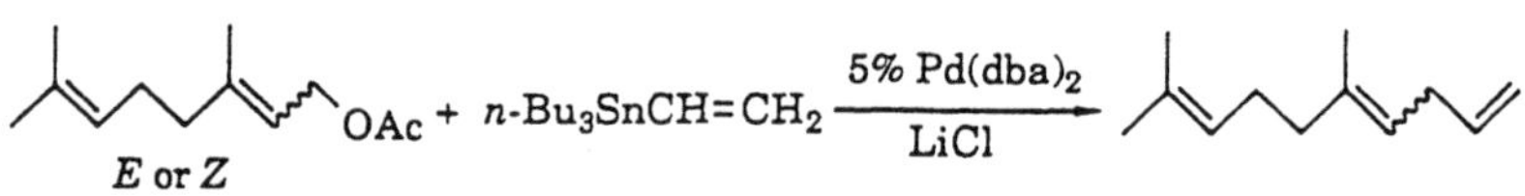

(473)

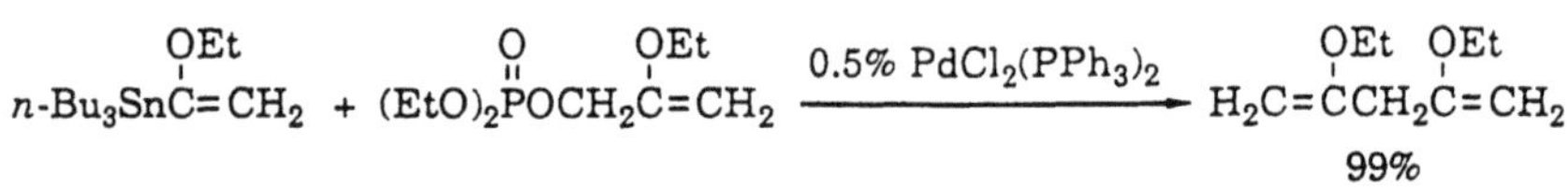

(474)

1,4-Dienes can also be obtained by cross-coupling vinylic stannanes and allylic stannanes (Eq. 475).[733] Crotyl tin reagents afford a mixture of head- and tail-coupled products.

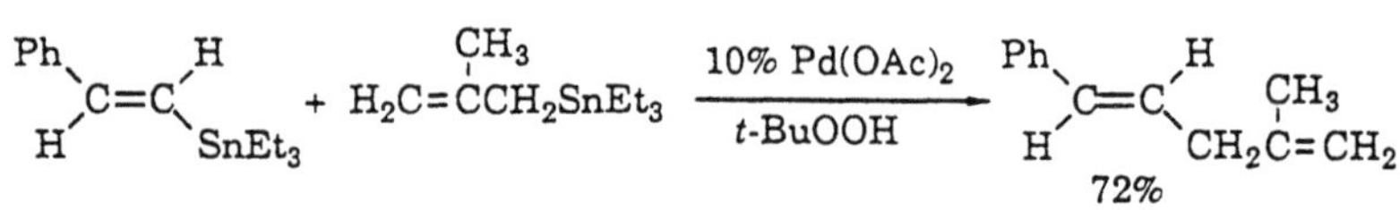

(475)

The palladium-catalyzed coupling of vinylic epoxides and vinylic stannanes provides dienols, presumably through π-allylpalladium intermediates (Eq. 476).[734,735]

$$(CH_3)_2C{=}CHSnMe_3 + H_2C{=}CHC(CH_3)\overset{O}{-}CH_2 \xrightarrow{\text{cat. } PdCl_2(CH_3CN)_2} \underset{72\%}{(CH_3)_2C{=}CHCH_2CH{=}C(CH_3)CH_2OH} \quad (476)$$

Symmetrical 1,3-dienes can be readily prepared by catalytically dimerizing vinylic stannanes (Eq. 477). Either 5% $PdCl_2(CH_3CN)_2$[736–738] or 10% $Pd(OAc)_2$ plus two equivalents of *t*-BuOOH[733] may be used as the catalyst.

$$2\ n\text{-}Bu_3SnCH{=}CHR \xrightarrow{\text{cat. Pd(II)}} (RCH{=}CH)_2 \quad (477)$$

1,3-Dienes have also been prepared by cross-coupling a vinyl stannane with a vinylic palladium intermediate generated by intramolecular alkyne insertion (Eq. 478).[739]

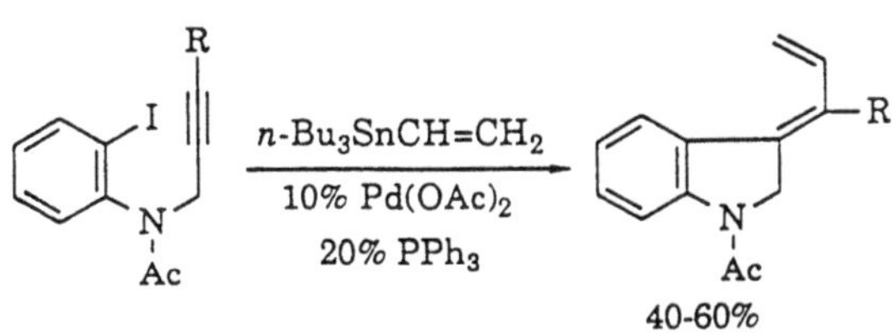

(478)

As noted earlier (Eqs. 469 and 470), benzylic halides can be catalytically cross-coupled with vinylic stannanes using palladium (Eq. 479).[413] The mechanism of this process has been examined.[740]

$$RX + n\text{-}Bu_3SnC(OEt){=}CH_2 \xrightarrow{\text{cat. } PdCl_2(PPh_3)_2} RC(OEt){=}CH_2 \quad (479)$$

R = benzyl, allyl, aryl, vinylic, acyl

One example of the coupling of a vinylic stannane and an alkynyl halide has been reported (Eq. 480).[416]

There are many examples of the cross-coupling of vinylic stannanes and aryl halides besides those already mentioned (Eqs. 469–471, 479, and 480). Various palladium catalysts and reaction conditions have been reported to effect this

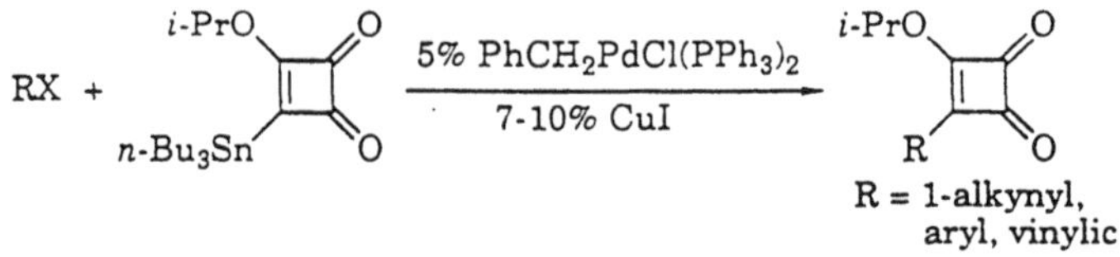

(480)

process.[464,741–743] Numerous functionally substituted aryl bromides and iodides have been utilized.[744] For example, this process has been utilized to prepare indoles (Eq. 481).[745] Aryl chlorides are unreactive, unless coordinated to a chromium tricarbonyl moiety.[746]

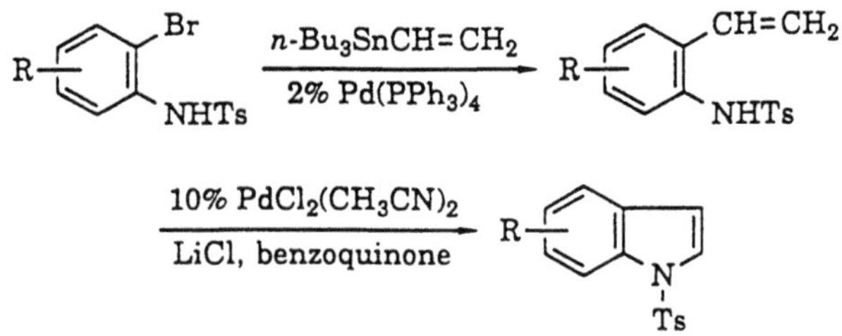

(481)

trans-1,2-Bis(tri-*n*-butylstannyl)ethylene has been employed in such coupling procedures to prepare *trans*-β-bromostyrenes,[747] stilbenes,[748] or polyheteroarenediylvinylenes[749] (Eq. 482).

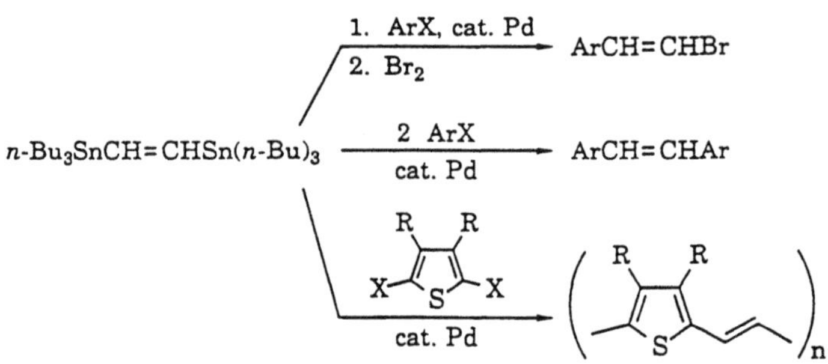

(482)

A wide variety of functionally-substituted vinylic stannanes have been employed in this overall arylation process. Vinylic stannanes bearing allylic silyl[750] or amino[751] groups, or homoallylic acetal functionality[752] couple efficiently.

Vinylic ethers bearing an α-stannyl moiety have been widely utilized.[412–414,753] This chemistry has been applied to furan-carbonyl adducts[754,755] and glycals[756–758] (Eqs. 483 and 484).

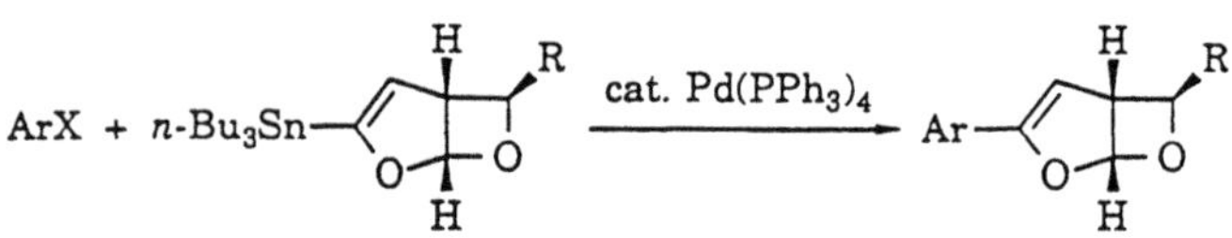

(483)

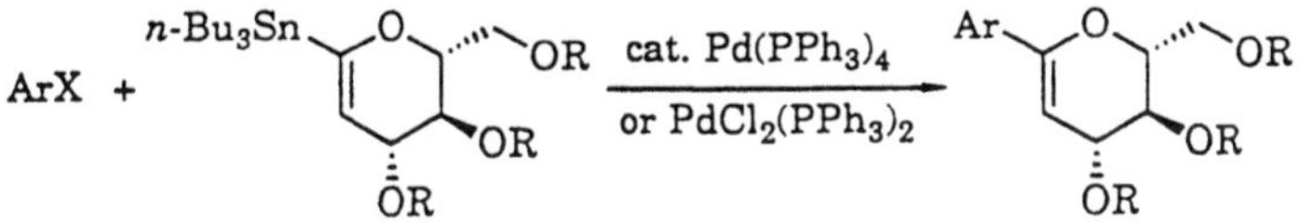

(484)

Lactones have been synthesized by such processes (Eqs. 485 and 486). A related intramolecular process has also been reported for the preparation of 2-benzylidene-benzocyclobutenones (Eq. 487).[760]

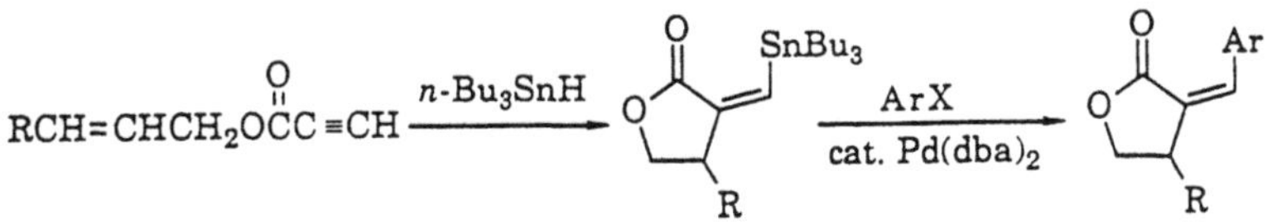

(485)[759]

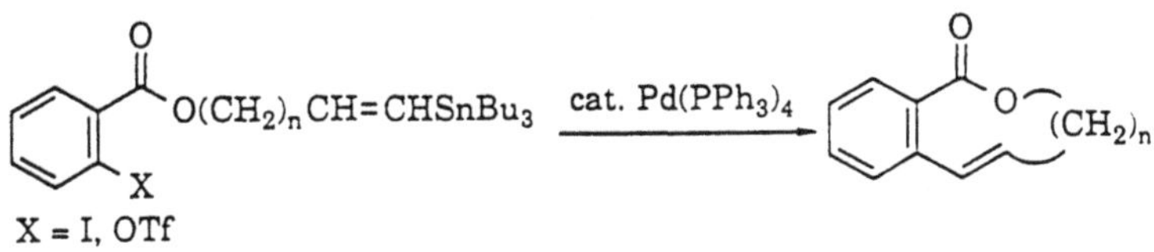

(486)[465]

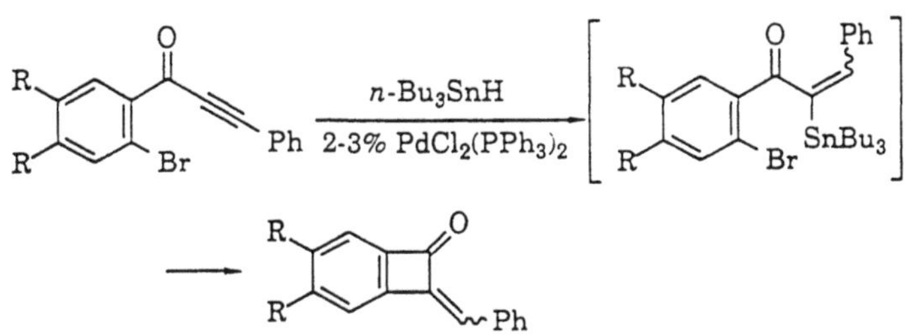

(487)

The palladium-catalyzed arylation of vinylic stannanes has been utilized as a key step in the synthesis of lacrimin A,[761,762] quinolone antibacterials,[763] and vinylated pyrimidines[407,408,764,765] and nucleosides.[766–769]

While little has been reported on the mechanism of this arylation process, it has been noted that rearrangements can occur, apparently by a process involving arylpalladium addition to the C–C double bond and subsequent rearrangement (Eq. 488).[550]

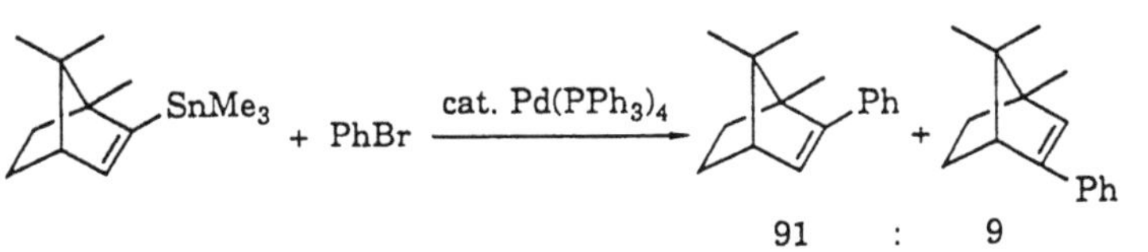

(488)

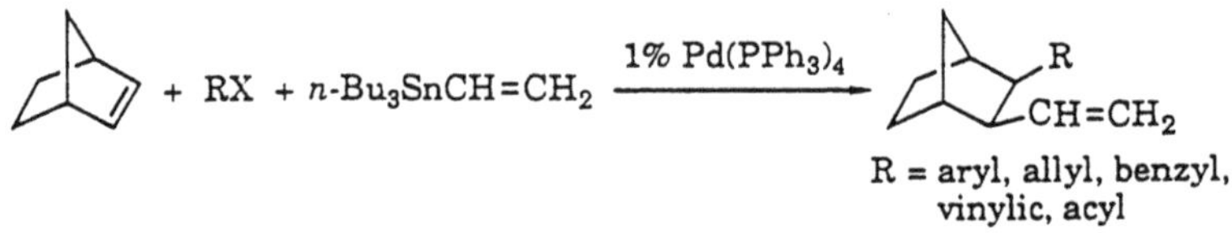

(489)

Aryl halides and vinylic stannanes also react with norbornene to afford alkene insertion products (Eq. 489).[136]

Aryl triflates also undergo palladium-catalyzed cross-coupling processes with vinylic stannanes. The presence of lithium chloride is required,[770] and tri(2-furyl)phosphine has proven useful as a ligand.[491] Besides simple aryl triflates,[770] this chemistry has been applied to triflates of naphthalene,[771] heteroaromatic systems,[745,763,772] and derivatives of resorcinol,[773] anthraquinones,[774] and carboalkoxyalkylphenylalanine.[775] It has been employed to synthesize aryl-substituted vinylic ethers,[412] and macrocyclic lactones[465] (see Eq. 486).

Vinylic stannanes couple with aryl fluorosulfonates at a rate comparable to aryl triflates (Eq. 490).[776]

$$ArOSO_2F + n\text{-}Bu_3SnCH{=}CHR \xrightarrow[\text{LiCl}]{5\%\ PdCl_2(PPh_3)_2} ArCH{=}CHR \qquad (490)$$

Aryl diazonium salts can also be employed in such arylation processes.[777] This reaction has been observed to proceed, however, by rearrangement (Eq. 491).[778]

$$R_3SnC(Ph){=}CH_2 + (ArN_2)BF_4 \xrightarrow{\text{cat. } Pd(dba)_2} PhCH{=}CHAr \qquad (491)$$

The acylation of vinylic stannanes is also catalyzed by palladium.[26,29,30] A number of different procedures and catalysts have been utilized,[743,779–783] and the mechanism has been examined.[784] Various functionally-substituted vinylic stannanes have been acylated, including those bearing allylic[750] and vinylic[723,785] (see Eq. 469) silyl groups, acetal functionality,[752] and α-alkoxy groups (see Eq. 479),[413,786] including 1,4-dioxins,[787] furan-carbonyl adducts,[755] and glycals.[406]

Stannyl-substituted diene iron tricarbonyl[788] and dihydropyridine (Eq. 492)[789] intermediates also can be acylated.

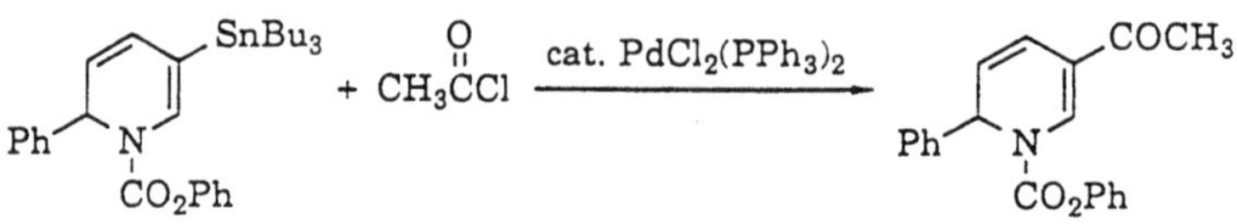

(492)

This acylation process has been employed in the synthesis of (±)-ar-tumerone,[790] pyrenophorin[780,791] and other macrolides, where both intramolecular and dimerization processes have been observed (Eq. 493).[792]

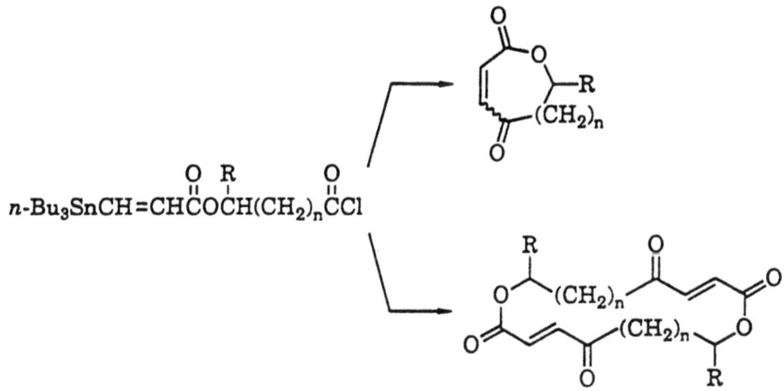

(493)

α,β-Unsaturated esters and amides may also be prepared via palladium-catalyzed acylation of vinylic stannanes (Eq. 494).[793]

$$(CH_3)_2C{=}CHSnBu_3 + ClC(O)X \xrightarrow{\text{cat. Pd(II)}} (CH_3)_2C{=}CHC(O)X \quad (494)$$

$X = n\text{-}C_8H_{17}O$ (70%), MePhN (60%)

As noted earlier (see Eq. 489), vinylic stannanes react with aryl halides and norbornene to afford norbornyl ketones.[136]

Palladium-catalyzed carbonylation of vinylic stannanes provides another useful route to ketones. In the presence of an aryl halide, aryl enones can be formed, but only in low yields (Eq. 495).[741] Much better results have been reported using (α-ethoxyvinyl)trimethylstannane (Eq. 496),[412] or aryl triflates (Eq. 497).[494]

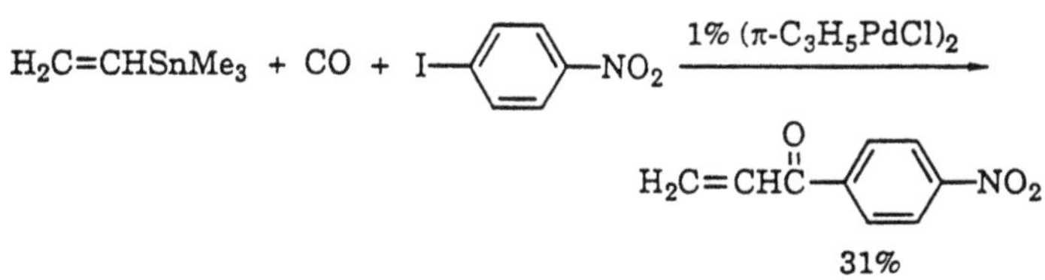

(495)

$$H_2C{=}C(OEt)SnMe_3 + CO + PhI \xrightarrow{2\%\ Pd(PPh_3)_4} H_2C{=}C(OEt){-}C(O){-}Ph \quad (496)$$

67%

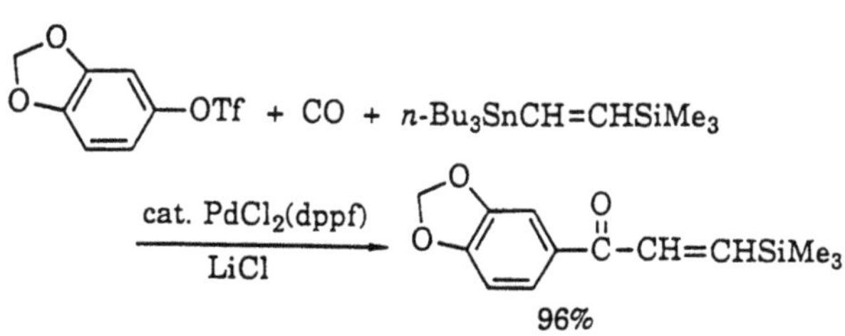

(497)

The catalytic carbonylation of vinylic stannanes in the presence of allylic halides also affords high yields of unsaturated ketones (Eq. 498).[721,722]

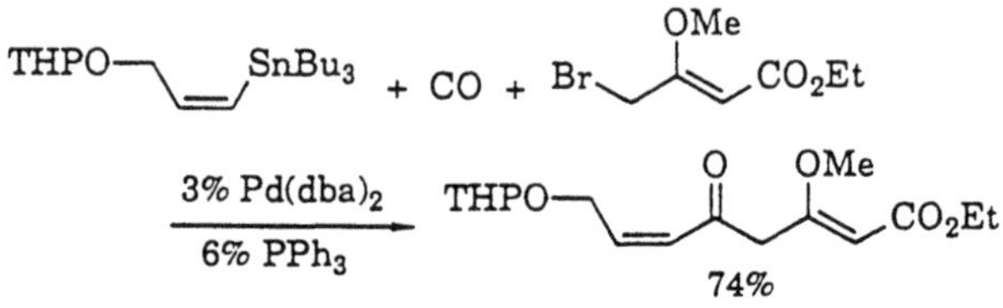

(498)

α,β-Unsaturated sulfones can be prepared from vinylic stannanes by reaction with sulfonyl chlorides (Eq. 499).[752,794]

$$p\text{-}MeC_6H_4SO_2Cl + n\text{-}Bu_3SnCH{=}CHCH_2CH(OEt)_2 \xrightarrow{2\%\ Pd(PPh_3)_4} \underset{85\%}{p\text{-}MeC_6H_4SO_2CH{=}CHCH_2CH(OEt)_2} \quad (499)$$

VIII. VINYLIC SULFONES

Vinylic sulfones can be either reduced or alkylated by Grignard reagents in the presence of a palladium catalyst. Alkylmagnesium reagents bearing β-hydrogens, such as EtMgBr or n-BuMgCl, afford reduction products (Eq. 500).[795]

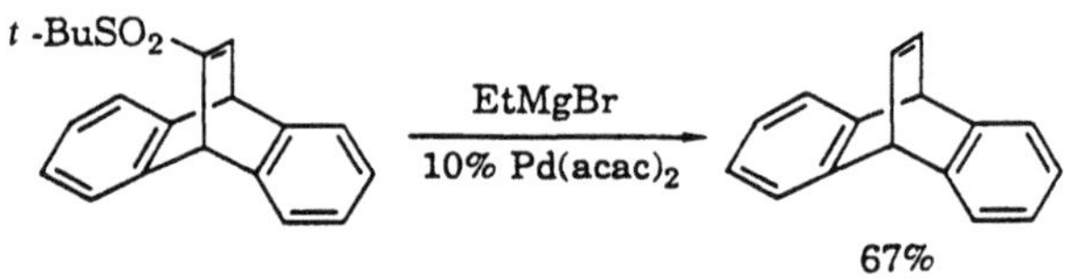

(500)

1,3-Dienes have been prepared in like manner.[796–798] Methyl or aryl Grignard reagents alkylate vinylic sulfones (Eq. 501).[795]

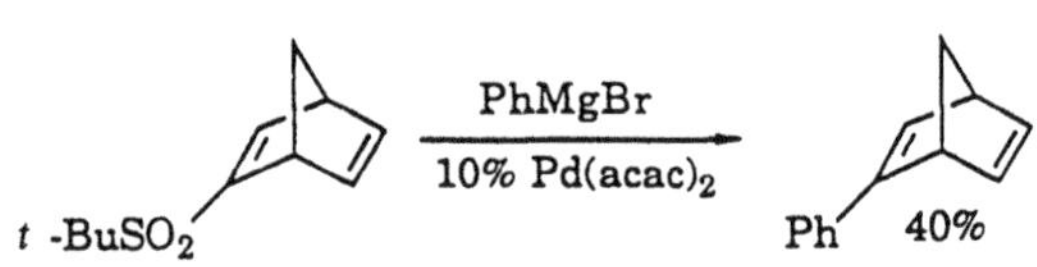

(501)

IX. VINYLIC TELLURIUM COMPOUNDS

Vinylic tellurium compounds can be carbonylated in the presence of a palladium catalyst to give either α,β-unsaturated carboxylic acids or esters (Eq. 502).[799–801]

$$\mathrm{RCH{=}CHTePh} \xrightarrow[\mathrm{LiCl}]{\mathrm{CO},\ \mathrm{PdCl_2}} \xrightarrow[\mathrm{2.\ CH_2N_2}]{\mathrm{1.\ H_3O^+}} \mathrm{RCH{=}CHCO_2CH_3} \qquad (502)$$

Some loss of stereochemistry is observed. This chemistry has been employed in the synthesis of butenolides from propargylic alcohols (Eq. 503).[801]

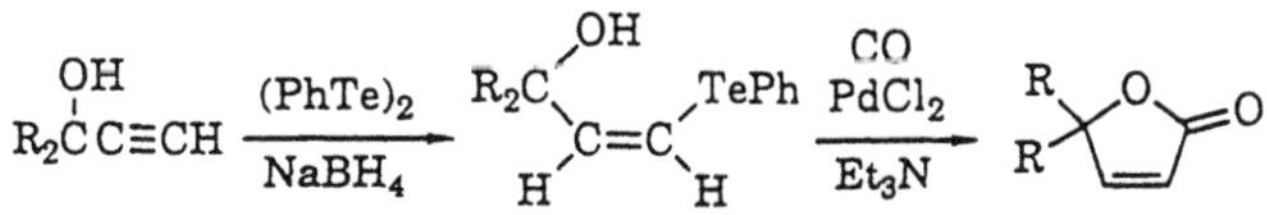

(503)

X. MISCELLANEOUS SUBSTRATES

A variety of other substrates react with palladium reagents to generate vinylic palladium intermediates useful in organic synthesis. For example, α,β-unsaturated acid chlorides react with alkenes and a palladium acetate catalyst to generate 1,3-dienes in low yield (Eq. 504).[802] Higher yields are reported for the analogous reaction of fumaryl chloride (Eq. 505).[803]

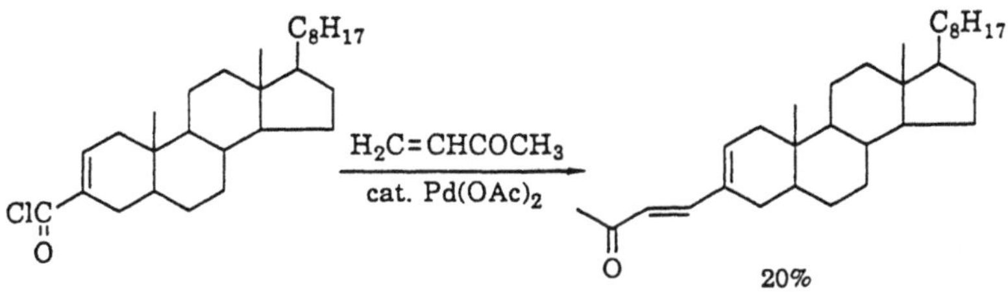

(504)

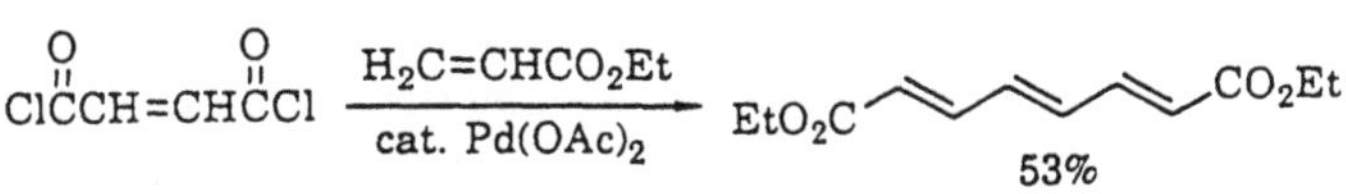

(505)

Unsaturated sulfonyl chlorides undergo a similar reaction (Eq. 506).[804]

$$\mathrm{PhCH{=}CHSO_2Cl} + \mathrm{H_2C{=}CHPh} \xrightarrow{2\%\ \mathrm{Pd(OAc)_2}} \underset{45\%}{\mathrm{PhCH{=}CHCH{=}CHPh}} \qquad (506)$$

Substituted cinnamaldehydes can be decarbonylated over palladium catalysts to produce styrenes (Eq. 507).[805]

$$\mathrm{Ph(H)C{=}C(R)CHO} \xrightarrow{\mathrm{cat.\ Pd/C}} \mathrm{PhCH{=}CHR} \qquad (507)$$

Certain alkenes react with palladium acetate to produce vinylpalladium intermediates capable of coupling with alkenes. Thus, one can dimerize certain alkenes using palladium acetate in NaOAc/HOAc (Eq. 508).[806] Analogous cross-coupling has been observed between uracil derivatives and alkenes (Eq. 509).[807,808]

$$2\ (CH_3)_2C{=}CH_2 \xrightarrow[\substack{NaOAc\\ HOAc}]{Pd(OAc)_2} \underset{54\%}{(CH_3)_2C{=}CHCH{=}C(CH_3)_2} \qquad (508)$$

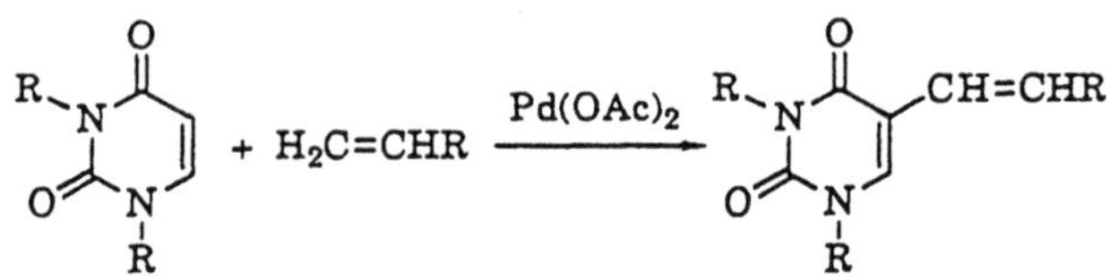

(509)

The carbonylation of such alkenes can also be accomplished (Eq. 510).[809]

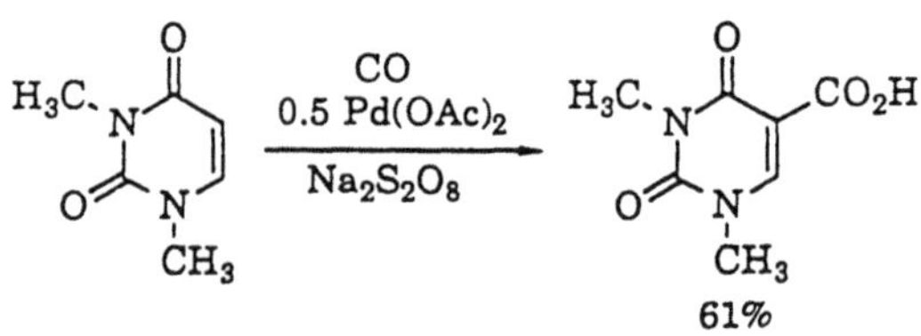

(510)

XI. CONCLUSION

Vinylic palladium intermediates are generated by a variety of vinylic substitution processes. They undergo a wide range of further reactions which have proven quite useful in organic synthesis. This chemistry can usually be effected under quite mild reaction conditions and accommodates most important organic functional groups. The yields are generally high and palladium can almost always be employed in only catalytic amounts. Increasing application of this methodology in the synthesis of complex molecules is evident.

REFERENCES

1. Mukhedkar, A. J.; Green, M.; Stone, F. G. A. *J. Chem. Soc. A* **1969**, 3023.
2. Mukhedkar, V. A.; Mukhedkar, A. J. *J. Inorg. Nucl. Chem.* **1981**, *43*, 2801.
3. Fitton, P.; McKeon, J. E. *J. Chem. Soc., Chem. Commun.* **1968**, 4.
4. Johnson, B. F. G.; Lewis, J.; Jones, J. D.; Taylor, K. A. *J. Chem. Soc., Dalton Trans.* **1974**, 34.
5. Mann, B. E.; Shaw, B. L.; Tucker, N. I. *J. Chem. Soc. A* **1971**, 2667.
6. Onishi, M.; Yamamoto, H.; Hiraki, K. *Bull. Chem. Soc. Jpn.* **1978**, *51*, 1856.
7. Oehme, G.; Pracejus, H. *Z. Chem.* **1974**, *14*, 24.
8. Fryzuk, M. D.; Bosnich, B. *J. Am. Chem. Soc.* **1979**, *101*, 3043.

9. Empsall, H. D.; Green, M.; Shakshooki, S. K.; Stone, F. G. A. *J. Chem. Soc. A* **1971**, 3472.
10. Christofides, A. *J. Organomet. Chem.* **1983**, *259*, 355.
11. Kudo, K.; Hidai, M.; Uchida, Y. *J. Organomet. Chem.* **1971**, *33*, 393.
12. Kudo, K.; Sato, M.; Hidai, M.; Uchida, Y. *Bull. Chem. Soc. Jpn.* **1973**, *46*, 2820.
13. Amatore, C.; Azzabi, M.; Jutand, A. *J. Am. Chem. Soc.* **1991**, *113*, 1670.
14. Oehme, G.; Baudisch, H. *Tetrahedron Lett.* **1974**, 4129.
15. Rybin, L. V.; Petrovskaya, E. A.; Rubinskaya, M. I.; Kuz'mina, L. G.; Struchkov, Yu. T.; Kaverin, V. V.; Koneva, N. Yu. *J. Organomet. Chem.* **1985**, *288*, 119.
16. Oehme, G.; Pracejus, H. *J. Prakt. Chem.* **1980**, *322*, 798.
17. Oehme, G. *J. Prakt. Chem.* **1984**, *326*, 779.
18. Carfagna, C.; Musco, A.; Pontellini, R. *J. Mol. Catal.* **1989**, *54*, L23.
19. Pri-Bar, I.; Buchman, O. *J. Org. Chem.* **1986**, *51*, 734.
20. Traas, P. C.; Takken, H. J.; Boelens, H. *Tetrahedron Lett.* **1977**, 2027.
21. Cook, G. K.; Hornback, W. J.; Jordan, C. L.; McDonald III, J. H.; Munroe, J. E. *J. Org. Chem.* **1989**, *54*, 5828.
22. Yamamura, K.; Murahashi, S.-I. *Tetrahedron Lett.* **1977**, 4429.
23. Schoenberg, A.; Heck, R. F. *J. Am. Chem. Soc.* **1974,** *96,* 7761.
24. Baillargeon, V. P.; Stille, J. K. *J. Am. Chem. Soc.* **1983**, *105,* 7175.
25. Baillargeon, V. P.; Stille, J. K. *J. Am. Chem. Soc.* **1986**, *108*, 452.
26. (a) Stille, J. K. *Pure Appl. Chem.* **1985**, *57*, 1771. (b) Stille, J. K. *Angew. Chem., Int. Ed. Engl.* **1986**, *25,* 508.
27. Negishi, E.; Zhang, Y.; Shimoyama, I.; Wu, G. *J. Am. Chem. Soc.* **1989**, *111,* 8018.
28. Wakita, Y.; Yasunaga, T.; Kojima, M. J. Organomet. Chem. **1985**, *288*, 261.
29. Beletskaya, I. P. *J. Organomet. Chem.* **1983**, *250*, 551.
30. Mitchell, T. N. *J. Organomet. Chem.* **1986**, *304*, 1.
31. Tanaka, M. *Tetrahedron Lett.* **1979**, 2601.
32. Goure, W. F.; Wright, M. E.; Davis, P. D.; Labadie, S. S.; Stille, J. K. *J. Am. Chem. Soc.* **1984**, *106*, 6417.
33. Dondoni, A.; Fantin, G.; Fogagnolo, M.; Mastellari, A.; Medici, A.; Negrini, E.; Pedrini, P. *Gazz. Chim. Ital.* **1988**, *118*, 211.
34. Negishi, E.; Miller, J. A. *J. Am. Chem. Soc.* **1983**, *105*, 6761.
35. Tour, J. M.; Negishi, E. *J. Am. Chem. Soc.* **1985**, *107,* 8289.
36. Shimoyama, I.; Zhang, Y.; Wu, G.; Negishi, E. *Tetrahedron Lett.* **1990**, *31*, 2841.
37. Kobayashi, T.; Tanaka, M. *J. Chem. Soc., Chem. Commun.* **1981**, 333.
38. Cassar, L.; Foa, M.; Gardano, A. *J. Organomet. Chem.* **1976**, *121*, C55.
39. Galamb, V.; Alper, H. *Tetrahedron Lett.* **1983**, *24*, 2965.
40. Galamb, V.; Gopal, M.; Alper, H. *Organometallics* **1983**, *2*, 801.
41. Tanaka, M.; Kobayashi, T.; Sakakura, T. *J. Chem. Soc., Chem. Commun.* **1985**, 837.
42. Schoenberg, A.; Bartoletti, I.; Heck, R. F. *J. Org. Chem.* **1974**, *39,* 3318.
43. Hidai, M.; Hikita, T.; Wada, Y.; Fujikura, Y.; Uchida, Y. *Bull. Chem. Soc. Jpn.* **1975**, *48*, 2075.
44. Woell, J. B.; Fergusson, S. B.; Alper, H. *J. Org. Chem.* **1985**, *50*, 2134.
45. Urata, H.; Maekawa, H.; Takahashi, S.; Fuchikami, T. *J. Org. Chem.* **1991**, *56*, 4320.
46. Alper, H.; Antebi, S.; Woell, J. B. *Angew. Chem., Int. Ed. Engl.* **1984**, *23,* 732.
47. Hashem, K. E.; Woell, J. B.; Alper, H. *Tetrahedron Lett.* **1984**, *25*, 4879.
48. Hegedus, L. S.; Holden, M. S. *J. Org. Chem.* **1986**, *51,* 1171.
49. Horino, H.; Inoue, N.; Asao, T. *Tetrahedron Lett.* **1981**, *22*, 741.
50. Fuchikami, T.; Ojima, I. *Tetrahedron Lett.* **1984**, *25*, 303.
51. Martin, L. D.; Stille, J. K. *J. Org. Chem.* **1982**, *47*, 3630.
52. Mori, M.; Washioka, Y.; Urayama, T.; Yoshiura, K.; Chiba, K.; Ban, Y. *J. Org. Chem.* **1983**, *48*, 4058.
53. Cowell, A.; Stille, J. K. *J. Am. Chem. Soc.* **1980**, *102,* 4193.

54. Cowell, A.; Stille, J. K. *Tetrahedron Lett.* **1979**, 133.
55. Marshall, J. A.; Lebreton, J.; DeHoff, B. S.; Jenson, T. M. *J. Org. Chem.* **1987,** *52*, 3883.
56. Marshall, J. A.; Lebreton, J.; DeHoff, B. S.; Jenson, T. M. *Tetrahedron Lett.* **1987**, *28*, 723.
57. Marshall, J. A.; Jenson, T. M.; DeHoff, B. S. *J. Org. Chem.* **1987**, *52*, 3860.
58. Schoenberg, A.; Heck, R. F. *J. Org. Chem.* **1974**, *39*, 3327.
59. Kobayashi, T.; Tanaka, M. *J. Organomet. Chem.* **1982**, *231*, C12.
60. Chiba, K.; Mori, M.; Ban, Y. *J. Chem. Soc., Chem. Commun.* **1980**, 770.
61. Mori, M.; Chiba, K.; Okita, M.; Ban, Y. *J. Chem. Soc., Chem. Commun.* **1979**, 698.
62. Mori, M.; Chiba, K.; Okita, M.; Kayo, I.; Ban, Y. *Tetrahedron* **1985,** *41*, 375.
63. Brickner, S. J.; Gaikema, J. J.; Torrado, J. T.; Greenfield, L. J.; Ulanowicz, D. A. *Tetrahedron Lett.* **1988, 29**, 5601.
64. Mori, M.; Chiba, K.; Ohta, N.; Ban, Y. *Heterocycles* **1979, 13**, 329.
65. Perry, R. J.; Turner, S. R. U.S. Patent 4,933,466 **1989**; *Chem. Abstr.* **1990**, *113*, 191156q.
66. Fuchikami, T.; Ojima, I. *Tetrahedron Lett.* **1982,** *23* , 4099.
67. Tanaka, M.; Kobayashi, T.; Sakakura, T.; Itatani, H.; Danno, S.; Zushi, K. *J. Mol. Catal.* **1985,** *32,* 115.
68. Son, T.; Yanagihara, H.; Ozawa, F.; Yamamoto, A. *Bull. Chem. Soc. Jpn.* **1988,** *61,* 1251.
69. Kobayashi, T.; Tanaka, M. *J. Organomet. Chem.* **1982**, *233*, C64.
70. Ozawa, F.; Soyama, H.; Yamamoto, T.; Yamamoto, A. *Tetrahedron Lett.* **1982**, *23*, 3383.
71. Tanaka, M.; Koyanagi, M.; Kobayashi, T. *Tetrahedron Lett.* **1981**, *22*, 3875.
72. Torii, S.; Tanaka, H.; Hamatani, T.; Morisaki, K.; Jutand, A.; Pfluger, F.; Fauvarque, J.-F. *Chem. Lett.* **1986**, 169.
73. Otsuka, S.; Ataka, K. *J. Chem. Soc., Dalton Trans.* **1976**, 327.
74. Heck, R. F.; Nolley, Jr., J. P. *J. Org. Chem.* **1972**, *37*, 2320.
75. Dieck, H. A.; Heck, R. F. *J. Am. Chem. Soc.* **1974**, *96*, 1133.
76. Dieck, H. A.; Heck, R. F. *J. Org. Chem.* **1975**, *40*, 1083.
77. Heck, R. F. *Pure Appl. Chem.* **1978**, *50*, 691.
78. Heck, R. F. *Pure Appl. Chem.* **1981**, *53*, 2323.
79. Heck, R. F. *ACS Adv. Chem.* **1982**, *196*, 213.
80. Heck, R. F. in *Fundamental Research in Organometallic Chemistry,* eds. M. Tsutsui, Y. Ishi and H. Yaozeng, Van Nostrand-Reinhold Co.: New York, 1982; p. 447.
81. Savoia, D.; Trombini, C.; Umani-Ronchi, A.; Verardo, G. *J. Chem. Soc., Chem. Commun.* **1981**, 541.
82. Jeffery, T. *J. Chem. Soc., Chem. Commun.* **1984**, 1287.
83. Jeffery, T. *Tetrahedron Lett.* **1985**, *26*, 2667.
84. Eddarir, S.; Mestdagh, H.; Rolando, C. *Tetrahedron Lett.* **1991**, *32*, 69.
85. Bergstrom, D. E.; Ogawa, M. K. *J. Am. Chem. Soc.* **1978**, *100*, 8106.
86. Hronowski, L. J. J.; Szarek, W. A. *J. Chem. Soc., Chem. Commun.* **1990**, 1547.
87. Stöbbe, M.; Reiser, O.; Näder, R.; de Meijere, A. *Chem. Ber.* **1987**, *120*, 1667.
88. Lansky, A.; Reiser, O.; de Meijere, A. *Synlett* **1990**, 405.
89. Ziegler, F. E.; Chakraborty, U. R.; Weisenfeld, R. B. *Tetrahedron* **1981**, *37*, 4035.
90. Kim, J. I.; Patel, B. A.; Heck, R. F. *J. Org. Chem.* **1981**, *46*, 1067.
91. Patel, B. A.; Kim, J. I.; Bender, D. D.; Kao, L.-C.; Heck, R. F. *J. Org. Chem.* **1981**, *46,* 1061.
92. Fuchikami, T.; Yatabe, M.; Ojima, I. *Synthesis* **1981**, 365.
93. Karabelas, K.; Hallberg, A. *J. Org. Chem.* **1988**, *53*, 4909.
94. Andersson, C.-M.; Hallberg, A. *J. Org. Chem.* **1989**, *54*, 1502.
95. Patel, B. A.; Dickerson, J. E.; Heck, R. F. *J. Org. Chem.* **1978**, *43*, 5018.
96. Mitsudo, T.; Fischetti, W.; Heck, R. F. *J. Org. Chem.* **1984**, *49*, 1640.
97. Jeffery, T. *Synth. Commun.* **1988**, *18*, 77.
98. Naora, H.; Ohnuki, T.; Nakamura, A. *Chem. Lett.* **1988**, 143.
99. Grigg, R.; Stevenson, P.; Worakun, T. *J. Chem. Soc., Chem. Commun.* **1984**, 1073.

100. O'Connor, B.; Zhang, Y.; Negishi, E. *Tetrahedron Lett.* **1988**, *29*, 3903.
101. Negishi, E.; Nguyen, T.; O'Connor, B. *Heterocycles* **1989**, *28*, 55.
102. Luly, J. R.; Rapoport, H. *J. Org. Chem.* **1984**, *49*, 1671.
103. Grigg, R.; Sridharan, V.; Sukirthalingam, S. *Tetrahedron Lett.* **1991**, *32*, 3855.
104. Kao, L.-C.; Stakem, F.G.; Patel, B. A.; Heck, R. F. *J. Org. Chem.* **1982**, *47*, 1267.
105. Jeffery, T. *J. Chem. Soc., Chem. Commun.* **1991**, 324.
106. Jeffery, T. *Tetrahedron Lett.* **1990**, *31*, 6641.
107. Mandai, T.; Hasegawa, S.; Fujimoto, T.; Kawada, M.; Nokami, J.; Tsuji, J. *Synlett* **1990**, 85.
108. Trost, B. M.; Walchli, R. *J. Am. Chem. Soc.* **1987**, *109*, 3487.
109. (a) Kosugi, M.; Hagiwara, I.; Migita, T. *Chem. Lett.* **1983**, 839. (b) Kosugi, M.; Hagiwara, I.; Sumiya, T.; Migita, T. *Bull. Chem. Soc. Jpn.* **1984**, *57*, 242.
110. Patel, B. A.; Heck, R. F. *J. Org. Chem.* **1978**, *43*, 3898.
111. Narula, C. K.; Mak, K. T.; Heck, R. F. *J. Org. Chem.* **1983**, *48*, 2792.
112. Shi, L.; Narula, C. K.; Mak, K. T.; Kao, L.; Xu, Y.; Heck, R. F. *J. Org. Chem.* **1983**, *48*, 3894.
113. Larock, R. C.; Leuck, D. J.; Harrison, L. W. *Tetrahedron Lett.* **1988**, *29*, 6399.
114. Balme, G.; Fournet, G.; Gore, J. *Tetrahedron Lett.* **1986**, *27*, 3855.
115. Fournet, G.; Balme, G.; Gore, J. *Tetrahedron* **1988**, *44*, 5809.
116. Fournet, G.; Balme, G.; Gore, J. *Tetrahedron* **1990**, *46*, 7763.
117. Larock, R. C.; Gong, W. H. *J. Org. Chem.* **1989**, *54*, 2047.
118. Torii, S.; Okumoto, H.; Kotani, T.; Akahoshi, F. *Chem. Lett.* **1989**, 1971.
119. Torii, S.; Okumoto, H.; Akahoshi, F.; Kotani, T. *J. Am. Chem. Soc.* **1989**, *111*, 8932.
120. Larock, R. C.; Kondo, F.; Narayanan, K.; Sydnes, L. K.; Hsu, M.-F. H. *Tetrahedron Lett.* **1989**, *30*, 5737.
121. Larock, R. C.; Song, H.; Baker, B. E.; Gong, W. H. *Tetrahedron Lett.* **1988**, *29*, 2919.
122. Negishi, E.; Zhang, Y.; O'Connor, B. *Tetrahedron Lett.* **1988**, *29*, 2915.
123. Zhang, Y.; O'Connor, B.; Negishi, E. *J. Org. Chem.* **1988**, *53*, 5588.
124. Rawal, V. H.; Michoud, C. *Tetrahedron Lett.* **1991**, *32*, 1695.
125. Grigg, R.; Sridharan, V.; Stevenson, P. *Tetrahedron* **1989**, *45*, 3557.
126. Sato, Y.; Sodeoka, M.; Shibasaki, M. *J. Org. Chem.* **1989**, *54*, 4738.
127. Sato, Y.; Sodeoka, M.; Shibasaki, M. *Chem. Lett.* **1990**, 1953.
128. Catellani, M.; Chiusoli, G. P. *J. Organomet. Chem.* **1982**, *233*, C21.
129. Catellani, M.; Chiusoli, G. P. *J. Organomet. Chem.* **1983**, *250*, 509.
130. Catellani, M.; Chiusoli, G. P.; Sgarabotto, P. *J. Organomet. Chem.* **1982**, *240*, 311.
131. Catellani, M.; Chiusoli, G. P.; Giroldini, W.; Salerno, G. *J. Organomet. Chem.* **1980**, *199*, C21.
132. Arcadi, A.; Marinelli, F.; Bernocchi, E.; Cacchi, S.; Ortar, G. *J. Organomet. Chem.* **1989**, *368*, 249.
133. Larock, R. C.; Johnson, P. L. *J. Chem. Soc., Chem. Commun.* **1989**, 1368.
134. Catellani, M.; Chiusoli, G. P. *J. Organomet. Chem.* **1984**, *275*, 257.
135. Kosugi, M.; Tamura, H.; Sano, H.; Migita, T. *Chem. Lett.* **1987**, 193.
136. Kosugi, M.; Tamura, H.; Sano, H.; Migita, T. *Tetrahedron* **1989**, *45*, 961.
137. Catellani, M.; Chiusoli, G. P. *Tetrahedron Lett.* **1982**, *23*, 4517.
138. Catellani, M.; Chiusoli, G. P.; Mari, A. *J. Organomet. Chem.* **1984**, *275*, 129.
139. Catellani, M.; Chiusoli, G. P.; Peloso, C. *Tetrahedron Lett.* **1983**, *24*, 813.
140. Torii, S.; Okumoto, H.; Ozaki, H.; Nakayasu, S.; Kotani, T. *Tetrahedron Lett.* **1990**, *31*, 5319.
141. Shimizu, I.; Sugiura, T.; Tsuji, J. *J. Org. Chem.* **1985**, *50*, 537.
142. Shimizu, I.; Tsuji, *J. Chem. Lett.* **1984**, 233.
143. Ahmar, M.; Cazes, B.; Gore, J. *Tetrahedron Lett.* **1984**, *25*, 4505.
144. Kopola, N.; Friess, B.; Cazes, B.; Gore, J. *Tetrahedron Lett.* **1989**, *30*, 3963.
145. Ahmar, M.; Barieux, J.-J.; Cazes, B.; Gore, J. *Tetrahedron* **1987**, *43*, 513.
146. Cazes, B.; Colovray, V.; Gore, J. *Tetrahedron Lett.* **1988**, *29*, 627.
147. Ahmar, M.; Cazes, B.; Gore, J. *Tetrahedron Lett.* **1985**, *26*, 3795.

148. Ahmar, M.; Cazes, B.; Gore, J. *Tetrahedron* **1987**, *43*, 3453.
149. Fischetti, W.; Mak, K. T.; Stakem, F. G.; Kim, J.; Rheingold, A. L.; Heck, R. F. *J. Org. Chem.* **1983**, *48*, 948.
150. Burns, B.; Grigg, R.; Ratananukul, P.; Sridharan, V.; Stevenson, P.; Worakun, T. *Tetrahedron Lett.* **1988**, *29*, 4329.
151. Patel, B. A.; Kao, L.-C.; Cortese, N. A.; Minkiewicz, J. V.; Heck, R. F. *J. Org. Chem.* **1979**, *44*, 918.
152. Bender, D. D.; Stakem, F. G.; Heck, R. F. *J. Org. Chem.* **1982**, *47*, 1278.
153. O'Connor, J. M.; Stallman, B. J.; Clark, W. G.; Shu, A. Y. L.; Spada, R. E.; Stevenson, T. M.; Dieck, H. A. *J. Org. Chem.* **1983**, *48*, 807.
154. Arcadi, A.; Bernocchi, E.; Burini, A.; Cacchi, S.; Marinelli, F.; Pietroni, B. *Tetrahedron Lett.* **1989**, *30*, 3465.
155. Zhang, Y.; Negishi, E. *J. Am. Chem. Soc.* **1989**, *111*, 3454.
156. Torii, S.; Okumoto, H.; Nishimura, A. *Tetrahedron Lett.* **1991**, *32*, 4167.
157. Burns, B.; Grigg, R.; Sridharan, V.; Stevenson, P.; Sukirthalingam, S.; Worakun, T. *Tetrahedron Lett.* **1989**, *30*, 1135.
158. Sonogashira, K.; Tohda, Y.; Hagihara, N. *Tetrahedron Lett.* **1975**, 4467.
159. Cassar, L. *J. Organomet. Chem.* **1975**, *93*, 253.
160. Dieck, H. A.; Heck, R. F. *J. Organomet. Chem.* **1975**, *93*, 259.
161. Bumagin, N. A.; Ponomarev, A. B.; Beletskaya, I. P. *Bull. Acad. Sci. USSR* **1984**, 1433.
162. Dougherty, T. K.; Lau, K. S. Y.; Hedberg, F. L. *J. Org. Chem.* **1983**, *48*, 5273.
163. Rossi, R.; Carpita, A.; Quirici, M. G.; Gaudenzi, M. L. *Tetrahedron* **1982**, *38*, 631.
164. Andreini, B. P.; Carpita, A.; Rossi, R.; Scamuzzi, B. *Tetrahedron* **1989**, *45*, 5621.
165. Weir, J. R.; Patel, B. A.; Heck, R. F. *J. Org. Chem.* **1980**, *45*, 4926.
166. (a) Nwokogu, G. C. *Tetrahedron Lett.* **1984**, *25*, 3263. (b) Nwokogu, G. C. *J. Org. Chem.* **1985**, *50*, 3900.
167. Eddarir, S.; Francesch, C.; Mestdagh, H.; Rolando, C. *Tetrahedron Lett.* **1990**, *31*, 4449.
168. Yang, Z.-Y.; Burton, D. J. *Tetrahedron Lett.* **1990**, *31*, 1369.
169. Ratovelomanana, V.; Hammoud, A.; Linstrumelle, G. *Tetrahedron Lett.* **1987**, *28*, 1649.
170. Ratovelomanana,V.; Linstrumelle, G. *Tetrahedron Lett.* **1981**, *22*, 315.
171. Ratovelomanana, V.; Linstrumelle, G. *Bull. Soc. Chim. Fr.* **1987**, 174.
172. Kende, A. S.; Smith, C. A. *J. Org. Chem.* **1988**, *53*, 2655.
173. Guillerm, D.; Linstrumelle, G. *Tetrahedron Lett.* **1985**, *26*, 3811.
174. Vollhardt, K. P. C.; Winn, L. S. *Tetrahedron Lett.* **1985**, *26*, 709.
175. Huynh, C.; Linstrumelle, G. *Tetrahedron* **1988**, *44*, 6337.
176. Ashe III, A. J.; Drone, F. J. *J. Am. Chem. Soc.* **1987**, *109*, 1879.
177. Arcadi, A.; Cacchi, S.; Marinelli, F. *Tetrahedron Lett.* **1989**, *30*, 2581.
178. Reisch, J.; Nordhaus, P. *J. Heterocycl. Chem.* **1991**, *28*, 167.
179. Gaudin, J.-M.; Morel, C. *Tetrahedron Lett.* **1990**, *31*, 5749.
180. Andreini, B. P.; Carpita, A.; Rossi, R. *Tetrahedron Lett.* **1986**, *27*, 5533.
181. Carpita, A.; Neri, D.; Rossi, R. *Gazz. Chim. Ital.* **1987**, *117*, 481.
182. Pattenden, G.; Robson, D. C. *Tetrahedron Lett.* **1987**, *28*, 5751.
183. Blade, R. J.; Robinson, J. E.; Peek, R. J.; Weston, J. B. *Tetrahedron Lett.* **1987**, *28*, 3857.
184. Crombie, L.; Horsham, M. A.; Blade, R. J. *Tetrahedron Lett.* **1987**, *28*, 4879 .
185. Crombie, L.; Hobbs, A. J. W.; Horsham, M. A.; Blade, R. J. *Tetrahedron Lett.* **1987**, *28*, 4875.
186. Takagaki, T.; Matsuo, N. Jpn. Kokai Tokkyo Koho JP 03 14,541 **1991**; *Chem. Abstr.* **1991**, *114*, 246862v.
187. Rossi, R.; Carpita, A.; Lezzi, A. *Tetrahedron* **1984**, *40*, 2773.
188. Butlin, R. J.; Holmes, A. B.; Mcdonald, E. *Tetrahedron Lett.* **1988**, *29*, 2989 .
189. Spencer, R. W.; Tam, T. F.; Thomas, E.; Robinson, V. J.; Krantz, A. *J. Am. Chem. Soc.* **1986**, *108*, 5589.

190. Tam, T. F.; Thomas, E.; Krantz, A. *Tetrahedron Lett.* **1987**, *28*, 1127.
191. Kabat, M.; Kiegiel, J.; Cohen, N.; Toth, K.; Wovkulich, P. M.; Uskoković, M. R. *Tetrahedron Lett.* **1991**, *32*, 2343.
192. Kende, A. S.; Kawamura, K.; DeVita, R. J. *J. Am. Chem. Soc.* **1990**, *112*, 4070.
193. Rossi, R.; Carpita, A. *Tetrahedron* **1983**, *39*, 287.
194. Michelot, D.; Guerrero, A.; Ratovelomanana, V. *J. Chem. Res. (S)*, **1982**, 93.
195. Rossi, R.; Carpita, A.; Piccardi, P. *Pesticide Chemistry: Human Welfare and the Environment* **1983**, *1*, 129.
196. Camps, F.; Chamorro, E.; Gasol, V.; Guerrero, A. *Synth. Commun.* **1989**, *19*, 3211.
197. Ratovelomana, V.; Linstrumelle, G. *Synth. Commun.* **1981**, *11*, 917.
198. Vu, M. H.; Ando, T.; Yoshida, S.; Takahashi, N.; Tatsuki, S.; Katagiri, K.; Yamane, A.; Ikeda, T. *Agric. Biol. Chem.* **1979**, *43*, 1615.
199. Ando, T.; Vu, M. H.; Yoshida, S.; Takahashi, N. *Agric. Biol. Chem.* **1982**, *46*, 717.
200. Ratovelomanana, V.; Linstrumelle, G. *Tetrahedron Lett.* **1984**, *25*, 6001.
201. Kobayashi, Y.; Shimazaki, T.; Taguchi, H.; Sato, F. *J. Org. Chem.* **1990**, *55*, 5324.
202. Kobayashi, Y.; Kato, N.; Shimazaki, T.; Sato, F. *Tetrahedron Lett.* **1988**, *29*, 6297.
203. Nicolaou, K. C.; Webber, S. E. *J. Am. Chem. Soc.* **1984**, *106*, 5734.
204. Nicolaou, K. C.; Veale, C. A.; Webber, S. E.; Katerinopoulos, H. *J. Am. Chem. Soc.* **1985**, *107*, 7515.
205. Nicolaou, K. C.; Zipkin, R. E.; Dolle, R. E.; Harris, B. D. *J. Am. Chem. Soc.* **1984**, *106*, 3548.
206. Nicolaou, K. C.; Ramphal, J. Y.; Palazon, J. M.; Spanevello, R. A. *Angew. Chem., Int. Ed. Engl.* **1989**, *28*, 587.
207. Kaye, A. D.; Pattenden, G.; Roberts, S. M. *Tetrahedron Lett.* **1986**, *27*, 2033.
208. Nicolaou, K. C.; Webber, S. E. *J. Chem. Soc., Chem. Commun.* **1986**, 1816.
209. Guillerm, D.; Linstrumelle, G. *Tetrahedron Lett.* **1986**, *27*, 5857.
210. Kobayashi, Y.; Okamoto, S.; Shimazaki, T.; Ochiai, Y.; Sato, F. *Tetrahedron Lett.* **1987**, *28*, 3959.
211. Bhalerao, U. T.; Dasaradhi, L.; Neelakantan, P.; Fadnavis, N. W. *J. Chem. Soc., Chem. Commun.* **1991**, 1197.
212. Corey, E. J.; d'Alarcao, M. *Tetrahedron Lett.* **1986**, *27*, 3589.
213. Robins, M. J.; Barr, P. J. *Tetrahedron Lett.* **1981**, *22*, 421.
214. Bergstrom, D. E. *Nucleosides & Nucleotides* **1982**, *1*, 1.
215. Robins, M. J.; Barr, P. J. *J. Org. Chem.* **1983**, *48*, 1854.
216. DeClercq, E.; Descamps, J.; Balzarini, J.; Giziewicz, J.; Barr, P. J.; Robins, M. J. *J. Med. Chem.* **1983**, *26*, 661.
217. Haralambidis, J.; Chai, M.; Tregear, G. W. *Nucleic Acids Res.* **1987**, *15*, 4857.
218. Cruickshank, K. A.; Stockwell, D. L. *Tetrahedron Lett.* **1988**, *29*, 5221.
219. Spaltenstein, A.; Robinson, B. H.; Hopkins, P. B. *J. Am. Chem. Soc.* **1988**, *110*, 1299.
220. Hobbs, Jr., F. W. *J. Org. Chem.* **1989**, *54*, 3420.
221. Casalnuovo, A. L.; Calabrese, J. C. *J. Am. Chem. Soc.* **1990**, *112*, 4324.
222. Robins, M. J.; Vinayak, R. S.; Wood, S. G. *Tetrahedron Lett.* **1990**, *31*, 3731.
223. Tanaka, H.; Haraguchi, K.; Koizumi, Y.; Fukui, M.; Miyasaka, T. *Can. J. Chem.* **1986**, *64*, 1560.
224. Haraguchi, K.; Itoh, Y.; Tanaka, H.; Miyasaka, T. *Tetrahedron Lett.* **1991**, *32*, 3391.
225. Doi, T.; Takahashi, T. *J. Org. Chem.* **1991**, *56*, 3465.
226. Porco, Jr., J. A.; Schoenen, F. J.; Stout, T. J.; Clardy, J.; Schreiber, S. L. *J. Am. Chem. Soc.* **1990**, *112*, 7410.
227. Isobe, M.; Hirose, Y.; Shimokawa, K.; Nishikawa, T.; Goto, T. *Tetrahedron Lett.* **1990**, *31*, 5499.
228. Nicolaou, K. C.; Maligres, P.; Shin, J.; de Leon, E.; Rideout, D. *J. Am. Chem. Soc.* **1990**, *112*, 7825.
229. Sakai, Y.; Nishiwaki, E.; Shishido, K.; Shibuya, M.; Kido, M. *Tetrahedron Lett.* **1991**, *32*, 4363.
230. Saito, I.; Yamaguchi, K.; Nagata, R.; Murahashi, E. *Tetrahedron Lett.* **1990**, *31*, 7469.

231. Magnus, P.; Pitterna, T. *J. Chem. Soc., Chem. Commun.* **1991**, 541.
232. Nicolaou, K. C.; Smith, A. L.; Wendeborn, S. V.; Hwang, C.-K. *J. Am. Chem. Soc.* **1991**, *113*, 3106.
233. Wender, P. A.; Harmata, M.; Jeffrey, D.; Mukai, C.; Suffert, J. *Tetrahedron Lett.* **1988**, *29*, 909.
234. Crévisy, C.; Beau, J.-M. *Tetrahedron Lett.* **1991**, *32*, 3171.
235. Wender, P. A.; Zercher, C. K. *J. Am. Chem. Soc.* **1991**, *113*, 2311.
236. Suffert, J. *Tetrahedron Lett.* **1990**, *31*, 7437.
237. Nishikawa, T.; Isobe, M.; Goto, T. *Synlett* **1991**, 393.
238. Magnus, P.; Fortt, S. M. *J. Chem. Soc., Chem. Commun.* **1991**, 544.
239. Myers, A. G.; Harrington, P. M.; Kuo, E. Y. *J. Am. Chem. Soc.* **1991**, *113*, 694.
240. Magnus, P.; Annoura, H.; Harling, J. *J. Org. Chem.* **1990**, *55*, 1709.
241. Krebs, A.; Wehlage, T.; Kramer, C.-P. *Tetrahedron Lett.* **1990**, *31*, 3533.
242. Nicolaou, K. C.; Hwang, C.-K.; Smith, A. L.; Wendeborn, S. V. *J. Am. Chem. Soc.* **1990**, *112*, 7416.
243. Wender, P. A.; McKinney, J. A.; Mukai, C. *J. Am. Chem. Soc.* **1990**, *112*, 5369.
244. Schoenen, F. J.; Porco, Jr., J. A.; Schreiber, S. L.; VanDuyne, G. D.; Clardy, J. *Tetrahedron Lett.* **1989**, *30*, 3765.
245. Tomioka, K.; Fujita, H.; Koga, K. *Tetrahedron Lett.* **1989**, *30*, 851.
246. Eaton, P. E.; Stössel, D. *J. Org. Chem.* **1991**, *56*, 5138.
247. Lee, G. C. M.; Tobias, B.; Holmes, J. M.; Harcourt, D. A.; Garst, M. E. *J. Am. Chem. Soc.* **1990**, *112*, 9330.
248. Silverberg, L. J.; Wu, G.; Rheingold, A. L.; Heck, R. F. *J. Organomet. Chem.* **1991**, *409*, 411.
249. (a) Danno, S.; Moritani, I.; Fujiwara, Y.; Teranishi, S. *Bull. Chem. Soc. Jpn.* **1970**, *43*, 3966. (b) Moritani, I.; Fujiwara, Y.; Danno, S. *J. Organomet. Chem.* **1971**, *27*, 279.
250. Grigg, R.; Coulter, T. *Tetrahedron Lett.* **1991**, *32*, 1359.
251. Grigg, R.; Stevenson, P.; Worakun, T. *J. Chem. Soc., Chem. Commun.* **1985**, 971.
252. Grigg, R.; Stevenson, P.; Worakun, T. *Tetrahedron* **1988**, *44*, 2049.
253. Mori, M.; Kaneta, N.; Shibasaki, M. *J. Org. Chem.* **1991**, *56*, 3486.
254. Piers, E.; Marais, P. C. *J. Org. Chem.* **1990**, *55*, 3454.
255. Yamamura, M.; Moritani, I.; Murahashi, S.-I. *J. Organomet. Chem.* **1975**, *91*, C39.
256. Murahashi, S.-I.; Yamamura, M.; Mita, N. *J. Org. Chem.* **1977**, *42*, 2870.
257. Murahashi, S.-I.; Yamamura, M.; Yanagisawa, K.; Mita, N.; Kondo, K. *J. Org. Chem.* **1979**, *44*, 2408.
258. Loar, M. K.; Stille, J. K. *J. Am. Chem. Soc.* **1981**, *103*, 4174.
259. Dang, H. P.; Linstrumelle, G. *Tetrahedron Lett.* **1978**, 191.
260. Michelot, D. *Synthesis* **1983**, 130.
261. Ratovelomanana, V.; Linstrumelle, G. *Synth. Commun.* **1984**, *14*, 179.
262. Hayashi, T.; Konishi, M.; Kobori, Y.; Kumada, M.; Higuchi, T.; Hirotsu, K. *J. Am. Chem. Soc.* **1984**, *106*, 158.
263. Negishi, E.; Valente, L. F.; Kobayashi, M. *J. Am. Chem. Soc.* **1980**, *102*, 3298.
264. Minato, A.; Suzuki, K.; Tamao, K. *J. Am. Chem. Soc.* **1987**, *109*, 1257.
265. Hayashi, T.; Konishi, M.; Fukushima, M.; Mise, T.; Kagotani, M.; Tajika, M.; Kumada, M. *J. Am. Chem. Soc.* **1982**, *104*, 180.
266. Hayashi, T.; Konishi, M.; Kumada, M. *Tetrahedron Lett.* **1979**, 1871.
267. Hayashi, T.; Konishi, M.; Yokota, K.; Kumada, M. *Chem. Lett.* **1980**, 767.
268. Rossi, R.; Carpita, A. *Tetrahedron Lett.* **1986**, *27*, 2529.
269. Terasawa, M.; Kaneda, K.; Imanaka, T.; Teranishi, S. *J. Organomet. Chem.* **1978**, *162*, 403.
270. Prestwich, G. D.; Wawrzeńczyk, C. *Tetrahedron Lett.* **1989**, *30*, 403.
271. Negishi, E.; Luo, F.-T.; Rand, C. L. *Tetrahedron Lett.* **1982**, *23*, 27.
272. Tamao, K.; Iwahara, T.; Kanatani, R.; Kumada, M. *Tetrahedron Lett.* **1984**, *25*, 1909.
273. Hayashi, T.; Konishi, M.; Okamoto, Y.; Kabeta, K.; Kumada, M. *J. Org. Chem.* **1986**, *51*, 3772.

274. Huynh, C.; Linstrumelle, G. *Tetrahedron Lett.* **1979**, 1073.
275. Fiandanese, V.; Marchese, G.; Naso, F.; Ronzini, L. *J. Chem. Soc., Chem. Commun.* **1982**, 647.
276. Fiandanese, V.; Marchese, G.; Naso, F.; Ronzini, L. *J. Chem. Soc., Perkin Trans 1* **1985**, 1115.
277. Naso, F. *Pure Appl. Chem.* **1988**, *60*, 79.
278. Zakharkin, L. I.; Anikina, E. V. *J. Org. Chem.* USSR **1986**, *22*, 2217.
279. Ratovelomanana,V.; Linstrumelle, G.; Normant, J.-F. *Tetrahedron Lett.* **1985**, *26*, 2575.
280. Lebedev, S. A.; Sorokina, R. S.; Berestova, S. S.; Petrov, E. S.; Beletskaya, I. P. *Bull. Acad. Sci. USSR* **1986**, 620.
281. Harada, T.; Kotani, Y.; Katsuhira, T.; Oku, A. *Tetrahedron Lett.* **1991**, *32*, 1573.
282. Harada, T.; Katsuhira, T.; Hattori, K.; Oku, A. *Tetrahedron Lett.* **1989**, *30*, 6039.
283. Piers, E.; Jean, M.; Marrs, P. S. *Tetrahedron Lett.* **1987**, *28*, 5075.
284. Negishi, E.; Matsushita, H.; Kobayashi, M.; Rand, C. L. *Tetrahedron Lett.* **1983**, *24*, 3823.
285. Knochel, P.; Normant, J. F. *Tetrahedron Lett.* **1986**, *27*, 4431.
286. McMurry, J. E.; Bosch, G. K. *J. Org. Chem.* **1987**, *52*, 4885.
287. Carpita, A.; Neri, D.; Rossi, R. *Gazz. Chim. Ital.* **1987**, *117*, 503.
288. Rand, C. L.; Van Horn, D. E.; Moore, M. W.; Negishi, E. *J. Org. Chem.* **1981**, *46*, 4093.
289. Martinet, P.; Sauvêtre, R.; Normant, J.-F. *J. Organomet. Chem.* **1989**, *367*, 1.
290. Negishi, E.; Swanson, D. R.; Rousset, C. J. *J. Org. Chem.* **1990**, *55*, 5406.
291. Kitazume, T.; Ishikawa, N. *Chem. Lett.* **1982**, 137.
292. Kitazume, T.; Ishikawa, N. *J. Am. Chem. Soc.* **1985**, *107*, 5186.
293. Asao, K.; Iio, H.; Tokoroyama, T. *Tetrahedron Lett.* **1989**, *30*, 6401.
294. Tamaru, Y.; Ochiai, H.; Nakamura, T.; Yoshida, Z. *Angew. Chem., Int. Ed. Engl.* **1987**, *26*, 1157.
295. Fauvarque, J. F.; Jutand, A. *J. Organomet. Chem.* **1981**, *209*, 109.
296. Nakamura, E.; Kuwajima, I. *Tetrahedron Lett.* **1986**, *27*, 83.
297. Tamaru, Y.; Ochiai, H.; Nakamura, T.; Yoshida, Z. *Tetrahedron Lett.* **1986**, *27*, 955.
298. Nakamura, E.; Aoki, S.; Sekiya, K.; Oshino, H.; Kuwajima, I. *J. Am. Chem. Soc.* **1987**, *109*, 8056.
299. Kobayashi, M.; Negishi, E. *J. Org. Chem.* **1980**, *45*, 5223.
300. Satoh, Y.; Serizawa, H.; Miyaura, N.; Hara, S.; Suzuki, A. *Tetrahedron Lett.* **1988**, *29*, 1811.
301. Hyuga, S.; Chiba, Y.; Yamashina, N.; Hara, S.; Suzuki, A. *Chem. Lett.* **1987**, 1757.
302. Hyuga, S.; Yamashina, N.; Hara, S.; Suzuki, A. *Chem. Lett.* **1988**, 809.
303. Yamashina, N.; Hyuga, S.; Hara, S.; Suzuki, A. *Tetrahedron Lett.* **1989**, *30*, 6555.
304. Chan, T. H.; Koumaglo, K. *Tetrahedron Lett.* **1986**, *27*, 883.
305. Ramiandrasoa, F.; Descoins, C. *Synth. Commun.* **1989**, *19*, 2703.
306. Millar, J. G. *Tetrahedron Lett.* **1989**, *30*, 4913.
307. (a) Miyaura, N.; Ishiyama, T.; Ishikawa, M.; Suzuki, A. *Tetrahedron Lett.* **1986**, *27*, 6369. (b) Miyaura, N.; Ishiyama, T.; Sasaki, H.; Ishikawa, M.; Satoh, M.; Suzuki, A. *J. Am. Chem. Soc.* **1989**, *111*, 314.
308. Soderquist, J. A.; Santiago, B. *Tetrahedron Lett.* **1990**, *31*, 5541.
309. Sato, M.; Miyaura, N.; Suzuki, A. *Chem. Lett.* **1989**, 1405.
310. Blumenkopf, T. A.; Bratz, M.; Castañeda, A.; Look, G. C.; Overman, L. E.; Rodriguez, D.; Thompson, A. S. *J. Am. Chem. Soc.* **1990**, *112*, 4386.
311. Hoshino, Y.; Ishiyama, T.; Miyaura, N.; Suzuki, A. *Tetrahedron Lett.* **1988**, *29*, 3983.
312. Stille, J. K.; Sweet, M. P. *Tetrahedron Lett.* **1989**, *30*, 3645.
313. Herdewijn, P.; Kerremans, L.; Wigerinck, P.; Vandendriessche, F.; Van Aerschot, A. *Tetrahedron Lett.* **1991**, *32*, 4397.
314. Banwell, M. G.; Collis, M. P.; Crisp, G. T.; Lambert, J. N.; Reum, M. E.; Scoble, J. A. *J. Chem. Soc., Chem. Commun.* **1989**, 616.
315. Férézou, J. P.; Julia, M.; Li, Y.; Liu, L. W.; Pancrazi, A. *Synlett* **1991**, 53.
316. Hatanaka, Y.; Hiyama, T. *J. Org. Chem.* **1988**, *53*, 918.
317. Yamamoto, Y.; Hatsuya, S.; Yamada, J. *J. Chem. Soc., Chem. Commun.* **1988**, 86.

318. Takayama, H.; Suzuki, T. *J. Chem. Soc., Chem. Commun.* **1988**, 1044.
319. Kreuzfeld, H.-J.; Döbler, C.; Abicht, H.-P. *J. Organomet. Chem.* **1987**, *336*, 287.
320. Hayashi, T.; Konishi, M.; Fukushima, M.; Kanehira, K.; Hioki, T.; Kumada, M. *J. Org. Chem.* **1983**, *48*, 2195.
321. Cross, G. A.; Kellogg, R. M. *J. Chem. Soc., Chem. Commun.* **1987**, 1746.
322. Hayashi, T.; Hagihara, T.; Katsuro, Y.; Kumada, M. *Bull. Chem. Soc. Jpn.* **1983**, *56*, 363.
323. Hayashi, T.; Konishi, M.; Ito, H.; Kumada, M. *J. Am. Chem. Soc.* **1982**, *104*, 4962.
324. Negishi, E.; Matsushita, H.; Okukado, N. *Tetrahedron Lett.* **1981**, *22*, 2715.
325. Negishi, E. *Pure Appl. Chem.* **1981**, *53*, 2333.
326. Tellier, F.; Sauvêtre, R.; Normant, J.-F. *J. Organomet. Chem.* **1985**, *292*, 19.
327. Negishi, E.; Takahashi, T.; Baba, S.; Van Horn, D. E.; Okukado, N. *J. Am. Chem. Soc.* **1987**, *109*, 2393.
328. Fiandanese, V.; Marchese, G.; Mascolo, G.; Naso, F.; Ronzini, L. *Tetrahedron Lett.* **1988**, *29*, 3705.
329. Crombie, L.; Horsham, M. A.; Jarrett, S. R. M. *Tetrahedron Lett.* **1989**, *30*, 4299.
330. Okukado, N.; Van Horn, D. E.; Klima, W. L.; Negishi, E. *Tetrahedron Lett.* **1978**, 1027.
331. Negishi, E.; Okukado, N.; King, A. O.; Van Horn, D. E.; Spiegel, B. I. *J. Am. Chem. Soc.* **1978**, *100*, 2254.
332. Wender, P. A.; Snapper, M. L. *Tetrahedron Lett.* **1987**, *28*, 2221.
333. Vincent, P.; Beaucourt, J.-P.; Pichat, L. *Tetrahedron Lett.* **1982**, *23*, 63.
334. Jabri, N.; Alexakis, A.; Normant, J. F. *Tetrahedron Lett.* **1981**, *22*, 959.
335. Jabri, N.; Alexakis, A.; Normant, J. F. *Bull. Soc. Chim. Fr. II* **1983**, 321.
336. Jabri, N.; Alexakis, A.; Normant, J. F. *Tetrahedron* **1986**, *42*, 1369.
337. Gardette, M.; Jabri, N.; Alexakis, A.; Normant, J. F. *Tetrahedron* **1984**, *40*, 2741.
338. Gardette, M.; Alexakis, A.; Normant, J. F. *Tetrahedron Lett.* **1982**, *23*, 5155.
339. Jabri, N.; Alexakis, A.; Normant, J. F. *Tetrahedron Lett.* **1982**, *23*, 1589.
340. Jabri, N.; Alexakis, A.; Normant, J. F. *Bull. Soc. Chim. Fr. II* **1983**, 332.
341. Negishi, E.; Takahashi, T.; Akiyoshi, K. *J. Organomet. Chem.* **1987**, *334*, 181.
342. Zhu, L.; Wehmeyer, R. M.; Rieke, R. D. *J. Org. Chem.* **1991**, *56*, 1445.
343. Andreini, B. P.; Benetti, M.; Carpita, A.; Rossi, R. *Tetrahedron* **1987**, *43*, 4591.
344. Duffault, J.-M.; Einhorn, J.; Alexakis, A. *Tetrahedron Lett.* **1991**, *32*, 3701.
345. Tellier, F.; Descoins, C. *Tetrahedron Lett.* **1990**, *31*, 2295.
346. Ramiandrasoa, F.; Tellier, F. *Synth. Commun.* **1990**, *20*, 333.
347. Björkling, F.; Norin, T.; Unelius, R. *Synth. Commun.* **1985**, *15*, 463.
348. Negishi, E.; Owczarczyk, Z. R.; Swanson, D. R. *Tetrahedron Lett.* **1991**, *32*, 4453.
349. Negishi, E.; Akiyoshi, K. *Chem. Lett.* **1987**, 1007.
350. Suzuki, A. *Pure Appl. Chem.* **1986**, *58*, 629.
351. Ogima, M.; Hyuga, S.; Hara, S.; Suzuki, A. *Chem. Lett.* **1989**, 1959.
352. Dolbier, Jr., W. R.; Koroniak, H.; Burton, D. J.; Heinze, P. *Tetrahedron Lett.* **1986**, *27*, 4387.
353. Telleri, F.; Sauvêtre, R.; Normant, J.-F. *J. Organomet. Chem.* **1989**, *364*, 17.
354. Tellier, F.; Sauvêtre, R.; Normant, J.-F. *J. Organomet. Chem.* **1986**, *303*, 309.
355. Gillet, J.-P.; Sauvêtre, R.; Normant, J.-F. *Tetrahedron Lett.* **1985**, *26*, 3999.
356. Isaka, M.; Matsuzawa, S.; Yamago, S.; Ejiri, S.; Miyachi, Y.; Nakamura, E. *J. Org. Chem.* **1989**, *54*, 4727.
357. Russell, C. E.; Hegedus, L. S. *J. Am. Chem. Soc.* **1983**, *105*, 943.
358. Negishi, E.; Luo, F.-T. *J. Org. Chem.* **1983**, *48*, 1560.
359. Sengupta, S.; Snieckus, V. *J. Org. Chem.* **1990**, *55*, 5680.
360. Gilchrist, T. L.; Healy, M. A. M. *Tetrahedron Lett.* **1990**, *31*, 5807.
361. Gilchrist, T. L.; Summersell, R. J. *J. Chem. Soc., Perkin Trans 1* **1988**, 2603.
362. Gilchrist, T. L.; Summersell, R. J. *Tetrahedron Lett.* **1987**, *28*, 1469.
363. Gilchrist, T. L.; Summersell, R. J. *J. Chem. Soc., Perkin Trans 1* **1988**, 2595.

364. Suzuki, A. *Acc. Chem. Res.* **1982,** *15,* 178.
365. Suzuki, A. Pure Appl. Chem. **1985,** *57,* 1749.
366. Miyaura, N.; Yamada, K.; Suzuki, A. *Tetrahedron Lett.* **1979,** 3437.
367. Miyaura, N.; Yamada, K.; Suginome, H.; Suzuki, A. *J. Am. Chem. Soc.* **1985,** *107,* 972.
368. Miyaura, N.; Suginome, H.; Suzuki, A. *Tetrahedron Lett.* **1981,** *22,* 127.
369. Miyaura, N.; Satoh, M.; Suzuki, A. *Tetrahedron Lett.* **1986,** *27,* 3745.
370. Rivera, I.; Soderquist, J. A. *Tetrahedron Lett.* **1991,** *32,* 2311.
371. Soderquist, J. A.; León-Colón, G. *Tetrahedron Lett.* **1991,** *32,* 43.
372. Miyaura, N.; Suginome, H.; Suzuki, A. *Tetrahedron* **1983,** *39,* 3271.
373. Miyaura, N.; Suginome, H.; Suzuki, A. *Tetrahedron Lett.* **1983,** *24,* 1527.
374. Rossi, R.; Carpita, A.; Quirici, M. G. *Tetrahedron* **1981,** *37,* 2617.
375. Kobayashi, Y.; Shimazaki, T.; Sato, F. *Tetrahedron Lett.* **1987,** *28,* 5849.
376. Miyaura, N.; Satoh, Y.; Hara, S.; Suzuki, A. *Bull. Chem. Soc. Jpn.* **1986,** *59,* 2029.
377. Satoh, M.; Miyaura, N.; Suzuki, A. *Chem. Lett.* **1986,** 1329.
378. Satoh, N.; Ishiyama, T.; Miyaura, N.; Suzuki, A. *Bull. Chem. Soc. Jpn.* **1987,** *60,* 3471.
379. Yanagi, T.; Oh-e, T.; Miyaura, N.; Suzuki, A. *Bull. Chem. Soc. Jpn.* **1989,** *62,* 3892.
380. Ishiyama, T.; Miyaura, N.; Suzuki, A. *Chem. Lett.* **1987,** 25.
381. Björkling, F.; Norin, T.; Unelius, C. R.; Miller, R. B. *J. Org. Chem.* **1987,** *52,* 292.
382. Miyaura, N.; Suginome, H.; Suzuki, A. Bull. Chem. Soc. Jpn. **1982,** *55,* 2221.
383. Haviv, F.; Ratajczyk, J. D.; DeNet, R. W.; Martin, Y. C.; Dyer, R. D.; Carter, G. W. *J. Med. Chem.* **1987,** *30,* 254.
384. Roush, W. R.; Riva, R. *J. Org. Chem.* **1988,** *53,* 710.
385. Roush, W. R.; Kageyama, M.; Riva, R.; Brown, B. B.; Warmus, J. S.; Moriarty, K. J. *J. Org. Chem.* **1991,** *56,* 1192.
386. Uenishi, J.; Beau, J.-M.; Armstrong, R. W.; Kishi, Y. *J. Am. Chem. Soc.* **1987,** *109,* 4756.
387. Roush, W. R.; Moriarty, K. J.; Brown, B. B. *Tetrahedron Lett.* **1990,** *31,* 6509.
388. Mavrov, M. V.; Urdaneta, N. A.; Serebryakov, E. P. *Bioorg. Khim.* **1990,** *16,* 711; *Chem. Abstr.* **1990,** *113,* 152777y.
389. Cassani, G.; Massardo, P.; Piccardi, P. *Tetrahedron Lett.* **1983,** *24,* 2513.
390. Miyaura, N.; Suzuki, A. *J. Organomet. Chem.* **1981,** *213,* C53.
391. Baba, S.; Negishi, E. *J. Am. Chem. Soc.* **1976,** *98,* 6729.
392. Negishi, E.; Takahashi, T.; Baba, S. Org. Syn. **1988,** *66,* 60.
393. Andreini, B. P.; Carpita, A.; Rossi, R. *Tetrahedron Lett.* **1988,** *29,* 2239.
394. Hatanaka, Y.; Hiyama, T. *J. Org. Chem.* **1989,** *54,* 268.
395. Tamao, K.; Kobayashi, K.; Ito, Y. *Tetrahedron Lett.* **1989,** *30,* 6051.
396. Hosomi, A.; Kohra, S.; Tominaga, Y. *Chem. Pharm. Bull.* **1988,** *36,* 4622.
397. Eaton, B.; King, Jr., J. A.; Vollhardt, K. P. C. *J. Am. Chem. Soc.* **1986,** *108,* 1359.
398. Stille, J. K.; Groh, B. L. *J. Am. Chem. Soc.* **1987,** *109,* 813.
399. Kende, A. S.; DeVita, R. J. *Tetrahedron Lett.* **1990,** *31,* 307.
400. Lee, J.; Snyder, J. K. *J. Org. Chem.* **1990,** *55,* 4995.
401. Forsyth, C. J.; Clardy, J. *J. Am. Chem. Soc.* **1990,** *112,* 3497.
402. Férézou, J. P.; Julia, M.; Liu, L. W.; Pancrazi, A. *Synlett* **1991,** 614.
403. Bestmann, H. J.; Attygalle, A. B.; Schwarz, J.; Garbe, W.; Vostrowsky, O.; Tomida, I. *Tetrahedron Lett.* **1989,** *30,* 2911.
404. Chan, C.; Cox, P. B.; Roberts, S. M. *J. Chem. Soc., Chem. Commun.* **1988,** 971.
405. Friesen, R. W.; Loo, R. W. *J. Org. Chem.* **1991,** *56,* 4821.
406. Dubois, E.; Beau, J.-M. *J. Chem. Soc., Chem. Commun.* **1990,** 1191.
407. Majeed, A. J.; Antonsen, O.; Benneche, T.; Undheim, K. *Tetrahedron* **1989,** *45,* 993.
408. Sandosham, J.; Benneche, T.; Moller, B. S.; Undheim, K. *Acta Chem. Scand. B* **1988,** *42,* 455.
409. Farina, V.; Hauck, S. I. *Synlett* **1991,** 157.
410. Crisp, G. T. *Synth. Commun.* **1989,** *19,* 2117.

411. Wang, L. R. R.; Benneche, T.; Undheim, K. *Acta Chem. Scand. B* **1990**, *44*, 726.
412. Kwon, H. B.; McKee, B. H.; Stille, J. K. *J. Org. Chem.* **1990**, *55*, 3114.
413. Kosugi, M.; Sumiya, T.; Obara, Y.; Suzuki, M.; Sano, H.; Migita, T. *Bull. Chem. Soc. Jpn.* **1987**, *60*, 767.
414. MacLeod, D.; Moorcroft, D.; Quayle, P.; Dorrity, M. R. J.; Malone, J. F.; Davies, G. M. *Tetrahedron Lett.* **1990**, *31*, 6077.
415. Liebeskind, L. S.; Wang, J. *Tetrahedron Lett.* **1990**, *31*, 4293.
416. Liebeskind, L. S.; Fengl, R. W. *J. Org. Chem.* **1990**, *55*, 5359.
417. Jeffery-Luong, T.; Linstrumelle, G. *Synthesis* **1982**, 738.
418. Ruitenberg, K.; Kleijn, H.; Meijer, J.; Oostveen, E. A.; Vermeer, P. *J. Organomet. Chem.* **1982**, *224*, 399.
419. Chauhan, Y. S.; Chandraratna, R. A. S.; Miller, D. A.; Kondrat, R. W.; Reischl, W.; Okamura, W. H. *J. Am. Chem. Soc.* **1985**, *107*, 1028.
420. Bumagin, N. A.; Kalinovskii, I. O.; Ponomarov, A. B.; Beletskaya, I. P. *Proc. Acad. Sci. USSR, Chem. Sec.* **1982**, *265*, 262.
421. King, A. O.; Okukado, N.; Negishi, E. *J. Chem. Soc., Chem. Commun.* **1977**, 683.
422. Stracker, E. C.; Zweifel, G. *Tetrahedron Lett.* **1991**, *32*, 3329.
423. Carpita, A.; Rossi, R. *Tetrahedron Lett.* **1986**, *27*, 4351.
424. Negishi, E.; Okukado, N.; Lovich, S. F.; Luo, F.-T. *J. Org. Chem.* **1984**, *49*, 2629.
425. Wudl, F.; Bitler, S. P. *J. Am. Chem. Soc.* **1986**, *108*, 4685.
426. Luo, F. T.; Lai, J. H.; Shaeh, J. C. *Bull. Inst. Chem., Acad. Sin.* **1987**, *34*, 17; *Chem. Abstr.* **1988**, *108*, 111757v.
427. Minato, A. *J. Org. Chem.* **1991**, *56*, 4052.
428. Tellier, F. *Synth. Commun.* **1990**, *20*, 371.
429. Collins, P. W.; Kramer, S. W.; Gasiecki, A. F.; Weier, R. M.; Jones, P. H.; Gullikson, G. W.; Bianchi, R. G.; Bauer, R. F. *J. Med. Chem.* **1987**, *30*, 193.
430. Vincent, P.; Beaucourt, J.-P.; Pichat, L. *Tetrahedron Lett.* **1981**, *22*, 945.
431. Tellier, F.; Sauvêtre, R.; Normant, J.-F. *Tetrahedron Lett.* **1986**, *27*, 3147.
432. Tellier, F.; Sauvêtre, R.; Normant, J.-F. *J. Organomet. Chem.* **1987**, *328*, 1.
433. Hatanaka, Y.; Matsui, K.; Hiyama, T. *Tetrahedron Lett.* **1989**, *30*, 2403.
434. Stille, J. K.; Simpson, J. H. *J. Am. Chem. Soc.* **1987**, *109*, 2138.
435. Rubin, Y.; Knobler, C. B.; Diederich, F. *J. Am. Chem. Soc.* **1990**, *112*, 1607.
436. Rudisill, D. E.; Castonguay, L. A.; Stille, J. K. *Tetrahedron Lett.* **1988**, *29*, 1509.
437. Hirama, M.; Tokuda, M.; Fujiwara, K. *Synlett* **1991**, 651.
438. Fujiwara, K.; Kurisaki, A.; Hirama, M. *Tetrahedron Lett.* **1990**, *31*, 4329.
439. Magriotis, P. A.; Doyle, T. J.; Kim, K. D. *Tetrahedron Lett.* **1990**, *31*, 2541.
440. Murahashi, S.-I.; Naota, T.; Tanigawa, Y. *Org. Syn.* **1984**, *62*, 39.
441. Minato, A.; Suzuki, K.; Tamao, K.; Kumada, M. *Tetrahedron Lett.* **1984**, *25*, 83.
442. Potter, G. A.; McCague, R. *J. Org. Chem.* **1990**, *55*, 6184.
443. Mathews, N.; Widdowson, D. A. *Synlett* **1990**, 467.
444. Kalinin, V. N.; Min, S. F. *J. Organomet. Chem.* **1988**, *352*, C34.
445. Negishi, E.; Luo, F.-T.; Frisbee, R.; Matsushita, H. *Heterocycles* **1982**, *18*, 117.
446. Ennis, D. S.; Gilchrist, T. L. *Tetrahedron* **1990**, *46*, 2623.
447. Miller, R. B.; Al-Hassan, M. I. *J. Org. Chem.* **1985**, *50*, 2121.
448. McCague, R.; Kuroda, R.; Leclercq, G.; Stoessel, S. *J. Med. Chem.* **1986**, *29*, 2053.
449. Al-Hassan, M. I. *Synth. Commun.* **1989**, *19*, 1619.
450. Al-Hassan, M. I. *Synth. Commun.* **1987**, *17*, 1247.
451. Al-Hassan, M. I. *J. Organomet. Chem.* **1987**, *321*, 119.
452. Al-Hassan, M. I. *Synth. Commun.* **1987**, *17*, 1787.
453. Al-Hassan, M. I. *Org. Prep. Proc. Int.* **1991**, *23*, 116.
454. Paterson, I.; Gardner, M.; Banks, B. J. *Tetrahedron* **1989**, *45*, 5283.

455. Iyoda, M.; Sato, K.; Oda, M. *Tetrahedron Lett.* **1987**, *28*, 625.
456. Vincent, P.; Beaucourt, J. P.; Pichat, L. *Tetrahedron Lett.* **1984**, *25*, 201.
457. Ennis, D. S.; Gilchrist, T. L. *Tetrahedron Lett.* **1989**, *30*, 3735.
458. Nair, V.; Powell, D. W.; Suri, S. C. *Synth. Commun.* **1987**, *17*, 1897.
459. Suri, S. C.; Nair, V. *Synthesis* **1990**, 695.
460. Hoshino, Y.; Miyaura, N.; Suzuki, A. *Bull. Chem. Soc. Jpn.* **1988,** *61*, 3008.
461. Yokoe, I.; Sugita, Y.; Shirataki, Y. *Chem. Pharm. Bull.* **1989**, *37*, 529.
462. Crisp, G. T.; Macolino, V. *Synth. Commun.* **1990**, *20*, 413.
463. Yoshida, J.; Tamao, K.; Yamamoto, H.; Kakui, T.; Uchida, T.; Kumada, M. *Organometallics* **1982**, *1*, 542.
464. Bumagin, N. A.; Bumagina, I. G.; Beletskaya, I. P. *Proc. Acad. Sci. USSR, Chem. Sec.* **1983**, *272*, 333.
465. Kalivretenos, A.; Stille, J. K.; Hegedus, L. S. *J. Org. Chem.* **1991**, *56*, 2883.
466. Yamamoto, Y.; Seko, T.; Nemoto, H. *J. Org. Chem.* **1989**, *54*, 4734.
467. Jin, H.; Uenishi, J.; Christ, W. J.; Kishi, Y. *J. Am. Chem. Soc.* **1986**, *108*, 5644.
468. Uozumi, Y.; Mori, M.; Shibasaki, M. *J. Chem. Soc., Chem. Commun.* **1991**, 81.
469. Li, P.; Alper, H. *J. Org. Chem.* **1986**, *51*, 4354.
470. Julia, M.; Blasioli, C. *Bull. Soc. Chim. Fr.* **1976**, 1941.
471. Foa, M.; Santi, R.; Garavaglia, F. *J. Organomet. Chem.* **1981**, *206*, C29.
472. Hirao, T.; Masunaga, T.; Ohshiro, Y.; Agawa, T. *Tetrahedron Lett.* **1980**, *21*, 3595.
473. Hirao, T.; Masunaga, T.; Yamada, N.; Ohshiro, Y.; Agawa, T. *Bull. Chem. Soc. Jpn.* **1982**, *55*, 909.
474. Xu, Y.; Li, Z. *Synthesis* **1986**, 240.
475. Xu, Y.; Wei, H.; Zhang, J.; Huang, G. *Tetrahedron Lett.* **1989**, *30*, 949.
476. Xu, Y.; Li, Z. *Tetrahedron Lett.* **1986**, *27*, 3017.
477. Xu, Y.; Xia, J.; Guo, H. *Synthesis* **1986**, 691.
478. Trost, B. M.; Yoshida, J. *Tetrahedron Lett.* **1983**, *24*, 4895.
479. Hatanaka, Y.; Hiyama, T. *Tetrahedron Lett.* **1987**, *28*, 4715.
480. Matsumoto, H.; Nagashima, S.; Kato, T.; Nagai, Y. *Angew. Chem., Int. Ed. Engl.* **1978**, *17*, 279.
481. Farina, V.; Hauck, S. I. *J. Org. Chem.* **1991**, *56*, 4317.
482. Moriarty, R. M.; Epa, W. R.; Awasthi, A. K. *J. Am. Chem. Soc.* **1991**, *113*, 6315.
483. Scott, W. J.; McMurry, J. E. *Acc. Chem. Res.* **1988**, *21*, 47.
484. Cacchi, S.; Morera, E.; Ortar, G. *Tetrahedron Lett.* **1984**, *25*, 4821.
485. Scott, W. J.; Stille, J. K. *J. Am. Chem. Soc.* **1986**, *108*, 3033.
486. Scott, W. J.; Crisp, G. T.; Stille, J. K. *J. Am. Chem. Soc.* **1984**, *106*, 4630.
487. Dupré, B.; Meyers, A. I. *J. Org. Chem.* **1991**, *56*, 3197.
488. Chou, W.-N.; White, J. B. *Tetrahedron Lett.* **1991**, *32*, 157.
489. Gyorkos, A. C.; Stille, J. K.; Hegedus, L. S. *J. Am. Chem. Soc.* **1990**, *112*, 8465.
490. Farina, V.; Baker, S. R.; Sapino, Jr., C. *Tetrahedron Lett.* **1988**, *29*, 6043.
491. Farina, V.; Baker, S. R.; Benigni, D. A.; Hauck, S. I.; Sapino, Jr., C. *J. Org. Chem.* **1990**, *55*, 5833.
492. Piers, E.; Fleming, F. F. *J. Chem. Soc., Chem. Commun.* **1989**, 756.
493. Crisp, G. T.; Scott, W. J.; Stille, J. K. *J. Am. Chem. Soc.* **1984**, *106*, 7500.
494. Echavarren, A. M.; Stille, J. K. *J. Am. Chem. Soc.* **1988**, *110*, 1557.
495. Huffman, J. W.; Zhang, X.; Wu, M.-J.; Joyner, H. H. *J. Org. Chem.* **1989**, *54*, 4741.
496. Huffman, J. W.; Zhang, X.; Wu, M.-J.; Joyner, H. H.; Pennington, W. T. *J. Org. Chem.* **1991**, *56*, 1481.
497. Tius, M. A.; Kannangara, G. S. K. *J. Org. Chem.* **1990**, *55*, 5711.
498. Cacchi, S.; Morera, E.; Ortar, G. *Tetrahedron Lett.* **1985**, *26*, 1109.
499. Agnel, G.; Negishi, E. *J. Am. Chem. Soc.* **1991**, *113*, 7424.
500. Hashimoto, H.; Furuichi, K.; Miwa, T. *J. Chem. Soc., Chem. Commun.* **1987**, 1002.

501. Rizzo, C. J.; Smith, III, A. B. *Tetrahedron Lett.* **1988**, *29*, 2793.
502. Metcalf, B. W.; Holt, D. A.; Levy, M. A. Eur. Pat. Appl. EP 414,529; US Appl. 398,192, **1989**; *Chem. Abstr.* **1991**, *114*, 247584t.
503. Aoki, S.; Nakamura, E. *Synlett* **1990**, 741.
504. Lan-Hargest, H.-Y.; Elliott, J. D.; Eggleston, D. S.; Holt, D. A.; Levy, M. A.; Metcalf, B. W. *Tetrahedron Lett.* **1987**, *28*, 6117.
505. Cacchi, S.; Morera, E.; Ortar, G. *Tetrahedron Lett.* **1984**, *25*, 2271.
506. Scott, W. J.; Peña, M. R.; Swärd, K.; Stoessel, S. J.; Stille, J. K. *J. Org. Chem.* **1985**, *50*, 2302.
507. Arcadi, A.; Marinelli, F.; Cacchi, S. *J. Organomet. Chem.* **1986**, *312*, C27.
508. Breining, T.; Schmidt, C.; Polos, K. *Synth. Commun.* **1987**, *17*, 85.
509. Arcadi, A.; Bernocchi, E.; Cacchi, S.; Cagliotti, L.; Marinelli, F. *Tetrahedron Lett.* **1990**, *31*, 2463.
510. Harnisch, W.; Morera, E.; Ortar, G. *J. Org. Chem.* **1985**, *50*, 1990.
511. Cacchi, S.; Ciattini, P. G.; Morera, E.; Ortar, G. *Tetrahedron Lett.* **1988**, *29*, 3117.
512. Peña, M. R.; Stille, J. K. *Tetrahedron Lett.* **1987**, *28*, 6573.
513. Peña, M. R.; Stille, J. K. *J. Am. Chem. Soc.* **1989**, *111*, 5417.
514. Hirota, K.; Kitade, Y.; Isobe, Y.; Maki, Y. *Heterocycles* **1987**, *26*, 355.
515. Ciattini, P. G.; Morera, E.; Ortar, G. *Tetrahedron Lett.* **1991**, *32*, 1579.
516. Cacchi, S.; Ciattini, P. G.; Morera, E.; Ortar, G. *Tetrahedron Lett.* **1987**, *28*, 3039.
517. Paquette, L. A.; Ra, C. S.; Edmonson, S. D. *J. Org. Chem.* **1990**, *55*, 2443.
518. Carpenter, N. E.; Kucera, D. J.; Overman, L. E. *J. Org. Chem.* **1989**, *54*, 5846.
519. Arcadi, A.; Bernocchi, E.; Cacchi, S.; Marinelli, F. *Tetrahedron* **1991**, *47*, 1525.
520. Hassan, M. E. *Can. J. Chem.* **1991**, *69*, 198.
521. Friess, B.; Cazes, B.; Gore, J. *Tetrahedron Lett.* **1988**, *29*, 4089.
522. Gauthier, V.; Cazes, B.; Gore, J. *Tetrahedron Lett.* **1991**, *32*, 915.
523. Kagechika, K.; Shibasaki, M. *J. Org. Chem.* **1991**, *56*, 4093.
524. Cacchi, S.; Morera, E.; Ortar, G. *Synthesis* **1986**, 320.
525. Arcadi, A.; Cacchi, S.; Delmastro, M.; Marinella, F. *Synlett* **1991**, 409.
526. Castedo, L.; Mouriño, A.; Sarandeses, L. A. *Tetrahedron Lett.* **1986**, *27*, 1523.
527. Castedo, L.; Mascareñas, J. L.; Mouriño, A.; Sarandeses, L. A. *Tetrahedron Lett.* **1988**, *29*, 1203.
528. Enas, J. D.; Shen, G.-Y.; Okamura, W. H. *J. Am. Chem. Soc.* **1991**, *113*, 3873.
529. Nakatani, K.; Arai, K.; Yamada, K.; Terashima, S. *Tetrahedron Lett.* **1991**, *32*, 3405.
530. Suffert, J.; Brückner, R. *Tetrahedron Lett.* **1991**, *32*, 1453.
531. Brückner, R.; Scheuplein, S. W.; Suffert, J. *Tetrahedron Lett.* **1991**, *32*, 1449.
532. Orsini, F.; Pelizzoni *Synth. Commun.* **1987**, *17*, 1389.
533. Orsini, F.; Pelizzoni, F.; Vallarino, L. M. *J. Organomet. Chem.* **1989**, *367*, 375.
534. Jackson, R. F. W.; Wythes, M. J.; Wood, A. *Tetrahedron Lett.* **1989**, *30*, 5941.
535. Oh-e, T.; Miyaura, N.; Suzuki, A. *Synlett* **1990**, 221.
536. Saulnier, M. G.; Kadow, J. F.; Tun, M. M.; Langley, D. R.; Vyas, D. M. *J. Am. Chem. Soc.* **1989**, *111*, 8320.
537. Hatanaka, Y.; Ebina, Y.; Hiyama, T. *J. Am. Chem. Soc.* **1991**, *113*, 7075.
538. Ciattini, P. G.; Morera, E.; Ortar, G. *Tetrahedron Lett.* **1990**, *31*, 1889.
539. Hatanaka, Y.; Hiyama, T. *Tetrahedron Lett.* **1990**, *31*, 2719.
540. McLaughlin, M. L.; McKinney, J. A.; Paquette, L. A. *Tetrahedron Lett.* **1986**, *27*, 5595.
541. Crisp, G. T.; Flynn, B. L. *Tetrahedron Lett.* **1990**, *31*, 1347.
542. Haiza, M.; Lee, J.; Snyder, J. K. *J. Org. Chem.* **1990**, *55*, 5008.
543. Corey, E. J.; Houpis, I. N. *J. Am. Chem. Soc.* **1990**, *112*, 8997.
544. Piers, E.; Friesen, R. W.; Keay, B. A. *J. Chem. Soc., Chem. Commun.* **1985**, 809.
545. Piers, E.; Friesen, R. W.; Keay, B. A. *Tetrahedron* **1991**, *47*, 4555.
546. Stille, J. K.; Tanaka, M. *J. Am. Chem. Soc.* **1987**, *109*, 3785.
547. Piers, E.; Friesen, R. W. *J. Org. Chem.* **1986**, *51*, 3405.

548. Piers, E.; Friesen, R. W. *J. Chem. Soc., Chem. Commun.* **1988**, 125.
549. Piers, E.; Llinas-Brunet, M. *J. Org. Chem.* **1989**, *54*, 1483.
550. Stork, G.; Isaacs, R. C. A. *J. Am. Chem. Soc.* **1990**, *112*, 7399.
551. McCague, R. *Tetrahedron Lett.* **1987**, *28*, 701.
552. Keenan, R. M.; Kruse, L. I. *Synth. Commun.* **1989**, *19*, 793.
553. Arcadi, A.; Burini, A.; Cacchi, S.; Delmastro, M.; Marinelli, F.; Pietroni, B. *Synlett* **1990**, 47.
554. Farina, V.; Roth, G. P. *Tetrahedron Lett.* **1991**, *32*, 4243.
555. Wattanasin, S. *Synth. Commun.* **1988**, *18*, 1919.
556. Rano, T. A.; Greenlee, M. L.; DiNinno, F. P. *Tetrahedron Lett.* **1990**, *31*, 2853.
557. Holt, D. A.; Erb, J. M. *Tetrahedron Lett.* **1989**, 30, 5393.
558. Kowalski, M. H.; Hinkle, R. J.; Stang, P. J. *J. Org. Chem.* **1989**, *54*, 2783.
559. Hinkle, R. J.; Stang, P. J.; Kowalski, M. H. *J. Org. Chem.* **1990**, *55*, 5033.
560. Wulff, W. D.; Peterson, G. A.; Bauta, W. E.; Chan, K.-S.; Faron, K. L.; Gilbertson, S. R.; Kaesler, R. W.; Yang, D. C.; Murray, C. K. *J. Org. Chem.* **1986**, *51*, 277.
561. Paquette, L. A.; Liang, S.; Galatsis, P. *Synlett* **1990**, 663.
562. Jarowicki, K.; Kocieński, P.; Marczak, S.; Willson, T. *Tetrahedron Lett.* **1990**, *31*, 3433.
563. Barber, C.; Jarowicki, K.; Kocienski, P. *Synlett* **1991**, 197.
564. Roth, G. P.; Sapino, C. *Tetrahedron Lett.* **1991**, *32*, 4073.
565. Hettrick, C. M.; Kling, J. K.; Scott, W. J. *J. Org. Chem.* **1991**, *56*, 1489.
566. Paley, R. S.; Snow, S. R. *Tetrahedron Lett.* **1990**, *31*, 5853.
567. Charbonnier, F.; Moyano, A.; Greene, A. E. *J. Org. Chem.* **1987**, *52*, 2303.
568. Sato, M.; Takai, K.; Oshima, K.; Nozaki, H. *Tetrahedron Lett.* **1981**, *22*, 1609.
569. Takai, K.; Sato, M.; Oshima, K.; Nozaki, H. *Bull. Chem. Soc. Jpn.* **1984**, *57*, 108.
570. Takai, K.; Oshima, K.; Nozaki, H. *Tetrahedron Lett.* **1980**, *21*, 2531.
571. Hashimoto, S.; Ban, M.; Yanagiya, Y.; Sakata, S.; Ikegami, S. *Tetrahedron Lett.* **1991**, *32*, 4027.
572. Fugami, K.; Oshima, K.; Utimoto, K. *Chem. Lett.* **1987**, 2203.
573. Hayashi, T.; Fujiwa, T.; Okamoto, Y.; Katsuro, Y.; Kumada, M. *Synthesis* **1981**, 1001.
574. Okuda, Y.; Sato, M.; Oshima, K.; Nozaki, H. *Tetrahedron Lett.* **1983**, *24*, 2015.
575. Rest, A. J.; Rosevear, D. T.; Stone, F. G. A. *J. Chem. Soc. A* **1967**, 66.
576. Wada, M.; Koyama, Y. *J. Organomet. Chem.* **1980**, *201*, 477.
577. Negishi, E.; Takahashi, T.; Akiyoshi, K. *J. Chem. Soc., Chem. Commun.* **1986**, 1338.
578. Fotiadu, F.; Cros, P.; Faure, B.; Buono, G. *Tetrahedron Lett.* **1990**, *31*, 77.
579. Bumagin, N. A.; Andryukhova, N. L.; Beletskaya, I. L. *Proc. Acad. Sci. USSR, Chem. Sec.* **1988**, *297*, 524.
580. Bumagin, N. A.; Andryukhova, N. P.; Beletskaya, I. P. *Metalloorg. Khim.* **1989**, *2*, 605; *Chem. Abstr.* **1990**, *112*, 55087v.
581. Bumagin, N. A.; Andryukhova, N. P.; Beletskaya, I. P. *Metalloorg. Khim.* **1989**, *2*, 893; *Chem. Abstr.* **1990**, *112*, 197696s.
582. Fauvarque, J.-F.; Jutand, A. *Bull. Soc. Chim. Fr.* **1976**, 765.
583. Murahashi, S.-I.; Tamba, Y.; Yamamura, M.; Yoshimura, N. *J. Org. Chem.* **1978**, *43*, 4099.
584. Nunomoto, S.; Kawakami, Y.; Yamashita, Y. *Bull. Chem. Soc. Jpn.* **1981**, *54*, 2831.
585. Hutzinger, M. W.; Oehlschlager, A. C. *J. Org. Chem.* **1991**, *56*, 2918.
586. Hayasi, Y.; Riediker, M.; Temple, J. S.; Schwartz, J. *Tetrahedron Lett.* **1981**, *22*, 2629.
587. Temple, J. S.; Riediker, M.; Schwartz, J. *J. Am. Chem. Soc.* **1982**, *104*, 1310.
588. Temple, J. S.; Schwartz, J. *J. Am. Chem. Soc.* **1980**, *102*, 7381.
589. Riediker, M.; Schwartz, J. *Tetrahedron Lett.* **1981**, 22, 4655.
590. McMurry, J. E.; Matz, J. R. *Tetrahedron Lett.* **1982**, *23*, 2723.
591. McMurry, J.; Matz, J. R.; Kees, K. L.; Bock, P. A. *Tetrahedron Lett.* **1982**, *23*, 1777.
592. Schwartz, J.; Loots, M. J.; Kosugi, H. *J. Am. Chem. Soc.* **1980**, *102*, 1333.
593. Jabri, N.; Alexakis, A.; Normant, J. F. *Tetrahedron Lett.* **1981**, *22*, 3851.
594. Jabri, N.; Alexakis, A.; Normant, J. F. *Tetrahedron Lett.* **1983**, *24*, 5081.

595. Larock, R. C.; Leach, D. R.; Bjorge, S. M. *J. Org. Chem.* **1986**, *51*, 5221.
596. Larock, R. C.; Leach, D. R.; Bjorge, S. M. *Tetrahedron Lett.* **1982**, *23*, 715.
597. Dunkerton, L. V.; Serino, A. J. *J. Org. Chem.* **1982**, *47*, 2812.
598. van der Louw, J.; van der Baan, J. L.; Bickelhaupt, F.; Klumpp, G. W. *Tetrahedron Lett.* **1987**, *28*, 2889.
599. van der Louw, J.; Komen, C. M. D.; Knol, A.; de Kanter, F. J. J.; van der Baan, J. L.; Bickelhaupt, F.; Klumpp, G. W. *Tetrahedron Lett.* **1989**, *30*, 4453.
600. Ruitenberg, K.; Kleijn, H.; Westmijze, H.; Meijer, J.; Vermeer, P. *Rec. Trav. Chim. Pays-Bas* **1982**, *101*, 405.
601. Keinan, E.; Bosch, E. *J. Org. Chem.* **1986**, *51*, 4006.
602. Kleijn, H.; Meijer, J.; Overbeek, G. C.; Vermeer, P. *Rec. Trav. Chim. Pays-Bas* **1982**, *101*, 97.
603. Kleijn, H.; Westmijze, H.; Meijer, J.; Vermeer, P. *Rec. Trav. Chim. Pays-Bas* **1983**, *102*, 378.
604. Bumagin, N. A.; Andryukhova, N. P.; Beletskaya, I. P. *Dokl. Akad. Nauk SSSR* **1989**, *307*, 375; *Chem. Abstr.* **1990**, *112*, 138656f.
605. Campbell, Jr., J. B.; Firor, J. W.; Davenport, T. W. *Synth. Commun.* **1989**, *19*, 2265.
606. Gillet, J. P.; Sauvêtre, R.; Normant, J. F. *Synthesis* **1986**, 538.
607. Heinze, P. L.; Burton, D. J. *J. Org. Chem.* **1988**, *53*, 2714.
608. Tellier, F.; Sauvêtre, R.; Normant, J. F.; Dromzee, Y.; Jeannin, Y. *J. Organomet. Chem.* **1987**, *331*, 281.
609. Heinze, P. L.; Burton, D. J. *J. Fluorine Chem.* **1986**, *31*, 115.
610. Tius, M. A.; Gu, X.; Gomez-Galeno, J. *J. Am. Chem. Soc.* **1990**, *112*, 8188.
611. Tius, M. A.; Gomez-Galeno, J.; Gu, X.; Zaidi, J. H. *J. Am. Chem. Soc.* **1991**, *113*, 5775.
612. Negishi, E.; Bagheri, V.; Chatterjee, S.; Luo, F.-T.; Miller, J. A.; Stoll, A. T. *Tetrahedron Lett.* **1983**, *24*, 5181.
613. Kaliya, O. L.; Temkin, O. N.; Flid, R. M.; Volkova, L. G. *Zh. Neorg. Khim.* **1970**, *15*, 2562; *Chem. Abstr.* **1970**, *73*, 109884z.
614. Larock, R. C.; Mitchell, M. A. *J. Am. Chem. Soc.* **1976**, *98*, 6718.
615. Larock, R. C.; Mitchell, M. A. *J. Am. Chem. Soc.* **1978**, *100*, 180.
616. Larock, R. C.; Leuck, D. J.; Harrison, L. W. *Tetrahedron Lett.* **1987**, *28*, 4977.
617. Larock, R. C.; Harrison, L. W.; Hsu, M. H. *J. Org. Chem.* **1984**, *49*, 3662.
618. Kim, J. I.; Lee, J. T. *Bull. Korean Chem. Soc.* **1986**, *7*, 142; *Chem. Abstr.* **1987**, *106*, 175732z.
619. Kim, J. I.; Lee, J. T.; Choi, C. K. *Bull. Korean Chem. Soc.* **1986**, *7*, 235; *Chem. Abstr.* **1987**, *106*, 83944s.
620. Dyke, S. F.; McCartney, M. J. *Tetrahedron* **1981**, *37*, 431.
621. Bigge, C. F.; Kalaritis, P.; Mertes, M. P. *Tetrahedron Lett.* **1979**, 1653.
622. Jones, A. S.; Verhelst, G.; Walker, R. T. *Tetrahedron Lett.* **1979**, 4415.
623. Bigge, C. F.; Lizotte, K. E.; Panek, J. S.; Mertes, M. P. *J. Carbohyd., Nucleosides, Nucleotides* **1981**, *8*, 295.
624. Bergstrom, D. E.; Inoue, H.; Reddy, P. A. *J. Org. Chem.* **1982**, *47*, 2174.
625. Dreyer, G. B.; Dervan, P. B. *Proc. Natl. Acad. Sci. USA* **1985**, *82*, 968.
626. Langer, P. R.; Waldrop, A. A.; Ward, D. C. *Proc. Natl. Acad. Sci. USA* **1981**, *78*, 6633.
627. Bergstrom, D. E.; Ruth, J. L. *J. Am. Chem. Soc.* **1976**, *98*, 1587.
628. Bigge, C. F.; Kalaritis, P.; Deck, J. R.; Mertes, M. P. *J. Am. Chem. Soc.* **1980**, *102*, 2033.
629. Larock, R. C.; Bernhardt, J. C.; Driggs, R. J. *J. Organomet. Chem.* **1978**, *156*, 45.
630. Bergstrom, D. E.; Ruth, J. L.; Warwick, P. *J. Org. Chem.* **1981**, *46*, 1432.
631. Ruth, J. L.; Bergstrom, D. E. *J. Org. Chem.* **1978**, *43*, 2870.
632. Larock, R. C.; Ilkka, S. J. *Tetrahedron Lett.* **1986**, *27*, 2211.
633. Larock, R. C.; Stolz-Dunn, S. K. *Tetrahedron Lett.* **1988**, *29*, 5069.
634. Larock, R. C.; Ding, S. *Tetrahedron Lett.* **1989**, *30*, 1897.
635. Larock, R. C.; Takagi, K.; Hershberger, S. S.; Mitchell, M. A. *Tetrahedron Lett.* **1981**, *22*, 5231.
636. Larock, R. C.; Takagi, K. *J. Org. Chem.* **1988**, *53*, 4329.

637. Kwok, D.-I.; Farr, R. N.; Daves, Jr., G. D. *J. Org. Chem.* **1991**, *56*, 3711.
638. Arai, I.; Daves, Jr., G. D. *J. Org. Chem.* **1978**, *43*, 4110.
639. Arai, I.; Daves, Jr., G. D. *J. Am. Chem. Soc.* **1978**, *100*, 287.
640. Arai, I.; Daves, Jr., G. D. *J. Am. Chem. Soc.* **1981**, *103*, 7683.
641. Arai, I.; Lee, T. D.; Hanna, R.; Daves, Jr., G. D. *Organometallics* **1982**, *1*, 742.
642. Lee, T. D.; Daves, Jr., G. D. *J. Org. Chem.* **1983**, *48*, 399.
643. Hacksell, U.; Daves, Jr., G. D. *J. Org. Chem.* **1983**, *48*, 2870.
644. Hacksell, U.; Daves, Jr., G. D. *Organometallics* **1983**, *2*, 772.
645. Kalinoski, H. T.; Hacksell, U.; Barofsky, D. F.; Barofsky, E.; Daves, Jr., G. D. *J. Am. Chem. Soc.* **1985**, *107*, 6476.
646. Hacksell, U.; Kalinkoski, H. T.; Barofsky, D. F.; Daves, Jr., G. D. *Acta Chem. Scand.* **1985**, *B39*, 469.
647. Cheng, J. C.-Y.; Hacksell, U.; Daves, Jr., G. D. *J. Org. Chem.* **1986**, *51*, 3093.
648. Cheng, J. C.-Y.; Daves, Jr., G. D. *J. Org. Chem.* **1987**, *52*, 3083.
649. Daves, Jr., G. D. *Acc. Chem. Res.* **1990**, *23*, 201.
650. Farr, R. N.; Daves, Jr., G. D. *J. Carbohydrate Chem.* **1990**, *9*, 653.
651. Cheng, J. C.-Y.; Daves, Jr., G. D. *Organometallics* **1986**, *5*, 1753.
652. Larock, R. C.; Kondo, F.; Narayanan, K.; Sydnes, L. K.; Hsu, M.-F. H. *Tetrahedron Lett.* **1989**, *30*, 5737.
653. Larock, R. C.; Hershberger, S. S.; Takagi, K.; Mitchell, M. A. *J. Org. Chem.* **1986**, *51*, 2450.
654. Larock, R. C.; Takagi, K.; Burkhart, J. P.; Hershberger, S. S. *Tetrahedron* **1986**, *42*, 3759.
655. Larock, R. C.; Hsu, M. H.; Narayanan, K. *Tetrahedron* **1987**, *43*, 2891.
656. Vedejs, E.; Salomon, M. F. *J. Chem. Soc., Chem. Commun.* **1971**, 1582.
657. Vedejs, E.; Weeks, P. D. *J. Chem. Soc., Chem. Commun.* **1974**, 223.
658. Larock, R. C.; Song, H. *Synth. Commun.* **1989**, *19*, 1463.
659. Larock, R. C.; Bernhardt, J. C. *J. Org. Chem.* **1978**, *43*, 710.
660. Kim, J. I.; Lee, K. H. *Taehan Hwahakhoe Chi* **1989**, *33*, 657; *Chem. Abstr.* **1990**, *112*, 157781j.
661. Larock, R. C. *J. Org. Chem.* **1975**, *40*, 3237.
662. Brailovskii, S. M.; Kabalina, G. A.; Shestakova, V. S.; Temkin, O. N. *J. Org. Chem. USSR* **1977**, *13*, 1066.
663. Walkup, R. D.; Park, G. *Tetrahedron Lett.* **1987**, *28*, 1023.
664. Walkup, R. D.; Park, G. *J. Am. Chem. Soc.* **1990**, *112*, 1597.
665. Larock, R. C.; Harrison, L. W. *J. Am. Chem. Soc.* **1984**, *106*, 4218.
666. Larock, R. C.; Riefling, B. *Tetrahedron Lett.* **1976**, 4661.
667. Larock, R. C.; Riefling, B.; Fellows, C. A. *J. Org. Chem.* **1978**, *43*, 131.
668. Larock, R. C.; Liu, C.-L. *J. Org. Chem.* **1983**, *48*, 2151.
669. Döbler, C.; Kreuzfeld, H.-J.; Krause, H. *Z. Chem.* **1971**, *11*, 147.
670. Larock, R. C. *J. Org. Chem.* **1976**, *41*, 2241.
671. Vedejs, E.; Weeks, P. D. *Tetrahedron Lett.* **1974**, 3207.
672. Elbe, H.-L.; Köbrich, G. *Chem. Ber.* **1974**, *107*, 1654.
673. Larock, R. C.; Riefling, B. *J. Org. Chem.* **1978**, *43*, 1468.
674. Arai, I.; Hanna, R.; Daves, Jr., G. D. *J. Am. Chem. Soc.* **1981**, *103*, 7684.
675. Larock, R. C.; Oertle, K.; Beatty, K. M. *J. Am. Chem. Soc.* **1980**, *102*, 1966.
676. Bergstrom, D.; Beal, P.; Husain, A.; Lind, R.; Jenson, J. *J. Am. Chem. Soc.* **1989**, *111*, 374.
677. Bergstrom, D. E.; Beal, P.; Jenson, J.; Lin, X. *J. Org. Chem.* **1991**, *56*, 5598.
678. Yatagai, H.; Yamamoto, Y.; Maruyama, K. *J. Chem. Soc., Chem. Commun.* **1978**, 702.
679. Yatagai, H. *Bull. Chem. Soc. Jpn.* **1980**, *53*, 1670.
680. Yatagai, H.; Yamamoto, Y.; Maruyama, K.; Sonoda, A.; Murahashi, S.-I. *J. Chem. Soc., Chem. Commun.* **1977**, 852.
681. Kim, J. I.; Lee, J. T.; Yeo, K. D. *Bull. Korean Chem. Soc.* **1985**, *6*, 366; *Chem. Abstr.* **1987**, *106*, 49558j.

682. Miyaura, N.; Tanabe, Y.; Suginome, H.; Suzuki, A. J. Organomet. Chem. **1982,** *233,* C13.
683. Miyaura, N.; Suzuki, A. *Yuki Gosei Kagaku Kyokaishi* **1988,** *46,* 848; *Chem. Abstr.* **1989,** *111,* 153263y.
684. Miyaura, N.; Yano, T.; Suzuki, A. *Tetrahedron Lett.* **1980,** *21,* 2865.
685. Miyaura, N.; Suginome, H.; Suzuki, A. *Tetrahedron Lett.* **1984,** *25,* 761.
686. Sasaya, F.; Miyaura, N.; Suzuki, A. *Bull. Korean Chem. Soc.* **1987,** *8,* 329; *Chem. Abstr.* **1988,** *108,* 111748t.
687. Sharma, S.; Oehlschlager, A. C. *Tetrahedron Lett.* **1988,** *29,* 261.
688. Miyaura, N.; Maeda, K.; Suginome, H.; Suzuki, A. *J. Org. Chem.* **1982,** *47,* 2117.
689. Satoh, M.; Miyaura, N.; Suzuki, A. *Synthesis* **1987,** 373.
690. Hegedus, L. S.; Toro, J. L.; Miles, W. H.; Harrington, P. J. *J. Org. Chem.* **1987,** *52,* 3319.
691. Miyaura, N.; Suzuki, A. *J. Chem. Soc., Chem. Commun.* **1979,** 866.
692. Cullen, K. E.; Sharp, J. T. *J. Chem. Soc., Chem. Commun.* **1991,** 658.
693. Miyaura, N.; Suzuki, A. *Chem. Lett.* **1981,** 879.
694. Ohe, T.; Ohe, K.; Uemura, S.; Sugita, N. *J. Organomet. Chem.* **1988,** *344,* C5.
695. Miyaura, N.; Suzuki, A. *Main Group Met. Chem.* **1987,** *10,* 295.
696. Rao, V. V. R.; Kumar, C. V.; Devaprabhakara, D. *J. Organomet. Chem.* **1979,** *179,* C7.
697. Matsushita, H.; Negishi, E. *J. Am. Chem. Soc.* **1981,** *103,* 2882.
698. Negishi, E.; Matsushita, H. Org. Syn. **1984,** *62,* 31.
699. Negishi, E.; Chatterjee, S.; Matsushita, H. *Tetrahedron Lett.* **1981,** *22,* 3737 .
700. Matsushita, H.; Negishi, E. *J. Chem. Soc., Chem. Commun.* **1982,** 160.
701. Chatterjee, S.; Negishi, E. J. Organomet. Chem. **1985,** *285,* C1.
702. Chatterjee, S.; Negishi, E. *J. Org. Chem.* **1985,** *50,* 3406.
703. Chorev, M.; Klausner, Y. S. *J. Chem. Soc., Chem. Commun.* **1976,** 596.
704. Al-Hassan, M. I. *Synth. Commun.* **1987,** *17,* 1413.
705. Al-Hassan, M. I. *Synthesis* **1987,** 816.
706. Chen, Q.; He, Y. *Acta Chim. Sin. (Engl. Ed.)* **1989,** 94; *Chem. Abstr.* **1990,** *112,* 138658h.
707. Wakamatsu, K.; Okuda, Y.; Oshima, K.; Nozaki, H. *Bull. Chem. Soc. Jpn.* **1985,** *58,* 2425.
708. Bumagin, N. A.; Ponomaryov, A. B.; Beletskaya, I. P. *Tetrahedron Lett.* **1985,** *26,* 4819.
709. Yamamoto, K.; Shinohara, K.; Ohuchi, T.; Kumada, M. *Tetrahedron Lett.* **1974,** 1153.
710. Ohta, T.; Hosokawa, T.; Murahashi, S.-I.; Miki, K.; Kasai, N. *Organometallics* **1985,** *4,* 2080.
711. Weber, W. P.; Felix, R. A.; Willard, A. K.; Koenig, K. E. *Tetrahedron Lett.* **1971,** 4701.
712. Ishikawa, M.; Fuchikami, T.; Kumada, M. *J. Chem. Soc., Chem. Commun.* **1977,** 352.
713. Ishikawa, M.; Sugisawa, H.; Kumada, M.; Higuchi, T.; Matsui, K.; Hirotsu, K. *Organometallics* **1982,** *1,* 1473.
714. Yoshida, J.; Tamao, K.; Takahashi, M.; Kumada, M. *Tetrahedron Lett.* **1978,** 2161.
715. Karabelas, K.; Hallberg, A. *J. Org. Chem.* **1989,** *54,* 1773.
716. Hatanaka, Y.; Fukushima, S.; Hiyama, T. *Heterocycles* **1990,** *30,* 303.
717. Somei, M.; Ohta, T.; Shinoda, J.; Somada, Y. *Heterocycles* **1989,** *29,* 653.
718. Tamao, K.; Kakui, T.; Kumada, M. *Tetrahedron Lett.* **1979,** 619.
719. Masters, J. J.; Hegedus, L. S.; Tamariz, J. *J. Org. Chem.* **1991,** *56,* 5666.
720. Matsubara, S.; Mitani, M.; Utimoto, K. *Tetrahedron Lett.* **1987,** *28,* 5857.
721. Sheffy, F. K.; Stille, J. K. *J. Am. Chem. Soc.* **1983,** *105,* 7173.
722. Sheffy, F. K.; Godschalz, J. P.; Stille, J. K. *J. Am. Chem. Soc.* **1984,** *106,* 4833.
723. Mitchell, T. N.; Wickenkamp, R.; Amamria, A.; Dicke, R.; Schneider, U. *J. Org. Chem.* **1987,** *52,* 4868.
724. Mitchell, T. N.; Kwetkat, K.; Rutschow, D.; Schneider, U. *Tetrahedron* **1989,** *45,* 969.
725. Farina, V.; Baker, S. R.; Benigni, D. A.; Sapino, Jr., C. *Tetrahedron Lett.* **1988,** *29,* 5739.
726. Miftakhov, M. S.; Lesnikova, E. T.; Tolstikov, G. A. *J. Org. Chem. USSR* **1986,** *22,* 2007.
727. Miftakhov, M. S.; Danilova, N. A.; Vel'der, Y. L.; Shitikova, O. V.; Tolstikov, G. A. *J. Org. Chem. USSR* **1988,** *24,* 1674.

728. Tolstikov, G. A.; Miftakhov, M. S.; Danilova, N. A.; Vel'der, Y. L.; Spirikhin, L. V. *Synthesis* **1989**, 625.
729. Beletskaya, I. P.; Kasatkin, A. N.; Lebedev, S. A.; Bumagin, N. A. *Bull. Acad. Sci. USSR* **1981**, 1994.
730. Bumagin, N. A.; Kasatkin, A. N.; Beletskaya, I. P. *Proc. Acad. Sci. USSR, Chem. Sec.* **1983**, *266,* 345.
731. Del Valle, L.; Stille, J. K.; Hegedus, L. S. *J. Org. Chem.* **1990**, *55*, 3019.
732. Kosugi, M.; Ohashi, K.; Akuzawa, K.; Kawazoe, T.; Sano, H.; Migita, T. *Chem. Lett.* **1987**, 1237.
733. Kanemoto, S.; Matsubara, S.; Oshima, K.; Utimoto, K.; Nozaki, H. *Chem. Lett.* **1987**, 5.
734. Echavarren, A. M.; Tueting, D. R.; Stille, J. K. *J. Am. Chem. Soc.* **1988**, *110,* 4039.
735. Tueting, D. R.; Echavarren, A. M.; Stille, J. K. *Tetrahedron* **1989**, *45*, 979.
736. Tolstikov, G. A.; Miftakhov, M. S.; Danilova, N. A.; Vel'der, Y. L.; Spirikhin, L. V. *Synthesis* **1989**, 633.
737. Tolstikov, G. A.; Miftakhov, M. S.; Danilova, N. A.; Vel'der, Y. A. *J. Org. Chem. USSR* **1988**, *24*, 396.
738. Danilova, N. A.; Vel'der, Y. L.; Miftakhov, M. S.; Spirikhin, L. V.; Khalilov, L. M.; Baikova, I. P.; Tolstikov, G. A. *Metalloorg Khim.* **1989**, *2*, 1079; *Chem. Abstr.* **1990**, *113*, 5312q.
739. Burns, B.; Grigg, R.; Ratananukul, P.; Sridharan, V.; Stevenson, P.; Sukirthalingam, S.; Worakun, T. *Tetrahedron Lett.* **1988**, *29,* 5565.
740. Milstein, D.; Stille, J. K. *J. Am. Chem. Soc.* **1979**, *101*, 4992.
741. Bumagin, N. A.; Bumagina, I. G.; Kashin, A. N.; Beletskaya, I. P. *Proc. Acad. Sci. USSR, Chem. Sec.* **1981**, *261,* 532.
742. Bumagin, N. A.; Bumagina, I. G.; Beletskaya, I. P. *Proc. Acad. Sci. USSR, Chem. Sec.* **1984**, *274*, 39.
743. Wright, M. E.; Lowe-Ma, C. K. *Organometallics* **1990**, *9*, 347.
744. McKean, D. R.; Parrinello, G.; Renaldo, A. F.; Stille, J. K. *J. Org. Chem.* **1987**, *52,* 422.
745. Krolski, M. E.; Renaldo, A. F.; Rudisill, D. E.; Stille, J. K. *J. Org. Chem.* **1988**, *53,* 1170.
746. Scott, W. J. *J. Chem. Soc., Chem. Commun.* **1987**, 1755.
747. Haack, R. A.; Penning, T. D.; Djurić, S. W.; Dziuba, J. A. *Tetrahedron Lett.* **1988**, *29,* 2783.
748. Zimmermann, E. K.; Stille, J. K. *Macromolecules* **1985**, *18,* 321.
749. Galarini, R.; Musco, A.; Pontellini, R.; Bolognesi, A.; Destri, S.; Catellani, M.; Mascherpa, M.; Zhuo, G. *J. Chem. Soc., Chem. Commun.* **1991**, 364.
750. Kang, K.-T.; Kim, S. S.; Lee, J. C. *Tetrahedron Lett.* **1991**, *32,* 4341.
751. Corriu, R. J. P.; Bolin, G.; Moreau, J. J. E. *Tetrahedron Lett.* **1991**, *32*, 4121.
752. Parrain, J.-L.; Duchene, A.; Quintard, J.-P. *Tetrahedron Lett.* **1990**, *31,* 1857 .
753. Sakamoto, T.; Kondo, Y.; Yasuhara, A.; Yamanaka, H. *Heterocycles* **1990**, *31,* 219.
754. Schreiber, S. L.; Desmaele, D.; Porco, Jr., J. A. *Tetrahedron Lett.* **1988**, *29,* 6689.
755. Schreiber, S. L.; Porco, Jr., J. A. *J. Org. Chem.* **1989**, *54,* 4721.
756. Dubois, E.; Beau, J.-M. *Tetrahedron Lett.* **1990**, *31*, 5165.
757. Friesen, R. W.; Sturino, C. F. *J. Org. Chem.* **1990**, *55,* 2572.
758. Friesen, R. W.; Sturino, C. F. *J. Org. Chem.* **1990**, *55*, 5808.
759. Lee, E.; Hur, C. U.; Jeong, Y. C.; Rhee, Y. H.; Chang, M. H. *J. Chem. Soc., Chem. Commun.* **1991**, 1314.
760. Bradley, J. C.; Durst, T. *J. Org. Chem.* **1991**, *56,* 5459.
761. Takle, A.; Kocieński, P. *Tetrahedron* **1990**, *46,* 4503.
762. Takle, A.; Kocieński, P. *Tetrahedron Lett.* **1989**, *30*, 1675.
763. Laborde, E.; Lesheski, L. E.; Kiely, J. S. *Tetrahedron Lett.* **1990**, *31*, 1837.
764. Solberg, J.; Undheim, K. *Acta Chem. Scand. B* **1987**, *41*, 712.
765. Kondo, Y.; Watanabe, R.; Sakamoto, T.; Yamanaka, H. *Chem. Pharm. Bull.* **1989**, *37,* 2814.
766. Nair, V.; Turner, G. A.; Chamberlain, S. D. *J. Am. Chem. Soc.* **1987**, *109,* 7223.
767. Nair, V.; Turner, G. A.; Buenger, G. S.; Chamberlain, S. D. *J. Org. Chem.* **1988**, *53,* 3051.

768. Moriarty, R. M.; Epa, W. R.; Awasthi, A. K. *Tetrahedron Lett.* **1990**, *31*, 5877.
769. Nair, V.; Lyons, A. G. *Tetrahedron* **1989**, *45*, 3653.
770. Echavarren, A. M.; Stille, J. K. *J. Am. Chem. Soc.* **1987**, *109*, 5478.
771. Crisp, G. T.; Papadopoulos, S. *Aust. J. Chem.* **1988**, *41*, 1711.
772. Crisp, G. T.; Papadopoulos, S. *Aust. J. Chem.* **1989**, *42*, 279.
773. Martorell, G.; García-Raso, A.; Saá, J. M. *Tetrahedron Lett.* **1990**, *31*, 2357.
774. Tamayo, N.; Echavarren, A. M.; Paredes, M. C.; Fariña, F.; Noheda, P. *Tetrahedron Lett.* **1990**, *31*, 5189.
775. Tilley, J. W.; Sarabu, R.; Wagner, R.; Mulkerins, K. *J. Org. Chem.* **1990**, *55*, 906.
776. Roth, G. P.; Fuller, C. E. *J. Org. Chem.* **1991**, *56*, 3493.
777. Kikukawa, K.; Kono, K.; Wada, F.; Matsuda, T. *J. Org. Chem.* **1983**, *48*, 1333.
778. Kikukawa, K.; Umekawa, H.; Matsuda, T. *J. Organomet. Chem.* **1986**, *311*, C44.
779. Milstein, D.; Stille, J. K. *J. Org. Chem.* **1979**, *44*, 1613.
780. Labadie, J. W.; Tueting, D.; Stille, J. K. *J. Org. Chem.* **1983**, *48*, 4634.
781. Renaldo, A. F. *Synth. Commun.* **1987**, *17*, 1823.
782. Bumagin, N. A.; Bumagina, I. G.; Kashin, A. N.; Beletskaya, I. P. *J. Org. Chem. USSR* **1982**, *18*, 977.
783. Kosugi, M.; Shimizu, Y.; Migita, T. *Chem. Lett.* **1977**, 1423.
784. Labadie, J. W.; Stille, J. K. *J. Am. Chem. Soc.* **1983**, *105*, 6129.
785. Chenard, B. L.; Van Zyl, C. M.; Sanderson, D. R. *Tetrahedron Lett.* **1986**, *27*, 2801.
786. Soderquist, J. A.; Leong, W. W.-H. *Tetrahedron Lett.* **1983**, *24*, 2361.
787. Blanchot, V.; Fétizon, M.; Hanna, I. *Synthesis* **1990**, 755.
788. Colson, P.-J.; Franck-Neumann, M.; Sedrati, M. *Tetrahedron Lett.* **1989**, *30*, 2393.
789. Comins, D. L.; Mantlo, N. B. *Tetrahedron Lett.* **1987**, *28*, 759.
790. Duchêne, A.; Quintard, J.-P. *Synth. Commun.* **1985**, *15*, 873.
791. Labadie, J. W.; Stille, J. K. *Tetrahedron Lett.* **1983**, *24*, 4283.
792. Baldwin, J. E.; Adlington, R. M.; Ramcharitar, S. H. *J. Chem. Soc., Chem. Commun.* **1991**, 940.
793. Balas, L.; Jousseaume, B.; Shin, H.; Verlhac, J.-B.; Wallian, F. *Organometallics* **1991**, *10*, 366.
794. Labadie, S. S. *J. Org. Chem.* **1989**, *54*, 2496.
795. Virgili, M.; Belloch, J.; Moyano, A.; Pericas, M. A.; Riera, A. *Tetrahedron Lett.* **1991**, *32*, 4583.
796. Fabre, J.-L.; Julia, M. *Tetrahedron Lett.* **1983**, *24*, 4311.
797. Cuvigny, T.; Fabre, J. L.; du Penhoat, C. H.; Julia, M. *Tetrahedron Lett.* **1983**, *24*, 4319.
798. Cuvigny, T.; du Penhoat, C. H.; Julia, M. *Tetrahedron* **1987**, *43*, 859.
799. Ohe, K.; Takahashi, H.; Uemura, S.; Sugita, N. *J. Organomet. Chem.* **1987**, *326*, 35.
800. Uemura, S.; Ohe, K.; Kim, J.-R.; Kudo, K.; Sugita, N. *J. Chem. Soc., Chem. Commun.* **1985**, 271.
801. Ohe, K.; Takahashi, H.; Uemura, S.; Sugita, N. *J. Org. Chem.* **1987**, *52*, 4859.
802. Kasahara, A.; Izumi, T.; Kyuda, K. *Chem. Ind.* **1988**, 467.
803. Kasahara, A.; Izumi, T.; Kudou, N. *Synthesis* **1988**, 704.
804. Kasahara, A.; Izumi, T.; Ogihara, T. *Chem. Ind.* **1988**, 728.
805. Hoffman, N. E.; Kanakkanatt, A. T.; Schneider, R. F. *J. Org. Chem.* **1962**, *27*, 2687.
806. Volger, H. C. *Rec. Trav. Chim. Pays-Bas* **1967**, *86*, 677.
807. Itahara, T. *Chem. Lett.* **1986**, 239.
808. Hirota, K.; Isobe, Y.; Kitade, Y.; Maki, Y. *Synthesis* **1987**, 495.
809. Itahara, T. *Chem. Lett.* **1983**, 127.

RUTHENIUM-CATALYZED OXIDATIVE TRANSFORMATIONS OF ALCOHOLS

Shun-Ichi Murahashi and Takeshi Naota

I. INTRODUCTION

Metal-catalyzed oxidation of alcohols has important synthetic and biological aspects.[1] During the course of studies on the simulation of enzymatic functions of

Advances in Metal-Organic Chemistry
Volume 3, pages 225–254.

ISBN: 1-55938-406-9

alcohol dehydrogenases[2] with metal catalysts, we found two types of novel catalytic oxidations of alcohols promoted by ruthenium complexes: (1) oxidative transformations of alcohols and aldehydes; and (2) oxidation of alcohols with *t*-butyl hydroperoxide.

II. OXIDATIVE CONDENSATION OF ALCOHOLS AND ALDEHYDES

Dehydrogenation process of alcohols by metal catalysts has importance in biological and industrial viewpoints. We have found that alcohols are activated efficiently with low valent ruthenium complexes to give the carbonyl dihydridoruthenium intermediates. Capture of the intermediates with nucleophiles provides novel

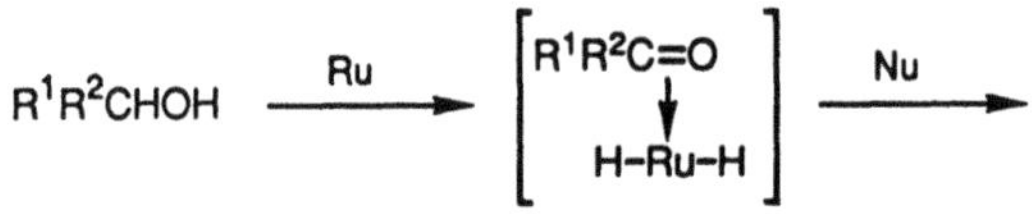

catalytic oxidative condensations of alcohols. Thus, primary alcohols undergo oxidative condensation upon treatment with dihydridotetrakis(triphenylphosphine)ruthenium ($RuH_2(PPh_3)_4$) catalyst to give esters along with evolution of molecular hydrogen. Similar treatment of diols affords the corresponding lactones. The ruthenium-catalyzed reaction of alcohols with amines gives unsymmetrical secondary and tertiary amines. Aminoalcohols can be converted into cyclic amines, while similar treatment of aminoalcohols in the presence of a hydrogen acceptor gives the corresponding lactams. Application of the present reaction provides novel catalytic reactions, which involve the oxidative condensation of aldehydes with alcohols to give esters and that with water to give acids or esters.

Details of these reactions are described with respect to scope, mechanism, and synthetic applications.

A. Catalytic Transformation of Alcohols to Esters and Lactones

The reported oxidative dehydrogenation of alcohols with homogeneous catalysts are limited to few reactions which involve formation of aldehydes along with evolution of molecular hydrogen,[3] alkylation of amines with alcohols,[4] and condensation of phenylacetonitriles with alcohols,[5] although hydrogen-transfer reaction of alcohols are well documented.[6]

During the course of the study on the simulation of the function of alcohol dehydrogenase[2] with metal complex catalysts, an efficient method for the oxidative condensations of alcohols to give esters and lactones catalyzed by $RuH_2(PPh_3)_4$ has been found as shown in Eq. 1.[7]

$$2\ RCH_2OH \xrightarrow{RuH_2(PPh_3)_4} RCO_2CH_2R + 2\ H_2 \qquad (1)$$

Preparation of esters[8] and lactones[9] by means of oxidative condensation of alcohols has been performed by using various stoichiometric oxidants. Heterogeneous metal-catalyzed transformation of alcohols to esters and lactones has been reported; however, these reactions require extremely high reaction temperatures and proceed nonselectively.[10] Homogeneous catalysts, such as $Pd(OAc)_2$ and $Ru_3(CO)_{12}$ catalyze the oxidative condensation of alcohols in the presence of a stoichiometric amount of bromobenzene[11] and diphenylacetylene,[12] respectively. Our method has advantage over the previous methods with respect to high efficiency, freedom from the need of stoichiometric amount of oxidants, and facile isolation of desired products. Recently, Shvo demonstrated that when (η^4-tetracyclone)$RuH_2(CO)_2$ is used as a catalyst, the oxidative condensation of alcohols proceeds without hydrogen acceptors.[13]

Treatment of a variety of primary alcohols with a suitable group 8 transition metal catalyst gives the corresponding esters along with evolution of molecular hydrogen. The catalytic activity of various metal complexes was examined with respect to the reaction of 1-hexanol. The products detected during the reaction are hexyl hexanoate, 1,1-bis(hexyloxy)hexane (**1**), and dihexyl ether (**2**). Ruthenium dihydride complexes, such as $RuH_2(PPh_3)_4$ and $RuH_2(CO)(PPh_3)_3$ give the best results. Chloride complexes, such as $RuCl_3{\bullet}H_2O$, $RuCl_2(PPh_3)_3$, $PdCl_2$, $RhCl_3{\bullet}H_2O$, and $RhCl(PPh_3)_3$ produce either **1** or **2** predominantly. Using a nonpolar solvent such as hexane, toluene, and mesitylene or in the absence of a solvent, satisfactory results are obtained. A reaction temperature higher than 140 °C is required.

Representative results of the oxidative condensation of primary alcohols with $RuH_2(PPh_3)_4$ catalyst are shown in Table 1. Alkyl and benzyl alcohols are readily converted into the corresponding esters with evolution of molecular hydrogen. The reaction proceeds efficiently at 180 °C either in mesitylene (method A) or in toluene in a sealed tube (method B). Generally, higher conversion of alcohols is obtained by method A rather than by method B, because the open system promotes the dissociation of molecular hydrogen. In method A, the reaction of the alcohols bearing a substituent at the β position, such as benzyl alcohol and 2-methyl-1-butanol, gives a considerable amount of the hydrocarbons (20–30%), which are derived from decarbonylation of the corresponding intermediate aldehydes. The closed system (method B) retards the dissociation of carbon monoxide. Aminoalcohols also undergo the esterification efficiently. Since the dehydrogenation step in the present esterification is reversible,[14] the reaction can be enhanced by

Table 1. Ruthenium-Catalyzed Reaction of Primary Alcohols

Alcohol	*Cond.*[a]	*Conv.(%)*	*Ester*	*Yield*[b]*(%)*
C_4H_9OH	A	96	$C_3H_7CO_2C_4H_9$	97
$C_6H_{13}OH$	A	95	$C_5H_{11}CO_2C_6H_{13}$	95
$C_6H_{13}OH$	B	75	$C_5H_{11}CO_2C_6H_{13}$	98
$C_8H_{17}OH$	A	100	$C_7H_{15}CO_2C_8H_{17}$	95[c]
$C_{14}H_{29}OH$	A	100	$C_{13}H_{27}CO_2C_{14}H_{29}$	82[c]
$PhCH_2OH$	A	100	$PhCO_2CH_2Ph$	60
$PhCH_2OH$	B	88	$PhCO_2CH_2Ph$	97
$C_2H_5CH(CH_3)CH_2OH$	B	52	$C_2H_5CH(CH_3)CO_2CH_2CH(CH_3)C_2H_5$	78
$(CH_3)_2CHCH_2CH_2OH$	B		$(CH_3)_2CHCH_2CO_2CH_2CH_2CH(CH_3)_2$	60[c]
$PhCH_2CH_2OH$	B		$PhCH_2CO_2CH_2Ph$	79[c]
$C_6H_{11}CH_2OH$	B	89	$C_6H_{11}CO_2CH_2C_6H_{11}$	93[d]
$(CH_3)_2NCH_2CH_2OH$	B	54	$(CH_3)_2NCH_2CO_2CH_2CH_2N(CH_3)_2$	55

Notes: [a]The reaction was carried out at 180°C in mesitylene (method A) or in toluene in a sealed tube (method B).
[b]Determined by GLC analysis based on alcohols.
[c]Isolated yield.
[d]Diphenylacetylene (2.5 mmol) was added.

accepting hydrogen with an appropriate hydrogen acceptor. Particularly, the reaction of less reactive, sterically hindered alcohols is enhanced remarkably by using a hydrogen acceptor. Thus, the reaction of cyclohexanemethanol in the presence of diphenylacetylene gave cyclohexylmethyl cyclohexanecarboxylate in 93% yield.

The usefulness of the present reaction is illustrated by the preparation of perfume esters[15] from readily available industrial materials. The reaction of 2-methyl-1-pentanol with $RuH_2(PPh_3)_4$ catalyst gives 2-methylpentyl 2-methylpentanoate (**3**, 60%), which is the fragrance component of perfume esters.[16] Similar treatment of 2-phenylethanol gives 2-phenylethyl 2-phenylacetate (**4**, hyacinth-like fragrance) in 79% isolated yield.

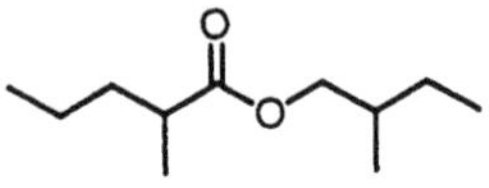

3

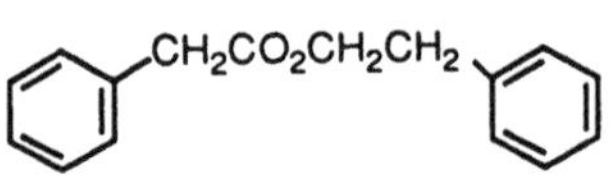

4

The oxidative condensation reaction can be applied to lactone synthesis from diols (Eq. 2). The addition of hydrogen acceptors such as diphenylacetylene, methyl vinyl ketone, and benzalacetone enhances the rate of the lactonization remarkably, however, these hydrogen acceptors are expensive and in some cases difficult to be removed from the reaction mixture. We have found that acetone is an excellent hydrogen acceptor for the lactonization because of its cheapness and easy separation from the reaction mixture.

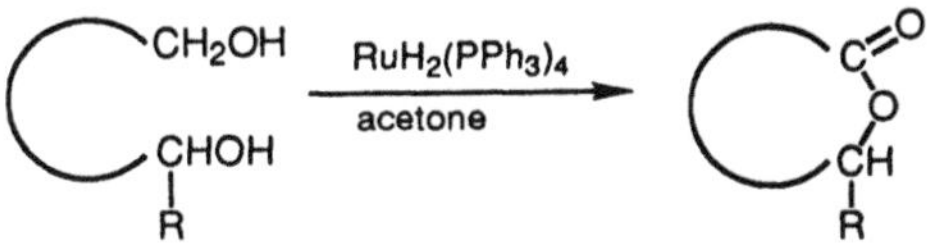

(2)

Table 2 summarizes the representative results of the ruthenium-catalyzed lactonization of diols. 1,4- and 1,5-Diols can be converted into five- and six-membered lactones in excellent yields. Allylic alcohols such as *cis*-2-butene-1,4-diol undergo the lactonization along with hydrogenation of the carbon-carbon double bonds. Phenols also undergo lactonization slowly. Benzylic alcohols are oxidized to give lactones of benzoic acid preferentially. Diethanolamines are also cyclized efficiently giving morpholine skeletons.

Interestingly, the addition of one equivalent of acetonitrile retards the intramolecular reaction of diols. Thus, the ruthenium-catalyzed reaction of 1,5-pentanediol in the presence of acetone gives δ-valerolactone in 82% yield, while the reaction in the presence of one equivalent of acetonitrile gives intermolecular esterification product, 5-hydroxypentyl 5-hydroxypentanoate (**5**) in 41% yield. δ-Valerolactone is not detected among the reaction products. This is due to the strong coordination of acetonitrile to the ruthenium complex in comparison with alcohols.[17] In this case bidentate chelation of diols and subsequent ring closure seems to be inhibited.

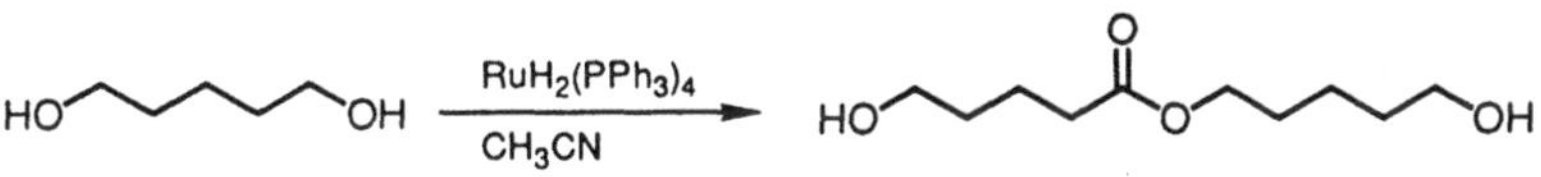

5

Since electronic stabilization resulting from hydrogen abstraction favors oxidation at the secondary position,[1] selective oxidation of primary alcohols continues to represent a notable synthetic challenge. Using the present reaction primary alcohols are oxidized chemoselectively in the presence of secondary alcohols. It is reported that primary alcohols are oxidized with a stoichiometric amount of $RuCl_2(PPh_3)_3$ 50 times faster than secondary alcohols.[18] The reaction of 1,4-pentanediol with $RuH_2(PPh_3)_4$ as catalyst in the presence of three equivalents of acetone at 180 °C for 3 h gave γ-valerolactone (**6**) in 91% yield. Similar treatment of 1,4-octanediol gave γ-octanolactone (**7**) in 98% yield. The reaction of *trans*-2-(2-hydroxyethyl)cyclohexanol gives *trans*-hexahydro-2-benzofuranone (**8**) in 90% yield. 1,2,6-Hexanetriol undergoes oxidative condensation to afford 6-hydroxy-5-hexanolide (**9**) chemoselectively in 71% yield. The favored reactivity of primary hydroxyl groups over secondary hydroxyl groups is ascribed to the difference in the stereochemical congestion between primary and secondary hydroxyls.

Table 2. Ruthenium-Catalyzed Lactonization of Diols

Diol	*Time (h)*	*Acetone*[a]	*Conv. (%)*	*Lactone*	*Yield*[b] *(%)*
	3	1	90	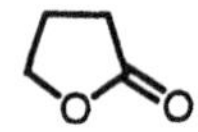	99
	2	3	86	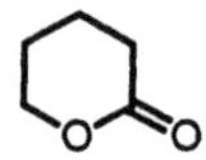	95
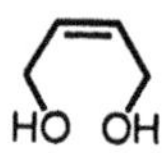	3	3	100	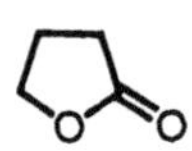	88
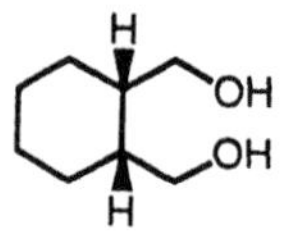	15	0	100	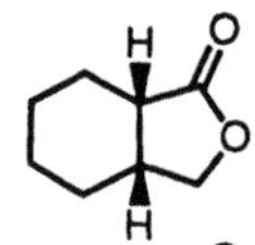	83[c)]
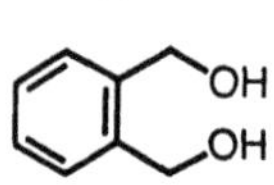	12	0	100	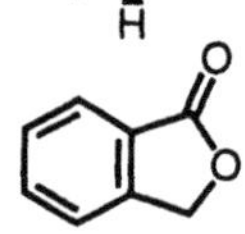	82[c)]
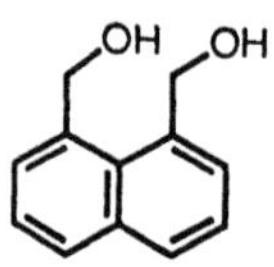	6	3	100	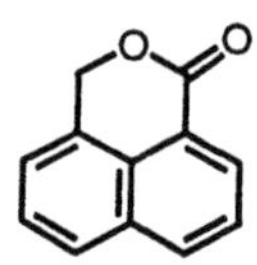	95[c)]
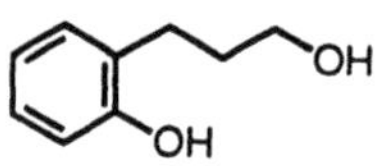	10	3	89	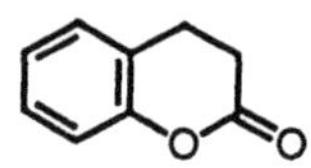	90
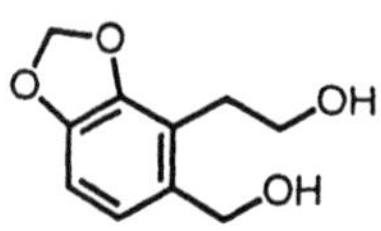	3	3		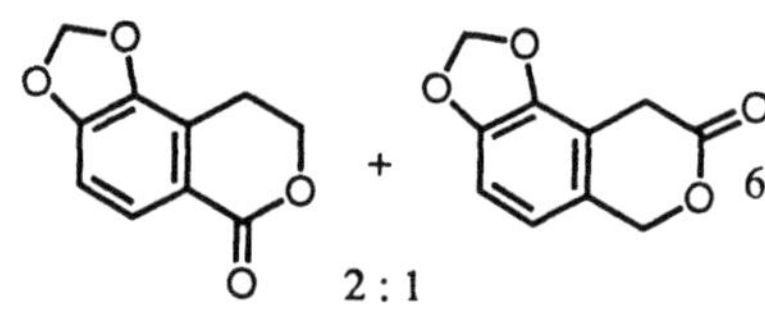	60[c)]
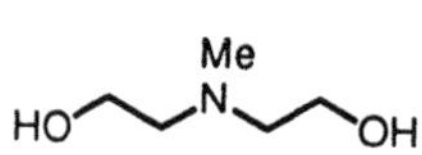	3	3	100	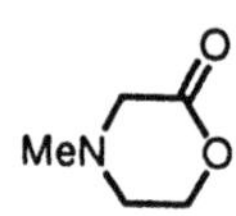	95[c)]

Notes: [a]The molar ratio of acetone/diol.
[b]Determined by GLC analysis based on diols converted.
[c]Isolated yield.

The lactone formation from unsymmetrical diols proceeds regioselectively. The unsymmetrical primary diols bearing bulky substituents are oxidized at the sterically less hindered position. The formation of **9** is such an example. The reaction of α-substituted diol **11**, derived from lactone **10**, with $RuH_2(PPh_3)_4$ catalyst gives

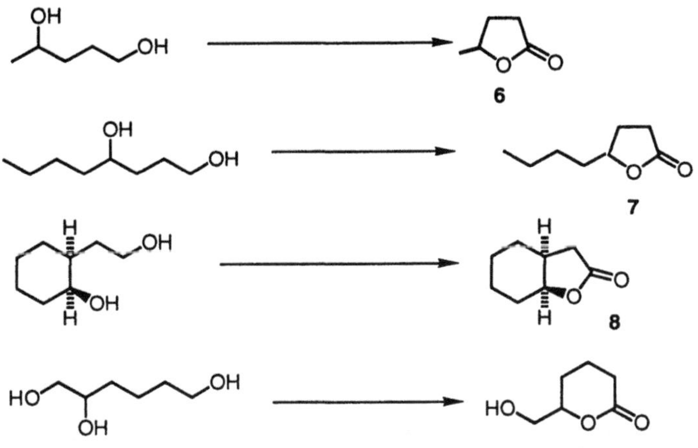

6 7 8 9

lactone **12** in 95% yield. The regioselectivity of the lactonization is up to 97:3. Since the starting unsymmetrical diols can be readily prepared by the α-substitution of lactones[19] followed by reduction, the present reaction provides an efficient method for the preparation of β-substituted-γ-butyrolactones from α-substituted-γ-butyrolactones. Similar regioselective oxidations using ruthenium phosphine complexes[20] and $NiBr_2/Bz_2O_2$[9d] have been reported.

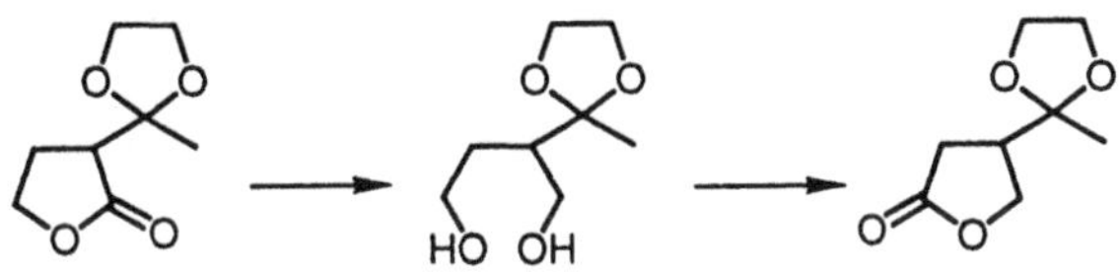

10 **11** **12**

α,ω-Diols except 1,4- and 1,5-diols undergo intermolecular condensation to give polyesters efficiently (Eq. 3). Typically, $RuH_2(PPh_3)_4$-catalyzed reaction of 1,12-dodecanediol gave polyester **13** (mp 71–72 °C) in 89% yield. The molecular weight (Mn) is determined to be 13,900 by cryoscopy in benzene. Similar treatment of 1,9-nonanediol and 1,4-cyclohexanedimethanol gives polyester **14** (87%, mp 56–57 °C, Mn 1600) and **15** (96%, mp 133–135 °C, Mn 1300), respectively. Shvo et al. reported polyester synthesis from diols by using $Ru_3(CO)_{12}$ catalyst and stoichiometric amount of diphenylacetylene.[21]

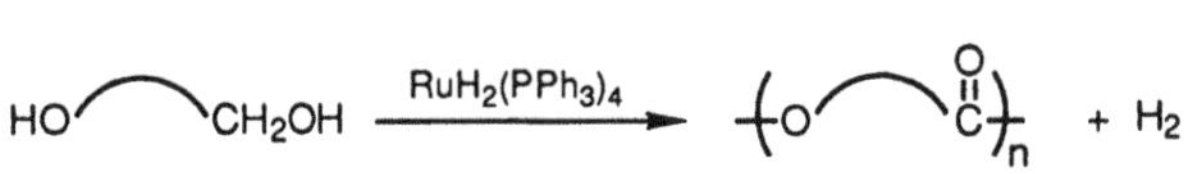

(3)

The ruthenium-catalyzed transformation of primary alcohols to esters can be rationalized by assuming the formation of the hemiacetal intermediate as shown in

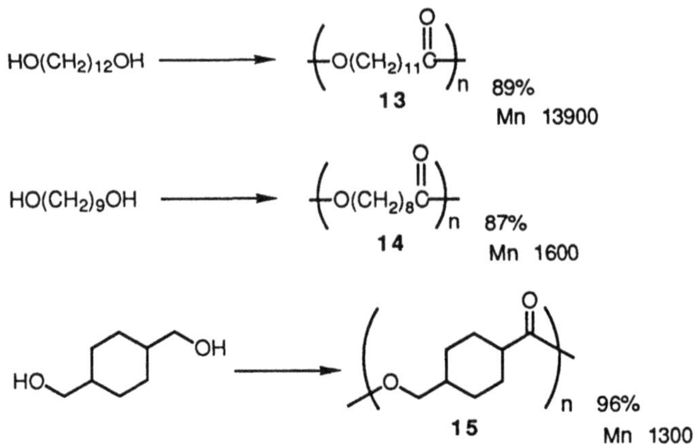

Scheme 1. The catalytically active species seems to be the coordinatively unsaturated ruthenium complex, (Ru)L_n (**16**), which is formed by reductive elimination of hydrogen from hydridoruthenium complex (Ru)H_2L_n. When the reaction is carried out in the presence of a hydrogen acceptor, zerovalent complex **16** is formed by hydrogen transfer from $RuH_2(PPh_3)_4$[22] to the hydrogen acceptor (A). It is also suggested that the active intermediate for the $RuH_2(PPh_3)_4$-catalyzed hydrogen transfer from 2-propanol to olefins[23] is not a Ru(II) species, but coordinatively unsaturated $Ru(PPh_3)_3$. Oxidative addition of **16** to the O–H bond of primary alcohols gives hydridoalkoxyruthenium complex **17**. The presence of alkoxyruthenium intermediates has been postulated from kinetic experiments.[6b] β-Elimination of **17** leads to complex **18**.[24,25] The evidence supporting β-elimination mechanism for the ruthenium-catalyzed oxidation of alcohols has been obtained

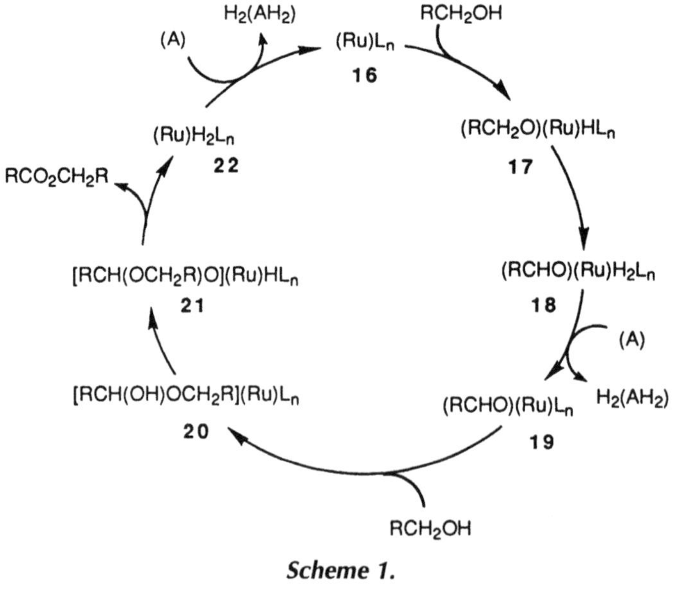

Scheme 1.

from both H-D exchange reactions and racemization reactions of optically active alcohols.[14] Such a pathway has been also suggested in the hydrogen-transfer reaction of alcohols[6b] and alkylation of amines with alcohols.[4] Dissociation of molecular hydrogen from **18** and subsequent reaction with an alcohol at the coordinated carbonyl carbon of **19** gives hemiacetal complex **20**. Further, oxidative addition of the ruthenium into the O-H bond of the coordinated hemiacetal gives alkoxyruthenium complex **21**. β-Elimination of **21** gives ester and dihydridoruthenium complex **22**, which undergoes elimination of molecular hydrogen to complete the catalytic cycle. In the presence of hydrogen acceptors (A) such as acetone and α,β-unsaturated ketones, transfer hydrogenations[22,23] from both complexes **18** and **22** proceed efficiently to enhance the condensation reaction. $Ru_3(CO)_{12}$-catalyzed esterification of alcohols needs a stoichiometric amount of a hydrogen acceptor such as diphenylacetylene;[12a,b] however, esters could not be obtained without a hydrogen acceptor, indicating that reductive elimination of molecular hydrogen from mononuclear complexes **18** and **22** proceeds more readily in comparison with that from hydridoruthenium clusters. A nonredox process has been proposed for the dehydrogenation of alcohols catalyzed by ruthenium(II) carboxylate complexes $Ru(OCOR^1)_2L_n$.[3b,c] Solvolysis of the carboxylate complexes with alcohols R^2CH_2OH gives acids R^1CO_2H and alkoxides $Ru(II)(OCOR^1)(OCH_2R^2)Ln$ which undergo β-elimination to give aldehydes and $Ru(II)H(OCOR^1)L_n$. Subsequent reaction of the hydrido complex with the acid R^1CO_2H liberates molecular hydrogen and regenerates the catalyst. However, the nonredox process seems not to be operative in the present reaction, because the observed acceleration caused by hydrogen acceptors can not be explained by this mechanism. $Ru(OCOCF_3)_2(CO)(PPh_3)_2$ catalyst[3b] shows lower catalytic activity for the esterification in comparison with the hydridoruthenium catalyst.

On working up the system after completion of the reaction, $Ru(CO)_3(PPh_3)_2$ (**23**) was isolated as the sole characterizable ruthenium complex. The formation of **23** can be accounted for by the pathway shown in Scheme 2. Thus, the aldehyde in complex **19** oxidatively adds to the ruthenium to give hydridoacylruthenium complex **24**. Such oxidative addition of metal into the C–H bond of aldehydes has been postulated in various metal-catalyzed decarbonylations of aldehydes,[26] and some hydridoacyl complexes have been isolated.[27] Acyl-alkyl rearrangement gives

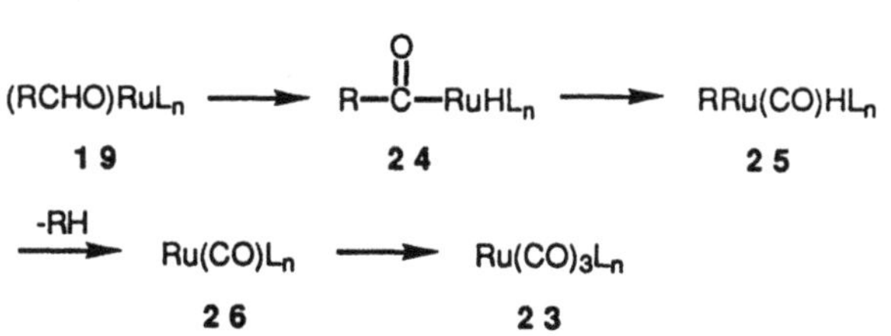

Scheme 2.

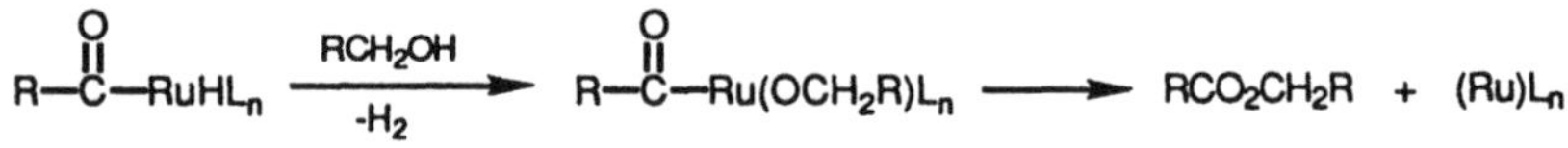

24 **27**

alkylcarbonylruthenium complex **25**, which undergoes reductive elimination of hydrocarbons to give carbonylruthenium complex **26**. Repetition of these oxidative addition and decarbonylation would give tricarbonylruthenium complex **23**. The formation of hydrocarbons by such a dehydrogenation-decarbonylation sequence of alcohols has been suggested.[3b,24,25,28] Eventually, GC-MASS analysis shows that a small amount of the dehydrogenation-decarbonylation product, heptane, is formed from the reaction of 1-octanol with $RuH_2(PPh_3)_4$ catalyst. These facts indicate the possibility of the alternative Tishchenko-type mechanism. That is, the reaction of **24** with alcohols gives acylalkoxyruthenium complex **27**, which undergoes reductive coupling to give esters. Recently, A. Yamamoto et al. reported that the ruthenium-catalyzed Tishchenko-type dimerization of aldehydes proceeds *in vacuo* without a solvent to give esters.[29] However, this mechanism seems unlikely in the present reaction, because initial step is not operative under the reaction condition. In fact, the control experiments for the $RuH_2(PPh_3)_4$-catalyzed reaction of 1,4-butanediol shows that the reaction proceeds via the formation of lactol **28**, which undergoes further dehydrogenation to give γ-butyrolactone. The corresponding dialdehyde is not detected during the reaction. These facts suggest strongly that hemiacetal formation is much faster than oxidative addition of ruthenium into an aldehyde under the present reaction conditions.

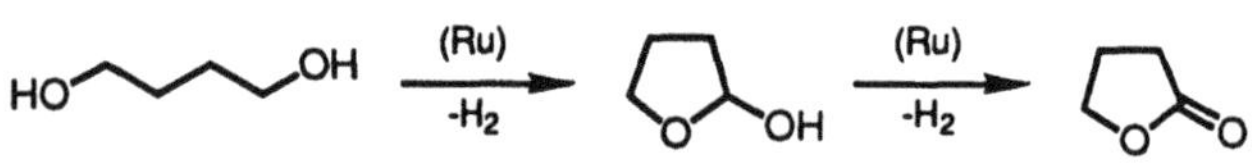

28

High reactivity of ruthenium towards alcohols has been applied to catalytic hydrogen transfer reactions of alcohols to various substrates. Ketones,[30] α,β-unsaturated ketones,[31] imines,[32] and even halides[33] are hydrogenated with alcohols in the presence of ruthenium catalysts.

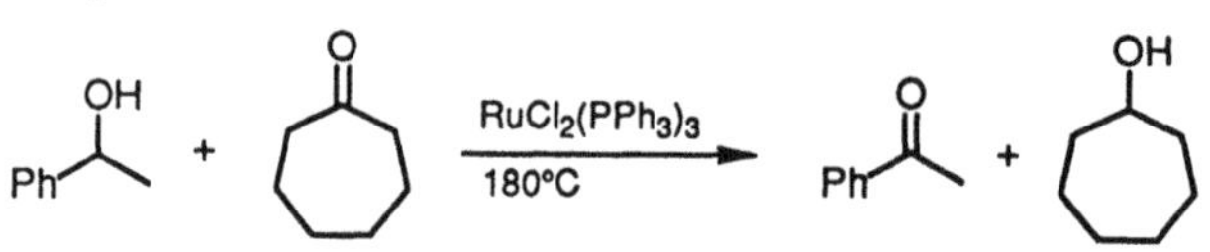

30a)

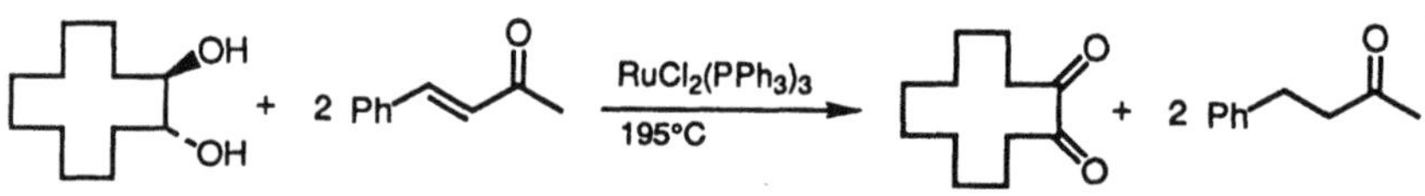

31c)

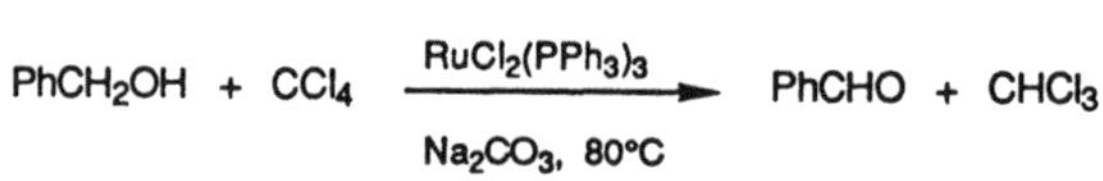

33)

B. Condensation of Alcohols with Amines

The above mentioned oxidative condensation of alcohols can be rationalized simply by assuming the reaction shown in Scheme 3. The activation of alcohols by means of ruthenium complex gives carbonyl compounds and ruthenium dihydride

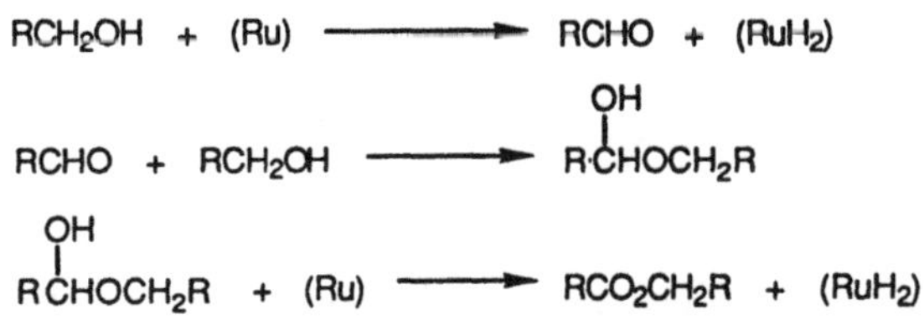

Scheme 3.

complex. Nucleophilic attack of alcohols to the intermediate aldehydes gives hemiacetals which undergo further dehydrogenation to afford esters. Capture of the carbonyl intermediates with other nucleophiles such as amines provides novel catalytic reactions of alcohols. We have found that ruthenium-catalyzed reaction of alcohols with amines gives unsymmetrical secondary and tertiary amines as depicted in Eq. 4.[4b] The reaction of various aliphatic amines with alcohols proceeds

$$R^1R^2CHOH + R^3R^4NH \xrightarrow{RuH_2(PPh_3)_4} R^1R^2CHNR^3R^4 \quad (4)$$

highly efficiently at 150–180 °C in the presence of $RuH_2(PPh_3)_4$ catalyst. Intramolecular version of the present reaction provides an efficient method for synthesis of cyclic amines. Aminoalcohols can be converted into secondary cyclic amines. Tertiary cyclic amines can be prepared by either the reaction of diols with primary amines or the reaction of aminoalcohols with alcohols. The representative results of these reactions are shown below. Efficient formation of 7-membered-ring compounds is due to the template effect of ruthenium complexes to the difunctional substrates.

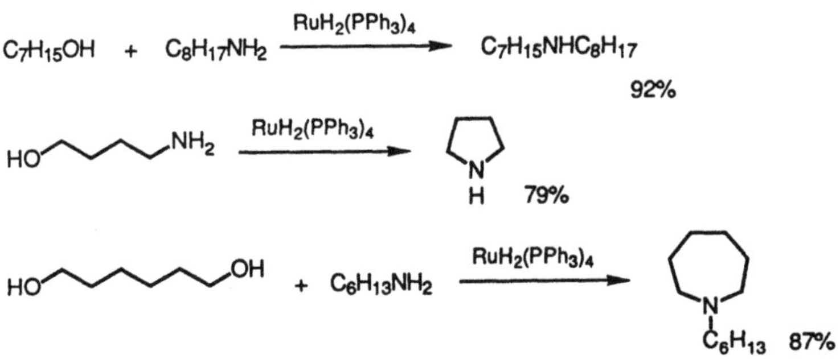

Further applications of the synthesis of cyclic amines confirm the efficiency of the present reaction. Tetrahydroisoquinolines can be readily prepared. Thus, treatment of diol **29** with benzylamine gives tetrahydroisoquinoline **30** in 62% isolated yield. Moreover, the cyclization reaction can be utilized for construction of indole derivatives. Thus, treatment of 1,5-pentanediol with tryptamine gives indolylethylpiperidine **31** (33%) along with indoloquinolizine **32** (6%).

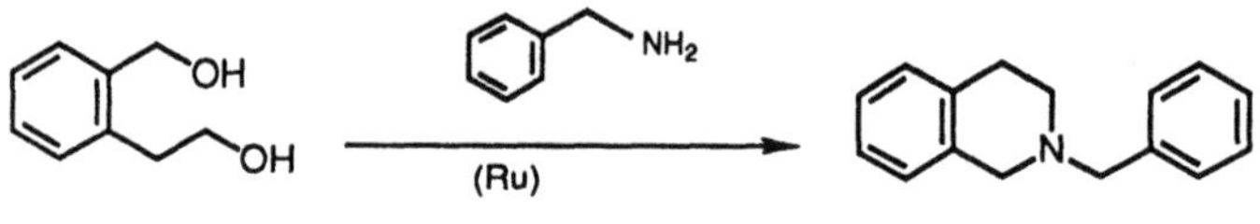

29 **30**

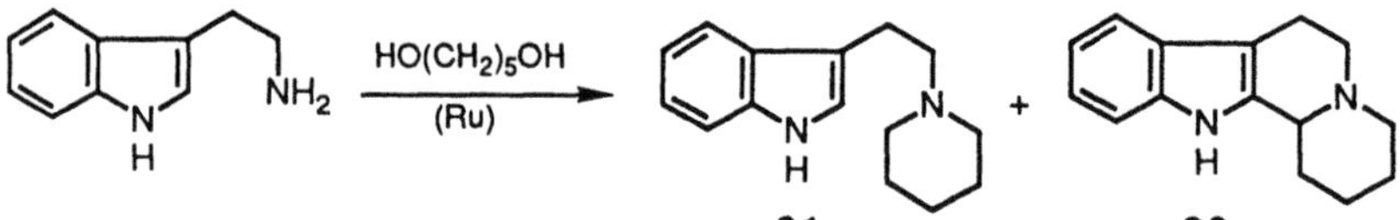

31 **32**

The reaction can be rationalized by assuming the mechanism as shown in Scheme 4. In a similar manner to the esterification of alcohols, dihydride complex **33** is formed by oxidative addition of ruthenium into the O–H bond of an alcohol and subsequent β-elimination. Nucleophilic reaction with an amine at the coordinated carbonyl carbon of **33** gives iminium ion complex **34**. Intramolecular hydride transfer of **34** gives a product amine and low valent ruthenium complex to complete catalytic cycle. The formation of quinolizine **32** indicates clearly the intermediacy of iminium ion **34**.

$$R^1R^2CHOH + (Ru) \rightleftharpoons \underset{\mathbf{33}}{(R^1R^2C{=}O)(Ru)(H_2)}$$

$$(R^1R^2C{=}O)(Ru)(H_2) + R^3R^4NH \rightleftharpoons \underset{\mathbf{34}}{(R^1R^2C{=}\overset{+}{N}R^3R^4)(Ru)(H^-)} + H_2O$$

$$(R^1R^2C{=}\overset{+}{N}R^3R^4)(Ru)(H^-) \rightleftharpoons R^1R^2CHNR^3R^4 + (Ru)$$

Scheme 4.

Watanabe reported that $RuCl_2(PPh_3)_3$ is a good catalyst for the *N*-alkylation of aromatic amines with alcohols.[34] The reaction is applied to the intramolecular cyclization reactions, which involved the reaction with diols to give *N*-aryl cyclic amines[35] and synthesis of indoles from 2-(2-hydroxyethyl)anilines.[36] Similar treatment of aromatic amines with 1,2- and 1,3-diol[37] and allylic alcohols[38] gives the corresponding quinoline derivatives by the similar *N*-alkylation and subsequent aromatic ring closure.

When the $RuH_2(PPh_3)_4$-catalyzed reaction of aminoalcohols is carried out in the presence of a hydrogen acceptor, the dehydrogenation reaction takes place exclusively to afford the corresponding lactams (Eq. 5).[39] This is in contrast to the result

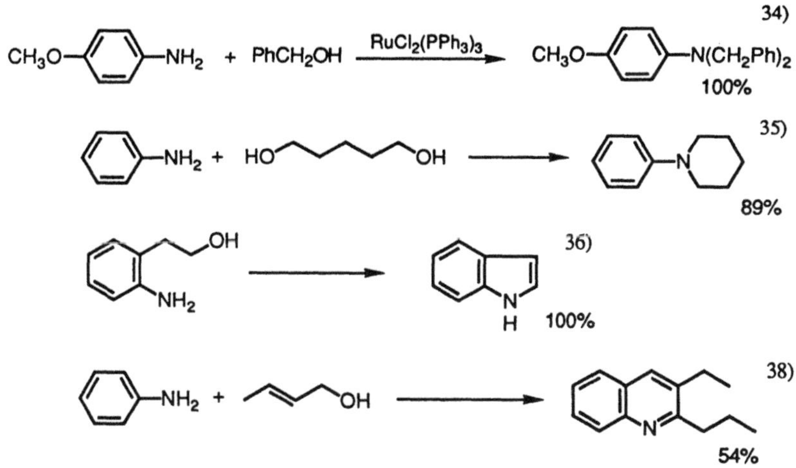

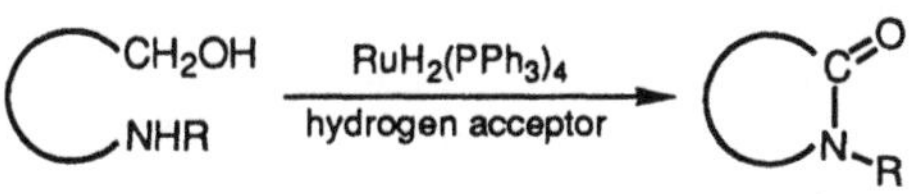

(5)

obtained by the cyclization reaction without a hydrogen acceptor. Various 1,4- and 1,5-aminoalcohols can be converted into the corresponding five- and six-membered lactams in good yields. In the case of the aminoalcohols bearing a primary amino group, two molar equivalents of water are required in order to prevent the formation of cyclic imines. Generally, α,β-unsaturated ketones and alkynes can be used as a hydrogen acceptor, and benzalacetone gives the best result among hydrogen acceptors examined.

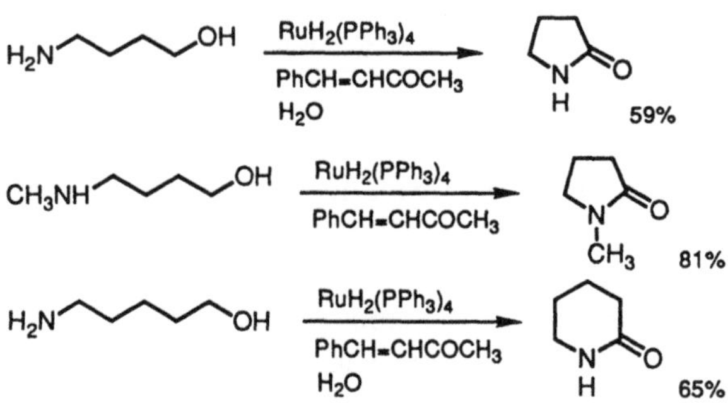

The present reaction can be rationalized by assuming the mechanism as shown in Scheme 5. This is principally similar to that of the ruthenium-catalyzed lactonization of diols. The oxidative addition of the low valent ruthenium complex to the O–H bond of an alcohol followed by elimination of (RuH_2) species gives aminoal-

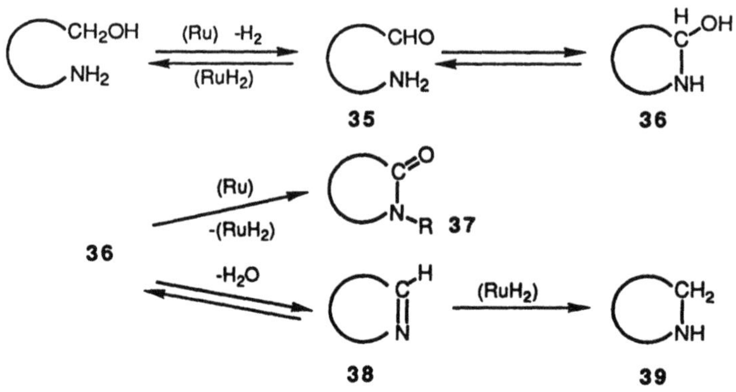

Scheme 5.

dehyde **35**, which undergoes condensation to give intermediate **36**. Further dehydrogenation of **36** in the presence of a hydrogen acceptor gives lactams **37**. In contrast, dehydration of **36** gives cyclic imine **38** which undergoes transfer hydrogenation to give amine **39**.

Catalytic alkylation with alcohols can be performed by using other nucleophiles. Primary amides undergo *N*-alkylation by the $RuCl_2(PPh_3)_3$-catalyzed reaction with alcohols.[40] $RuH_2(PPh_3)_4$-catalyzed reaction of phenylacetonitrile with ethanol proceeds in the presence of inorganic base to give the corresponding α-ethylated product.[41]

$$C_9H_{19}OH + PhCONH_2 \xrightarrow[Na_2CO_3]{RuCl_2(PPh_3)_3} PhCONHC_9H_{19} \quad 76\% \qquad 40)$$

$$C_2H_5OH + PhCH_2CN \xrightarrow{RuH_2(PPh_3)_4} PhCH(C_2H_5)CN \quad 92\% \qquad 41)$$

C. Ruthenium-Promoted Oxidative Condensation of Aldehydes

The principle of the catalytic dehydrogenation process of alcohols can be extended to a novel catalytic condensation of aldehydes. With assumption of the formation of hemiacetals as shown in Scheme 3, the oxidative condensation of aldehydes with alcohols can be performed to afford esters as depicted in Eq. 6.[7b] Since oxidative transformations of aldehydes to esters are important, much effort has been devoted to find efficient methods for this type of oxidation, and few

$$RCHO + RCH_2OH \xrightarrow{RuH_2(PPh_3)_4} RCO_2CH_2R + H_2 \qquad (6)$$

Table 3. Ruthenium-Catalyzed Reaction of Aldehydes with Either Alcohols or Water

Aldehyde	*Alcohol (Water)*	*Conv (%) Aldehyde*	*Product*	*Yield*[a] *(%)*
C_3H_7CHO	C_4H_9OH	100	$C_3H_7CO_2C_4H_9$	96
$C_7H_{15}CHO$	$C_8H_{17}OH$	98	$C_7H_{15}CO_2C_8H_{17}$	80
PhCOH	$PhCH_2OH$	93	$PhCO_2CH_2Ph$	85
C_3H_7CHO	H_2O	99	$C_3H_7CO_2H$	91[b]
$C_5H_{11}CHO$	H_2O	99	$C_5H_{11}CO_2H$	67[b]
$C_7H_{15}CHO$	H_2O	99	$C_7H_{15}CO_2H$	70[b]
PhCHO	H_2O	93	$PhCO_2H$	75[b]

Notes: [a]Determined by GLC analysis based on aldehydes.
[b]Benzalacetone (1 equiv.) was added.

methods[42] have been explored. The reaction of aldehydes with an equimolar amount of alcohols at 180 °C in the presence of $RuH_2(PPh_3)_4$ catalyst gives the corresponding esters directly in high yields. The results are shown in Table 3. Typically, the ruthenium-catalyzed reaction of butanal with 1-butanol gave butyl butanoate in 96% yield. The cross condensation of aldehydes (R^1CHO) with alcohols (R^2CH_2OH) proceeds nonselectively. Thus, the reaction of benzaldehyde with 1-octanol at 180 °C gave a mixture of octyl octanoate (21%), benzyl benzoate (19%), octyl benzoate (15%), and benzyl octanoate (24%). The scrambling of the product distribution is due to the formation of alcohol (R^1CH_2OH) by the hydrogen transfer to the starting aldehyde (R^1CHO). Therefore, when a hydrogen acceptor is used, the hydrogen transfer is completely retarded and the unsymmetrical esters are obtained selectively. Typically, the reaction of octanal with methanol in the presence of mesityl oxide at 140 °C gave methyl octanoate (**40**, 66%) without any formation of octyl octanoate and octyl formate.

$$C_7H_{15}CHO + CH_3OH \xrightarrow[(CH_3)_2C{=}CHCOCH_3]{RuH_2(PPh_3)_4} \underset{\textbf{40}}{C_7H_{15}CO_2CH_3}$$

By analogy to the reaction of aldehydes with alcohols, we next investigated the reaction of aldehydes with water. Treatment of aldehydes with water in the presence of a hydrogen acceptor gave the corresponding carboxylic acid (Eq. 7). The results are listed in Table 3. Typically, the $RuH_2(PPh_3)_4$-catalyzed reaction of butanal with water (2.0 equiv) in 1,2-dimethoxyethane in the presence of benzalacetone (1.0 equiv) at 180 °C gave butyric acid (91%). In the absence of a hydrogen acceptor, the reaction of aldehydes with water gave the corresponding ester preferentially

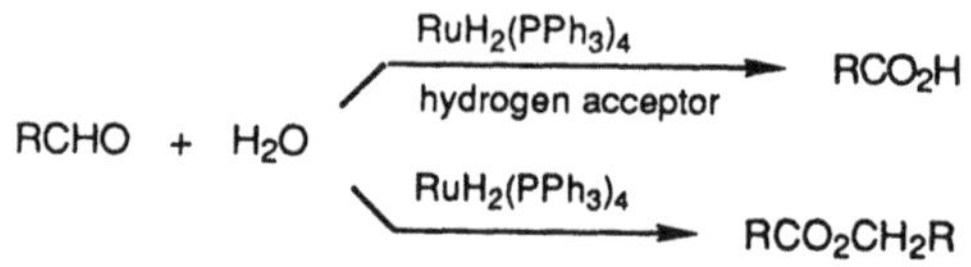

(7)

(8)

(Eq. 8). Maitlis reported that transition-metal-promoted disproportionation of acetaldehyde gave acetic acid and ethanol.[43]

α,β-Unsaturated aldehydes undergo the condensation reaction with water in the absence of a hydrogen acceptor, giving saturated carboxylic acids *via* intramolecular hydrogen transfer. Typically, the reaction of crotonaldehyde with water in the presence of $RuH_2(PPh_3)_4$ catalyst gives butyric acid in 68% yield.

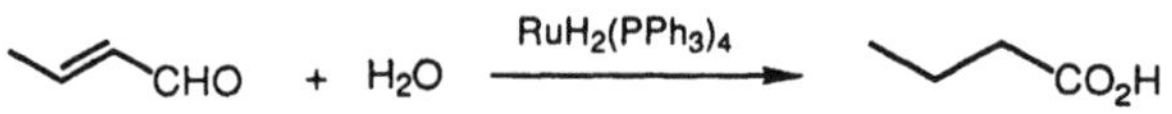

The ruthenium-catalyzed reaction of aldehydes with water is rationalized by assuming the similar pathway shown in Scheme 6. Aldehydes react with water in the presence of **16** to give *gem*-diol complex **41**. Oxidative addition of ruthenium into O–H bond gives complex **42**. β-Elimination gives carboxylic acid and hydridoruthenium complex **22**, which reacts with hydrogen acceptor (A) to complete the catalytic cycle. A similar pathway has been postulated for the metal-catalyzed

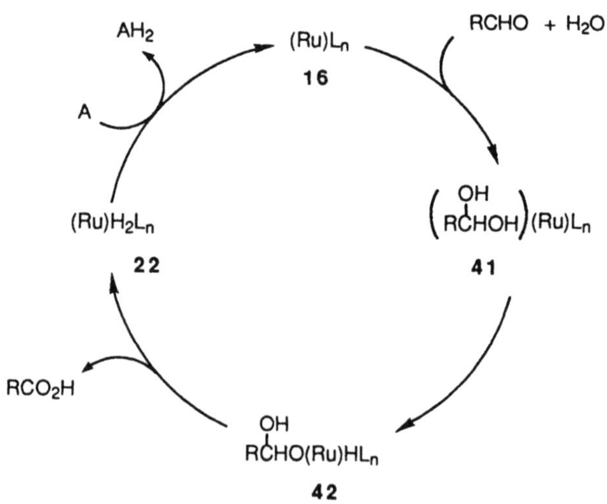

Scheme 6.

Cannizzaro reaction of acetaldehyde.[43] Furthermore, it has been reported that the reaction of $RuH_2(PPh_3)_4$ with aldehydes in the presence of water gives carboxylatocarbonyl complex, $Ru(O_2CR)_2(CO)(PPh_3)_2$.[29] In the absence of a hydrogen acceptor, the starting aldehyde is hydrogenated with ruthenium hydride **22** to give the alcohol. Esters are then formed by three pathways; the condensation of carboxylic acids with alcohols, the catalytic reaction of aldehydes with alcohols, and the catalytic oxidative condensation of alcohols.

III. RUTHENIUM-CATALYZED OXIDATION OF ALCOHOLS AND CYANOHYDRINS WITH *t*-BUTYL HYDROPEROXIDE

The dehydrogenation reactions are extremely accelerated by using an appropriate hydrogen acceptor. This is due to the smooth reproduction of active species (Ru) by the reaction of (RuH_2) with a hydrogen acceptor. During the course of searching for a suitable oxidant to trap the hydrogen of (RuH_2) species, a novel catalytic oxidation of alcohols with *t*-butyl hydroperoxide has been found. Thus, the $RuCl_2(PPh_3)_3$-catalyzed oxidation of alcohols with *t*-BuOOH proceeds under mild conditions to give the corresponding ketones efficiently (Eq. 9).

$$R^1{-}CH(OH){-}R^2 \xrightarrow[t\text{-BuOOH}]{RuCl_2(PPh_3)_3} R^1{-}C(=O){-}R^2 \qquad (9)$$

The present catalytic system is highly efficient for the synthesis of acyl cyanides from cyanohydrins (Eq. 10).[44] As a consequence of the utility of the present oxidation reaction, novel reactions of acyl cyanides have been explored. Thus, acyl cyanides are highly useful reagents for selective acylation of amino compounds. Chemoselective acylation of aminoalcohols proceeds to give *N*-acylated aminoalcohols.[44] Primary amines are selectively acylated in the presence of secondary amines.[45] Palladium-catalyzed decarbonylation of acyl cyanides proceeds highly efficiently under mild conditions to give the corresponding nitriles.[46]

$$R{-}CH(OH)CN \xrightarrow[t\text{-BuOOH}]{RuCl_2(PPh_3)_3} R{-}C(=O){-}CN \qquad (10)$$

A. Oxidation of Alcohols

Ruthenium-catalyzed oxidation of alcohols with *t*-BuOOH proceeds under mild conditions to give the corresponding carbonyl compounds. The catalytic activity of various ruthenium complexes has been examined for the transformation of 1-phenylethanol to acetophenone (**43**) with *t*-BuOOH in benzene. $RuCl_2(PPh_3)_3$

Table 4. Conversion of Alcohols to Ketones

Alcohol	*Product*	*Time(h)*	*Conv. (%)*	*Yield*[a] *(%)*
$C_6H_{13}CH(OH)CH_3$	$C_6H_{13}C(O)CH_3$	3	73	67[b)]
$PhCH(OH)CH_3$	$PhC(O)CH_3$ **(43)**	3	94	100[b)]
$CH_3CH{=}CHCH(OH)C_6H_{13}$	$CH_3CH{=}CHC(O)C_6H_{13}$	3		57
$PhCH(OH){-}C(O)Ph$	$PhC(O){-}C(O)Ph$	18		66
$PhCH(OH)CO_2Et$	$PhC(O)CO_2Et$	17		96

Notes: [a]Isolated yield.
[b]Determined by GLC analysis based on the starting alcohol converted.

has been proved to be the most effective catalyst (convn. of alcohol 94%, yield of **43** 100%). Other catalysts such as $RuCl_3 \bullet nH_2O$ and $RuH_2(PPh_3)_4$ gives satisfactory results (83%, 100%; 82%, 97%). RuO_2 shows low catalytic activity (13%, 100%). When other transition metal catalysts such as $RhCl(PPh_3)_3$ and $Pd(OAc)_2$ are used, the starting alcohol is recovered. The oxidation in an aprotic solvent such as benzene, acetone, and CH_2Cl_2 gives satisfactory results. A solution of *t*-BuOOH in a dry solvent such as benzene, toluene, and dichloromethane has proved to be the most effective oxidant. The representative results of the oxidation of various alcohols with *t*-butyl hydroperoxide are shown in Table 4. Aliphatic, aromatic, and allylic alcohols are converted into the corresponding carbonyl compounds. Benzoin and methyl mandelate also undergo the oxidation to give the corresponding 1,2-dicarbonyl compounds.

Ruthenium-catalyzed oxidation of alcohols has been performed by using various oxidants such as $NaIO_4$,[47] $NaBrO_3$,[48] *N*-methylmorpholine *N*-oxide,[49] iodosylben-

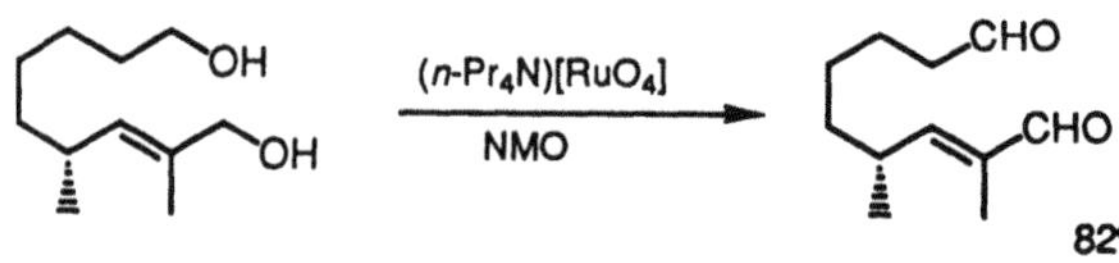

54)

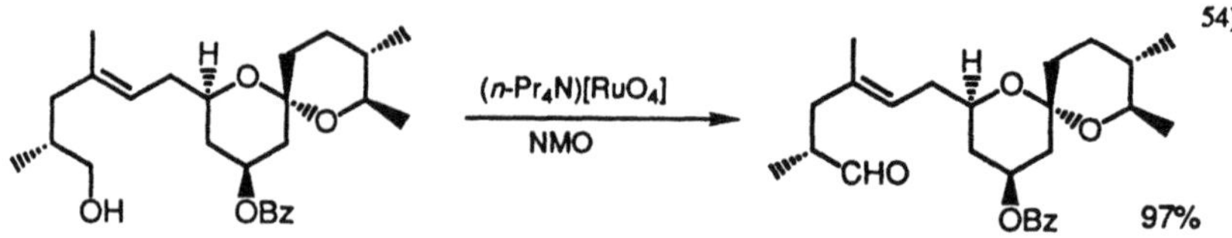

54)

zene,[50] $Me_3SiOOSiMe_3$,[51] O_2,[52] and allyl methyl carbonate.[53] Our method is advantageous over the previous methods with respect to simple operation, mild reaction conditions, and high efficiency. Recently, Griffith et al. reported that a combination of tetra-*n*-propylperruthenate and *N*-methylmorpholine *N*-oxide (NMO), is a mild and convenient catalytic system for the oxidation of alcohols to carbonyl compounds.[54] A wide range of alcohols are oxidized efficiently under mild reaction conditions.

B. Oxidation of Cyanohydrins to Acyl Cyanides

The present oxidation can be applied to the synthesis of acyl cyanides from the corresponding aldehyde cyanohydrins. Acyl cyanides are versatile synthetic intermediates and have been utilized in a variety of transformation of CO and CN functions,[55] nucleophilic reactions with enolates,[56] and Michael reactions.[5] Acyl cyanides have been prepared by condensation of acid halides with a variety of metal cyanides[58] or $(CH_3)_3SiCN$.[59] Alternative method is the oxidation of cyanohydrins with oxidants such as CrO_3[60] and manganese dioxide.[61] The last method is convenient for the conversion of sensitive aldehydes into esters, although acyl cyanides are not isolated.[61] *O*-Silylated cyanohydrins[62,63] and thionitriles[64] are also oxidized to give acyl cyanides with bromine, pyridine dichromate, and *N*-bromosuccinimide. The present oxidation reaction is advantageous over the previous methods because of its high efficiency, and facile isolation of the desired products.

Table 5 summarizes the representative results of the synthesis of acyl cyanides. Aromatic and α,β-unsaturated cyanohydrins are converted into acyl cyanides efficiently. Irrespective of the substituents on the aryl group, acyl cyanides are obtained in high yields. The oxidation of α,β-unsaturated aldehyde cyanohydrins proceeds slowly, and hence *t*-BuOOH should be added slowly (5–6 h) to complete the oxidation. Aliphatic aldehyde cyanohydrins are converted into the corresponding carboxylic acids because of the facile hydrolysis of unstable acyl cyanides.

Since acyl cyanides can be smoothly converted into a variety of carboxylic acid derivatives,[55] the reaction provides a convenient method for the synthesis of acid derivatives from sensitive aldehydes. Typically, the ruthenium-catalyzed oxidation of 2-hydroxy-3-heptenenitrile (**44**) followed by treatment with methanol in the

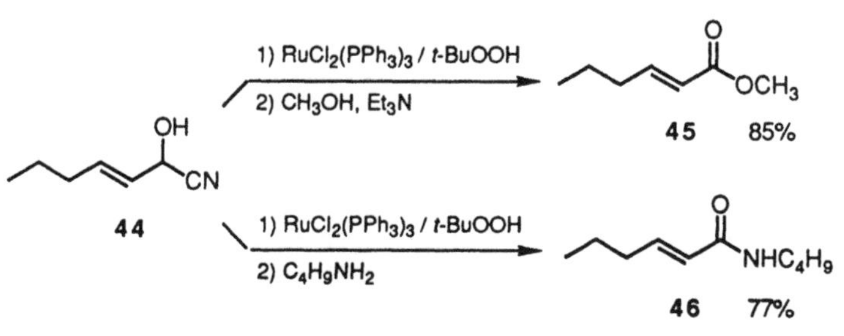

Table 5. Conversion of Cyanohydrins to Acyl Cyanides

Acyl Cyanide	*Method*[a]	*Time (h)*	*Yield*[b] *(%)*
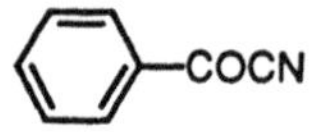 **(49)**	A	3	87
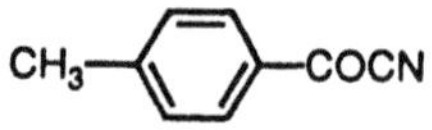	A	3	99
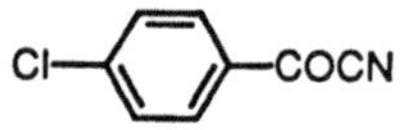	A	3	77
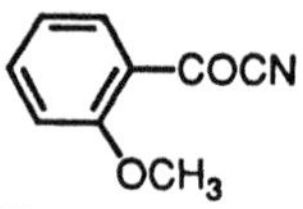	A	11	82
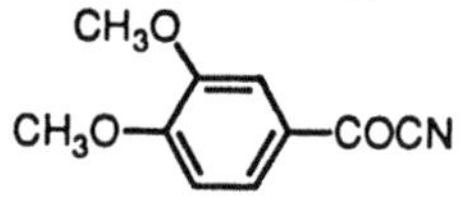 **(57)**	A	2	89
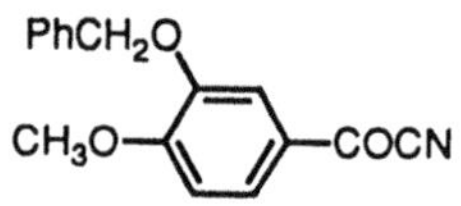	B	15	87
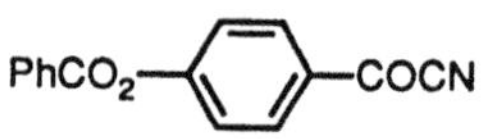	B	5	80
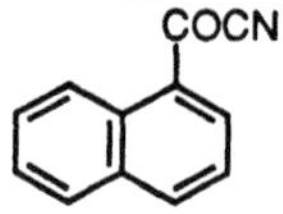	B	24	65
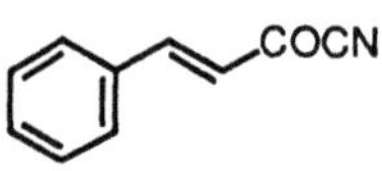 **(60)**	B	9	92
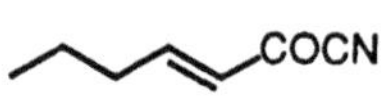	B	8	81

Notes: [a]Cyanohydrins were prepared in 90–95% yields, according to the following methods: A, $NaHSO_3$–KCN; B, $(CH_3)_3SiCN/ZnI_2$–2N HCl.
[b]Isolated yields.

presence of triethylamine afforded the corresponding methyl ester **45** in 85% yield. Similar treatment with butylamine gave the *N*-butylamide **46** in 77% yield.

Oxoruthenium intermediates have been postulated for the ruthenium-catalyzed oxidation of amines with *tert*-butyl hydroperoxide.[65] The present oxidation can be rationalized by assuming the similar mechanism as shown in Scheme 7. The reaction of Ru(II) complex with *t*-BuOOH would give Ru(II)OO-*t*-Bu which undergoes heterolytic cleavage of the O–O bond to give oxoruthenium(IV) intermediate **47**.[66] Complex **47** abstracts α-hydrogen atom of alcohols or cyanohydrins to afford [R^1R^2COH Ru(III)(OH)] species, which undergoes one electron transfer

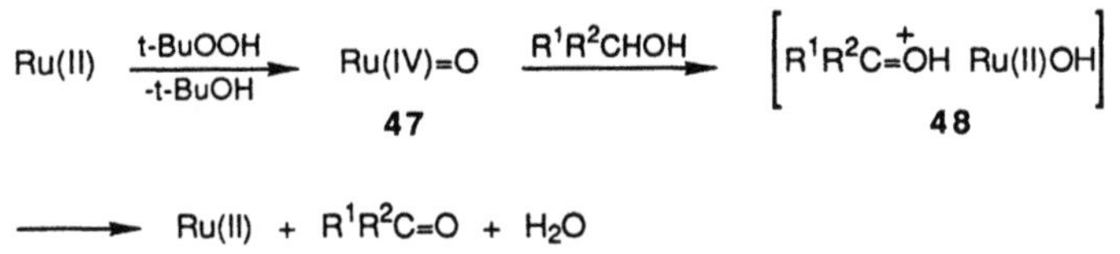

Scheme 7.

to give intermediate **48**. Alternatively, the oxidation of alcohols would occur by two electron, hydride transfer to produce **48**.[66a] The complex **48** is converted into Ru(II), carbonyl compounds, and water to complete catalytic cycle. The possibility of homolytic cleavage of *t*-butyl hydroperoxide followed by direct hydrogen abstraction by alkoxy radical and others[67] seems unlikely, because: (1) oxidation of benzylic positions does not occur (Table 5); and (2) no radical coupling product is detected among the products. Alternative mechanism which involves the dehydrogenation of alcohols by ruthenium complex[7] followed by oxidation of the resulting dihydridoruthenium intermediate with *t*-BuOOH seems unlikely, because the oxidation of alcohols in the presence of a hydrogen acceptor such as benzalacetone does not occur under the reaction conditions.

C. Synthetic Applications of Acyl Cyanides

1. Selective Acylation of Amino Compounds with Acyl Cyanides

As a consequence of exploring synthetic application of the present oxidation reaction, we have found that acyl cyanides thus obtained are highly useful reagents for selective acylation of amino compounds. Thus, the reaction of aminoalcohols with acyl cyanides gives *N*-acylated products selectively (Eq. 11).[44] Furthermore, primary amines are selectively acylated in the presence of secondary amines (Eq. 12).[45]

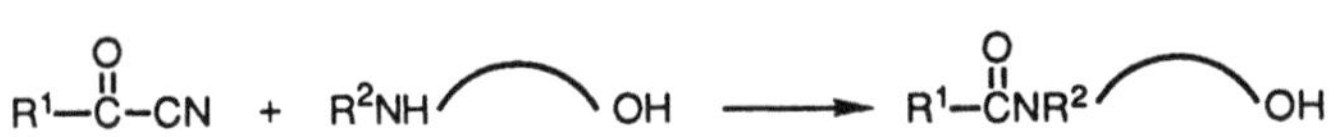

(11)

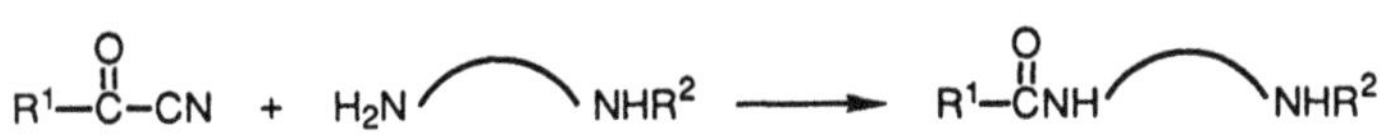

(12)

Since selective protection of amines is important for synthesis of nitrogen compounds, much effort has been devoted to find new reagents for selective protection of amine functions.[68] In the chemistry of aminosugars[69,70] and oligonucleotides,[71] many methods for selective *N*-acylation in the presence of hydroxy groups have been reported; however, the reported methods are specific and often lack generality of applicable substrates. The direct methods for selective protection of aminoalcohols are limited to a few reactions, which involve *N*-acylation with pentafluorophenyl acetate,[72] poly(3-acyl-2-oxazolone),[73] 2-oxo-3-oxazolinylphosphonate,[74] and 3-acyl-1,3-thiazolidine-2-thione[75], and *N*-tosylation with *N*-methyl-*N*-tosylpyrrolidium perchlorate.[76]

Selective protection and functionalization of polyamines are of importance in view of synthesis of naturally occurring polyamines such as spermidine and spermine alkaloids[77] which have potent antibiotic[78] and antineoplustic[79] properties. Several direct[73–75,80] and multi-step[77a,b,81] methods for selective protection of polyamines have been reported. The present acylation reaction is advantageous over the previous methods, because: (1) the reaction proceeds generally and highly selectively under mild conditions; (2) the products can be isolated simply without any washing process; and (3) a variety of acyl cyanides can be readily prepared by either the ruthenium-catalyzed oxidation of cyanohydrins or the condensation of acyl halides with metal cyanides.[58] It is noteworthy that $RuH_2(PPh_3)_4$-catalyzed acylation of primary amines with nitriles proceeds chemoselectively in the presence of secondary amines.[17]

The reaction of aminoalcohols with acyl cyanides gives *N*-acylated products selectively. The amino groups of various aminoalcohols are selectively acylated under neutral conditions. The detectable by-products in these reactions are small amounts of *N,O*-dibenzoylated products (0–3%), which can be readily removed by recrystallization or chromatographic separation. Typically, the reaction of 3-amino-1-propanol with benzoyl cyanide (**49**) gives *N*-(3-hydroxypropyl)benzamide (**50**) in 93% yield. Other examples are formation of *N*-benzoylated products **51** (100%) and **52** (95%).

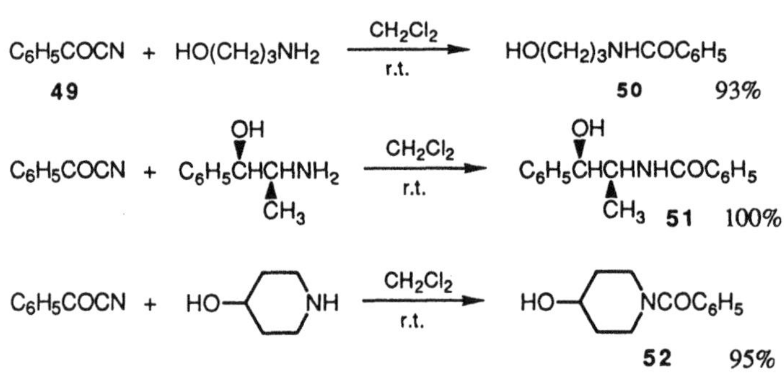

Chemoselective acylation of primary amines of polyamines can be performed in the presence of secondary amines under the similar reaction conditions. Linear polyamines such as spermidine and spermine are selectively acylated with various acyl cyanides. The reactions of benzoyl cyanide (**49**) with spermidine (**53**) and spermine (**54**) give N^1,N^8-dibenzoylspermidine (**55**) and N^1,N^{12}-dibenzoylspermine (**56**) in 92% and 91% yields, respectively. The only detectable by-products in these reactions are a trace amount of fully acylated products (0–3%), which can be

$$\underset{\mathbf{49}}{C_6H_5COCN} + \underset{\mathbf{53}}{H_2N(CH_2)_3NH(CH_2)_4NH_2} \xrightarrow[\text{r.t.}]{CH_2Cl_2} \underset{\mathbf{55}}{C_6H_5CONH(CH_2)_3NH(CH_2)_4NHCOC_6H_5} \quad 92\%$$

$$\mathbf{49} + \underset{\mathbf{54}}{H_2N(CH_2)_3NH(CH_2)_4NH(CH_2)_3NH_2} \xrightarrow[\text{r.t.}]{CH_2Cl_2} \underset{\mathbf{56}}{C_6H_5CONH(CH_2)_3NH(CH_2)_4NH(CH_2)_3NHCOC_6H_5} \quad 91\%$$

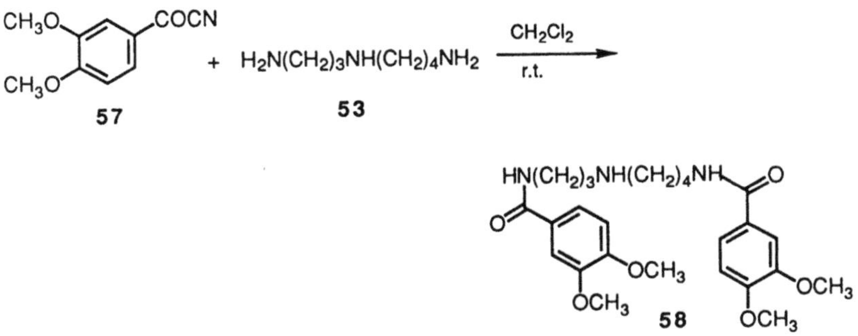

57 **53** **58** 83%

readily removed by recrystallization or chromatography. The reaction of spermidine (**53**) with 3,4-dimethoxybenzoyl cyanide (**57**) (2.0 equiv.) gives N^1,N^8-bis(3,4-dimethoxybenzoyl)spermidine (**58**) (83%), which is a precursor of pistillarin [N^1,N^8-bis(3,4-dihydroxybenzoyl)spermidine].[82] The acylation with alkyl acyl cyanides proceeds quite rapidly even at –78 °C and results in lower chemoselectivity (70–80%).

The efficiency of the present reaction is illustrated by the short-step synthesis of maytenine [N^1,N^8-bis(*trans*-cinnamoyl)spermidine][83] (**61**), which has been isolated from *Maytenus chuchuhuasha*. Thus, the $RuCl_2(PPh_3)_3$-catalyzed oxidation of *trans*-cinnamaldehyde cyanohydrin (**59**) with *t*-butyl hydroperoxide gives *trans*-cinnamoyl cyanide (**60**) in 92% yield. The reaction of spermidine with two equivalents of **60** in dry dichloromethane at room temperature gives **61** in 92% yield (mp. 156–157 °C). Although maytenine has been synthesized by several

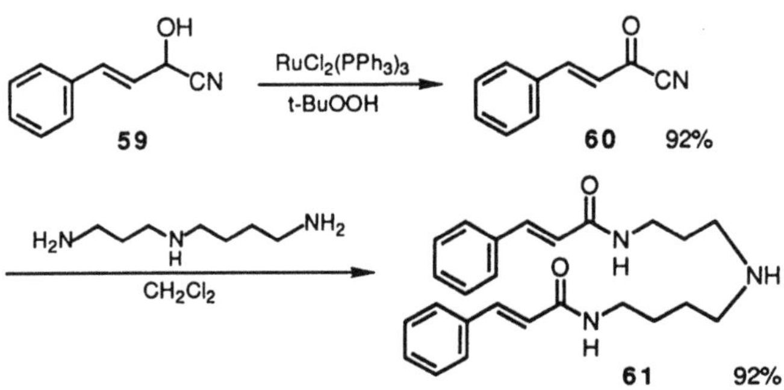

59 **60** 92% **61** 92%

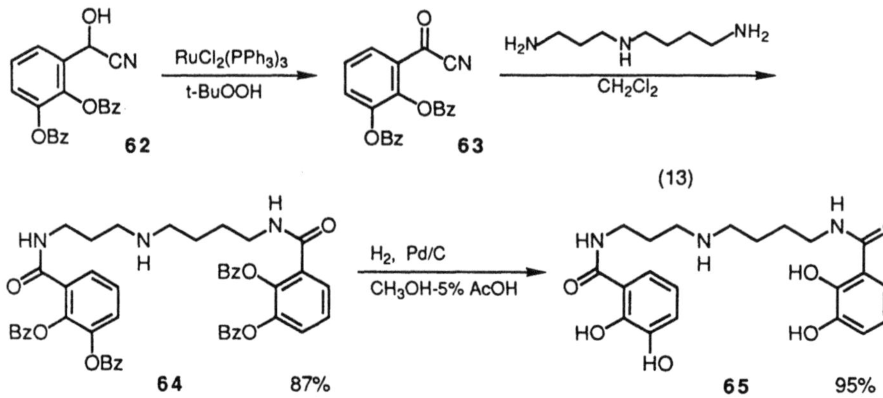

methods,[75,77b,80c,d,83] this approach has the advantages of easy operation, high selectivity, and mildness of the reaction conditions.

Most importantly, the present method is useful for the synthesis of spermidine siderophores,[77a] which are biologically important class of microbially produced iron transport compounds.[84] A typical example is the synthesis of siderophore, N^1,N^8-bis(2,3-dihydroxybenzoyl)spermidine[85] (**65**) which has been isolated from *Micrococcus denitrificans* and is an important precursor of the spermidine catecholamides such as agrobactin[86] and parabactin.[87] The reaction of spermidine with two equivalents of 2,3-dibenzyloxybenzoyl cyanide (**63**) derived from cyanohydrin **62** gives the corresponding bis(benzoyl)spermidine **64** in 87% yield. Removal of benzyl groups by catalytic hydrogenation over palladium on charcoal in methanol-5% acetic acid affords siderophore **65** in 95% yield.

2. Palladium-Catalyzed Decarbonylation of Acyl Cyanides

During the course of examining the reactivity of acyl cyanides toward transition metal complexes, we have found that acyl cyanides undergo palladium-catalyzed decarbonylation readily to give the corresponding nitriles in excellent yields (Eq. 13).[88]

$$R{-}\overset{\overset{\displaystyle O}{\|}}{C}{-}CN \xrightarrow{Pd(PPh_3)_4} R{-}CN + CO \qquad (13)$$

The transition-metal-catalyzed decarbonylation of carbonyl compounds has been reported to proceed at elevated temperature.[26,89] Aldehydes undergo decarbonylation in the presence of rhodium and palladium catalysts to give hydrocarbons.[90–92] Aromatic acyl halides (X = Cl,[90,93] Br,[90] I,[94]) undergo similar decarbonylation, while decarbonylation of the aliphatic acyl cyanides (X = Cl, Br) having a β-hydrogen gives olefins.[95] Blum et al. reported that aromatic acyl cyanides undergo decarbonylation using $RhCl(PPh_3)_3$ catalyst;[96] however, the decarbonylation required extremely drastic reaction conditions (300 °C) and the applicable

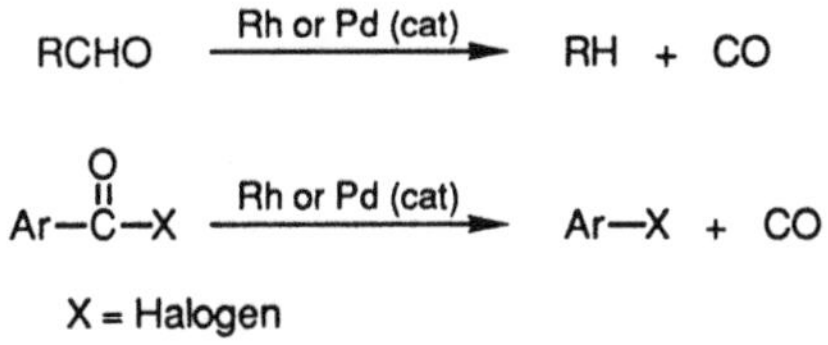

substrates are limited to reactive acyl cyanides such as naphthoyl and *p*-chlorobenzoyl cyanides.

The present palladium-catalyzed decarbonylation proceeds generally with aromatic and heteroaromatic acyl cyanides at 120 °C. Since acyl cyanides are readily prepared by the ruthenium-catalyzed oxidation of cyanohydrins with *t*-BuOOH, the present decarbonylation reaction provides an efficient method for the conversion of aromatic and heteroaromatic aldehydes to the corresponding nitriles under mild and neutral conditions.

$Pd(PPh_3)_4$ in benzene or toluene is the best catalyst among the catalysts examined. Using well-known catalysts for the decarbonylation of aldehydes and acyl halides such as $RhCl(PPh_3)_3$,[96] $RhCl(CO)(PPh_3)_3$,[90] Pd-C (5%),[93,95] and $PdCl_2$,[95] benzoyl cyanides are recovered. Typically, the treatment of a solution of benzoyl cyanide (**49**) with 3 mol% of $Pd(PPh_3)_4$ in benzene at 120 °C for 12 h gives benzonitrile (**66**) in quantitative yield. Irrespective of the substituent on the aryl group, acyl cyanides are obtained in excellent yields. Heteroaromatic acyl cyanides such as 2-furoyl cyanide (**67**) also undergo the decarbonylation. Thus, treatment of 2-furoyl cyanide with Pd(0) catalyst gives 1-cyanofuran (**68**) in 98% yield.

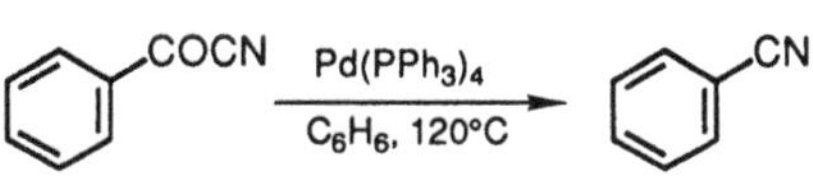

49 **66** 100%

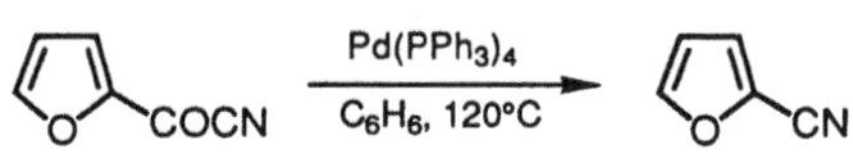

67 **68** 98%

Because of easy preparation of acyl cyanides by the oxidation of cyanohydrins, the present decarbonylation reaction provides an efficient method for the conversion of aldehydes to the corresponding nitriles. Generally, nitriles have been prepared from aldehydes by the acid-induced dehydration of the corresponding aldoximes.[97] Because the present reaction proceeds efficiently under mild and neutral conditions, the reactions are highly useful for the synthesis of acid-sensitive nitriles. The ruthenium-catalyzed oxidation of 3-(benzyloxy)-4-methoxybenzaldehyde cyanohydrin (**69**) with *t*-BuOOH gives the corresponding acyl cyanide **70** in

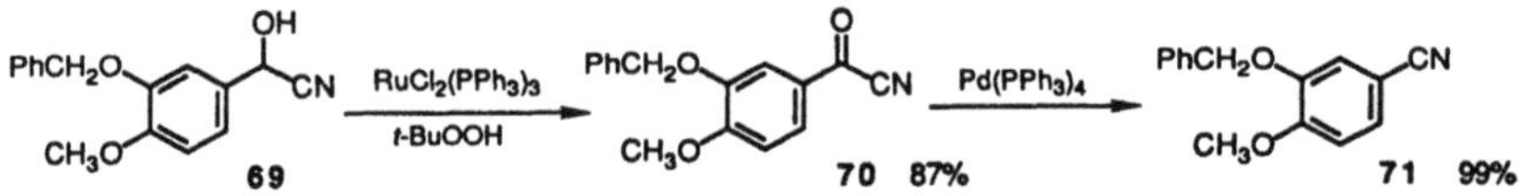

87% yield. The decarbonylation of **70** with $Pd(PPh_3)_4$ catalyst gives nitrile **71** in 99% yield. The similar treatment of 4-(benzoyloxy)benzaldehyde cyanohydrin (**72**) gives acyl cyanide **73** in 80% yield. The palladium-catalyzed reaction of **73** affords the corresponding nitrile **74** in 92% yield. In these reactions, benzyloxy and benzoyloxy groups are retained under the reaction conditions. Since acyl cyanides can be prepared by the reaction of acyl halides with metal cyanides[58] or trimethylsilyl cyanide,[59] the present reaction also provides a convenient method for the synthesis of nitriles from carboxylic acids.

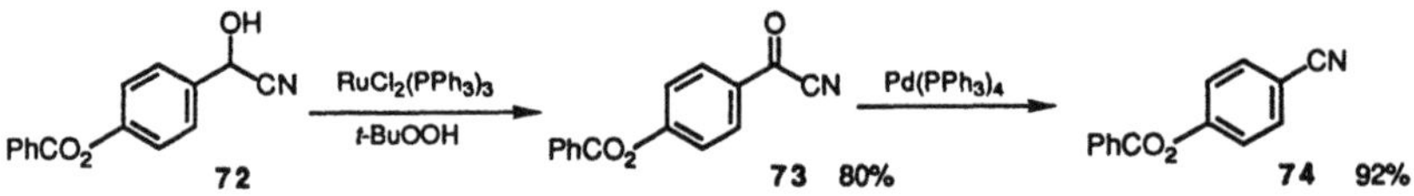

The palladium-catalyzed decarbonylation can be rationalized by assuming the pathway, which is quite similar to the mechanism for the decarbonylation of acyl halides and aldehydes.[98,99] The acyl cyanide undergoes the oxidative addition to $Pd(PPh_3)_4$ to give complex $Pd(CN)(COAr)(PPh_3)_n$. Acyl-aryl rearrangement gives complex $Pd(CN)(Ar)(CO)(PPh_3)_n$, whose reductive elimination gives aryl cyanide and $Pd(CO)(PPh_3)_n$, which undergoes dissociation of carbon monoxide.

REFERENCES

1. Sheldon, R. A.; Kochi, J. K. *Metal-Catalyzed Oxidations of Organic Compounds*; Academic Press: New York, 1981.
2. Bridges, A. J.; Raman, P. S.; Ng, G. S. Y.; Jones, J. B. *J. Am. Chem. Soc.* **1984**, *106*, 1461, and references cited therein.
3. (a) $RhCl_3$-$SnCl_2$: Chairman, H. B. *J. Chem. Soc. B* **1970**, 584; (b) $Ru(OCOCF_3)_2(CO)L_n$: Dobson, A.; Robinson, S. D. *Inorg. Chem.* **1977**, *16*, 137. Jung, C. W.; Garrou, P. E. *Organometallics* **1982**, *1*, 658. (c) $Ru_2(OAc)_4Cl$-PR_3: Shinoda, S.; Itagaki, H.; Saito, Y. *J. Chem. Soc., Chem. Commun.* **1985**, 860. (d) $[Rh(2,2'\text{-bipyridyl})_2]Cl$: Morton, D.; Cole-Hamilton, D. J. *J. Chem. Soc., Chem. Commun.* **1987**, 248.
4. (a) Pd: Murahashi, S.-I.; Shimamura, T.; Moritani, I. *J. Chem. Soc., Chem. Commun.* **1974**, 931. (b) $RuH_2(PPh_3)_4$: Murahashi, S.-I.; Kondo, K.; Hakata, T. *Tetrahedron Lett.* **1982**, *23*, 229. (c) $RuCl_2(PPh_3)_3$: Watanabe, Y.; Tsuji, Y.; Ige, H.; Ohsugi, Y.; Ohta, T. *J. Org. Chem.* **1984**, *49*, 3359. Tsuji, Y.; Huh, K.-T.; Ohsugi, Y.; Watanabe, Y. *Ibid.* **1985**, *50*, 1365. Arcelli, A.; Khai, B.-T.; Porzi, G. *J. Organomet. Chem.* **1982**, *235*, 93. (d) $RhH(PPh_3)_4$: Grigg, R; Mitchell, T. R. B.; Sutthivaiyakit, S.; Tongpenyai, N. *J. Chem. Soc., Chem. Commun.* **1981**, 611. (e) $(t\text{-}BuO)_3Al/Ni$: Botta, M.; De Angelis, F.; Nicoletti, R. *Synthesis* **1977**, 722. (f) Cu: Baiker, A.; Richarz, W. *Tetrahedron Lett.* **1977**, 1937.
5. Grigg, R.; Mitchell, T. R. B.; Sutthivaiyakit, S. *Tetrahedron Lett.* **1981**, 22, 4107.
6. (a) Review: Brieger, G.; Nestrick, T. J. *Chem. Rev.* **1974**, *74*, 567. Kolomnikov, I. S.; Kukolev, V. P.; Vol'pin, M. E. *Russ. Chem. Rev., Engl. Transl.* **1974**, *43*, 399. (b) Ru: Regen, S. L.;

Whitesides, G. M. *J. Org. Chem.* **1972**, *37*, 1832. Sasson, Y.; Blum, J. *J. Org. Chem.* **1975**, *40*, 1887. Imai, H.; Nishiguchi, T.; Fukuzumi, K. *Ibid.* **1976**, *41*, 665. Smith, T. A.; Maitlis, P. M. *J. Organomet. Chem.* **1985**, *289*, 385. (c) Fe: Landesberg, J. M.; Katz, L.; Olsen, C. *J. Org. Chem.* **1972**, *37*, 930. (d) Zr: Ishii, Y.; Nakano, T.; Inada, A.; Kishigami, Y.; Sakurai, K.; Ogawa, M. *J. Org. Chem.* **1986**, *51*, 240. (e) Rh: Imai, H.; Nishiguchi, T.; Fukuzumi, K. *J. Org. Chem.* **1974**, *39*, 1622. (f) Ir: Trocha-Grimshaw, J.; Henbest, H. B. *J. Chem. Soc., Chem. Commun.* **1967**, 544. (g) Pt: Bailar, J. C., Jr.; Itatani, H. *J. Am. Chem. Soc.* **1967**, *89*, 1592.

7. (a) Murahashi, S.-I.; Ito, K.; Naota, T.; Maeda, Y. *Tetrahedron Lett.* **1981**, *22*, 5327. (b) Murahashi, S.-I.; Naota, T.; Ito, K.; Maeda, Y.; Taki, H. *J. Org. Chem.* **1987**, *52*, 4319.
8. (a) $Na_2Cr_2O_7$: Robertson, G. R. *Org. Synth., Coll. Vol. I* **1944**, 138. (b) $NaBrO_2$: Kageyama, T.; Kawahara, S.; Kitamura, K.; Ueno, Y.; Okawara, M. *Chem. Lett.* **1983**, 1097. (c) Bu_3SnBr-NBS: Ogawa, T.; Matsui, M. *J. Am. Chem. Soc.* **1976**, *98*, 1629. (d) $Ca(OCl)_2$: Nwaukwa, S. O .; Keehn, P. M. *Tetrahedron Lett.* **1982**, *23*, 35. (e) Br_2-HMPT: Neirabeyeh, M. A.; Ziegler, J.-C.; Gross, B. *Synthesis* **1976**, 811. (f) Electrooxidation: Shono, T.; Matsumura, Y.; Hayashi, J.; Mizoguchi, M. *Tetrahedron Lett.* **1979**, 165.
9. (a) CrO_3: Stenberg, V. I.; Perkins, R. J. *J. Org. Chem.* **1963**, *28*, 323. (b) MnO_2: Marshall, J. A.; Cohen, N.; Hochstetler, A. R. *J. Am. Chem. Soc.* **1966**, *88*, 3408. (c) Ag_2CO_3-Celite: Fetizon, M.; Golfier, M.; Louis, J.-M. *Tetrahedron* **1975**, *31*, 171. (d) $NiBr_2$-Bz_2O_2: Doyle, M. P.; Dow, R. L.; Bagheri, V.; Patrie, W. J. *Tetrahedron Lett.* **1980**, *21*, 2795. (e) $Ni(OOCR)_2$-Br_2: Doyle, M. P.; Dow, R. L.; Bagheri, V.; Patrie, W. J. *J. Org. Chem.* **1983**, *48*, 476. (f) Pt-O_2: Lansbury, P. T.; Hangauer, D. G., Jr.; Vacca, J. P. *J. Am. Chem. Soc.* **1980**, *102*, 3964. (g) Oxoammonium salt: Semmelhack, M. F.; Chou, C. S.; Cortes, D. A. *J. Am. Chem. Soc.* **1983**, *105*, 4492. (h) $(C_6H_5)_3C^+BF_4^-$: Ref. 9d.
10. (a) Cu(0): Takeshita, K.; Nakamura, S.; Kawamoto, K. *Bull. Chem. Soc. Jpn.* **1978**, *51*, 2622. (b) CuO: Berthon, B.; Forestiere, A.; Leleu, G.; Sillion, B. *Tetrahedron Lett.* **1981**, *22*, 4073.
11. Pd-PhBr: Tamaru, Y.; Yamada, Y.; Inoue, K.; Yamamoto, Y.; Yoshida, Z. *J. Org. Chem.* **1983**, *48*, 1286.
12. (a) Blum, Y.; Reshef, D.; Shvo, Y. *Tetrahedron Lett.* **1981**, *22*, 1541. (b) Shvo, Y.; Blum, Y.; Reshef, D.; Menzin, M. *J. Organomet. Chem.* **1982**, *226*, C21. (c) Cottier, L.; Descotes, G.; Sabadie, J. *J. Mol. Catal.* **1980**, *7*, 337.
13. Blum, Y.; Shvo, Y. *J. Organomet. Chem.* **1985**, *282*, C7.
14. Sasson, Y.; Blum, J. *J. Chem. Soc., Chem. Commun.* **1974**, 309.
15. *The Condensed Chemical Dictionary*, 10th Ed.; Howley, G. G. Ed.; Van Nostrand Reinhold: New York, 1981.
16. Schaper, U. A.; Bruns, K. *Ger. Offen.* 3225293.
17. Murahashi, S.-I.; Naota, T.; Saito, E. *J. Am. Chem. Soc.* **1986**, *108*, 7846.
18. Tomioka, H.; Takai, K.; Oshima, K.; Nozaki, H. *Tetrahedron Lett.* **1981**, *22*, 1605.
19. Grieco, P. A. *Synthesis* **1975**, 67. Newaz, S. S. *Aldrichimica Acta* **1977**, *10*, 64.
20. Ishii, Y.; Osakada, K.; Ikariya, T.; Saburi, M.; Yoshikawa, S. *J. Org. Chem.* **1986**, *51*, 2034.
21. Shvo, Y.; Blum, Y.; Reshef, D.; Menzin, M. *J. Organomet. Chem.* **1982**, *226*, C21.
22. Komiya, S.; Yamamoto, A. *Bull. Chem. Soc. Jpn.* **1976**, *49*, 2553.
23. Imai, H.; Nishiguchi, T.; Kobayashi, M.; Fukuzumi, K. *Bull. Chem. Soc. Jpn.* **1975**, *48*, 1585.
24. Chatt, J.; Shaw, B. L.; Field, A. E. *J. Chem. Soc.* **1964**, 3466.
25. Chaudret, B. N.; Cole-Hamilton, D. J.; Nohr, R. S.; Wilkinson, G. *J. Chem. Soc., Dalton Trans.* **1977**, 1546.
26. Tsuji, J. In *Organic Synthesis via Metal Carbonyls;* Wender, I.; Pino, P. Eds., Wiley: New York, 1977; Vol. 2, p. 595.
27. Abeysekera, A. M.; Grigg, R.; Trocha-Grimshaw, J.; Viswanatha, V. *J. Chem. Soc., Perkin Trans. 1* **1977**, 1395. Suggs, J. W. *J. Am. Chem. Soc.* **1978**, *100*, 640. Rauchfuss, T. B. *Ibid.* **1979**, *101*, 1045.
28. Lupin, M. S.; Shaw, B. L. *J. Chem. Soc. A* **1968**, 741. Vaska, L.; DiLuzio, J. W. *J. Am. Chem. Soc.* **1961**, *83*, 1262. Chaudret, B.; Poilblanc, R. *Inorg. Chim. Acta* **1979**, *24*, L209.

29. Ito, T., Horino, H.; Koshiro, Y.; Yamamoto, A. *Bull. Chem. Soc. Jpn.* **1982**, *55*, 504.
30. (a) Sasson, Y.; Blum, J. *J. Org. Chem.* **1975**, *40*, 1887. (b) Watanabe, Y.; Ohta, T.; Tsuji, Y. *Bull. Chem. Soc. Jpn.* **1982**, *55*, 2441.
31. (a) Smith, T. A.; Maitlis, P. M. *J. Organomet. Chem.* **1985**, *289*, 385. (b) Bar, R.; Sasson, Y. *Tetrahedron Lett.* **1981**, *22*, 1709. (c) Regen, S. L.; Whitesides, G. M. *J. Org. Chem.* **1972**, *37*, 1832.
32. Basu, A.; Bhaduri, S.; Sharma, K.; Jones, P. G. *J. Chem. Soc., Chem. Commun.* **1987**, 1126.
33. Sasson, Y.; Wiener, H.; Bashir, S. *J. Chem. Soc., Chem. Commun.* **1987**, 1574.
34. Watanabe, Y.; Tsuji, Y.; Ige, H.; Ohsugi, Y.; Ohta, T. *J. Org. Chem.* **1984**, *49*, 3359.
35. (a) Tsuji, Y.; Huh, K.-T.; Ohsugi, Y.; Watanabe, Y. *J. Org. Chem.* **1985**, *50*, 1365. (b) Tsuji, Y.; Yokoyama, Y.; Huh, K.-T.; Watanabe, Y. *Bull. Chem. Soc. Jpn.* **1987**, *60*, 3456.
36. Tsuji, Y.; Huh, K.-T.; Yokoyama, Y.; Watanabe, Y. *J. Chem. Soc., Chem. Commun.* **1986**, 1575.
37. Tsuji, Y.; Huh, K.-T.; Watanabe, Y. *J. Org. Chem.* **1987**, *52*, 1673.
38. Watanabe, Y.; Tsuji, Y.; Ohsugi, Y.; Shida, J. *Bull. Chem. Soc. Jpn.* **1983**, *56*, 2452.
39. Naota, T.; Murahashi, S.-I.; *Synlett.* **1991**, 693.
40. Watanabe, Y.; Ohta, T.; Tsuji, Y. *Bull. Chem. Soc. Jpn.* **1983**, *56*, 2647.
41. Grigg, R.; Mitchell, T. R. B.; Sutthivaiyakit, S. *Tetrahedron Lett.* **1981**, *22*, 4107.
42. (a) Corey, E. J.; Gilman, N. W.; Ganem, B. E. *J. Am. Chem. Soc.* **1968**, *90*, 5616. (b) Ogawa, T.; Matsui, M. *Ibid.* **1976**, *98*, 1629. (c) Wilson, S. R.; Tofigh, S.; Misra, R. N. *J. Org. Chem.* **1982**, *47*, 1360. Shono, T.; Matsumura, Y.; Hayashi, J.; Inoue, K.; Iwasaki, F.; Itoh, T. *Ibid.* **1985**, *50*, 4967.
43. Cook, J.; Hamlin, J. E.; Nutton, A.; Maitlis, P. M. *J. Chem. Soc., Dalton Trans.* **1981**, 2342.
44. (a) Murahashi, S.-I.; Naota, T.; Nakajima, N. *Tetrahedron Lett.* **1985**, *26*, 925. (b) Murahashi, S.-I.; Naota, T. *Synthesis* **1993**, 433.
45. Murahashi, S.-I.; Naota, T.; Nakajima, N. *Chem. Lett.* **1987**, 879.
46. Murahashi, S.-I.; Naota, T.; Nakajima, N. *J. Org. Chem.* **1986**, *51*, 898.
47. Carlsen, P. H. J.; Katsuki, T.; Martin, V. S.; Sharpless, K. B. *J. Org. Chem.* **1981**, *46*, 3936.
48. (a) Yamamoto, Y.; Suzuki, H.; Moro-oka, Y. *Tetrahedron Lett.* **1985**, *26*, 2107. (b) Kanemoto, S.; Tomioka, H.; Oshima, K.; Nozaki, H. *Bull. Chem. Soc. Jpn.* **1986**, *59*, 105.
49. Sharpless, K. B.; Akashi, K.; Oshima, K. *Tetrahedron Lett.* **1976**, 2503.
50. Müller, P.; Godoy, J. *Tetrahedron Lett.* **1981**, *22*, 2361.
51. Kanemoto, S.; Oshima, K.; Matsubara, S.; Takai, K.; Nozaki, H. *Tetrahedron Lett.* **1983**, *24*, 2185.
52. (a) Matsumoto, M.; Watanabe, N. *J. Org. Chem.* **1984**, *49*, 3435. (b) Matsumoto, M.; Ito, S. *J. Chem. Soc., Chem. Commun.* **1981**, 907. (c) Tovrog, B. S.; Diamond, S. E.; Mares, F. *J. Am. Chem. Soc.* **1979**, *101*, 5067. (d) Tang, R.; Diamond, S. E.; Neary, N.; Mares, F. *J. Chem. Soc., Chem. Commun.* **1978**, 562.
53. Minami, I.; Yamada, M.; Tsuji, J. *Tetrahedron Lett.* **1986**, *27*, 1805.
54. Griffith, W. P.; Ley, S. V. *Aldrichimica Acta* **1990**, *23*, 13.
55. Hünig, S.; Schaller, R. *Angew. Chem. Int. Ed. Engl.* **1982**, *21*, 36.
56. (a) Hertenstein, U.; Hünig, S.; Reichelt, H.; Schaller, R. *Chem. Ber.* **1982**, *115*, 261. (b) Oku, A.; Arita, S. *Bull. Chem. Soc. Jpn.* **1979**, *52*, 3337.
57. (a) El-abed, P. D.; Jellal, A.; Santelli, M. *Tetrahedron Lett.* **1984**, *25*, 1463. (b) Jellal, A.; Zahra, J.-P.; Santelli, M. *Ibid.* **1983**, *24*, 1395. (c) Jellal, A.; Santelli, M. *Ibid.* **1980**, *21*, 4487.
58. (a) Ando, T.; Kawate, T.; Yamawaki, J.; Hanafusa, T. *Synthesis* **1983**, 637. (b) Haase, K.; Hoffman, H. M. R. *Angew. Chem. Int. Ed. Engl.* **1982**, *21*, 83. (c) Hoffman, H. M. R.; Haase, K.; Ismail, Z. M.; Preftitsi, S.; Weber, A. *Chem. Ber.* **1982**, *115*, 3880. (d) Tanaka, M.; Koyanagi, M. *Synthesis* **1981**, 973.
59. (a) Herrmann, K.; Simchen, G. *Synthesis* **1979**, 204. (b) Olah, G. A.; Arvanaghi, M.; Surya Prakash, G. K. *Ibid.* **1983**, 636.
60. Kindler, K.; Peschke, W. *Arch. Pharm.* **1931**, *269*, 581.

61. Corey, E. J.; Gilman, N. W.; Ganem, B. E. *J. Am. Chem. Soc.* **1968**, *90*, 5616.
62. Corey, E. J.; Schmidt, G. *Tetrahedron Lett.* **1980**, *21*, 731.
63. Härle, H.; Jochims, J. C. *Chem. Ber.* **1986**, *119*, 1400.
64. Pochat, F. *Tetrahedron Lett.* **1977**, 3813.
65. Murahashi, S.-I.; Naota, T.; Yonemura, K. *J. Am. Chem. Soc.* **1988**, *110*, 8256.
66. (a) Roecker, L.; Meyer, T. J. *J. Am. Chem. Soc.* **1987**, *109*, 746. (b) Marmion, M. E.; Takeuchi, K. J. *J. Am. Chem. Soc.* **1986**, *108*, 510.
67. Sosnovsky, G.; Rawlinson, D. J. In *Organic Peroxides*; Swern, D. Ed.; Wiley: New York, 1971; Vol. 2, p. 595.
68. Greene, T. W. *Protective Groups in Organic Synthesis*; Wiley: New York, 1981, Chapter 7.
69. (a) 2-Nitrobenzyloxycarbonyl chloride: Amit, B.; Zehavi, U.; Patchornik, A. *J. Org. Chem.* **1974**, *39*, 192. (b) 1,2-Diphenylmaleic anhydride: Zehavi, U. *Ibid.* **1977**, *42*, 2819. (c) *N*-[(2*S*)-4-benzyloxycarbonylamino-2-hydroxybutanoyloxy]succinimide: Cron, M. J.; Keil, J. G.; Lin, J.-S.; Ruggeri, M. V.; Walker, D. *J. Chem. Soc., Chem. Commun.* **1979**, 266. (d) N-Acylation via metal complexing of an amino-hydroxy group pair: Hanessian, S.; Patil, G. *Tetrahedron Lett.* **1978**, 1035. Nagabhushan, T. L.; Cooper, A. B.; Turner, W. N.; Tsai, H.; McCombie, S.; Mallams, A. K.; Rane, D.; Wright, J. J.; Reichert, P.; Boxler, D. L.; Weinstein, J. *J. Am. Chem. Soc.* **1978**, *100*, 5253. (e) O to N-Acetyl migration: Tsuchiya, T.; Takagi, Y.; Umezawa, S. *Tetrahedron Lett.* **1979**, 4951. Thomas, M. B.; Williams, M. T. *Tetrahedron Lett.* **1980**, *21*, 4981.
70. Acyl cyanides have been used for protection of carbohydrates and hydroxysteroids: (a) Abbas, S. A.; Haines, A. H. *Carbohydrate Res.* **1975**, *39*, 358. (b) Havel, M.; Velek, J.; Pospíšek, J.; Souček, M. *Collect. Czech. Chem. Commun.* **1979**, *44*, 2443. (c) Goto, J.; Goto, N.; Shamsa, F.; Saito, M.; Komatsu, S.; Suzaki, K.; Nambara, T. *Anal. Chim. Acta* **1983**, *147*, 397. (d) Kubo, I.; Komatsu, S. *J. Chromatogr.* **1986**, *362*, 61.
71. (a) Review, Zhdanov, R. I.; Zhenodarova, S. M. *Synthesis* **1975**, 222. (b) Pentafluorophenyl benzoate: Igolen, J.; Morin, C. *J. Org. Chem.* **1980**, *45*, 4802. (c) Butyl chlorothioformate: Fujii, M.; Yamakage, S.; Takaku, H.; Hata, T. *Tetrahedron Lett.* **1987**, *28*, 5713. (d) Phenoxyacetic anhydride: Wu, T.; Ogilvie, K. K.; Pon, R. T. *Tetrahedron Lett.* **1988**, *29*, 4249. (e) 9-Fluorenylmethoxycarbonyl chloride: Koole, L. H.; Moody, H. M.; Broeders, N. L. H. L.; Quaedflieg, P. J. L. M.; Kuijpers, W. H. A.; van Genderen, M. H. P.; Coenen, A. J. J. M.; van der Wal, S.; Buck, H. M. *J. Org. Chem.* **1989**, *54*, 1657. (f) α,α-Dimethoxyamines: McBride, L. J.; Kierzek, R.; Beaucage, S. L.; Caruthers, M. H. *J. Am. Chem. Soc.* **1986**, *108*, 2040. (g) Multi-step method: Schulhof, J. C.; Molko, D.; Teoule, R. *Tetrahedron Lett.* **1987**, *28*, 51. Ti, G. S.; Gaffney, B. L.; Jones, R. A. *J. Am. Chem. Soc.* **1982**, *104*, 1316.
72. Kisfaludy, L.; Mohacsi, T.; Low, M.; Drexler, F. *J. Org. Chem.* **1979**, *44*, 654.
73. Kunieda, T.; Higuchi, T.; Abe, Y.; Hirobe, M. *Tetrahedron Lett.* **1982**, *23*, 1159.
74. Kunieda, T.; Higuchi, T.; Abe, Y.; Hirobe, M. *Tetrahedron* **1983**, *39*, 3253.
75. Nagao, Y.; Seno, K.; Kawabata, K.; Miyasaka, T.; Takao, S.; Fujita, E. *Chem. Pharm. Bull.* **1984**, *32*, 2687.
76. Oishi, T.; Kamata, K.; Kosuda, S.; Ban, Y. *J. Chem. Soc., Chem. Commun.* **1972**, 1148.
77. (a) Bergeron, R. J. *Acc. Chem. Res.* **1986**, *19*, 105. (b) Ganem, B. *Ibid.* **1982**, *15*, 290. (c) Badawi, M. M.; Bernauer, K.; van den Broek, P.; Groger, D.; Guggisberg, A.; Johne, S.; Kompis, I.; Schneider, F.; Veith, H.-J.; Hesse, M.; Schmid, H. *Pure Appl. Chem.* **1973**, *33*, 81.
78. Hlavka, J. J.; Bitha, P.; Boothe, J.; Fields, T. *J. Antibiot.* **1978**, *31*, 477.
79. Schmitz, F. J.; Hollenbeak, K. H.; Prasad, R. S. *Tetrahedron Lett.* **1979**, 3387.
80. (a) 1-Hydroxypiperidine esters: Husson, A.; Besselievre, R.; Husson, H.-P. *Tetrahedron Lett.* **1983**, *24*, 1031. (b) Acylimidazoles: Joshua, A. V.; Scott, J. R. *Tetrahedron Lett.* **1984**, *25*, 5725. (c) Carboxylic acids with benzo- or naphthosultones: Acher, F.; Wakselman, M. *J. Org. Chem.* **1984**, *49*, 4133. (d) *t*-Butylchlorodiphenylsilane: Overman, L. E.; Okazaki, M. E.; Mishra, P. *Tetrahedron Lett.* **1986**, *27*, 4391. (e) *N*-Methoxydiacetamide: Kikugawa, Y.; Mitsui, K.; Sakamoto, T.; Kawase, M.; Tamiya, H. *Tetrahedron Lett.* **1990**, *31*, 243.

81. (a) Wälchli-Schaer, E.; Eugster, C. H. *Helv. Chim. Acta* **1978**, *61*, 928. (b) Humora, M.; Quick, J. *J. Org. Chem.* **1979**, *44*, 1166. (c) Tice, C. M.; Ganem, B. *J. Org. Chem.* **1983**, *48*, 2106. (d) Nordlander, J. E.; Payne, M. J.; Balk, M. A.; Gress, J. L.; Harris, F. D.; Lane, J. S.; Lewe, R. F.; Marshall, S. E.; Nagy, D.; Rachlin, D. J. *Ibid.* **1984**, *49*, 133. (e) Sundaramoorthi, R.; Marazano, C.; Fourrey, J.-L.; Das, B. C. *Tetrahedron Lett.* **1984**, *25*, 3191.
82. Steglich, W.; Steffan, B.; Stroech, K.; Wolf, M. *Z. Naturforsch., C: Biosci.* **1984**, *39C*, 10.
83. Englert, G.; Klinga, K.; Raymond-Hamet; Schlittler, E.; Vetter, W. *Helv. Chim. Acta* **1973**, *56*, 474.
84. Raymond, K. N.; Muller, G.; Matzanke, B. F. *Top. Curr. Chem.* **1984**, *123*, 50.
85. Tait, G. H. *Biochem. J.* **1975**, *146*, 191.
86. Bergeron, R. J.; McManis, J. S.; Dionis, J. B.; Garlich, J. R. *J. Org. Chem.* **1985**, *50*, 2780.
87. Bergeron, R. J.; Kline, S. J. *J. Am. Chem. Soc.* **1982**, *104*, 4489.
88. Murahashi, S.-I.; Naota, T.; Nakajima, N. *J. Org. Chem.* **1986**, *51*, 898.
89. Tsuji, J.; Ohno, K. *Synthesis* **1969**, 157.
90. Ohno, K.; Tsuji, J. *J. Am. Chem. Soc.* **1968**, *90*, 99.
91. Hanthorn, J. O.; Wilt, M. H. *J. Org. Chem.* **1960**, *25*, 2215.
92. Doughty, D. H.; Pignolet, L. H. *J. Am. Chem. Soc.* **1978**, *100*, 7083.
93. Verbicky, J. W., Jr.; Dellacoletta, B. A.; Williams, L. *Tetrahedron Lett.* **1982**, *23*, 371.
94. Blum, J.; Rosenman, H.; Bergmann, E. D. *J. Org. Chem.* **1968**, *33*, 1928.
95. Tsuji, J.; Ohno, K. *J. Am. Chem. Soc.* **1968**, *90*, 94.
96. Blum, J.; Oppenheimer, E.; Bergmann, E. D. *J. Am. Chem. Soc.* **1967**, *89*, 2338.
97. (a) Friedrich, K.; Wallenbells, K. *The Chemistry of the Cyano Group*; Patai, S., Ed.; Wiley: New York, 1970; p. 67. (b) Other methods: Hydrazone formation followed by oxidation. Smith, R. F.; Albright, J. A.; Waring, A. M. *J. Org. Chem.* **1966**, *31*, 4100. Binkley, R. W. *Tetrahedron Lett.* **1970**, 2085. Condensation with NH_3 followed by oxidation. Nakagawa, K.; Mineo, S.; Kawamura, S.; Horikawa, M.; Tokumoto, T.; Mori, O. *Synth. Commun.* **1979**, *9*, 529 and references cited therein.
98. Lau, K. S. Y.; Becker, Y.; Huang, F.; Baenziger, N.; Stille, J. K. *J. Am. Chem. Soc.* **1977**, *99*, 5664.
99. Egglestone, D. L.; Baird, M. C.; Lock, C. J. L.; Turner, G. *J. Chem. Soc., Dalton Trans.* **1977**, 1576.

PALLADIUM-CATALYZED CARBONYL ALLYLATION VIA π-ALLYLPALLADIUM COMPLEXES

Yoshiro Masuyama

Advances in Metal-Organic Chemistry
Volume 3, pages 255–303.

ISBN: 1-55938-406-9

I. INTRODUCTION

π-Allylpalladium complex has been usually applied as an allylic cation in organic syntheses.[1] The rapid development of the chemistry of π-allylpalladium complex is due to the ease of preparation and storage of starting materials, allylic esters, and due to the application to regio- and stereoselection that are one of the important subjects in organic synthesis. Allylic ester is therefore superior to allylic halide as a synthon of allylic cation. However, from the viewpoint of the wide variety of the formation of allylic skeleton, the chemistry of π-allylpalladium complex derived from allylic ester has not always been superior to that of allylic halide, because allylic halide can be utilized not only as a synthon of allylic cation, but also as that of allylic anion with metal reagents.[2–20] Therefore, umpolung of π-allylpalladium complex, namely allylic ester, will be an important theme.[21,22] Here, we shall focus on the transformation of allylic alcohol or ester **1** to synthon of allylic anion **3** via π-allylpalladium complex **2**,[23–28] followed by carbonyl allylation (Scheme 1).[2–20]

1, X:H, Ac, COOMe   3

Carbonyl allylation can be applied to acyclic stereocontrol, which is the most attractive subject in modern organic synthesis.[29] Diastereoselection has been studied in detail for the addition reaction of 2-butenylmetal compounds **4** (R^1:CH_3) to aldehydes; it is effected by coordination of either carbonyl group to metal (M) or carbonyl group to Lewis acid (Scheme 2).

The carbonyl allylation by allylic alcohol or ester via the formation of π-allylpalladium complex has high possibilities in acyclic stereocontrol, since various allylic alcohols and esters, such as (*E*)- and (*Z*)-3-substituted allylic alcohols and

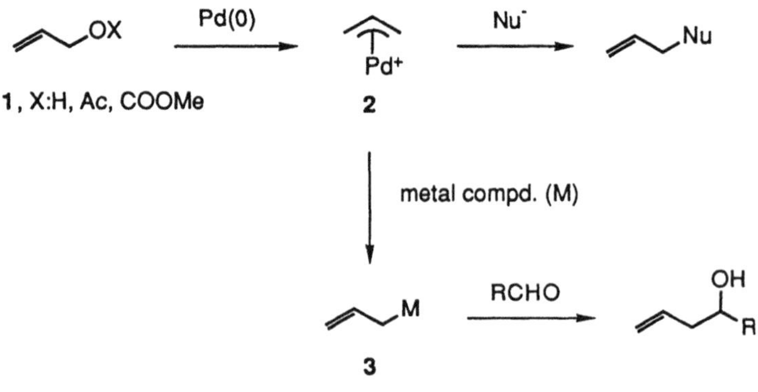

Scheme 1.

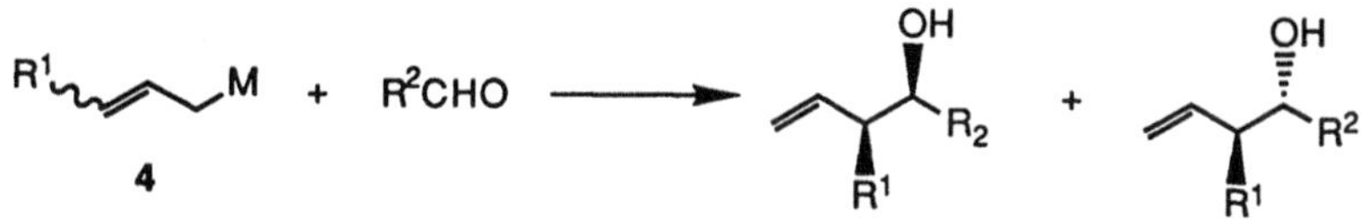

4

without Lewis acid			
(E)	*syn*	<	*anti*
(Z)	*syn*	>	*anti*
with Lewis acid			
(E) and *(Z)*	*syn*	>	*anti*

Scheme 2.

esters and 1-substituted ones, are prepared regio- and stereoselectively.[30] Further, the advantage of this carbonyl allylation is that 1-substituted allylic alcohols **6** (X:H), easily prepared from aldehydes and vinyl lithium, probably function as well as (*E*)-3-substituted allylic ones **5** (X:H), since the actual allylating agents are probably allylic metal compounds **8** derived from same π-allylpalladium complexes **7** (Scheme 3). Palladium-catalyzed carbonyl allylation by 1-substituted allylic alcohols may be made to occur with either *syn*- or *anti*-selection by the choice of reaction conditions.

In this review we shall first discuss some of our initial studies in charge reversal of π-allylpalladium complexes followed by carbonyl allylation with allylic esters, while explaining some background of the chemistry of the charge reversal followed by the carbonyl allylation with allylic metal compounds. Next, the recent progress of palladium-catalyzed carbonyl allylation by allylic alcohols with stannous chloride,[31] based on the same conception as that by allylic esters, will be demonstrated; we shall try to show the originality of this carbonyl allylation, that is, how allylic alcohols function as synthons of allylic carbanions, to elucidate the mechanism of this carbonyl allylation by NMR spectroscopic investigation, and to establish a solvation-controlled diastereoselection and a chelation-controlled one.

II. CARBONYL ALLYLATION VIA CHARGE REVERSAL OF π-ALLYLPALLADIUM COMPLEX

We shall first focus on apparent charge reversal of π-allylpalladium complex with various metal reagents (regiocontrolled transmetallation) followed by regio- and stereocontrolled carbonyl allylation, using allylic esters as starting materials.

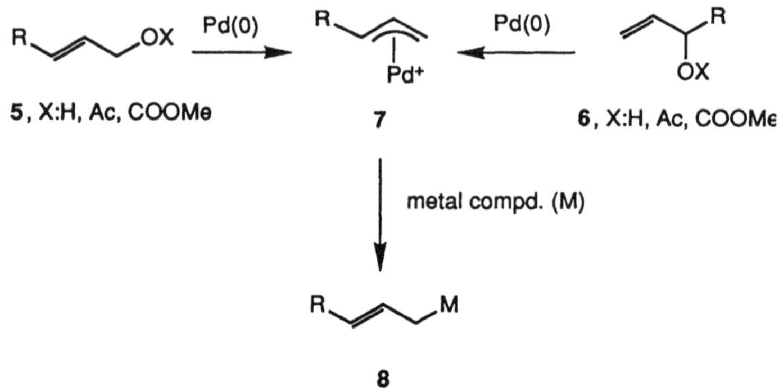

Scheme 3.

A. Apparent Charge Reversal of π-Allylpalladium Complex (Transmetallation)

1. Fundamental Conceptions and Short History

Is apparent charge reversal of π-allylpalladium complex with metal reagent, namely transmetallation, possible? Can metal reagents that have been used in the allylmetallation of allylic halide be applied to the transmetallation? The choice of metal reagent is important: (1) One electron transfer (reduction) must not take place from the metal reagent to π-allylpalladium complex to obtain allylic homocoupling product,[32] (2) the metal reagent has to attack π-allylpalladium complex regioselectively, and (3) the accompanying nucleophilic addition of prepared allylmetal intermediate to carbonyl compound has to occur in the same pot. In initial stages of our studies, a few transmetallations of this kind had already been reported. Stannylaluminum reagents have been utilized for carbonyl allylation via the formation of π-allylpalladium complex without allylic homocoupling, using allylic acetates in 1984[23] and allylic phosphates in 1985 (Scheme 4).[24] The carbonyl allylation by allylic acetates is not a real one-pot method. That is, allylic tin intermediate **9** has actually been isolated and then used in the carbonyl allylation. (*E*)-Crotyl phosphate is allowed to cause the carbonyl allylation without isolation of (*E*)-crotyl tin intermediate **10** in one pot. Regiocontrol is preferable for the investigation of diastereocontrolled carbonyl allylation, since carbonyl compounds apparently attack the more substituted allylic position of π-allylpalladium complexes. However, the diastereoselectivity is not so high. This stannylaluminum reagent has been used for the cross coupling of allylic acetates with aromatic or vinyl bromides to lead to cyclization (Scheme 5).[28] One pot carbonyl allylation by available allylic acetates has been carried out with samarium iodide in 1986 (Scheme 6).[25] Regiocontrol cannot usually be controlled because of high reactivity

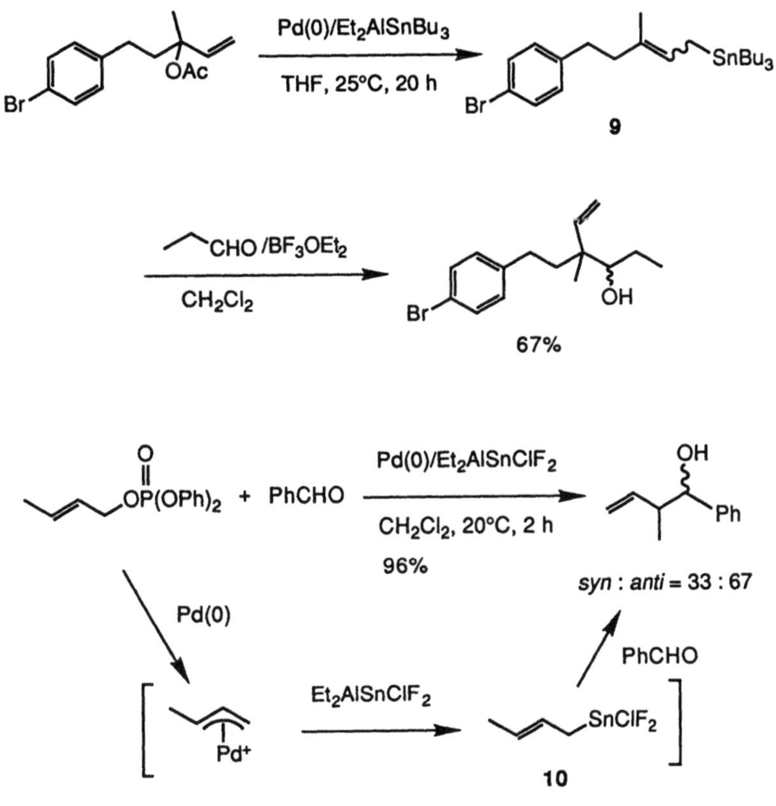

Scheme 4.

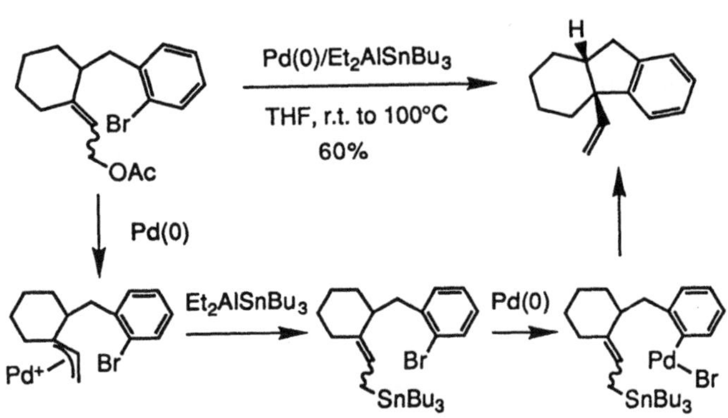

Scheme 5.

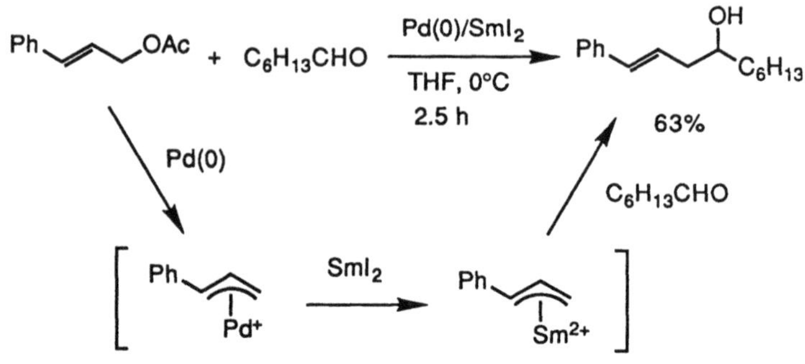

11

Scheme 6.

of η^3-allylsamarium complex such as **11**. Samarium iodide cannot be used in the allylation of aromatic or α,β-unsaturated carbonyl compounds, because it causes pinacol-type self-coupling. Using Pd(0)-samarium iodide, allylic acetates can be allowed to react with chlorosilane to give allylic silanes.[27]

2. Carbonyl Allylation by Allylic Acetates with $Pd(PPh_3)_4$-Zn[33]

Are the limitations suggested above solvable by using other metal reagents or are they essential in this kind of carbonyl allylation? The possibilities of the carbonyl allylation via the formation of π-allylpalladium complex led us to examine other metal reagents. The suitability of metal reagents that are available in the reaction of allylic halides[2–20] was investigated under several reaction conditions. Zinc powder is useful in palladium-catalyzed carbonyl allylation by allylic acetates in dry dioxane at room temperature. The carbonyl allylations with the low-valent metal compounds, such as $SnCl_2$,[5,6] Sn,[11] $CrCl_3$-$LiAlH_4$,[9] $MnCl_2$-$LiAlH_4$,[10] and VCl_3-$LiAlH_4$,[34] instead of Zn, do not occur significantly under the same conditions. Benzene, tetrahydrofuran (THF), acetonitrile, and N,N-dimethylformamide (DMF) as solvents inhibit the reaction dramatically. This carbonyl allylation requires more than 5 equimolar amounts of allylic acetates to aldehydes, owing to the enhancement of the yields based on the aldehydes. This demerit is because of the homocoupling of the acetates via one electron transfer to π-allylpalladium complex **12** as mentioned above (Scheme 7).[32]

Addition reaction of 2-(E)-butenyl acetate to benzaldehyde with $Pd(PPh_3)_4$. To a suspension of zinc powder (0.65 g, 10 mg atom), benzaldehyde (0.21 g, 2.0 mmol), and $Pd(PPh_3)_4$ (46 mg, 0.04 mmol) in dioxane (3 mL) is added 2-(*E*)-butenyl acetate (1.1 g, 10 mmol) dropwise for 24 h at room temperature under a nitrogen atmosphere. After being stirred for an additional 3 days at room temperature, the reaction mixture is poured into water (30 mL) and extracted with ether (2 × 25 mL). The combined extracts are dried over anhydrous $MgSO_4$. Evaporation of ether and purification by column chromatography on silica gel (hexane/ethyl acetate = 5/1) afford 0.23 g (71%, *syn*/*anti* = 55/45) of 2-methyl-1-phenyl-3-butenol as a colorless oil.

The results of the allylation of aldehydes by various allylic acetates with $Pd(PPh_3)_4$- Zn are summarized in Table 1. Aromatic and α,β-unsaturated aldehydes can be employed in this Pd(0)-Zn system without pinacol-type self-coupling.[25] Aldehydes regioselectively attack the more substituted allylic position of the π-allylpalladium complex **12** to give a single regioisomer (Scheme 7). This carbonyl allylation seems to proceed via the formation of allylzinc intermediates **13**.[2,18,20] The same regioselection can be observed in the carbonyl allylation by both geranyl acetate (**14**) and linalyl acetate (**15**) via the formation of 1,1-disubstituted π-allylpalladium complex (Scheme 8). This carbonyl allylation consequently exhibits slight diastereoselectivity (*syn* selectivity), except for the cases of 1-phenyl-

Table 1. Carbonyl Allylation by Allylic Acetates with Pd(0)-Zn[a]

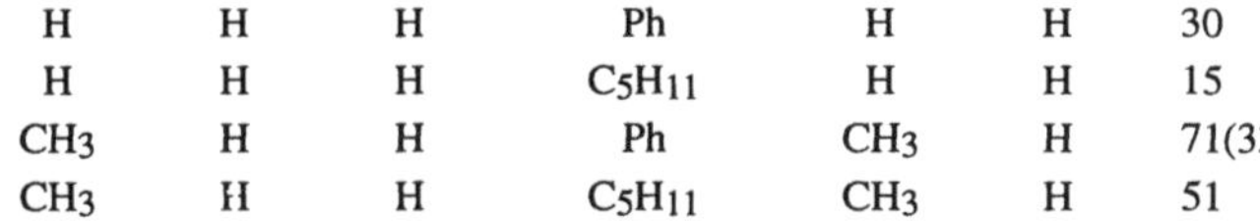

	Acetate			Aldehyde	Product			
Entry	R^1	R^2	R^3	R^4	*R*	*R'*	*Yield(%)*[b]	*Syn:Anti*[c]
1	H	H	H	Ph	H	H	30	—
2	H	H	H	C_5H_{11}	H	H	15	—
3	CH_3	H	H	Ph	CH_3	H	71(33[d])	55:45
4	CH_3	H	H	C_5H_{11}	CH_3	H	51	58:42
5	CH_3	H	H	$CH_3CH{=}CH$	CH_3	H	20	58:42
6	H	H	CH_3	Ph	CH_3	H	37[d]	53:47
7	CH_3	CH_3	H	Ph	CH_3	CH_3	99	—
8	Ph	H	H	Ph	Ph	H	70	13:87
9	Ph	H	H	$CH_3CH{=}CH$	Ph	H	90	34:66
10	H	H	Ph	Ph	Ph	H	57[d]	17:83
11	H	H	C_7H_{15}	Ph	C_7H_{15}	H	26[d]	73:27
12	H	H	C_7H_{15}	C_5H_{11}	C_7H_{15}	H	33[d]	58:42

Notes: [a]Addition reaction of allylic acetates (10 mmol) to aldehydes (2 mmol) with $Pd(PPh_3)_4$ (0.04 mmol) and Zn (10 mmol) is carried out in dioxane at room temperature for 4–5 days.

[b]The yields of isolated pure products based on aldehydes. No other regioisomer is detected by 270-MHz ^1H NMR.

[c]The ratio is determined by GLPC or 270-MHz ^1H NMR.

[d]Two equivalents of the acetate are used.

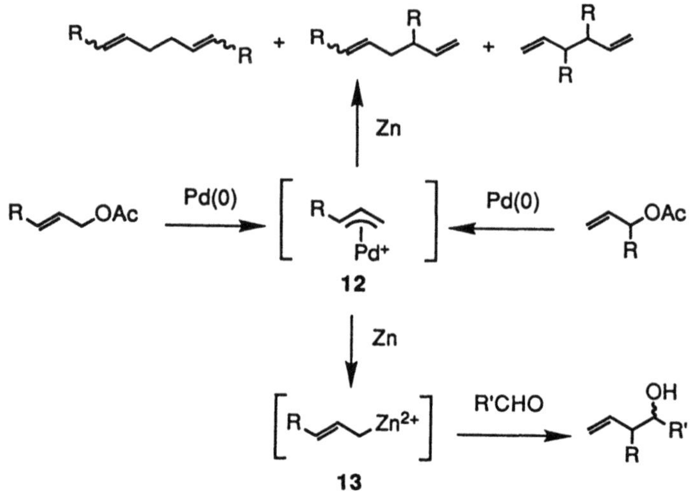

Scheme 7.

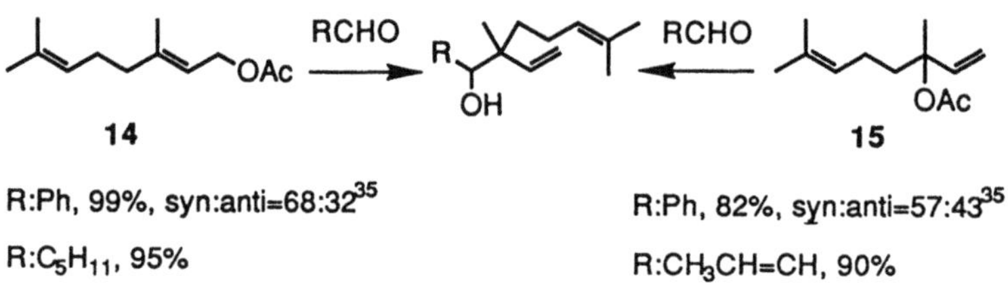

Scheme 8.

2-propenyl acetate and 3-phenyl-2-propenyl acetate that cause *anti*-addition. Ketones such as 4-*t*-butylcyclohexanone and acetophenone don't react under the same conditions. As shown in Scheme 9, the allylation of an aldehyde is chemoselectively performed in the presence of a ketone group or an ester group.[9,16]

3. *Carbonyl Allylation by Allylic Acetates with $PdCl_2(PhCN)_2$-$SnCl_2$*[36]

The $Pd(PPh_3)_4$-Zn system can lead to the satisfactory results for regio- and chemoselection. However, the problems of suppression of allylic homocoupling and enhancement of diastereoselectivity still remain. Therefore, we continued to try to find other Pd-metal systems which avoid those demerits of the $Pd(PPh_3)_4$-Zn system. Stannous chloride, which has been employed in carbonyl allylation by allylic bromides,[5,6] can be an active reagent for this palladium-catalyzed carbonyl allylation with neither allylic homocoupling nor pinacol-type self-coupling even at

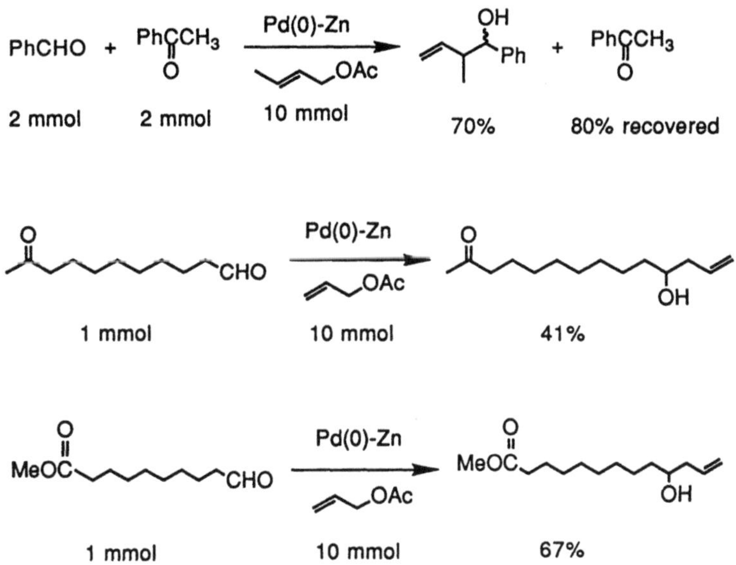

Scheme 9.

60 °C.[37] 1,3-Dimethylimidazolidin-2-one (DMI) (50 °C, 20 h, 82%) is a better solvent in the allylation of benzaldehyde by 2-propenyl acetate with $Pd(PPh_3)_4$-$SnCl_2$, than are THF (reflux, 72 h, 14%), dioxane (50 °C, 72 h, 0%), and DMF (50 °C, 35 h, 69%). Insoluble stannous fluoride does not serve for this carbonyl allylation in any polar solvent such as THF, DMF, and DMI. $PdCl_2(PhCN)_2$, which is easily prepared and is highly stable to air, has the same activity as that of $Pd(PPh_3)_4$ (60 °C, 15 h, 82%). Therefore, $PdCl_2(PhCN)_2$ will be employed in the allylation of various aldehydes by various allylic acetates with $SnCl_2$ in DMI as shown below. The Pd(II) catalyst, $PdCl_2(PhCN)_2$, may be reduced to Pd(0) complex by $SnCl_2$ to form π-allylpalladium complex in this system.[1]

The palladium-catalyzed carbonyl allylations by 2-propenyl acetate are summarized in Table 2. These reaction rates are faster than those with $Pd(PPh_3)_4$-Zn. The carbonyl allylation with $PdCl_2(PhCN)_2$-$SnCl_2$ system exhibits the same chemoselectivity to aldehyde as that with $Pd(PPh_3)_4$-Zn system (entries 7–9). The allylation of acetophenone does not occur at 60 °C in DMI. However, the addition of cesium fluoride or tetrabutylammonium bromide (TBA) realizes the allylation of acetophenone at 60 °C for 48 h in 19% and 48% yields respectively.[38] 2-Propenyltrihalotin intermediate **16**, which is more active than 2-propenylacetoxydichlorotin in the absence of TBA, may be prepared from π-allylpalladium complex and stannous chloride with TBA, or pentacoordinated 2-propenyltin compound **17**, which is effective for the allylation, may be formed from **16** and another TBA (Scheme 10).[39]

Table 2. Carbonyl Allylation by 2-Propenyl Acetate with $PdCl_2(PhCN)_2$-$SnCl_2$[a]

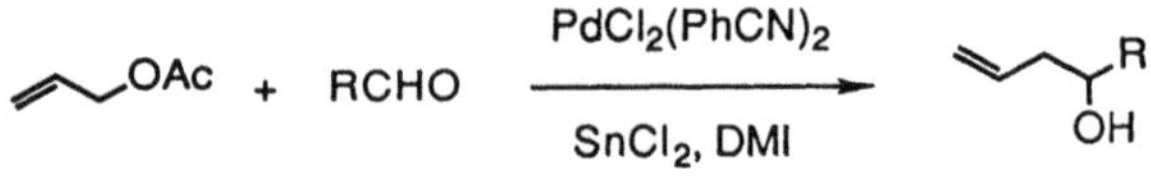

Entry	*Aldehyde*	*Temp.(°C)*	*Time(h)*	*Yield(%)*[b]
1	hexanal	60	19	54
2	undec-10-enal	60	17	76
3	benzaldehyde	60	15(25)[c]	82(75)[c]
4	piperonal	25	60	63
5	cinnamaldehyde	25	57	60
6	citral	25	59	50
7	10-oxoundecanal	50	20	68
8	$MeOOC(CH_2)_4CHO$	25	48	84
9	$MeOOC(CH_2)_8CHO$	50	20	70

Notes: [a]Addition reaction of 2-propenyl acetate (3 mmol) to aldehydes (1 mmol) with $PdCl_2(PhCN)_2$ (0.02 mmol) and $SnCl_2$ (3 mmol) is carried out in DMI (5 ml).

[b]The yields of isolated pure products are based on the aldehydes.

[c]One equimolar amount of 2-propenyl acetate is used.

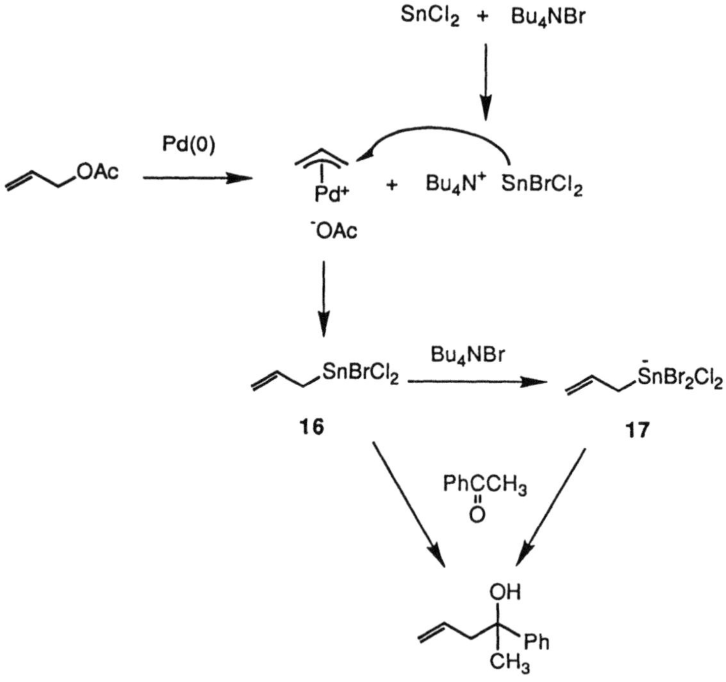

Scheme 10.

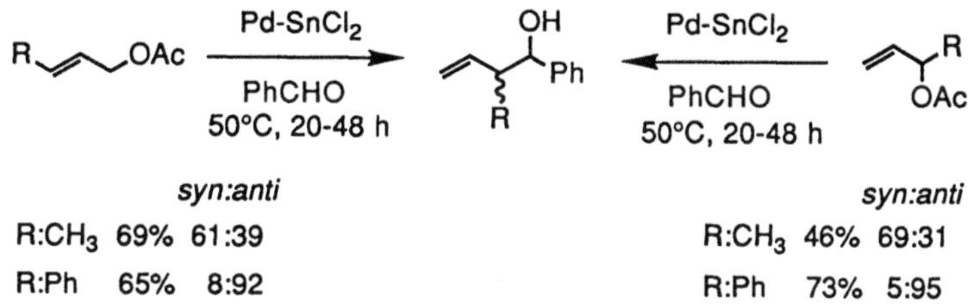

Scheme 11.

Chemoselective carbonyl allylation by 2-propenyl acetate with $PdCl_2(PhCN)_2$-$SnCl_2$. To a solution of stannous chloride (0.57 g, 3 mmol), 2-propenyl acetate (0.30 g, 3 mmol), and 10-oxoundecanal (0.18 g, 1 mmol) in DMI (5 ml) is added $PdCl_2(PhCN)_2$ (8 mg, 0.02 mmol) at room temperature under a nitrogen atmosphere. After being stirred at 50 °C for 20 h, the reaction mixture is diluted with ether (20 ml). After removal of precipitate, the mixture is poured into water (30 ml), and extracted with ether-methylene chloride (3:2, 100 ml). The solution is washed with 10% aqueous HCl sol. followed by brine, and dried over anhydrous $MgSO_4$. Evaporation of solvent and purification by column chromatography on silica gel (hexane/ethyl acetate = 2/1) afford 0.14 g (62%) of 13-oxotetradec-1-en-4-ol as a white solid.

This carbonyl allylation with $PdCl_2(PhCN)_2$-$SnCl_2$ possesses the same regioselectivity as that in the carbonyl allylation with $Pd(PPh_3)_4$-Zn. That is to say, aldehydes attack γ-position of 3-substituted allyltin intermediates derived from 1- and 3-substituted π-allylpalladium complexes with $SnCl_2$.[37] The representative results of the addition reaction to benzaldehyde are illustrated in Scheme 11. Regioselection and even diastereoselection in the reaction of geranyl acetate (**14**) are the same as those of linalyl acetate (**15**) (Scheme 12). Neryl acetate (**18**) also causes *syn* addition similar to **14**. The reaction of **14** and **18**, bearing a trisubstituted olefinic moiety, is slower than that of monosubstituted olefinic **15**.

4. Carbonyl Allylation by 2-Propenyl Carbonate with $PdCl_2(PhCN)_2$-$SnCl_2$[40]

The carbonyl allylation by allylic acetates with $PdCl_2(PhCN)_2$-$SnCl_2$ leaves diastereocontrol unsolved. The poor diastereoselectivity is presumably ascribable to the relatively high reaction temperature (50–60 °C), which the formation of π-allylpalladium complex from allylic acetate requires. 2-Propenyl carbonate, which is more reactive than 2-propenyl acetate for the complexation with Pd(0) catalyst,[41] can be employed at relatively low temperatures (10–25 °C) in the

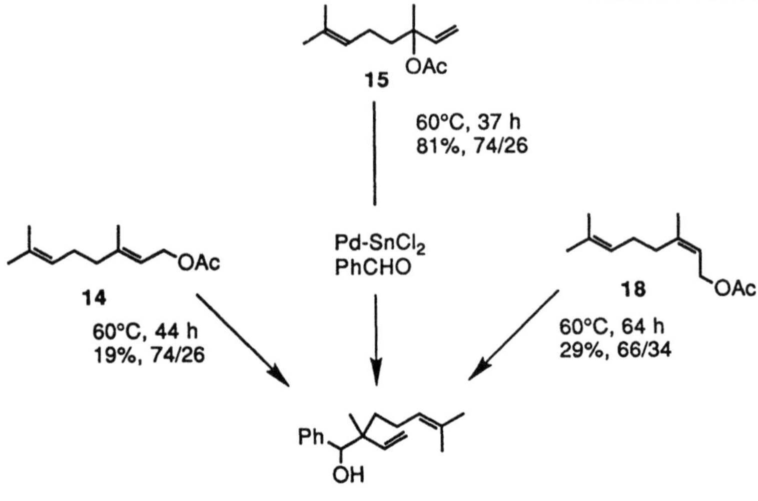

syn/anti

Scheme 12.

carbonyl allylation with $PdCl_2(PhCN)_2$-$SnCl_2$ (Table 3). The reactivity of a ketone, namely 4-*t*-butylcyclohexanone, is very low even at 60 °C (entry 6).

High chemoselectivity toward a carbonate group is observed in the reaction of 4-acetoxy-(*Z*)-2-butenyl methyl carbonate (**19**) with benzaldehyde under the same conditions, similarly to the palladium-catalyzed reaction of **19** with nucleophile (Scheme 13).[41] The diastereoselectivity is confirmed by ^{1}H NMR investigation of 1,3-dioxanes **21a** and **21b** prepared by reduction of **20a** and **20b** with DIBAH

Table 3. Carbonyl Allylation by 2-Propenyl Carbonate with $PdCl_2(PhCN)_2$-$SnCl_2$

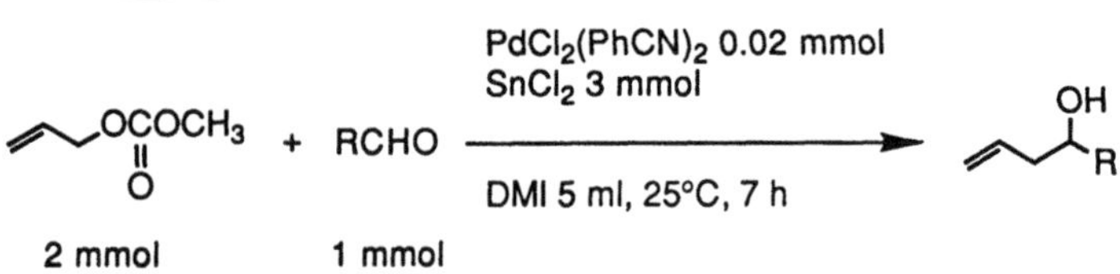

Entry	*Aldehyde*	*Yield(%)*
1	hexanal	49
2	octanal	57
3	10-undecenal	75
4	piperonal	95
5	cinnamaldehyde	92
6	4-*t*-butylcyclohexanone	8[a]

Note: [a]The reaction is carried out at 60 °C for 48 h.

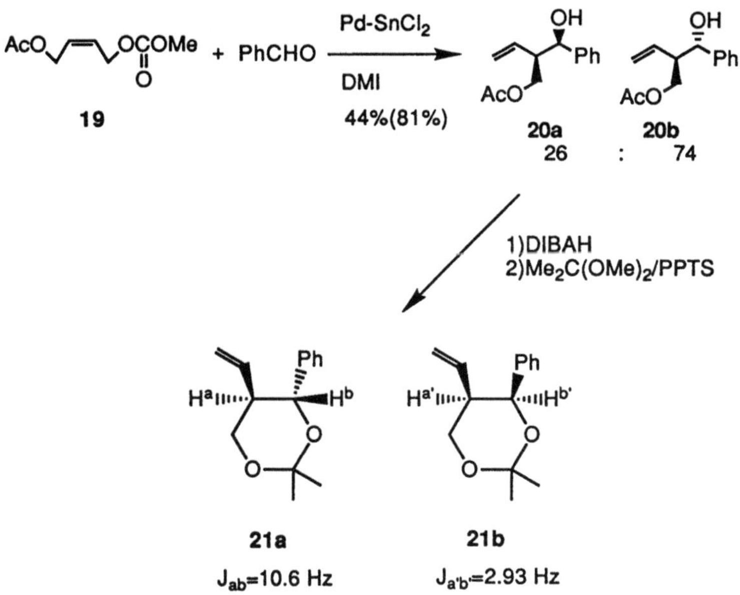

Scheme 13.

followed by acetalization with 2,2-dimethoxypropane.[42] Diastereoselection in the carbonyl allylation by 2-butenyl carbonates will be described in the next section along with that by 2-(*E*)-butenyl acetate under various conditions.

Chemoselection between acetate and carbonate in palladium-catalyzed carbonyl allylation with $SnCl_2$. To a solution of stannous chloride (0.29 g, 1.5 mmol), acetoxycarbonate **19** (0.19 g, 1.0 mmol), benzaldehyde (0.12 g, 1.2 mmol) in DMI (3 ml) is added $PdCl_2(PhCN)_2$ (8 mg, 0.02 mmol). After stirring for 21 h at 25 °C under a nitrogen atmosphere, ether (20 ml) is poured into the reaction mixture and then the precipitate is removed. The mixture is poured into water (30 ml), and is extracted with ether-methylene chloride (3:2, 100 ml). The solution is washed with 10% HCl sol. followed by brine, and is dried over anhydrous $MgSO_4$. Evaporation of solvent and purification by column chromatography on silica gel (hexane/ethyl acetate = 2/1) afford 97 mg (44%) of **20a** and **20b** as a colorless oil.

B. Possibility of Diastereoselection in Carbonyl Allylation via Charge Reversal

Using both the Pd-Zn system and the Pd-$SnCl_2$ one, we tried to control a diastereoselection in the addition reaction of 2-(*E*)- and 2-(*Z*)-butenyl acetate or carbonate to benzaldehyde.

Table 4. Effects of Metal, Ligand, and Temperature on Diastereoselection

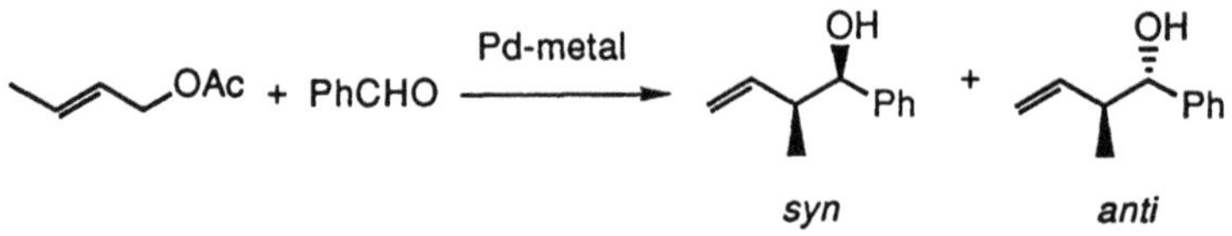

Entry	Catalyst	Metal	Solvent	Temp. (°C)	Time (h)	Yield (%)	Syn:Anti
1	$Pd(PPh_3)_4$	Zn	dioxane	25	120	71	55:45
2	$Pd(OAc)_2$-$4PPh_3$	Zn	dioxane	25	120	82	52:48
3	$PdCl_2(PPh)_2$	Zn	dioxane	25	120	82	48:52
4	$Pd(PPh_3)_4$-4dppm	Zn	dioxane	25	120	81	57:43
5	$Pd(PPh_3)_4$-4dppe	Zn	dioxane	25	120	71	46:54
6	$Pd(PPh_3)_4$-4dppp	Zn	dioxane	25	120	50	46:54
7	$Pd(PPh_3)_4$	$SnCl_2$	DMI	100	19	49	50:50
8	$Pd(PPh_3)_4$	$SnCl_2$	DMI	50	20	71	59:41
9	$PdCl_2(PPh_3)_2$	$SnCl_2$	DMI	100	18	51	53:47
10	$PdCl_2(PPh_3)_2$	$SnCl_2$	DMI	60	20	71	59:41
11	$PdCl_2(PhCN)_2$	$SnCl_2$	DMI	100	16	64	50:50
12	$PdCl_2(PhCN)_2$	$SnCl_2$	DMI	60	20	69	61:39
13	$PdCl_2(PhCN)_2$	$SnCl_2$	DMI	25	90	61	43:57
14	$PdCl_2(PhCN)_2$	$SnCl_2$	DMI	–20[a]	100	10	36:64

Note: [a]2-(*E*)-Butenyl acetate is stirred with $PdCl_2(PhCN)_2$ and $SnCl_2$ in DMI-THF(1/1) at 50°C for 10 h, and then benzaldehyde is added to the solution at –20 °C.

The effects of phosphine ligand on palladium(0), metal, and temperature were investigated in the addition reaction of 2-(*E*)-butenyl acetate to benzaldehyde, as summarized in Table 4.[43] The yield and the diastereoselectivitiy do not change with bulkiness of phosphine ligand of palladium catalyst. Therefore, those should not be dependent upon the structure of π-allylpalladium complexes. Since the *syn*/*anti* ratio is about 1:1 using the Pd(0)-Zn system, this carbonyl allylation does not seem to clearly proceed either via a chair form of six-membered cyclic transition state or via an acyclic antiperiplanar transition state (entries 1–6).[29] In the Pd(0)-$SnCl_2$ system, the reactions at 60 °C exhibit slight *syn* selectivity, in contrast to the nonselective reactions at 100 °C (entries 7–12). The reaction at 25 °C leads to *anti* selection, which is opposite to that at 60 °C (entry 13). The reaction does not occur below 10 °C. Thus, we tried the formation of 2-butenyltin intermediate at 50 °C followed by the addition reaction to benzaldehyde at –20 °C. However, the reaction is very slow and the *anti* selectivity is similar to that in the simple one-pot reaction at 25 °C (entry 14). Addition of bidentate phosphine ligands such as dppe and dppp inhibits the carbonyl allylation with $SnCl_2$, in contrast with the results for the Pd(0)-Zn system.

Table 5. Effect of Olefin Geometry on Diastereoselection

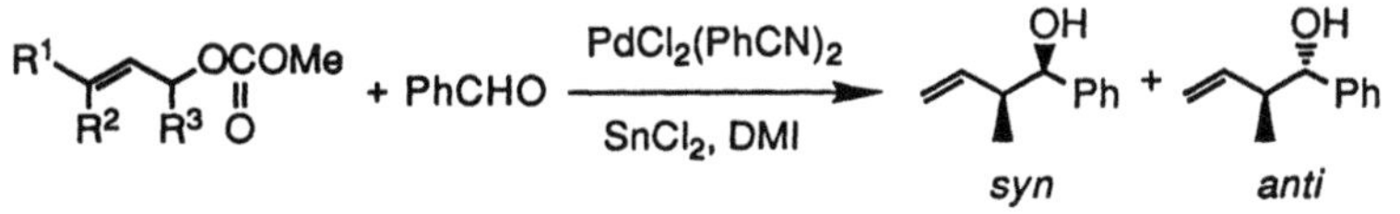

Entry	*Allylic Carbonate* R^1	R^2	R^3	*Temp.* (°C)	*Time* (h)	*Yield* (%)	*Syn:Anti*
1	CH_3	H	H	25	48	80	33:67
2	CH_3	H	H	10	72	90	27:73
3	H	CH_3	H	25	60	42	48:52
4	H	H	CH_3	10	84	95	31:69

2-(*E*)-Butenyl **22**, 2-(*Z*)-butenyl **23**, and 1-methyl-2-propenyl carbonate **24**, which are more reactive than the corresponding acetates, are available for the investigation of effect of olefin geometry in the diastereoselection below 25 °C.[40] The results are summarized in Table 5. Addition of **22** at 10 °C leads to high *anti* selection compared to that of 2-(*E*)-butenyl acetate (entry 2). (*Z*)-Carbonate **23** does not cause diastereoselective addition (entry 3). Addition of **24** exhibits the same diastereoselectivity value as that of **22** (entry 4). Therefore, 1-substituted allyl carbonate can be used in place of (*E*)-3-substituted allyl carbonate.

As a result, *syn*-π-allylpalladium complex **25a**, which is prepared from both **22** and **24** with Pd(0), reacts with $SnCl_2$ to produce 2-(*E*)-butenyltin intermediate **26a**, which causes nucleophilic addition at γ-position to benzaldehyde, as illustrated in Scheme 14. The *anti* selection can be explained by a chair form of six-membered cyclic transition state.[29] The minor *syn* addition seems to be based on an acyclic transition state, since coordination of DMI instead of benzaldehyde to Sn(IV) disturbs the formation of the six-membered cyclic transition state. On the other hand, 2-(*Z*)-butenylmetal compounds cause *syn* addition either via the cyclic transition state or via the acyclic transition state.[29] However, the carbonyl allylation by 2-(*Z*)-butenyl carbonate **23** leads to nondiastereoselection (entry 3). This result can be demonstrated by the isomerization of *anti*-π-allylpalladium complex **25b**, which is initially prepared from **23** with Pd(0), to stable *syn* complex **25a** (Scheme 14).

Even if reactive allylic carbonate is used, the carbonyl allylation requires higher reaction temperature than 10 °C. Therefore, the development of diastereoselection in this kind of carbonyl allylation needs either: (1) the formation of π-allylpalladium complex followed by that of allylmetal intermediate at lower temperature (–80 °C); or (2) other control methods such as chelation-control.

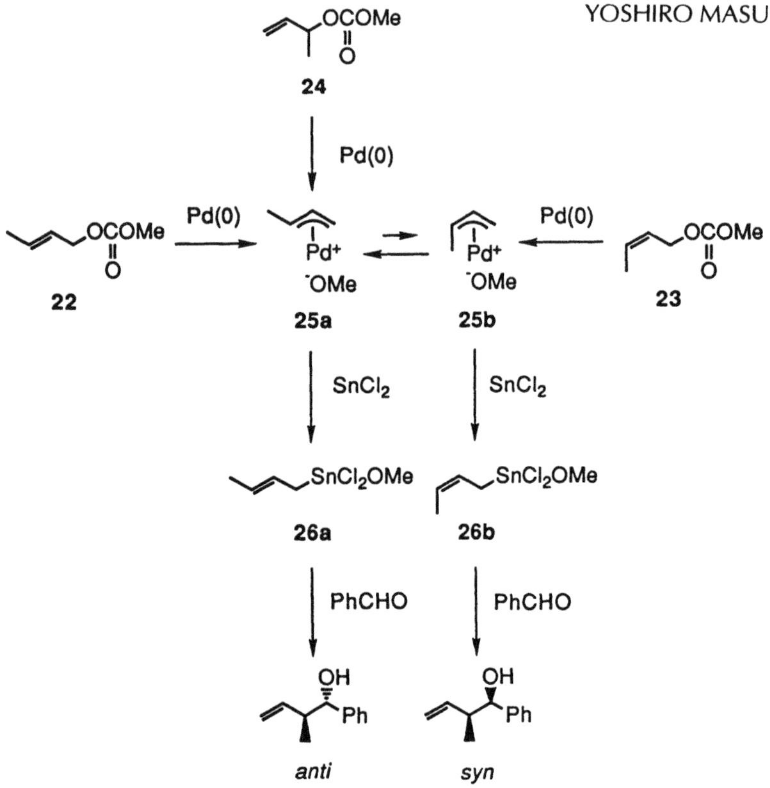

Scheme 14.

III. PALLADIUM-CATALYZED CARBONYL ALLYLATION BY ALLYLIC ALCOHOLS

We shall establish (1) the application of allylic alcohols to the palladium-catalyzed carbonyl allylation, (2) mechanism of the carbonyl allylation by NMR spectroscopic investigation, and (3) diastereoselection by means of solvation-control and chelation-control method. It will be very important that this palladium-catalyzed carbonyl allylation by allylic alcohols can be carried out *in aqueous solution under air.*

A. Allylic Alcohol as Synthon of Allylic Carbanion

Can allylic alcohol be used as a synthon of allylic carbanion? The achievement of this procedure may overthrow the nucleophilic allylation method using allylic halide, which is one of the established organic synthetic methods.[2–20] If π-allylpalladium complex could be formed from allylic alcohol with Pd(0), allylic alcohol would be employed as a nucleophilic allylating agent, namely allylic carbanion, instead of allylic ester in palladium-catalyzed carbonyl allylation. Allylic alcohol

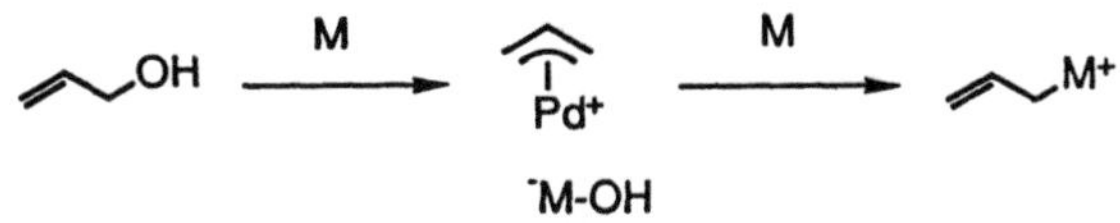

Scheme 15.

has great advantages as to (1) stability and tractability, and/or (2) ease of preparation, compared with allylic halides and esters. However, π-allylpalladium complex can not usually be prepared from allylic alcohol under the same conditions as can be prepared from allylic ester or halide. The reaction requires either high reaction temperature or a reagent for inducing the elimination of hydroxy group.[44] Thus, palladium-catalyzed carbonyl allylation by allylic alcohol needs metal reagents which possess two functions; that of assisting elimination of hydroxy group and that of reducing π-allylpalladium complex (Scheme 15). Stannous chloride, which has been used in the carbonyl allylation by allylic esters, possesses both functions and can be applied to the carbonyl allylation by various allylic alcohols.[45]

Table 6. Addition Reaction of 2-(*E*)-Butenol to Benzaldehyde under Various Conditions

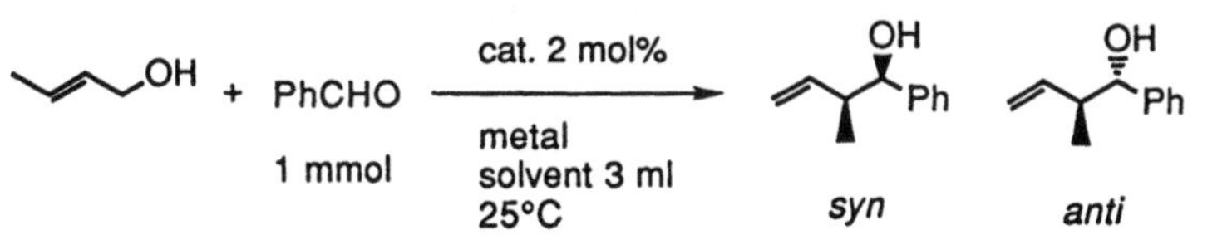

Entry	Alcohol (mmol)	Catalyst (mmol)	Metal Reagent		Solvent	Time (h)	Yield[a] (%)	Ratio[b] Syn:Anti
1	1.0	$PdCl_2(PhCN)_2$	$SnCl_2$	1.0	DMF	216	40	25:75
2	1.0	$PdCl_2(PhCN)_2$	$SnCl_2$	1.5	DMF	216	62	26:74
3	1.0	$PdCl_2(PhCN)_2$	$SnCl_2$	2.0	DMF	216	80	24:76
4	1.5	$PdCl_2(PhCN)_2$	$SnCl_2$	3.0	DMI	25	74	29:71
5	1.5	$PdCl_2(PPh_3)_2$	$SnCl_2$	3.0	DMI	90	34	45:55
6	1.5	$Pd(PPh_3)_4$	$SnCl_2$	3.0	DMI	90	65	29:71
7	1.5	$Pd_2(dba)_3CHCl_3$	$SnCl_2$	3.0	DMI	24	98	37:63
8	1.5	$Pd(dppe)_2$	$SnCl_2$	3.0	DMI	48	69	29:71
9	1.5	$PdCl_2(PhCN)_2$	SnF_2	3.0	DMI	72	0	---
10	1.5	$PdCl_2(PhCN)_2$	Sn	3.0	DMI	72	0	---
11	1.5	$PdCl_2(PhCN)_2$	Zn	3.0	DMI	72	0	---
12	1.5	$NiBr_2(PPh_3)_2$	$SnCl_2$	3.0	DMI	72	0	---
13	1.5	$NiBr_2(PBu_3)_2$	$SnCl_2$	3.0	DMI	72	0	---

Notes: [a]Isolated yield based on benzaldehyde.
[b]The ratio was determined by ^1H NMR (GX-270) and GC (PEG 20M, 0.25 mm × 30 m).

Addition reaction of 2-(*E*)-butenyl alcohol to benzaldehyde was first examined under various conditions. The results are summarized in Table 6. Pd catalysts such as $PdCl_2(PhCN)_2$, $PdCl_2(PPh_3)_2$, $Pd(PPh_3)_4$, $Pd_2(dba)_3CHCl_3$, and $Pd(dppe)_2$ can be used in this carbonyl allylation by allylic alcohol with $SnCl_2$ (entries 4-8). No catalytic activity of Ni(II) complexes is observed for this carbonyl allylation (entries 12 and 13). $PdCl_2(PhCN)_2$ and $Pd_2(dba)_3CHCl_3$ are superior to other Pd complexes bearing phosphine ligands with regard to stability to air and reactivity to this allylation. $PdCl_2(PhCN)_2$ is better than $Pd_2(dba)_3CHCl_3$, because dba ligand can often not be separated from allylated products by column chromatography. SnF_2, Sn, and Zn do not serve as reducing agents for this allylation under the same conditions as $SnCl_2$ (entries 9–11). This allylation needs more than two equimolar amounts of $SnCl_2$ (entries 1–3). Thus, $SnCl_2$ may function during two steps; elimination of hydroxy group and reduction of π-allylpalladium complex. Addition of butenyltin intermediate, derived from 2-butenol, occurs regioselectively at γ-position, similarly to cases of 2-butenyl esters. The structure of allyltin intermediate was estimated by NMR spectroscopic investigations described in the last section (III-D). The reaction rate suggests that this carbonyl allylation needs to use

Table 7. Allylation of Various Carbonyl Compounds by 2-Propenol

Entry	*Carbonyl Compounds*	*Solvent*	*Temp.(°C)*	*Time(h)*	*Yield(%)*
1	hexanal	DMI	25	41	62
2	10-undecenal	DMI	25	39	65
3	crotonaldehyde	DMI	25	26	36
4	cinnamaldehyde	DMI	25	24	81
5	piperonal	DMI	25	39	88
6	benzaldehyde	DMI	25	25	74
7	benzaldehyde	THF	25	24	83
8	acetophenone	DMI	25	96	36
9	acetophenone	DMI	50	45	59[a]
10	cyclohexanone	DMF	25	68	98
11	2-methylcycohexanone	DMF	25	67	51
12	4-*t*-butylcyclohexanone	DMF	25	142	89[b]
13	cycloheptanone	DMF	25	67	45
14	cyclododecanone	DMF	25	118	55

Notes: [a]4-Phenyl-1,3-pentadiene was obtained as a by-product (6%), and acetophenone was recovered (10%).
[b]*syn:anti* = 94:6.

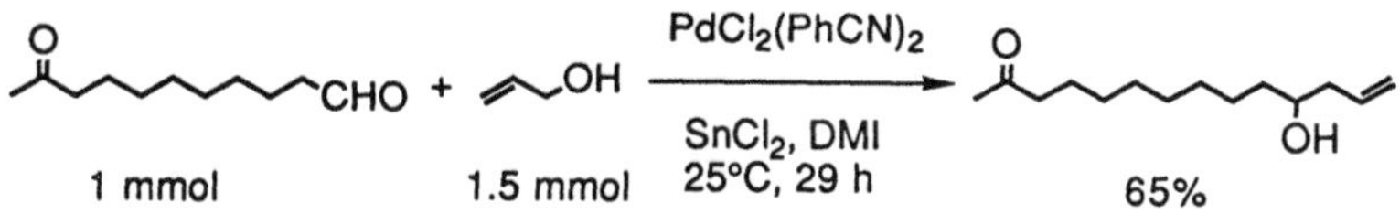

Scheme 16.

1.5 equimolar amounts of 2-(*E*)-butenol and 3.0 equimolar amounts of stannous chloride to benzaldehyde (entry 4).

The allylation of various carbonyl compounds by allylic alcohol can be carried out with $PdCl_2(PhCN)_2$-$SnCl_2$ in DMI, DMF, or THF. The results are summarized in Table 7. This allylation by 2-propenol can be applied not only to aldehydes, but also to ketones at 25 °C without additives such as cesium fluoride and tetrabutylammonium bromide, in contrast with allylic esters (entries 8–13). However, addition to ketones is slower than that to aldehydes. Thus, chemoselective addition to an aldehyde can be realized in the presence of a ketone group similarly to that of 2-propenyl acetate (Scheme 16). Addition to 4-*t*-butylcyclohexanone occurs predominantly from an equatorial direction (*syn*:*anti* = 99:1, entry 12).[9]

Differences in reactivity between allylic alcohol and allylic esters were directly investigated using 2-(*Z*)-butene-1,4-diol and its ester derivatives. The results are shown in Table 8. The examination of the chemoselectivity between an allylic alcohol and an allylic acetate with 4-acetoxy-2-(*Z*)-butenol demonstrates the high reactivity of allylic alcohol (entry 3). The product is the same as that found in the

Table 8. Chemoselection to Allylic Derivatives in Allylation of Benzaldehyde

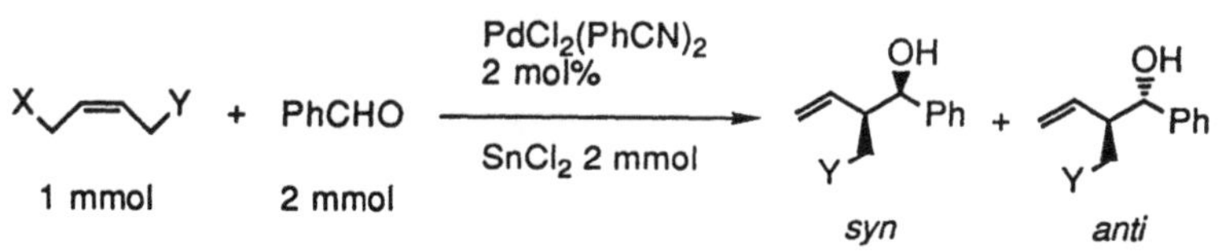

	Allylic Derivative		*Time*	*Product*	*Yield*	
Entry	*X*	*Y*	*(h)*	*Y*	*(%)*	*Syn:Anti*[a]
1	OH	OH	43	OH	60	35:65
2	OAc	OAc	123	OAc	61	28:72
3	OH	OAc	50	OAc	57	31:69
4	OCOOMe	OH	25	OH	56	26:74
5	OCOOMe	OAc	21	OAc	44	32:68

Note: [a]After the adduct is converted to 1,3-dioxane derivative as shown in Scheme 13, the configuration is determined by following the results reported in ref. 42.

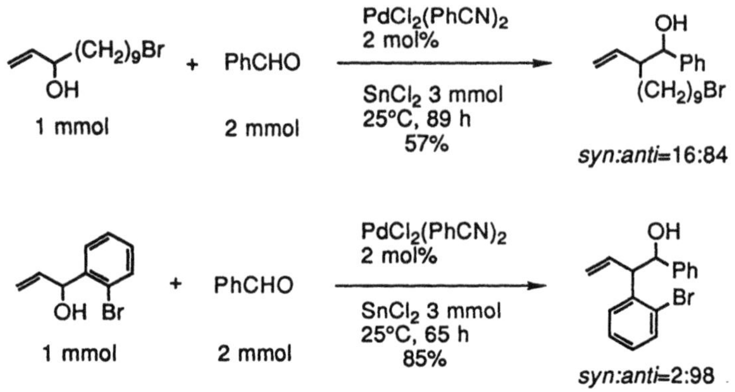

Scheme 17.

addition of 1,4-diacetoxy-2-(*Z*)-butene (entry 2). On the other hand, an allylic carbonate is more reactive than an allylic alcohol (entry 4). As a result, the order of the reactivity of leaving groups is OCOOMe > OH > OAc in the palladium-catalyzed carbonyl allylation with $SnCl_2$.

Aryl, vinyl, and alkyl bromide do not cause palladium-catalyzed homo- or cross-coupling reactions with $SnCl_2$ under the same conditions as the carbonyl allylation by allylic alcohol. Therefore, allylic alcohols bearing either an aliphatic or an aromatic bromide can be applied chemoselectively to produce the corresponding homoallylic alcohols as shown in Scheme 17. This result demonstrates that oxidative addition of allylic alcohol to Pd(0) is faster than that of aryl bromide under these conditions.[28]

We expect that this carbonyl allylation will become a method of considerable synthetic utility, complementing the known allylmetal method,[29,46] since the starting allylating agents, allylic alcohols, are easily prepared and are relatively stable.[30]

B. Solvation-Controlled Diastereoselection

In this section we demonstrate that diastereoselection in palladium-catalyzed carbonyl allylation by allylic alcohols is changed by the solvation modes of various polar solvents.

1. Solvation-Controlled Diastereoselection by 2-Butenol Derivatives[47]

The diastereoselection in the allylation of benzaldehyde with 2-(*E*)-butenol was investigated in various solvents at many different temperatures (Scheme 18). The results are summarized in Table 9. The allylation takes place in any polar solvent in which $SnCl_2$ dissolves, to give homoallylic alcohols in good yields. Regio- and diastereoselectivity reveal a solvent dependency. The reactions in DMI, DMF, ethylene glycol (EG), or DMSO possess high regioselectivity (γ-addition) at 25 °C (entries 13, 14, 24, and 26). Use of EG and DMSO leads to *syn* selection, contrary to the *anti* selection found in DMI and DMF. The regio- and diastereoselectivity of the reactions in these solvents can be demonstrated by the dielectric constant and Swain's parameter of the solvents (entries 2, 13, 14, 17, 19, 21–24, and 26):

1. The larger the dielectric constant is, the higher the regioselectivity (γ-addition) is.
2. If the solvent possesses a weaker solvation effect for cation than benzaldehyde does, *anti* addition occurs. On the contrary, if the effect is stronger, *syn* addition is preferred.
3. A solvent with a stronger solvation effect for anion leads to relatively high *syn* selection.

THF cannot be applied to regio- and diastereoselection at 25 °C, but can be at lower temperatures (< –10 °C) (entries 1–6). The reaction does not occur at all at –40 °C in THF (entry 7). Although carbonyl allylations using allylic organometallic reagents are usually inhibited by the presence of H_2O, the palladium-catalyzed carbonyl allylation with $SnCl_2$ is accelerated by the addition of H_2O in any solvent used.[48] Furthermore, in the presence of H_2O, the reaction is more selective with respect to both γ-addition (regiocontrol) and *syn*/*anti* addition (diastereocontrol). Thus, use of THF-H_2O medium leads to highly selective *anti* addition without α-addition even at 25 °C (entry 9). On the other hand, surprisingly, the control of the amount of H_2O in DMSO medium permits both *syn* selection (a small amount of H_2O) and *anti* selection (a large amount of H_2O) (entries 27–34).

In solvents such as THF-H_2O and DMSO-H_2O (a large amount), 3-buten-2-ol (**29**) and 2-methylbut-2-(*E*)-enol (**30**) as well as **27** cause *anti* addition (entries 42, 46–49, and 54), and 2-(*Z*)-butenol (**28**) and 2-methylbut-2-(*Z*)-enol (**31**) cause *syn* addition (entries 35–37, 41, 55, 57, and 58). These results can be explained by chair forms of six-membered cyclic transition states **32** and **33** (Scheme 19).[9,29] Less *syn* selectivity in the addition of (*Z*)-**28** at 25 °C in THF should be dependent on the preparation of an 2-(*E*)-butenyltin intermediate via the rapid isomerization of *anti*-π-allylpalladium complex to *syn* complex, similarly to the case of 2-(*Z*)-butenyl carbonate (**23**) as illustrated in Scheme 14. On the contrary, (*Z*)-**31** bearing 2-methyl group exhibits high *syn* selectivity, which is probably due to the suppression of the isomerization of *anti*-π-allylpalladium complex to *syn* complex by steric hindrance of the 2-methyl group. On the other hand, additions of both (*E*)-**27** and (*Z*)-**28** in DMSO-H_2O (a small amount) medium lead to *syn* selection without depending upon double-bond geometry (entries 28 and 40). Thus, this *syn* selection suggests the existence of an acyclic antiperiplanar transition state **35**, as shown in Scheme 19.[29] Both (*E*)-**30** and (*Z*)-**31** cause *anti* addition in DMSO-H_2O medium (a small amount of H_2O; entries 52 and 56). This *anti* addition cannot be accounted for by either the six-membered cyclic transition state or the acyclic antiperiplanar

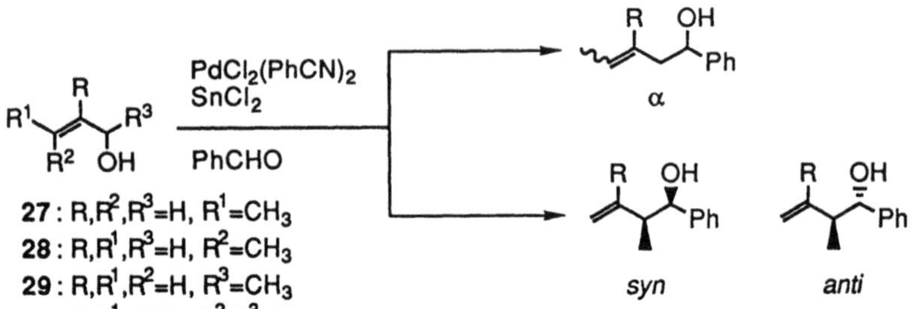

27 : R,R²,R³=H, R¹=CH₃
28 : R,R¹,R³=H, R²=CH₃
29 : R,R¹,R²=H, R³=CH₃
30 : R,R¹=CH₃, R²,R³=H
31 : R,R²=CH₃, R¹,R³=H

Scheme 18.

Table 9. Dependence of Diastereoselection on Solvent[a]

Entry	*Alcohol*	*Solvent, ml*	*H_2O(mmol)*	*Temp.(°C)*	*Time(h)*	*Yield(%)*[b]	*Syn:Anti:α*[c]
1	**27**	THF 5	0	50	24	67	23:23:54
2	**27**	THF 3	0	25	24	72	22:25:53
3	**27**	THF 3	0	0	95	80	16:78:6
4	**27**	THF 3	0	–10	130	81	10:90:0
5	**27**	THF 3	0	–15	135	62	7:93:0
6	**27**	THF 3	0	–20	120	21	3:97:0
7	**27**	THF 3	0	–40	120	0	—
8	**27**	THF 4	5	25	7	76	20:79:1
9	**27**	THF 4	10	25	6	74	16:84:0
10	**27**	THF 4	25	25	13	84	23:77:0
11	**27**	THF 4	65	25	16	71	34:66:0
12	**27**	DMI 3	0	50	7	65	49:51:0
13	**27**	DMI 3	0	25	36	63	29:71:0
14	**27**	DMF 3	0	25	63	89	30:70:0
15	**27**	DMF 3	4	25	39	71	25:75:0
16	**27**	DMF 3	64	25	15	77	18:82:0
17	**27**	acetone 8	0	25	121	31	3:39:58
18	**27**	acetone 5	3	25	73	42	9:52:39
19	**27**	DME 6	0	25	142	22	41:3:56
20	**27**	DME 8	6	25	90	67	31:64:5
21	**27**	MeOH 5	0	25	135	9	60:38:2
22	**27**	EtOH 3	0	25	72	79	38:41:21
23	**27**	BuOH 3	0	25	64	72	36:58:6
24	**27**	EG 3	0	25	37	78	58:42:0
25	**27**	EG 3	5	25	36	85	43:57:0
26	**27**	DMSO 3	0	25	136	34	65:35:0
27	**27**	DMSO 3	2	25	91	45	74:26:0
28	**27**	DMSO 3	3	25	88	84	86:14:0
29	**27**	DMSO 3	10	25	88	75	83:17:0

(continued)

Table 9. (*continued*)

Entry	*Alcohol*	*Solvent, ml*	H_2O*(mmol)*	*Temp.(°C)*	*Time(h)*	*Yield(%)*[b]	*Syn:anti:α*[c]
30	**27**	DMSO 3	20	25	69	77	68:32:0
31	**27**	DMSO 3	28	25	70	65	49:51:0
32	**27**	DMSO 3	43	25	50	71	39:61:0
33	**27**	DMSO 3	63	25	43	42	22:78:0
34	**27**	DMSO 3	170	25	70	70	16:84:0
35	**28**	THF 3	0	−10	139	71	77:23:0
36	**28**	THF 4	10	25	6	80	56:44:0
37	**28**	THF 4	12	0	76	73	70:30:0
38	**28**	DMI 3	0	25	47	67	66:34:0
39	**28**	DMSO 3	0	25	138	24	75:25:0
40	**28**	DMSO 3	3	25	95	51	84:16:0
41	**28**	DMSO 3	170	25	95	48	79:21:0
42	**29**	THF 4	10	25	6	91	27:73:0
43	**29**	DMI 3	0	25	24	75	29:71:0
44	**29**	DMSO 3	3	25	92	95	84:16:0
45	**29**	DMSO 3	6	25	106	64	75:25:0
46	**29**	DMSO 3	54	25	53	76	34:66:0
47	**29**	DMSO 3	170	25	55	60	24:76:0
48	**30**	THF 4	11	25	24	60	10:80:10
49	**30**	THF 4	20	25	47	63	11:89:0
50	**30**	DMI 3	0	25	90	64	11:89:0[d]
51	**30**	DMSO 3	0	25	134	15	10:90:0
52	**30**	DMSO 3	3	25	109	83	27:73:0[d]
53	**30**	DMSO 3	6	25	100	46	9:91:0
54	**30**	DMSO 3	54	25	68	67	3:97:0
55	**31**	THF 4	20	25	35	79	87:13:0
56	**31**	DMSO 3	3	25	113	87	38:62:0[d]
57	**31**	DMSO 3	54	25	71	88	81:19:0
58	**31**	DMSO 3	170	25	130	34	91:9:0

Notes: [a]The allylation of benzaldehyde (1 mmol) by allylic alcohols (1.5 mmol) is carried out with $PdCl_2(PhCN)_2$ (0.02 mmol) and $SnCl_2$ (3 mmol).
[b]Isolated yield based on benzaldehyde.
[c]The ratio is determined by GC (capillary column: PEG 20 M, 0.25 mm × 30 m, 155 °C).
[d]Three mmol of allylic alcohols were used.

transition state. Since (1) an *anti*-π-allylpalladium complex derived from (*Z*)-**31** has isomerized to a *syn* complex very little in THF-H_2O and DMSO-H_2O (a large amount) and (2) the reactions of (*E*)-**27** and (*Z*)-**28** have proceeded via an acyclic transition state **35** in DMSO-H_2O (a small amount), we propose one possible acyclic transition state, namely the acyclic synclinal transition state **34** in the reactions of (*E*)-**30** and (*Z*)-**31** in DMSO-H_2O a small amount) (Scheme 19).[49,50] Steric hin-

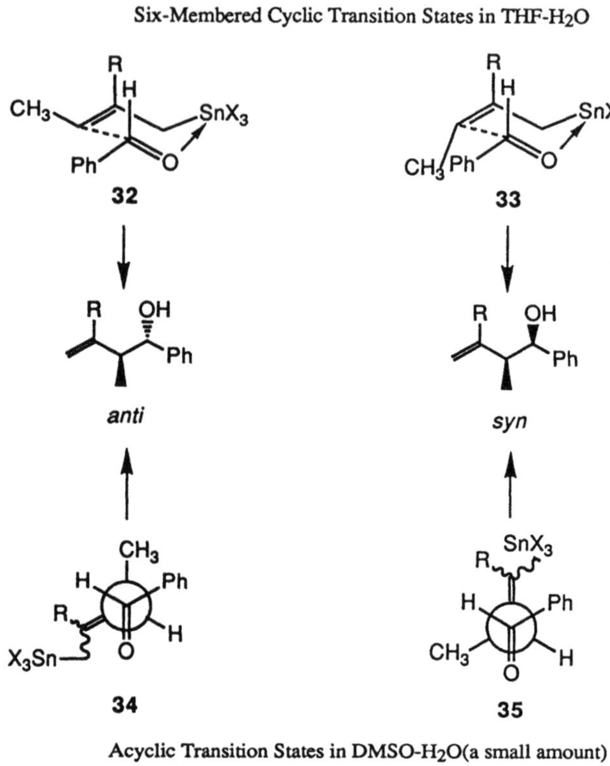

Scheme 19.

drance due to a 2-methyl group in (*E*)-**30** and (*Z*)-**31**, which is more bulky than a 2-proton in (*E*)-**27** and (*Z*)-**28**, presumably leads to the more favorable synclinal transition state **34** rather than to the antiperiplanar transition state **35**.

In contrast with solvents such as THF, DMI, and DMF, DMSO should strongly coordinate to Sn(IV) in the allylic tin intermediates, thus disturbing the formation of the six-membered cyclic transition states **32** and **33**. The small amount of H_2O seems to cause the hydrolysis of Sn^{IV}-Cl bonds to activate the allylic tin intermediates,[48] and the large amount of H_2O in DMSO seems not only to hydrolyze the Sn^{IV}–Cl bonds, but also to rupture the coordination of DMSO to the Sn(IV).

2. Steric and Electronic Effect of Substituents for Solvation-Controlled Diastereoselection[51]

As mentioned above, desirable diastereoselection in the 2-butenols/benzaldehyde system can be easily achieved by the choice of solvent.[52] We next investigated

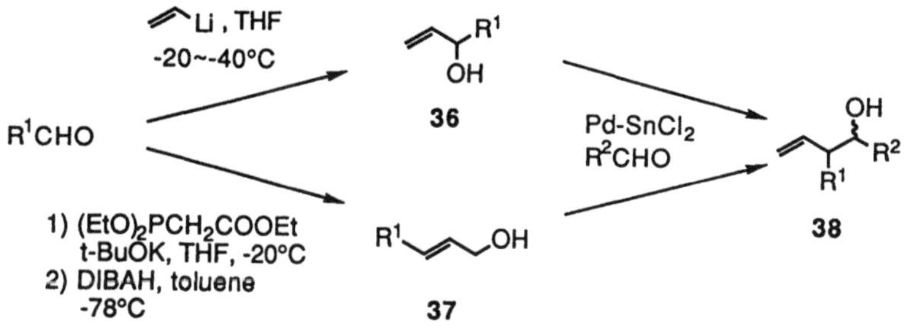

Scheme 20.

steric and electronic effect of 1- or 3-(*E*)-substituent of allylic alcohol and substituent of aldehyde for the regio- and diastereoselection. The representative results are summarized in Table 10. The rate of allylation is more affected by the steric hindrance of 1- or 3-(*E*)-substituent of allylic alcohols than that of aldehydes. 2-Propenols bearing a bulky substituent, such as isopropyl or cyclohexyl group, at 1- or 3-position do not cause the addition reaction in a poor DMSO-H_2O medium. The rate is also dependent upon the electrophilicity of the aldehyde (entries 1–8). Both 1-substituted **36** and 3-(*E*)-substituted allylic alcohol **37** react with the aldehyde at 25 °C to give one regioisomer **38**, of which the diastereomer ratio prepared from **36** is almost the same as that prepared from **37** (Scheme 20). Therefore, 1-substituted allylic alcohol **36**, which can be prepared more easily than 3-(*E*)-**37**, can be generally applied to this carbonyl allylation. 3-Cyclohexyl-2-propenol (**39**) bearing a bulky substituent at 3-position, as an exception, causes an apparent α-addition to benzaldehyde even in DMI at 25 °C to produce (*E*)-homoallylic alcohol **44**, as shown in Scheme 21 (entries 23 and 24). The reaction of π-allylpalladium complex **40** with $SnCl_2$ does not produce 1-cyclohexyl-2-propenyltin intermediate **42**, but a 3-cyclohexyl-2-propenyltin one **41**. However, γ-addition of **41** to benzaldehyde should be suppressed by steric hindrance between a cyclohexyl group and a phenyl group in transition state **45**. 1-Cyclohexyl intermediate **42**, which slightly exists in an equilibrium state, may consequently cause γ-addition via transition state **46** to afford (*E*)-homoallylic alcohol **44**.[48,53,54] The preparation of (*Z*)-geometric homoallyl alcohols **48** via transition state **47a**, in which R^1 group is at pseudo-axial position, has been reported in the carbonyl allylation by 1-substituted allylic tin compounds (Scheme 22).[48,53,55] It has been demonstrated by the steric hindrance between pseudo-equatorial 1-substituent (R^1) and a butyl group on Sn(IV) in transition state **47b**. In contrast with these results, our results suggest that the transition state **46** is favorable; in other words, a substituent (X) on Sn(IV) of allyltin intermediate **42** is not so bulky under our conditions.

Table 10. Addition of 1- or 3-(*E*)-Substituted Allylic Alcohols to Aldehydes

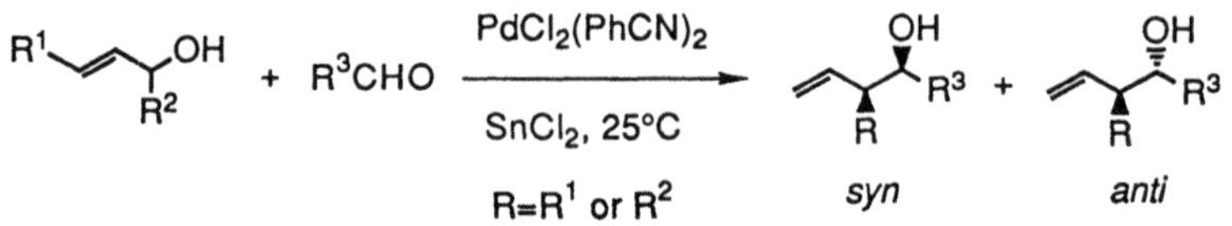

Entry	Allylic Alcohol R^1	R^2	Aldehyde R^3	Solvent[a]	Time (h)	Yield[b] (%)	Ratio[c] Syn:Anti
1	CH_3	H	*p*-MeOOCPh	DMF	38	99	21:79
2	CH_3	H	*p*-ClPh	DMF	125	50	22:78
3	CH_3	H	*p*-MeSPh	DMF	148	68	34:66
4	CH_3	H	*p*-MeOPh	DMF	173	70	36:64
5	CH_3	H	*p*-MeOOCPh	DMSO-H_2O	43	90	77:23
6	CH_3	H	*p*-ClPh	DMSO-H_2O	120	84	70:30
7	CH_3	H	*p*-MeSPh	DMSO-H_2O	216	65	75:25
8	CH_3	H	*p*-MeOPh	DMSO-H_2O	240	79	82:18
9	CH_3	H	c-C_6H_{11}	THF-H_2O	72	66(10)	14:86
10	CH_3	H	c-C_6H_{11}	DMI	42	83	35:65
11	CH_3	H	c-C_6H_{11}	DMSO-H_2O	116	13	29:71
12	CH_3	H	n-C_6H_{13}	THF-H_2O	26	66	51:49
13	CH_3	H	n-C_6H_{13}	DMI	72	65	45:55
14	CH_3	H	n-C_6H_{13}	DMSO-H_2O	117	27	60:40
15	H	CH_3	n-C_6H_{13}	DMI	96	75	45:55
16	H	CH_3	n-C_6H_{13}	THF-H_2O	72	90	54:46
17	H	CH_3	c-C_6H_{11}	THF-H_2O	72	65	32:68
18	n-Pr	H	Ph	DMI	96	57	12:88
19	n-Pr	H	Ph	THF-H_2O	48	82	20:80
20	n-Pr	H	Ph	DMSO-H_2O	120	83	70:30
21	i-Pr	H	Ph	THF-H_2O	50	36	3:97
22	i-Pr	H	n-C_6H_{13}	THF-H_2O	50	47	33:67
23	c-C_6H_{11}	H	Ph	DMI	70	(36)	—
24	c-C_6H_{11}	H	Ph	THF-H_2O	96	(45)	—
25	c-C_6H_{11}	H	n-C_6H_{13}	DMI	96	56	33:67
26	c-C_6H_{11}	H	n-C_6H_{13}	THF-H_2O	70	55(7)	40:60
27	H	i-Pr	Me	DMI	89	45	36:64
28	H	c-C_6H_{11}	Me	DMI	68	25	46:54
29	Ph	H	Me	DMI	72	73	6:94
30	Ph	H	Me	THF-H_2O	72	67	8:92
31	Ph	H	Ph	DMI	73	79	7:93
32	Ph	H	Ph	THF-H_2O	42	67	0:100
33	Ph	H	Ph	DMSO-H_2O	169	57	3:97
34	Ph	H	c-C_6H_{11}	DMI	70	59	0:100
35	Ph	H	c-C_6H_{11}	THF-H_2O	64	(66)	—
36	Ph	H	c-C_6H_{11}	DMSO-H_2O	94	37	1:99
37	H	*p*-MeOPh	CH_3	DMI	75	71	8:92
38	H	*p*-MeOPh	Ph	DMSO-H_2O	69	23	1:99
39	H	*p*-NCPh	CH_3	DMI	87	75	6:94
40	H	*o*-HOPh	CH_3	DMI	54	58	5:95
41	H	*o*-BrPh	CH_3	DMI	67	58	5:95

Notes: [a]DMSO-H_2O: DMSO (3 ml) and H_2O (3 mmol) were used. THF-H_2O: THF (3 ml) and H_2O (10 mmol) are used.

[b]Figures in parentheses are yields of α-adducts.

[c]The ratio was determined by ^1H NMR (Jeolco GX-270) and GC (capillary column; PEG 20M, 0.25 mm × 30 m).

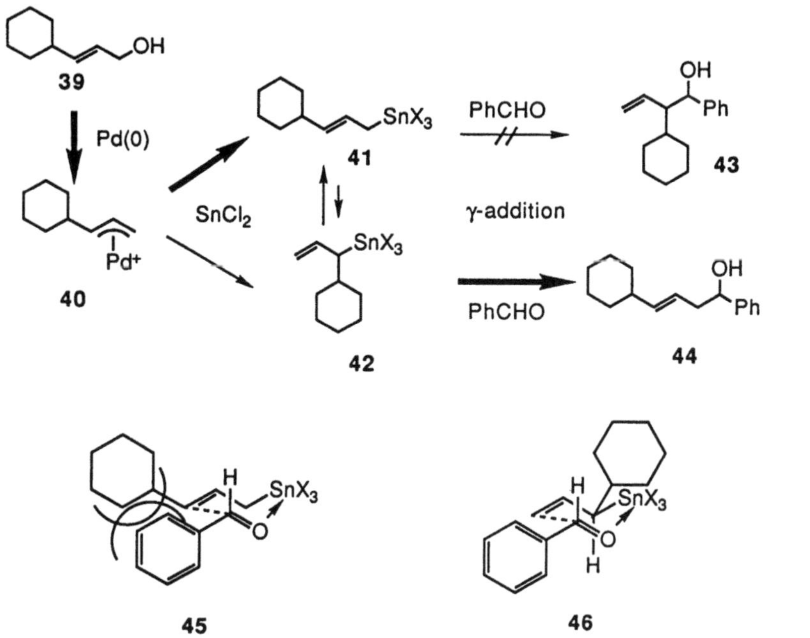

Scheme 21.

Steric and electronic effects of substituent R^3 of aldehyde for diastereoselection in the carbonyl allylation were investigated using 2-(*E*)-butenol and 1-methyl-2-propenol. In the cases of benzaldehyde derivatives, the diastereoselection can be changed by the choice of solvents without the electronic effect of *p*-substituents, as in the reaction of benzaldehyde (entries 1–8). Heptanal exhibits no diastereoselectivity in either THF-H_2O or DMSO-H_2O (entries 12, 14, and 16). Bulky cyclohexanecarboxaldehyde causes anti addition in both of them (entries 9, 11, and 17). The relationship between the steric effect of 1- (R^2) or 3-aliphatic substituent (R^1) of allylic alcohol and the diastereoselection cannot be discussed from our results, because (1) 2-propenols, having a bulky substituent at 1- or 3-position, do not react in DMSO-H_2O, and (2) 1- or 3-cyclohexyl-2-propenol causes α-addition in DMI and THF-H_2O. 2-Propenols having an aromatic substituent at 1- or 3-position exhibit high *anti* selectivity even in DMSO-H_2O (entries 29–41). Less reactive 3-phenyl-2-propenyltin intermediate, due to an electron-withdrawing phenyl group, may be able to cause the γ-addition only via six-membered cyclic transition state, which is favorable in entropy.

The steric effect of (*Z*)-geometric allylic alcohols cannot be examined, since *anti*-π-allylpalladium complex, derived from (*Z*)-allylic alcohol bearing a bulky

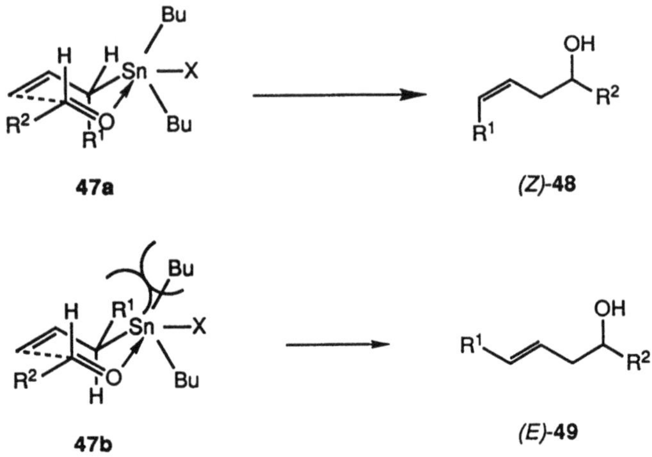

Scheme 22.

substituent at 3-position, rapidly isomerizes to a stable *syn* complex to afford (*E*)-allylic tin intermediate.

3. Application of Pd-SnCl₂ System to 1,2-Asymmetric Induction[56]

We tried the application of carbonyl allylation to 1,2-asymmetric induction, as realized in the addition reaction of allylic alcohols to 2-phenylpropanal, which has a stereocenter at 2-position, with $PdCl_2(PhCN)_2$-$SnCl_2$ at 25 °C in DMI or THF-H_2O. 2-Propenol causes *syn* addition, which is explained by Cram's rule (Scheme 23).[57] 2-Methyl-2-propenol exhibits the same diastereoselectivity as that of 2-propenol (Scheme 23). Steric hindrance of 2-methyl group consequently does not affect the diastereoselection. The bulkiness of 3-substituent in allylic alcohol dramatically enhances 1,2-diastereoselectivity, in contrast with the addition of 2-butenyltrimethyltin with BF_3OEt_2.[57] That is, the addition of 2-(*E*)-butenol to

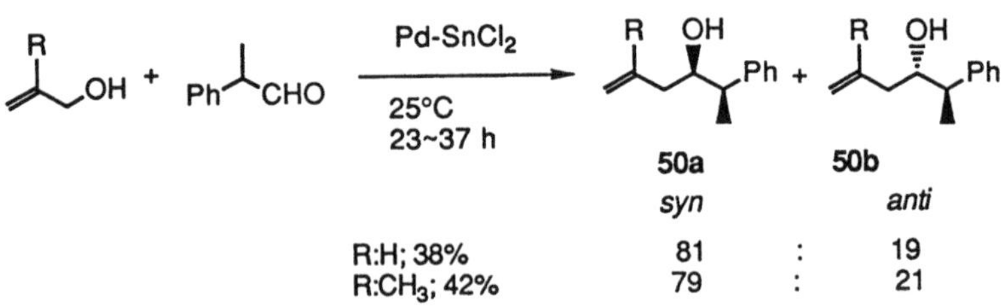

Scheme 23.

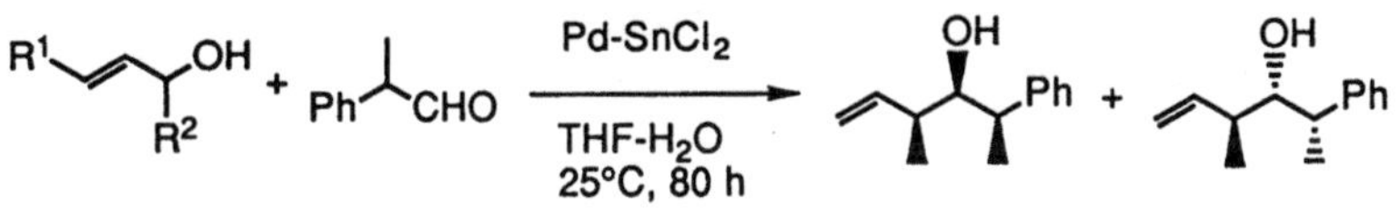

51a *syn* — **51b** *anti*

	51a		51b
R^1:CH_3, R^2:H; 77%	15	:	85
R^1:H, R^2:CH_3; 74%	18	:	82

Scheme 24.

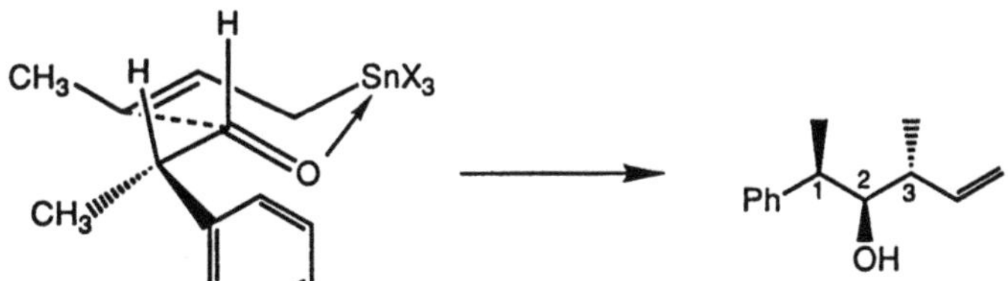

52 **51b**

Scheme 25.

2-phenylpropanal exhibits higher *syn* selectivity (1,2-*syn*/1,2-*anti* = 99/1) than that of 2-propenol to give major products **51a** and **51b** (2,3-*syn*/2,3-*anti* = 15/85) (Scheme 24). This 1,2-*syn* 2,3-*anti* selection, which produces diastereomer **51b**, can be explained by the combination of Felkin's model and six-membered cyclic transition state model, namely a transition state **52** (Scheme 25). In general, the trajectory of attack of nucleophile Nu^- on aldehyde R*CHO shifts from the normal plane (the plane perpendicular to the R*–C–H and containing the C═O bond) to the opposite side of bulky stereocenter R* (route **a** in Figure 1).[58] Hence, the

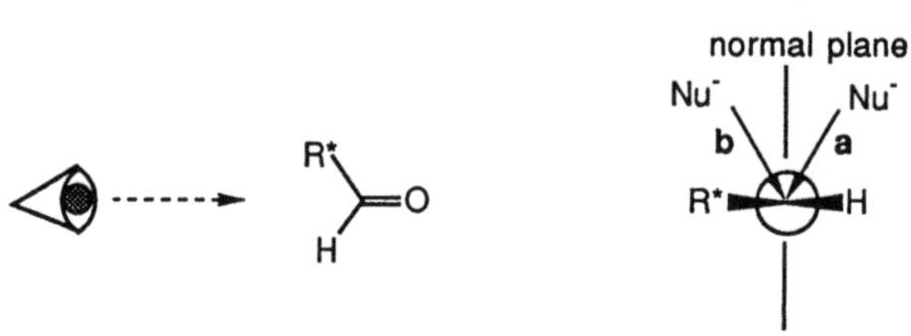

Figure 1. Route **a** is the usual trajectory of attack of nucleophile Nu^- on carbonyl compound R*CHO, and route **b** is a proposed trajectory in carbonyl allylation via a six-membered cyclic transition state.

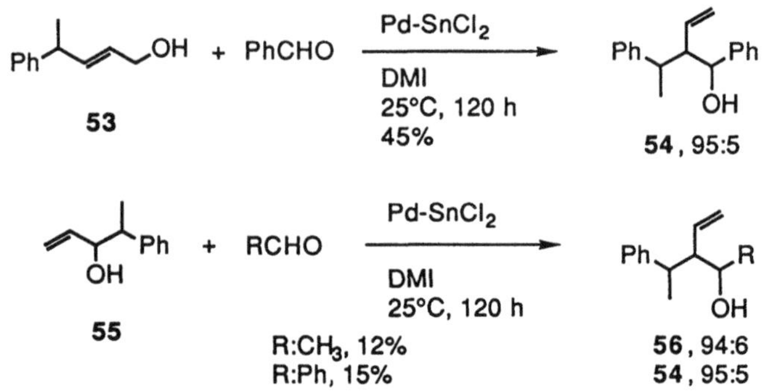

Scheme 26.

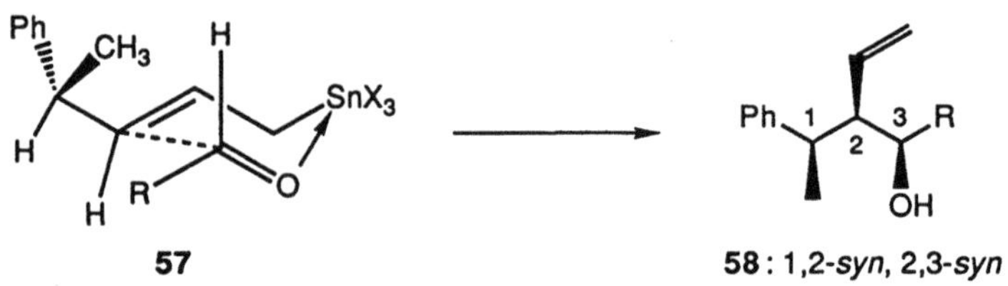

Scheme 27.

diastereofacial selection is not so strongly affected by the stereocenter R*, which is far from the trajectory **a**. On the contrary, high 1,2-*syn* selectivity in this palladium-catalyzed carbonyl allylation suggests that the diastereofacially selective attack of allyltin intermediate is strongly affected by the stereocenter R*. Six-membered cyclic transition state **52** seems to serve for a fixation of trajectory **b** in the direction of the sterically more demanding R*.

We also tried 1,2,3-asymmetric induction in the carbonyl allylation by allylic alcohol having a stereocenter (Scheme 26). The addition reaction of **53** and **55** to benzaldehyde or acetoaldehyde occurs diastereoselectively to afford two diastereomers; their structure cannot be confirmed yet. If transition state model **57**, which is the same as **52** in the allylation of aldehyde having a stereocenter, could be applied to the carbonyl allylation by allylic alcohol **53** and **55**, the diastereomer would be 1,2-*syn,* 2,3-*syn* **58** (Scheme 27).

C. Chelation-Controlled Diastereoselection

We now present chelation-controlled diastereoselection in palladium-catalyzed carbonyl allylation by allylic alcohol with $SnCl_2$ in polar solvents.

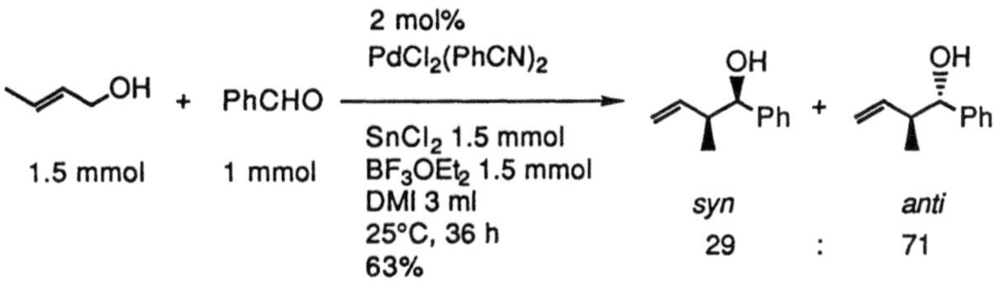

Scheme 28.

The coordination of Lewis acids to aldehydes plays an important role in the diastereocontrol of carbonyl allylation by allylic tin compounds: (1) The addition of BF_3OEt_2 leads to *syn* selection via an acyclic antiperiplanar transition state in the reaction of 2-butenyltrialkyltin with aldehydes, regardless of the geometry of the 2-butenyl unit,[29] and (2) the chelation of α-alkoxy-α-alkylacetoaldehyde to $MgBr_2$ or $TiCl_4$ in the reaction with allyltrialkyltin leads to *syn* selection, in contrast with the *anti* selection used in Cram's model.[59,60,61] Indeed, the palladium-catalyzed allylation of benzaldehyde by 2-(*E*)-butenol with $SnCl_2$, in any polar solvent in which $SnCl_2$ is dissolved, does not exhibit any effect due to BF_3OEt_2; the diastereomer ratio is the same as that without BF_3OEt_2 (Scheme 28). No chelation of $MgBr_2$ is recognized in the allylation of 2-benzyloxypropanal by 2-propenol with $PdCl_2(PhCN)_2$-$SnCl_2$ in THF, since the formation of chelate intermediate **59** ought to have led to *syn* selection (Scheme 29). Hence, chelation-controlled diastereoselection in polar solvents such as THF, DME, DMF, and DMI may need a stronger metal-oxygen bond than the coordination of ether oxygen to metal, as illustrated in Scheme 30.

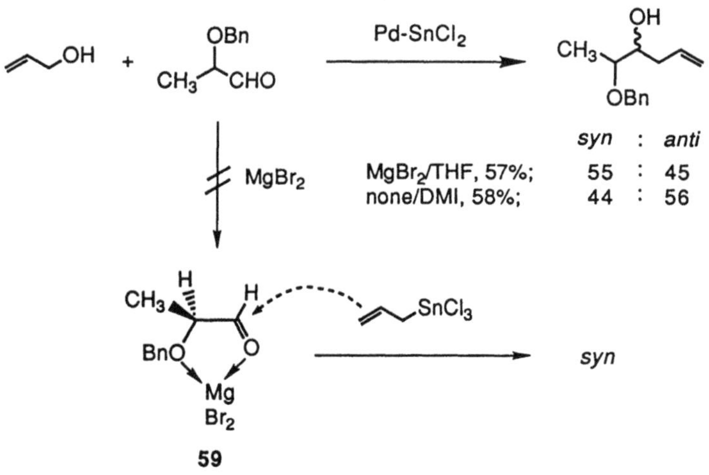

Scheme 29.

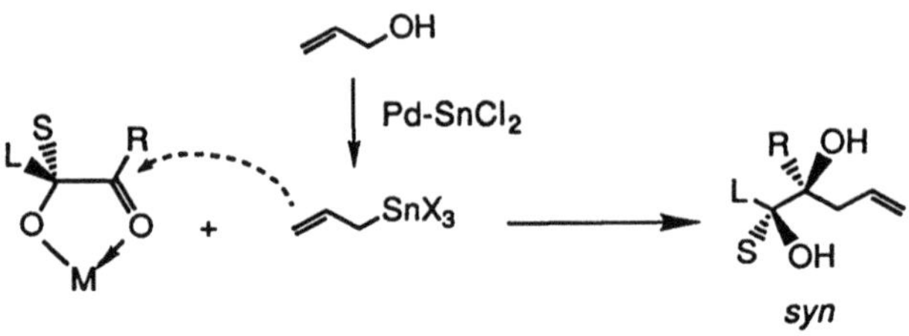

Scheme 30.

1. *Diastereoselective Diallylation of Diketo Compounds*[62]

We first tried to apply a tin-oxygen bond formed *in situ* to chelation-controlled diastereoselection.[63,64] Indeed, diallylation of 1,2-diketones by 2-propenol was carried out with Pd-$SnCl_2$ in DMI at 25 °C. The results are summarized in Table 11. Dialkyl-substituted 1,2-diketones do not react with even one equimolar amount of 2-propenol without undergoing diallylation (entries 1–6). As shown in Scheme 31, the second allylation should be promoted by the coordination of a second

Table 11. Diallylation of 1,2-Diketones by 2-Propenol with Pd-$SnCl_2$

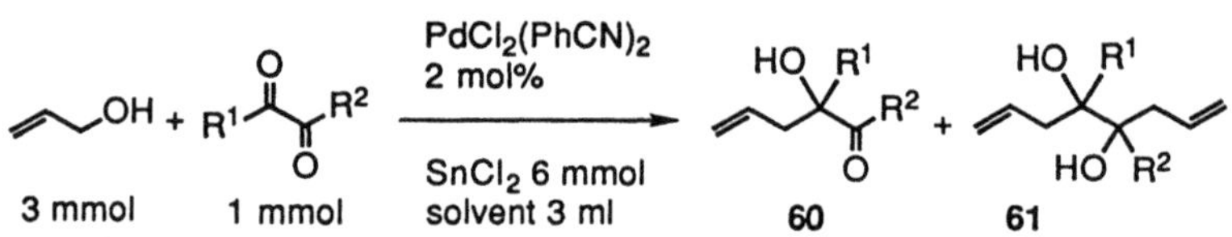

Entry	Diketone R^1	R^2	Solvent	Temp. (°C)	Time (h)	Yield(%) 60	61	Ratio of **61**[a] Syn:Anti
1	CH_3	CH_3	DMI	25	48	0	59	70:30
2	CH_3	CH_3	THF[b]	25	48	0	53	61:39
3	CH_3	CH_3	DMSO[b]	25	88	0	33	89:11
4	CH_3	CH_3CH_2	DMI	25	48	0	47	62:38
5	CH_3	CH_3CH_2	THF[b]	25	48	0	51	55:45
6	$(CH_2)_4$		DMI	25	48	0	46	100:0[c]
7	Ph	Ph	DMI	50	55	22	61	0:100
8	Ph	Ph	THF[b]	50	19	19	36	0:100
9	Ph	Ph	DMSO[b]	50	57	27	55	0:100

Notes: [a]The ratio was determined by GC (capillary column: PEG 20M, 0.25 mm × 30 m) and 270 MHz ^1H NMR (Jeolco GX-270). 1,2-Diol structures were confirmed by ^1H NMR observation of ketals derived from 1,2-diols and 2,2-dimethoxypropane.

[b]H_2O (5 mmol) was added.

[c]See ref. 66.

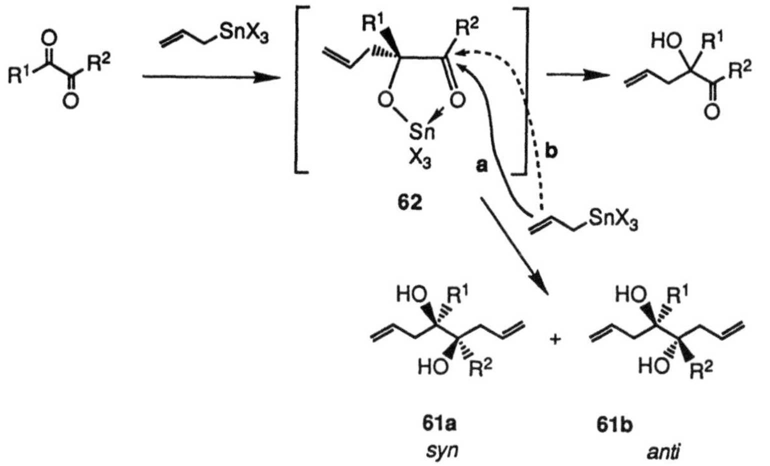

Scheme 31.

carbonyl group to Sn(IV) in monoallylated tin alkoxide intermediate **62**. In the case of phenyl-substituted 1,2-diketone, benzil, steric hindrance of the phenyl group probably leads to isolation of monoallylated product (entries 7–9). Diallylation of 2,3-butanedione and 2,3-pentanedione exhibits *syn* selectivity (entries 1–5), and that of benzil leads to complete *anti* selection (entries 7–9). These results also establish the existence of a chelation of the second carbonyl group to Sn(IV) in the monoallylated tin alkoxide intermediate **62** (Scheme 31). That is, the allylation of **62**, where R^1 is the phenyl group more bulky than the allyl group, presumably proceeds on path **b** to lead to *anti* selection. In a case of $R^1 = CH_3$, which is less bulky than the allyl group, path **a** is superior to path **b**. In contrast with the allylation of benzil, carbonyl allylation of monoallylated α-hydroxyketone **60** (R^1,R^2 = Ph), derived from benzil and 2-propenol, with $PdCl_2(PhCN)_2$-$SnCl_2$ in DMI at 50 °C exhibits slight *syn* selectivity (*syn:anti* = 59:41).[65] The existence of the chelate intermediate **62** is further supported by this result. In a case of 1,2-cyclohexanedione, the first allylation probably occurs on equatorial approach to a carbonyl group, followed by the second allylation on axial approach to another carbonyl group via a chelate intermediate **63** to produce *syn* 1,2-diol **64** (Scheme 32).[66]

Diallylation of phthalaldehyde, which is a kind of 1,4-diketo compound, also seems to cause a diastereofacial selection via a 7-membered chelate intermediate to give *anti*-rich diol **65** (*syn:anti* = 18:82) (Scheme 33).[63]

In addition, methyl pyruvate is monoallylated at only keto position, and the allylation exhibits no diastereoselectivity (Scheme 34). The addition of 2-butene-1,4-diol to methyl pyruvate leads to cyclization, without the formation of 2-hydroxy ester **66**, to produce α-hydroxy-γ-butyrolactone **67** under the same conditions.

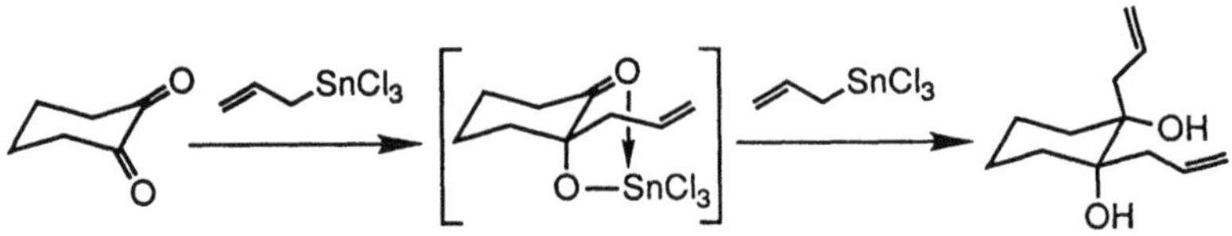

63 64

Scheme 32.

Intramolecular diallylation of 1,2-diketones by 2-methylene-1,3-propanediol (**68**) can be achieved to give only *syn* diol **69** under the same conditions (Scheme 35).[65] This reaction should proceed via the formation of 5-membered chelate intermediate **70** followed by **71**, similarly to intermolecular diallylation of 1,2-diketones as illustrated in Scheme 31. As mentioned above, the diallylation of 1,2-diketone suggests the existence of a chelation of carbonyl group to Sn(IV) of tin-oxygen bond in polar solvents, which allows us to consider possibilities of chelation-controlled diastereoselective allylation of ω-hydroxy carbonyl compounds in polar solvents.

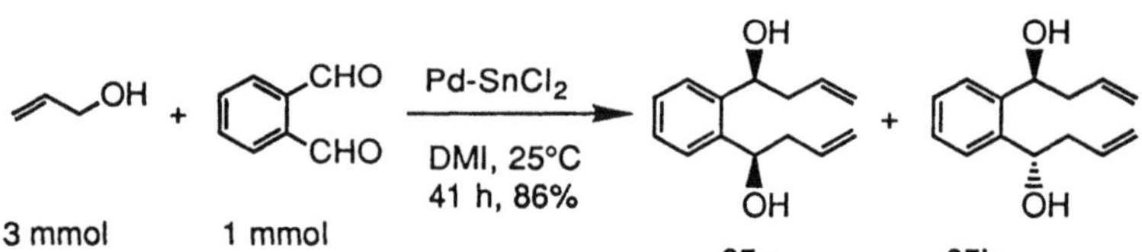

65a syn 18 : 65b anti 82

Scheme 33.

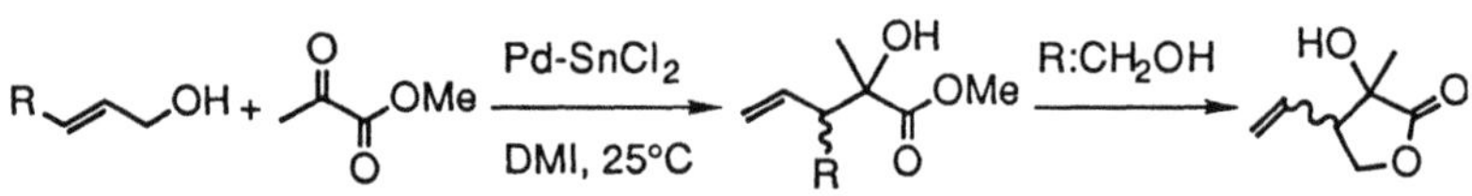

1.5 mmol 1 mmol 66 67

R: H, 85 h; 42%
R: CH_3, 89 h; 52%(50:50)
R: Ph, 89 h; 32%(54:46)
R: CH_2OH, 133 h; 32%(54:46)

Scheme 34.

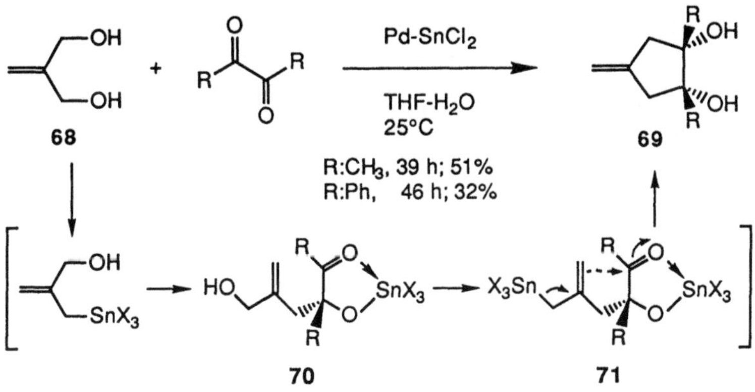

Scheme 35.

2. *Diastereofacial Selection to α-Hydroxyketone*[67]

The chelation of carbonyl group to Sn(IV) of tin alkoxide formed *in situ* proved to be effective for diastereoselective carbonyl allylation by allylic alcohol with Pd-$SnCl_2$ (Scheme 31). We here describe the effect of a chelation of ω-hydroxy carbonyl compounds to *Sn(IV) of allylic tin intermediates* in palladium-catalyzed diastereoselective carbonyl allylation.[68] The addition reaction of 2-(*E*)-butenol (**27**) to benzaldehyde derivatives with Pd-$SnCl_2$ in THF-H_2O usually exhibits *anti* selectivity (Section III-B). As had been expected, *o*-methoxybenzaldehyde also undergoes *anti* addition (**72a**:**72b** = 6:94) in the carbonyl allylation by **27** (Scheme 36). On the contrary, the reaction of *o*-hydroxybenzaldehyde with **27** leads to *syn*

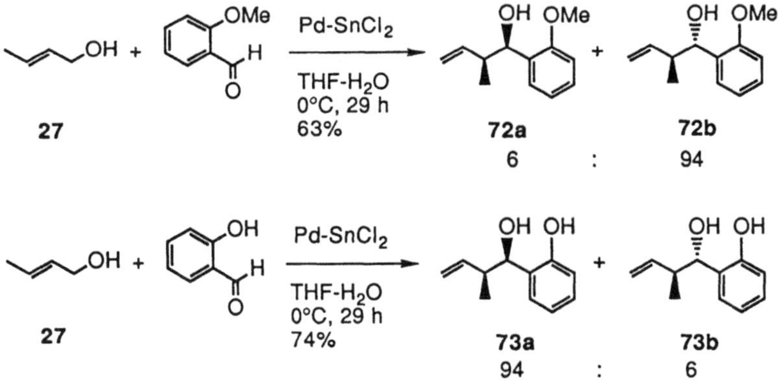

Scheme 36.

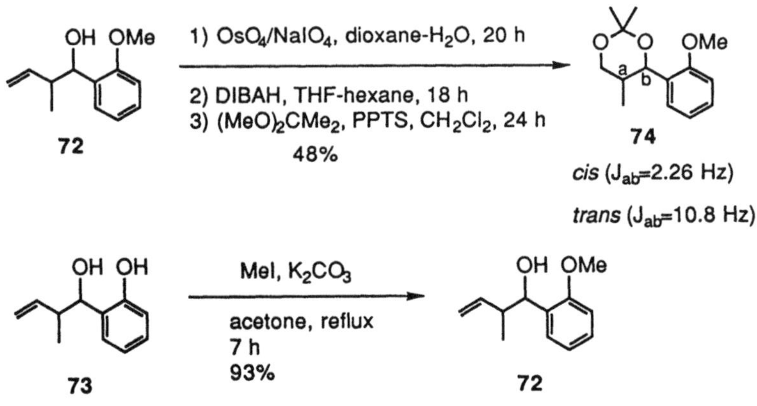

Scheme 37.

selection (**73a**:**73b** = 94:6) in THF-H_2O (Scheme 36). The configuration of **72** was determined by the conversion of **72** into 1,3-dioxane derivative **74** followed by coupling constants between H_a and H_b of **74** in ^1H NMR spectrum (Scheme 37). The configuration of **73** was determined by the conversion into **72** (Scheme 37). These results allowed us to presume a bicyclic transition state **75** containing a chelation of *o*-hydroxybenzaldehyde to Sn(IV) of 2-(*E*)-butenyltin intermediate (Scheme 38), and led us to an application of this kind of chelation of α-hydroxyketones.

The effect of a chelation of hydroxyacetone (**76**) to Sn(IV) of 2-(*E*)-butenyltin intermediate was first investigated. Palladium-catalyzed carbonyl allylation of **76** by both **27** and **29** predominantly produces *syn*-**77a** at 0 °C in THF (Scheme 39). This *syn* selectivity can be explained by the same chelated bicyclic transition state as **75** in the addition of **27** to *o*-hydroxybenzaldehyde in THF-H_2O. Use of THF-H_2O as a solvent lowers the *syn* selectivity (0 °C, **77a**:**77b** = 64:36). This result

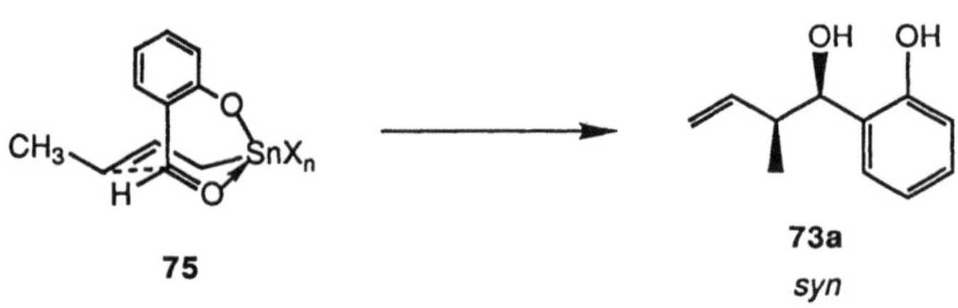

Scheme 38.

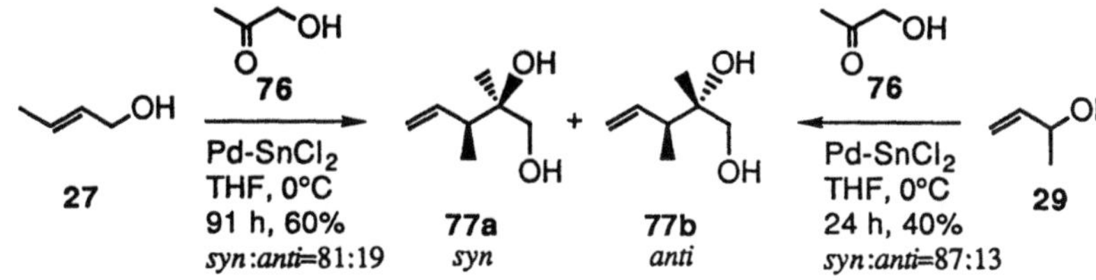

Scheme 39.

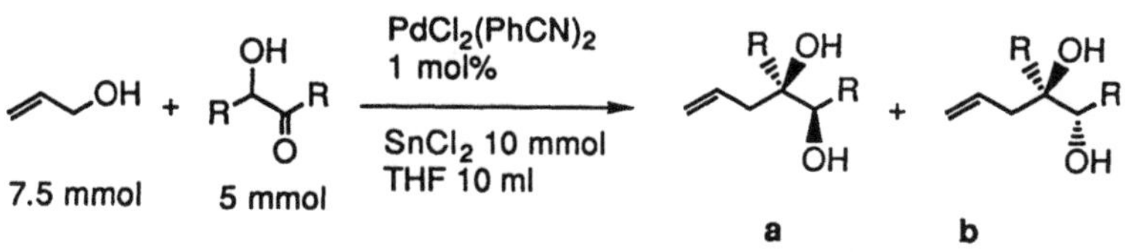

			a syn		b anti
78; R: CH_3	0°C, 40 h	**80**;75%	82	:	18
79; R: Ph	25°C, 20 h	**81**; ~100%	99	:	1

Scheme 40.

suggests that H_2O disturbs the coordination of the hydroxy group, which is needed for the formation of a chelated bicyclic transition state like **75**, as distinct from that of phenolic hydroxy group to the Sn(IV). Hence, this kind of chelation should be dependent upon the acidity of the hydroxy groups.

The addition reaction of 2-propenol to α-hydroxyketones **78** and **79** produces diastereoselectively 1,2-*syn*-diol **80** (**a**:**b** = 82:18) and **81** (**a**:**b** = 99:1) in 75% and ~100% yields, respectively (Scheme 40). This selectivity can be explained by a bicyclic transition state **82** (R′ = H) to afford *syn* 1,2-diol **83**, similarly to **75** in the addition of **27** to *o*-hydroxybenzaldehyde (Scheme 41). Nucleophilic attack of 2-propenyltin intermediate occurs predominantly from the side opposite to a bulky substituent R. Benzoin **79**, bearing a more bulky substituent, phenyl group, exhibits higher 1,2-*syn* selectivity than acetoin **78**, bearing a less bulky substituent, methyl group. The addition reaction of 2-methyl-2-propenol to **78** demonstrates that 1,3-interaction in a bicyclic transition state **82** (R′ = CH_3) enhances 1,2-*syn* selectivity (*syn*:*anti* = 89:11), even if R is methyl (Scheme 42).

syn Selective allylation of α-hydroxyketones via chelated bicyclic transition state. To a solution of stannous chloride (1.9 g, 10 mmol), 2-propenol (0.29 g, 5 mmol), and acetoin (**78**, 0.88 g, 10 mmol) in THF (10 ml) is added $PdCl_2(PhCN)_2$ (0.1 mmol). After stirring for 40 h at 0 °C, the reaction mixture

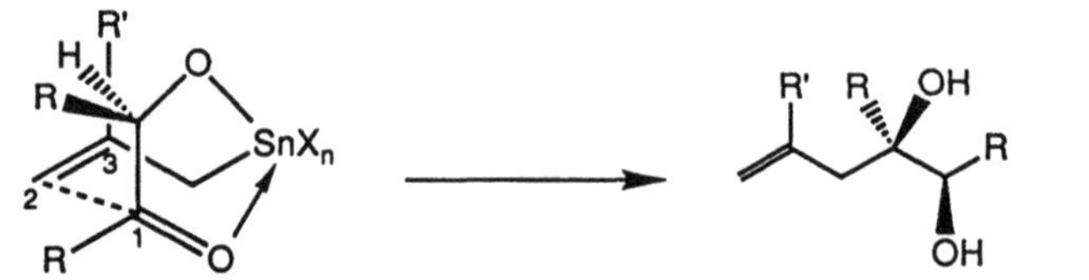

82 → **83** *syn*

Scheme 41.

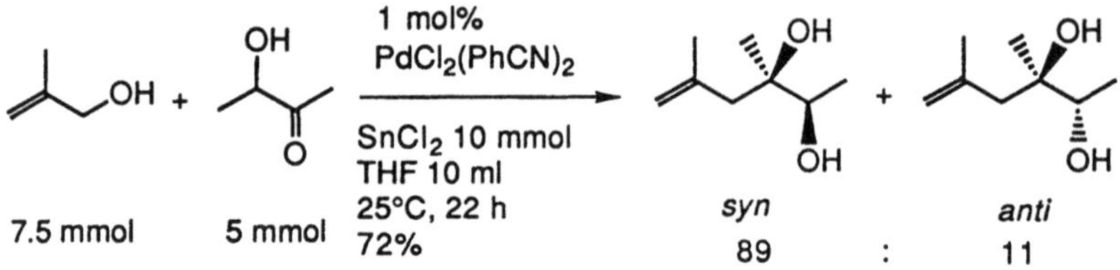

Scheme 42.

is poured into water (20 ml). The mixture, to which 10% aqueous HCl (20 ml) is added, is extracted for 68 h by continuous liquid-liquid extraction method [ether (300 ml)]. Evaporation of ether and purification by column chromatography on silica gel (hexane/ethyl acetate = 2/1) affords 0.48 g (75%) of 3-methylhex-5-ene-2,3-diol (**80**) as a colorless oil. The ^{1}H and ^{13}C NMR spectra (GX-270) corresponded with those of an authentic sample.[68] The diastereomer ratio was determined by ^{1}H NMR and capillary GC (PEG 20M, 0.25 mm × 30 m).[68]

We carried out 1,2,3-asymmetric induction in the addition of 2-(*E*)-butenol to **78** or **79** for the investigation of the influence of H_2O on this kind of coordination of hydroxy group (Scheme 43). 1,2-*syn* Selection was observed in both THF and

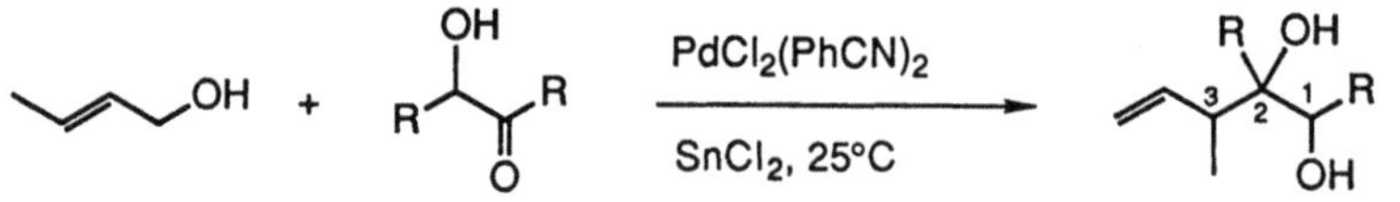

				C^1C^2 *syn/anti*,	C^2C^3 *syn/anti*
78	THF	96 h	**85**, 64%	81/19	65/35
78	THF-H_2O	69 h	**85**, 98%	78/22	44/56
79	THF	72 h	**86**, 33%	100/0	60/40
79	THF-H_2O	72 h	**86**, 89%	100/0	11/89

Scheme 43.

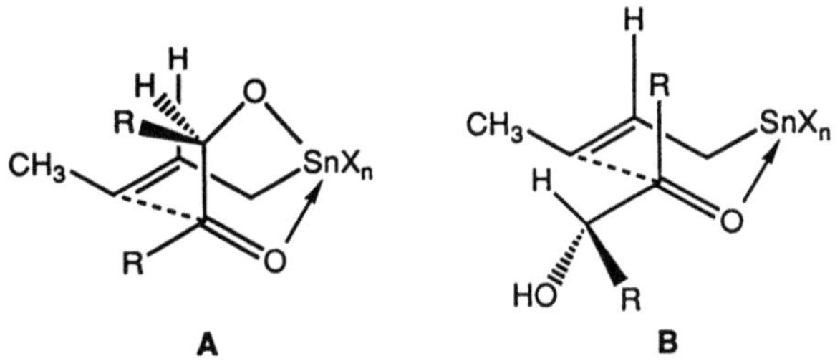

Figure 2. A chelated bicyclic model **A** and a six-membered cyclic Felkin model **B**.

THF-H_2O, while diastereoselectivity between C^2 and C^3 was *syn*-rich in THF and *anti*-rich in THF-H_2O. Thus, these results can be explained by a chelated bicyclic transition state **A** in THF and a six-membered cyclic Felkin transition state **B** in THF-H_2O (Figure 2). The bulkiness of substituent R seems to be disadvantageous for the formation of the chelated bicyclic transition state **A**.

3. 2-Alkoxycarbonylallylation with 2-(Hydroxymethyl)acrylate: Coordination of Alkoxycarbonyl Group to Sn(IV)[69]

We here describe the effect of a third kind of chelation, namely an intramolecular chelation to Sn(IV) of allylic tin intermediate in palladium-catalyzed diastereoselective carbonyl allylation. We investigated an intramolecular chelation of 2-alkoxycarbonyl group to Sn(IV) in allylic tin intermediates **88** derived from available and stable 2-(hydroxymethyl)acrylates **87**, after accomplishing a diastereoselective synthesis of α-methylene-γ-butyrolactones **89** (Scheme 44).[70]

Metal-mediated 2-alkoxycarbonylallylation of aldehydes is one of the most attractive methods for preparation of α-methylene-γ-butyrolactones, and 2-(bromomethyl)acrylates are usually employed as starting materials.[70,71] However, 2-(bromomethyl)acrylates cannot be so easily prepared, and are not so stable. The development of 2-alkoxycarbonylallylating reagents, which are available and stable, is therefore desired.[72] We first tried the synthesis of α-methylene-γ-butyrolactones **92** via palladium-catalyzed carbonyl allylation by ethyl 2-(hy-

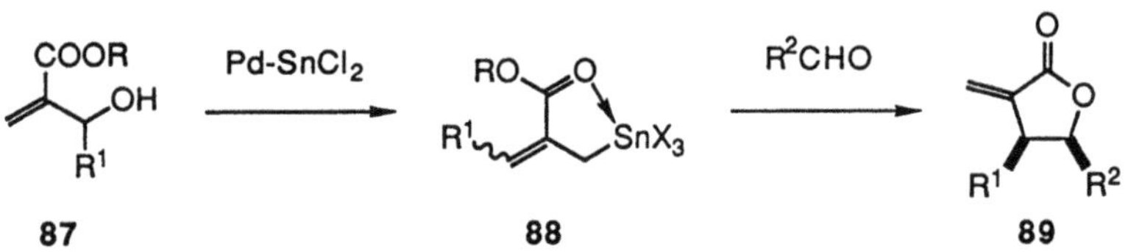

Scheme 44.

Table 12. Palladium-Catalyzed Carbonyl Allylation by Ethyl 2-(Hydroxymethyl)acrylate (**90**) and Ethyl 2-(Methoxycarbonyloxymethyl)acrylate (**91**)

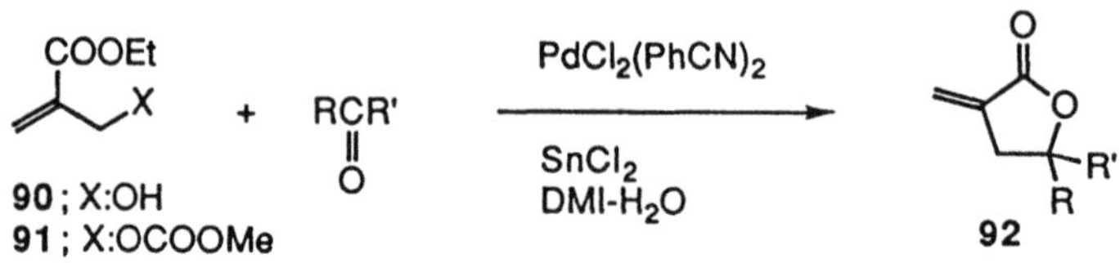

Entry	*Acrylate*	*Carbonyl Compound*	*Temp.(°C)*	*Time(h)*	**92**, *Yield(%)*[a]
1	**90**	benzaldehyde	80	47	33
2	**90**	piperonal	80	25	47
3	**90**	octanal	80	70	36
4	**90**	cyclohexanecarboxaldehyde	80	90	30
5	**90**	cinnamaldehyde	80	120	44
6	**90**	cyclohexanone	80	110	23
7	**91**	benzaldehyde	50	52	36
8	**91**	piperonal	50	25	52
9	**91**	octanal	50	60	51
10	**91**	cyclohexanecarboxaldehyde	50	48	61
11	**91**	cinnamaldehyde	50	80	47
12	**91**	cyclohexanone	80	90	36

Note: [a]Isolated yields based on the carbonyl compounds.

droxymethyl)acrylate (**90**),[73] which is more available and more stable than 2-(bromomethyl)acrylate. Some representative results are summarized in Table 12.

> *α-Methylene-γ-butyrolactones: 2-Ethoxycarbonylallylation of carbonyl compounds.* To a solution of stannous chloride (0.76 g, 4.0 mmol), ethyl 2-(hydroxymethyl)acrylate (**90**, 0.26 g, 2.0 mmol), and piperonal (0.18 g, 1.2 mmol) in DMI (3 ml)-H_2O (0.5 ml) is added $PdCl_2(PhCN)_2$ (16 mg, 0.04 mmol). After stirring for 25 h at 80 °C under a nitrogen atmosphere, ether (20 ml) is poured into the reaction mixture. The mixture is poured into water (30 ml), and is extracted with ether-methylene chloride (3:2, 100 ml). The solution is washed with water followed by brine, and is dried over anhydrous $MgSO_4$. Evaporation of solvent and purification by column chromatography on silica gel (hexane/ethyl acetate = 5/1) afford 0.21 g (47%) of 2-methylene-4-(3,4-methylenedioxyphenyl)butyrolactone as a colorless oil.

Aliphatic, aromatic, and α,β-unsaturated aldehydes can be used in this palladium-catalyzed carbonyl allylation (entries 1–5). A ketone such as cyclohexanone also reacted under the same conditions to give spiro[cyclohexane-1,4′-(2′-methylene)butyrolactone] in 23% yield (entry 6). This reaction did not occur at less

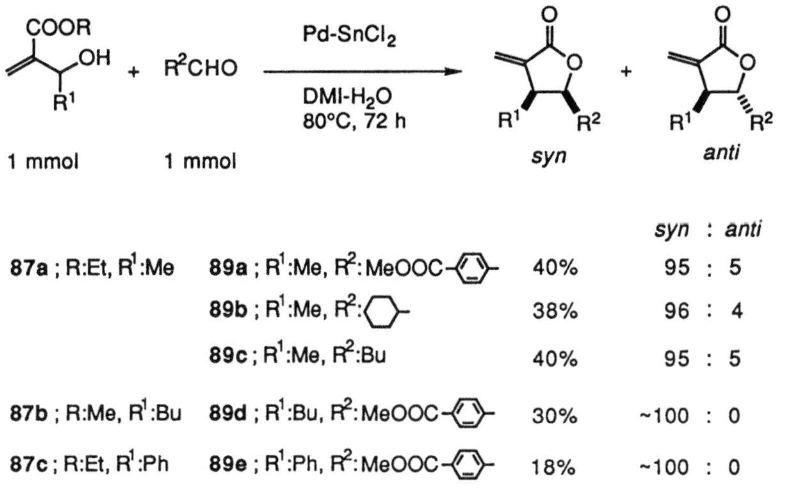

Scheme 45.

than 50 °C. In spite of the use of two equimolar amounts of **90** with aldehydes, the yields are generally low. The use of ethyl 2-(methoxycarbonyloxymethyl)acrylate (**91**) instead of **90** made it possible to lower the reaction temperature (50 °C);[40] this allowed the yields of **92** to enhance (entries 7–12). The polymerization of **91** at 50 °C might be depressed, compared with that of **90** at 80 °C. However, taking account of the troublesomeness of the preparation of allylic carbonate **91** from allylic alcohol **90**, the reaction of **90** might be superior to that of **91**.

Diastereoselection was investigated in the allylation of aldehydes with 2-(hydroxymethyl)acrylates **87a–c**[74] bearing a substituent (R^1) at an allylic position at 80 °C in DMI-H_2O (Scheme 45). An addition reaction of **87a** to aldehydes, such as 4-methoxycarbonylbenzaldehyde, cyclohexanecarboxaldehyde, and pentanal, led to *syn* selection (>95%) to afford *cis*-lactones **89a-c** without any influence of the bulkiness of R^2. The reaction of **87b** or **87c**, bearing a more bulky substituent (R^1 = Bu or Ph) than that (R^1 = Me) in **87a**, with 4-methoxycarbonylbenzaldehyde exhibited complete *syn* selectivity. This *syn* addition probably proceeds via an acyclic *anti*periplanar transition state **94**,[70] which suggests the existence of chelation of oxygen of ester group to Sn(IV) of allylic tin intermediate **88** to disturb the formation of a six-membered cyclic transition state (Scheme 46).[47] The palladium-catalyzed carbonyl allylation by 1-substituted allylic alcohol **87** proves itself a powerful diastereoselective allylation method, since **87** is prepared more easily than 3-substituted allylic alcohol and is regioselectively converted into 3-substituted allylic tin intermediate **88** via the formation of π-allylpalladium complex **93**.[45,47]

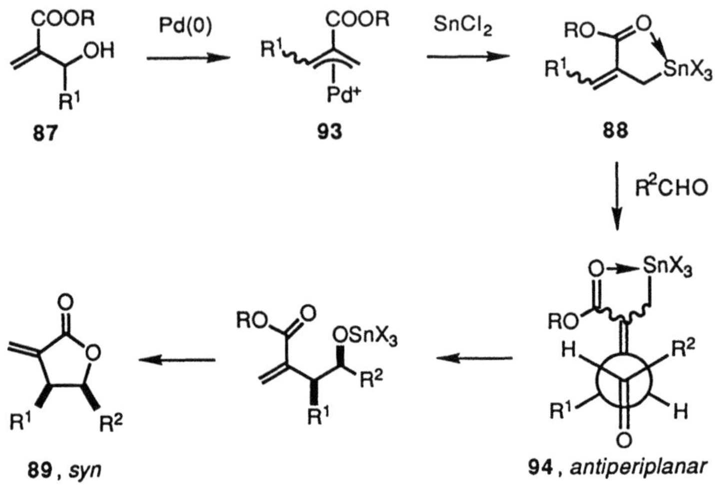

Scheme 46.

D. NMR Spectroscopic Investigation[75]

This palladium-catalyzed carbonyl allylation by allylic alcohol has contained some enigmas;

1. this reaction requires two equimolar amounts of $SnCl_2$ to allylic alcohol;
2. the diastereoselection can be changed by the choice of solvents (*syn* selection in DMSO and *anti* selection in DMI, DMF and THF); and
3. the addition of H_2O accelerates the allylation and enhances the diastereoselectivity.

Thus, we have considered (1) detection of an actual allylating agent, namely allyltin intermediate and (2) estimation of the transition state in the reaction of the allyltin intermediate with aldehyde by NMR spectroscopic investigation; these will contribute to the elucidation of the enigmas.

We first carried out direct ^1H and ^{13}C NMR observation of the reaction of 2-propenol (1 mmol) and $SnCl_2$ (2 mmol) with $PdCl_2(PhCN)_2$ (0.04 mmol) in DMF-d_7 (0.5 ml) at 25 °C (sample **I**; spectra **A** and **B** in Figure 3). After 48 h, the starting 2-propenol is completely consumed. ^1H and ^{13}C NMR spectra of a major detectable product is almost the same as those of the reaction of 3-chloropropene (1 mmol) and $SnCl_2$ (1 mmol) with $PdCl_2(PhCN)_2$ (0.01 mmol) in DMF-d_7 (0.5 ml) (sample **II**; spectra **C** and **D** in Figure 4). The ^{13}C NMR spectrum is the same as that of 2-propenyltrichlorotin.[76] 2-Propenyltrichlorotin **96** should therefore be prepared from 2-propenol via the formation of π-allylpalladium trichlorostannate

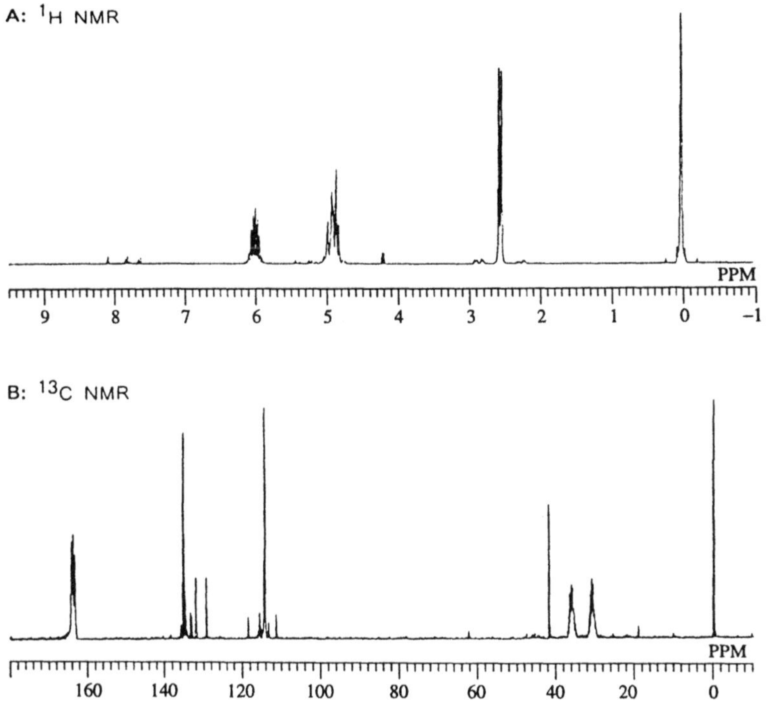

Figure 3. The reaction of allyl alcohol with cat. Pd-2$SnCl_2$in DMF-d_7.

95 (Scheme 47).[77] A minor product, having broad ^{13}C peaks as shown on spectrum **F** in Figure 5, is obtained in the reaction of 2-propenol, and the ^{13}C chemical shifts resemble those of 2-propenyltrichlorotin **96**, respectively. Thus, the minor product, which is not detected in the reaction of 3-chloropropene, also seems to consist of a 2-propenyltin moiety.

^{119}Sn NMR spectra of the above samples **I** and **II** were measured to confirm the structure of the major product (DMF-d_7, 25 °C, Me_4Sn as an external reference compound). Sample **I** shows two signals (–491.5 and –504.1 ppm). On the other hand, sample **II** shows a fairly sharp signal (–490.1 ppm). In both samples, **I** and **II**, ^{119}Sn peak (–326.7 ppm) of $SnCl_2$ almost disappears. Adding six equimolar amounts of D_2O to each sample, sample **I** changes to one signal (–503.6 ppm) and the signal of sample **II** shifts to upfield (–495.6 ppm).

To support the above assumption, we observed the alteration of the ^{13}C chemical shifts of the allyl species, while adding a controlled amount of D_2O to sample **I** and **II**. The ^{13}C NMR spectra between 110 and 140 ppm are shown in Figure 5; spectra **E–J**. Adding one equimolar amount of D_2O to 2-propenyltrichlorotin [spectrum

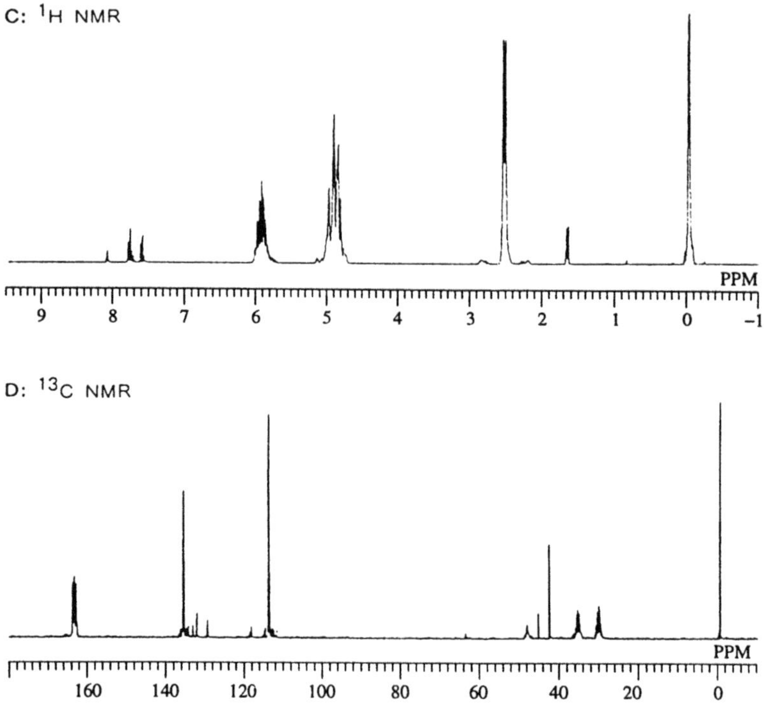

Figure 4. The reaction of allyl chloride with cat. Pd-$SnCl_2$ in DMF-d_7.

H], C_2 (136.0 ppm) and C_3 (114.2 ppm) shifts to upfield and downfield to form each two peaks C_2' (135.9 ppm) and C_2'' (135.6 ppm), and C_3' (114.5 ppm) and C_3'' (114.7 ppm) for two different allyl species, respectively [spectrum **I**]. The spectrum **I** exhibits the same signals as those on spectrum **F**. On continual addition of D_2O to the samples showing the spectra **F** and **I**, the sharp peaks C_2' and C_3' become small and the broad peaks C_2'' and C_3'' become large. The peaks C_2' and C_3' finally disappear. The final allyl species, which is similar to the minor allyl species in the reaction of 2-propenol, may be a 2-propenyltin compound having no Sn^{IV}-Cl bond.

From the results mentioned above, signals for the major allyl species on spectrum **B** are assigned to 2-propenyltrichlorotin, and the minor allyl species seems to be either a 2-propenylhydroxytin derivative **97** or its secondary one.[78] A reaction path is illustrated in Scheme 47. The detection of 2-propenyltrichlorotin supports that $SnCl_2$ possesses two functions: that of assisting elimination of hydroxy group and that of reducing π-allylpalladium complex. The addition of D_2O produces the hydrated allylic tin species **97**, which is active for the carbonyl allylation.

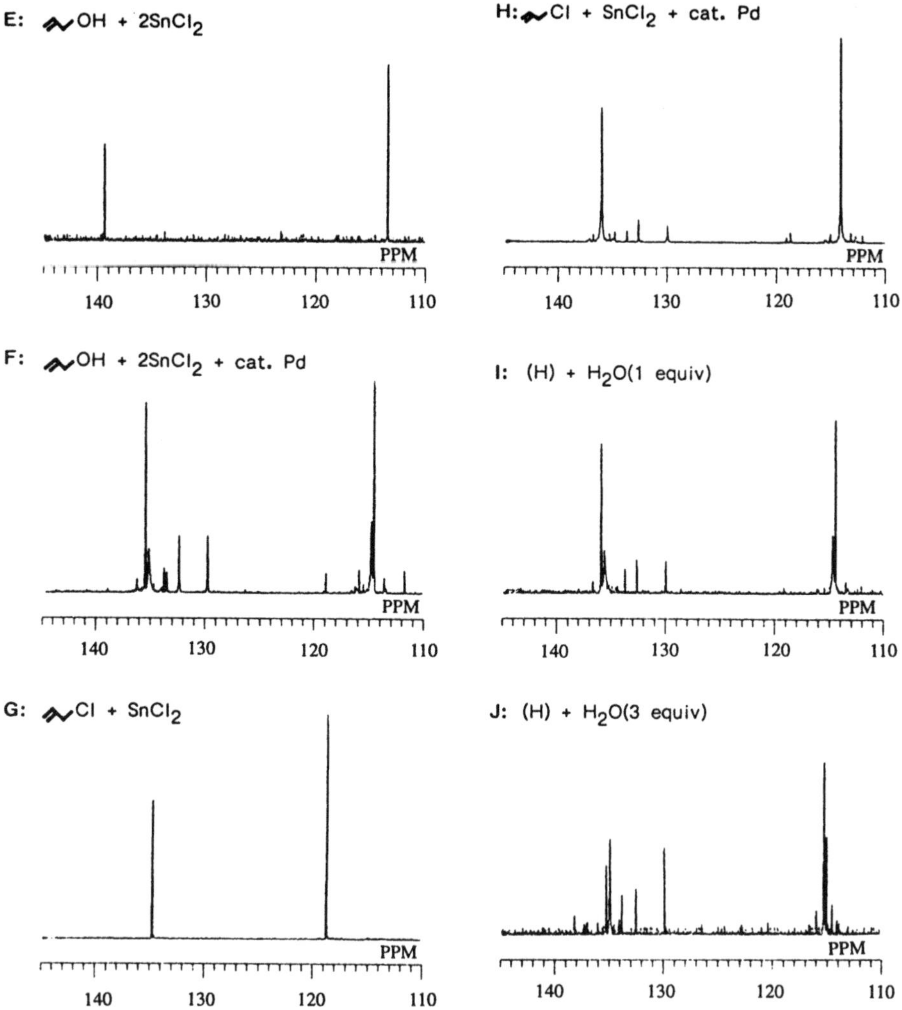

Figure 5. ^{13}C NMR in DMF-d_7.

We tried to observe a change of the coordination of carbonyl group to Sn(IV) of 2-propenyltin intermediate at –80 °C in DMF-$CDCl_3$ or DMSO-$CDCl_3$, while adding controlled amounts of carbonyl compounds such as benzaldehyde and 4-*t*-butylcyclohexanone, for the estimation of the assumed transition state in Scheme 19. To our regret, no change of the coordination can be confirmed for the present, because a suitable carbonyl compound, which is soluble at –80 °C in those mixed solvents and does not react with 2-propenyltin intermediate at –80 °C, is not found.

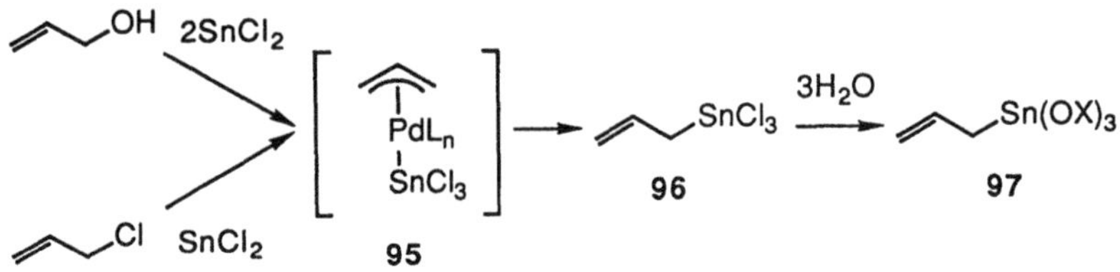

Scheme 47.

IV. CONCLUSION

This palladium-catalyzed carbonyl allylation by allylic alcohols and esters with $SnCl_2$ smoothly proceeds *in aqueous medium*, as shown in Scheme 48. $PdCl_2(PhCN)_2$ acts on the carbonyl allylation as a catalyst *in aqueous medium under air.* This reaction implies that allylic alcohols and esters function as synthons of allylic carbanions. *It is particularly noteworthy that allylic alcohols can be applied as synthons of allylic carbanions.* The actual allylating agents in dry medium are *allylic trichlorotins*, which are prepared from allylic alcohols via the formation of π-allylpalladium trichlorostannates. The carbonyl allylation by allylic alcohols can achieve *high chemo-, regio-, and diastereoselection by the choice of solvent*, and can be applied to *chelation-controlled diastereoselection in polar solvents.*

We expect that this carbonyl allylation will become a method of considerable synthetic utility, complementing the known allylmetal method,[29,37,46] since (1) the starting allylating agents, allylic alcohols, are easily prepared and are relatively stable, and (2) all operations are simple.

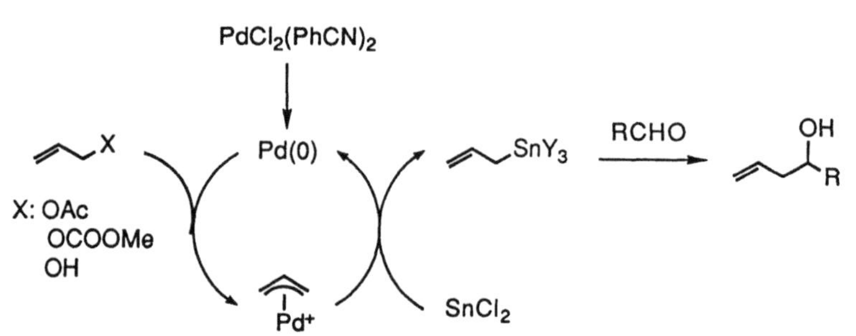

Scheme 48.

ACKNOWLEDGMENTS

This research was carried out at Sophia University in Tokyo, Japan. I am grateful to Nobuyoshi Kinugawa, Kiyotaka Otake, Ryoji Hayashi, Tomoyoshi Tsunoda, Jun. P. Takahara, and Yumiko Nimura for their enthusiastic contributions to this research. I would like to thank Professor Yasuhiko Kurusu of Sophia University for his valuable comments. I would also like to thank Professor Mitsuo Kira of Tohoku University for the determination of the structure of allylated products in diastereofacial selection of α-hydroxyketones. We acknowledge the Ministry of Education, Science and Culture, Japan and Mitsubishi Kasei Corporation for financial support.

REFERENCES AND NOTES

1. (a) Trost, B. M.; Verhoeven, T. R. In *Comprehensive Organometallic Chemistry*; Wilkinson, G., Ed.; Pergamon Press: Oxford, 1982; Vol. 8, p. 802. (b) Hegedus, L. S. In *Comprehensive Carbanion Chemistry*; Buncel, E.; Durst, T., Eds.; Elsevier: Amsterdam, 1984; Vol. p. 30. (c) Heck, R. F. In *Palladium Reagents in Organic Syntheses*; Academic Press: London, 1985; Vol. p. 117. (d) Tsuji, J. In *The Chemistry of the Metal-Carbon Bond*; Hartley, F. R.; Patai, S., Eds.; Wiley: New York, 1985; Vol. 3, p. 163. (e) Consiglio, G.; Waymouth, R. M.; *Chem. Rev.* **1989**, *89*, 257.
2. Courtois, G.; Miginiac, L. *J. Organomet. Chem.* **1974**, *69*, 1.
3. Benkeser, R. A.; Siklosi, M. P.; Mozdzen, E. C. *J. Am. Chem. Soc.* **1978**, *100*, 2134.
4. Luche, J.-L.; Damiano, J.-C. *J. Am. Chem. Soc.* **1980**, *102*, 7926.
5. Mukaiyama, T.; Harada, T.; Shoda, S. *Chem. Lett.* **1980**, 1507.
6. Mukaiyama, T.; Harada, T. *Chem. Lett.* **1981**, 1527.
7. Sato, F.; Iida, K.; Iijima, S.; Morita, H.; Sato, M. *J. Chem. Soc., Chem. Commun.* **1981**, 1140.
8. Souppe, J.; Namy, J. L.; Kagan, H. B. *Tetrahedron Lett.* **1982**, *23*, 3497.
9. Hiyama, T.; Okude, Y.; Kimura, K.; Nozaki, H. *Bull. Chem. Soc. Jpn.* **1982**, *55*, 561.
10. Hiyama, T.; Obayashi, M.; Nakamura, A. *Organomatallics* **1982**, *1*, 1249.
11. Nokami, J.; Otera, J.; Sudo, T.; Okawara, R. *Organometallics* **1983**, *2*, 191.
12. Yamamoto, Y.; Maruyama, K. *J. Org. Chem.* **1983**, *48*, 1564.
13. Imamoto, T.; Kusumoto, T.; Tawarayama, Y.; Sugiura, Y.; Mita, T.; Hatanaka, Y.; Yokoyama, M. *J. Org. Chem.* **1984**, *49*, 3904.
14. Uneyama, K.; Kamaki, N.; Moriya, A.; Torii, S. *J. Org. Chem.* **1985**, *50*, 5396.
15. Wada, M.; Akiba, K. *Tetrahedron Lett.* **1985**, *26*, 4211.
16. Tanaka, H.; Yamashita, S.; Hamatani, T.; Ikemoto, Y.; Torii, S. *Chem. Lett.* **1986**, 1611.
17. Uneyama, K.; Nanbu, H.; Torii, S. *Tetrahedron Lett.* **1986**, *27*, 2395.
18. Einhorn, C.; Luche, J.-L. *J. Organomet. Chem.* **1987**, *322*, 177.
19. Durrandetti, S.; Sibille, S.; Perichon, J. *J. Org. Chem.* **1989**, *54*, 2198.
20. Shono, T.; Ishifune, M.; Kashimura, S. *Chem. Lett.* **1990**, 449.
21. Seebach, D. *Angew. Chem., Int. Ed. Engl.* **1979**, *18*, 239.
22. (a) Trost, B. M. *Bull. Chem. Soc. Jpn.* **1988**, *61*, 107. (b) Masuyama, Y.; Yamada, K.; Shimizu, S.; Kurusu, Y. *Bull. Chem. Soc. Jpn.* **1989**, *62*, 2913.
23. Trost, B. M.; Herndon, J. W. *J. Am. Chem. Soc.* **1984**, *106*, 6835.
24. Matsubara, S.; Wakamatsu, K.; Morizawa, Y.; Tsuboniwa, N.; Oshima, K.; Nozaki, H. *Bull. Chem. Soc. Jpn.* **1985**, *58*, 1196.
25. Tabuchi, T.; Inanaga, J.; Yamaguchi, M. *Tetrahedron Lett.* **1986**, *27*, 1195.
26. Qiu, W.; Wang, Z. *J. Chem. Soc., Chem. Commun.* **1989**, 356.
27. Tabuchi, T.; Inanaga, J.; Yamaguchi, M. *Tetrahedron Lett.* **1987**, *28*, 215.

28. Trost, B. M.; Walchli, R. *J. Am. Chem. Soc.* **1987**, *109*, 3487.
29. (a) Hoffmann, R. W. *Angew. Chem., Int. Ed. Engl.* **1982**, *21*, 555. (b) Yamamoto, Y.; Maruyama, K. *Heterocycles* **1982**, *18*, 357. (c) Reetz, M. T. *Book Organotitanium Reagents in Organic Synthesis*; Springer-Verlag: Berlin Heidelberg, 1986; Vol. p. 162. (d) Yamamoto, Y. *Acc. Chem. Res.* **1987**, *20*, 243.
30. Wilkinson, S. G. In *Comprehensive Organic Chemistry*, Barton, D. H. R.; Ollis, W. D.; and Stoddart, J. F. Pergamon: Oxford, 1979; Vol. 1, p. 579.
31. (a) Mukaiyama, T. *Pure Appl. Chem.* **1986**, *58*, 505. (b) Mukaiyama, T.; Iwasawa, N. *Nippon Kagaku Kaishi* **1987**, 1099.
32. Sasaoka, S.; Yamamoto, T.; Kinoshita, H.; Inomata, K.; Kotake, H. *Chem. Lett.* **1985**, 315.
33. Masuyama, Y.; Kinugawa, N.; Kurusu, Y. *J. Org. Chem.* **1987**, *52*, 3702.
34. Ho, T.-L.; Olah, G. A. *Synthesis* **1977**, 170.
35. Koreeda, M.; Tanaka, Y. *Chem. Lett.* **1982**, 1299.
36. Masuyama, Y.; Hayashi, R.; Otake, K.; Kurusu, Y. *J. Chem. Soc., Chem. Commun.* **1988**, 44.
37. (a) Kumar Das, V. G.; Chu, C.-K. In *The Chemistry of the Metal-Carbon Bond*; Wiley: New York, 1985; Vol. 3, p. 41. (b) Tagliavini, G. *Rev. Si, Ge, Sn, Pb* **1985**, *8*, 237. (c) Yamamoto, Y. *Aldrichimica Acta* **1987**, *20*, 45. (d) Pereyre, M.; Quintard, J.-P.; Rahm, A. *Tin in Organic Synthesis*; Butterworths: London, 1987, p. 211.
38. Masuyama, Y.; Haratani, Y.; Kurusu, Y. *Chem. Express* **1988**, *3*, 127.
39. Sakurai, H. *Synlett* **1989**, *1*, 1.
40. Masuyama, Y.; Otake, K.; Kurusu, Y. *Tetrahedron Lett.* **1988**, *29*, 3563.
41. Tsuji, J.; Shimizu, I.; Minami, I.; Ohashi, Y. *Tetrahedron Lett.* **1982**, *23*, 4809.
42. Koga, K.; Yamada, S. *Chem. Pharm. Bull.* **1972**, *20*, 526.
43. Masuyama, Y.; Kinugawa, N.; Otake, K.; Kurusu, Y. unpublished results.
44. (a) Atkins, K. E.; Walker, W. E.; Manyik, R. M. *Tetrahedron Lett.* **1970**, 3821. (b) Lu, X.; Lu, L.; Sun, J. *J. Mol. Catal.* **1987**, *41*, 245. (c) Lu, X.; Jiang, X.; Tao, X. *J. Organomet. Chem.* **1988**, *344*, 109. (d) Bergbreiter, D. E.; Weatherford, D. A. *J. Chem. Soc., Chem. Commun.* **1989**, 883.
45. Masuyama, Y.; Takahara, J. P.; Kurusu, Y. *J. Am. Chem. Soc.* **1988**, *110*, 4473.
46. (a) Wilkinson, G.; Stone, F. G. A.; Abel, E. W. In *Comprehensive Organometallic Chemistry*; Pergamon: Oxford, 1982; Vol. 7. (b) Hartley, F. R.; Patai, S., Eds.; *The Chemistry of the Metal-Carbon Bond*; Wiley: New York, 1985; Vol. 3.
47. Masuyama, Y.; Takahara, J. P.; Kurusu, Y. *Tetrahedron Lett.* **1989**, *30*, 3437.
48. Furlani, D.; Marton, D.; Tagliavini, G.; Zordan, M. *J. Organomet. Chem.* **1988**, *341*, 345.
49. (a) Denmark, S. E.; Weber, E. *J. Am. Chem. Soc.* **1987**, *109*, 2512. (b) Denmark, S. E.; Weber, E. J. *J. Am. Chem. Soc.* **1984**, *106*, 7970. (c) Denmark, S. E.; Weber, J. E. *Helv. Chim. Acta* **1983**, *66*, 1655.
50. Denmark, S. E.; Weber, E. J.; Wilson, T. M.; Wilson, T. M. *Tetrahedron* **1989**, *45*, 1053.
51. Takahara, J. P.; Masuyama, Y.; Kurusu, Y. *J. Am. Chem. Soc.* **1992**, *114*, 2577.
52. Keck, G. E.; Abbott, D. E.; Boden, E. P.; Enholm, E. J. *Tetrahedron Lett.* **1984**, *25*, 3927.
53. (a) Boaretto, A.; Marton, D.; Tagliavini, G. *J. Organomet. Chem.* **1987**, *321*, 199. (b) Gambaro, A.; Ganis, P.; Marton, D.; Peruzzo, V.; Tagliavini, G. *J. Organomet. Chem.* **1982**, *231*, 307.
54. Yamamoto, Y.; Maeda, N.; Maruyama, K. *J. Chem. Soc., Chem. Commun.* **1983**, 742.
55. Hull, C.; Mortlock, S. V.; Thomas, E. J. *Tetrahedron Lett.* **1987**, *28*, 5343.
56. Takahara, J. P.; Masuyama, Y.; Kurusu, Y. *J. Am. Chem. Soc.* **1992**, *114*, 2577.
57. Yamamoto, Y.; Yatagai, H.; Ishihara, Y.; Maeda, N.; Maruyama, K. *Tetrahedron* **1984**, *40*, 2239.
58. Mori, I.; Bartlett, P. A.; Heathcock, C. H. *J. Am. Chem. Soc.* **1987**, *109*, 7199.
59. Keck, G. E.; Boden, E. P. *Tetrahedron Lett.* **1984**, *25*, 265.
60. Yamamoto, Y.; Komatsu, K.; Maruyama, K. *J. Organomet. Chem.* **1985**, *285*, 31.
61. Keck, G. E.; Abbott, D. E.; Wiley, M. R. *Tetrahedron Lett.* **1987**, *28*, 139.
62. Masuyama, Y.; Tsunoda, T.; Kurusu, Y. *Chem. Lett.* **1989**, 1647.

63. Reetz, M. T.; Steinbach, R.; Westermann, J.; Urz, R.; Wenderoth, B.; Peter, R. *Angew. Chem., Int. Ed. Engl.* **1982**, *21*, 135.
64. Molander, G. A.; Shubert, D. C. *J. Am. Chem. Soc.* **1986**, *108*, 4683.
65. Masuyama, Y.; Tsunoda, T.; Kurusu, Y. unpublished results.
66. Marvell, E. N.; Cheng, J. C.-P. *J. Org. Chem.* **1980**, *45*, 4511.
67. Takahara, J. P.; Masuyama, Y.; Kurusu, Y. *Chem. Lett.* **1991**, 879.
68. Sato, K.; Kira, M.; Sakurai, H. *J. Am. Chem. Soc.* **1989**, *111*, 6429.
69. Masuyama, Y.; Nimura, Y.; Kurusu, Y. *Tetrahedron Lett.* **1991**, *32*, 225.
70. Okuda, Y.; Nakatsukasa, S.; Oshima, K.; Nozaki, H. *Chem. Lett.* **1985**, 481.
71. (a) Hegedus, L.S.; Wagner, S. D.; Waterman, E. L.; Siirala-Hansen, K. *J. Org. Chem.* **1975**, *40*, 593. (b) Semmelhack, M. F.; Wu E. S. C. *J. Am. Chem. Soc.* **1976**, *98*, 3384. (c) Öhler, E.; Reininger, K.; Schmidt, U. *Angew. Chem.* **1970**, *82*, 480. (d) Boldrini, G. P.; Savoia, D.; Tagliavini, E.; Trombini, C.; Umani-Ronchi, A. *J. Org. Chem.* **1983**, *48*, 4108. (e) Mattes, H.; Benezra, C. *Tetrahedron Lett.* **1985**, *26*, 5697. (f) Nokami, J.; Tamaoka, T.; Ogawa, H.; Wakabayashi, S. *Chem. Lett.* **1986**, 541. (g) K. Uneyama, K. Ueda, and S. Torii, *Chem. Lett.* 1986, 1201.
72. (a) Semmelhack, M. F.; Yamashita, A.; Tomesch, J. C.; Hirotsu, K. *J. Am. Chem. Soc.* **1978**, *100*, 5565. (b) Hosomi, A.; Hashimoto, H.; Sakurai, H. *Tetrahedron Lett.* **1980**, *21*, 951.
73. Villieras, J.; Rambaud, M. *Org. Synth.* **1987**, *66*, 220.
74. Hoffmann, H. M. R.; Rabe, J. *Helv. Chim. Acta* **1984**, *67*, 413.
75. Takahara, J. P.; Masuyama, Y.; Kurusu, Y. *J. Am. Chem. Soc.* **1992**, *114*, 2577.
76. Denmark, S. E.; Wilson, T.; Willson, T. M. *J. Am. Chem. Soc.* **1988**, *110*, 984.
77. Musco, A.; Pontellini, R.; Grassi, M.; Sironi, A.; Meille, S. V.; Rüegger, H.; Ammann, C.; Pregosin, P. S. *Organometallics* **1988**, *7*, 2130.
78. Poller, R. C. In *Comprehensive Organic Chemistry*; Jones, D. N.; Pergamon: Oxford, 1979; Vol. 3, p. 1073.

INDEX

www.ingramcontent.com/pod-product-compliance
Ingram Content Group UK Ltd.
Pitfield, Milton Keynes, MK11 3LW, UK
UKHW020238290926
15470UKWH00002B/607

9 781559 384063